CRUSTACEAN ISSUES

T0315450

General editor

RONALD VONK

Institute for Biodiversity and Ecosystem Dynamics
University of Amsterdam, The Netherlands

1   Crustacean Phylogeny
    Schram, F.R. (ed.)
    1983   ISBN 90 6191 231 8 Sold out

2   Crustacean Growth: Larval Growth
    Wenner, A. (ed.)
    1985   ISBN 90 6191 294 6

3   Crustacean Growth: Factors in Adult Growth
    Wenner, A. (ed.)
    1985   ISBN 90 6191 535 X

4   Crustacean Biogeography
    Gore, R.H. & Heck, K.L. (eds.)
    1986   ISBN 90 6191 593 7

5   Barnacle Biology
    Southward, A.J. (ed.)
    1987   ISBN 90 6191 628 3

6   Functional Morphology of Feeding and Grooming in Crustacea
    Felgenhauer, B.E., Thistle, A.B. & Watling, L. (eds.)
    1989   ISBN 90 6191 777 8

7   Crustacean Egg Production
    Wenner, A. & Kuris, A. (eds.)
    1991   ISBN 90 6191 098 6

8   History of Carcinology
    Truesdale, F.M. (ed.)
    1993   ISBN 90 5410 137 7

The Biology and Fisheries of the Slipper Lobster

Kari L. Lavalli
College of General Studies
Boston University
Boston, Massachusetts, U.S.A.

Ehud Spanier
The Leon Recanati Institute for Maritime Studies
and Department of Maritime Civilizations
University of Haifa
Haifa, Israel

CRC Press
Taylor & Francis Group
Boca Raton London New York

CRC Press is an imprint of the
Taylor & Francis Group, an **informa** business

Cover image courtesy of Megan Elizabeth Stover of the College of General Studies, Boston University, Boston, Massachusetts.

CRC Press
Taylor & Francis Group
6000 Broken Sound Parkway NW, Suite 300
Boca Raton, FL 33487-2742

First issued in paperback 2019

© 2007 by Taylor & Francis Group, LLC
CRC Press is an imprint of Taylor & Francis Group, an Informa business

No claim to original U.S. Government works

ISBN-13: 978-0-8493-3398-9 (hbk)
ISBN-13: 978-0-367-38952-9 (pbk)

Library of Congress Cataloging-in-Publication Data

The biology and fisheries of the slipper lobster / edited by Kari L. Lavalli and Ehud Spanier.
 p. cm.
 Includes bibliographical references and index.
 ISBN-13: 978-0-8493-3398-9 (alk. paper)
 ISBN-10: 0-8493-3398-9 (alk. paper)
 1. Scyllaridae. 2. Lobster fisheries. I. Lavalli, Kari L. II. Spanier, Ehud. III. Title.

QL444.M33B555 2007
595.3'84--dc22   2006026391

**Visit the Taylor & Francis Web site at
http://www.taylorandfrancis.com**

**and the CRC Press Web site at
http://www.crcpress.com**

Contents

Preface

Slipper or shovel-nosed lobsters are a fascinating family within the order Decapoda and yet, because of their lesser economic importance, have been virtually ignored in biological research. This situation contrasts with the well-known and commercially important nephropid and palinurid lobsters about which much has been and continues to be written. The first collection of scientific knowledge on clawed lobster biology was compiled and written by Francis Hobart Herrick in his monograph, "The American Lobster: A Study of Its Habits and Development," published in 1895 in the *Bulletin of the U.S. Fisheries Commission*, volume 15, and republished and updated in 1909 in the *Bulletin of the U.S. Bureau of Fisheries*, volume 29. These two editions came at a time when much information was being gathered on the early life stages (larvae) via culturing activities and when much information had been gathered about the adult life history via both laboratory work and fisheries sampling. Little information was available on the early benthic stages of the clawed lobster and that was to remain the case for nearly another century.

In the late 1970s, the first International Workshop on Lobster and Rock Lobster Ecology and Physiology was organized by Bruce F. Phillips and J. Stanley Cobb and held in Perth, Western Australia. Out of that conference, the first two-volume set (*The Biology and Management of Lobsters*, B.F. Phillips and J.S. Cobb, eds., Academic Press, New York, 1980) of books devoted to the biology and management of both clawed and spiny lobsters was published, with many of the conference participants acting as contributors. These volumes mentioned the slipper lobsters in the general biology chapter, but otherwise, no attention was focused on them.

As more and more attention was directed toward a better understanding of clawed lobster and spiny lobster fisheries around the world, and these fisheries became more and more profitable in the early 1990s, biological research on basic life history and physiology increased, and with spiny lobsters, concerted effort was focused on the potential of aquaculture. Additional international workshops yielded conference proceedings published in several journals: *Canadian Journal of Fisheries and Aquatic Sciences*, *Crustaceana*, and *Marine and Freshwater Research*. In none of these proceedings was any attempt made to tie together and update the 1980s volumes, which were sadly outdated. Thus, in 1994, Bruce Phillips, Stanley Cobb, and Jiro Kittaka edited a multi-authored volume focusing solely on spiny lobster biology as it pertained to aquaculture and management (*Spiny Lobster Management*, Fishing News Books, Oxford, 1994). This volume was followed by one focused solely on different aspects of the biology and fisheries management of the American clawed lobster, *Homarus americanus*, edited by Jan Robert Factor (*Biology of the Lobster Homarus americanus*, Academic Press, New York, 1995). Then in 2000, Bruce Phillips and Jiro Kittaka updated their previous volume (*Spiny Lobsters: Fisheries and Culture*). Most recently, in 2006 Bruce Phillips edited a volume entitled *Lobsters: Biology, Management, Fisheries and Aquaculture* (Blackwell Publishing, Oxford) that once again attempted to talk about both spiny and clawed lobster biology with separate sections on commercially important, well-studied species, including a chapter on the genus *Scyllarides*.

It is clear that scyllarid lobsters are not as commercially important as nephropid or palinurid lobsters, but increasingly they are being targeted as a saleable by-product in spiny lobster or shrimp fisheries and, in some regions of the world, are the focus of directed fisheries (the Galápagos Islands, Hawaii, India, Australia, and the Mediterranean). Often, commercial interest in these lobsters arises as the result of fisheries failures with other species (e.g., spiny lobsters or urchins). In the cases where slipper lobsters have been targeted directly, their numbers have declined rapidly and fisheries failures have occurred or are likely to occur in the near future. Sadly, these fisheries have arisen at a time when little to nothing is known about the biology of the targeted species — clearly an unwise manner of approaching fisheries management. Thus, we believe that it is now time to provide the first volume on the biology and fisheries of these lobsters to compile and synthesize available biological information and identify gaps in our knowledge that need to be addressed. We know that this volume will provide a significant resource to those interested in slipper lobsters in general, as well as those involved in exploitation and management of this resource. It is our hope that this volume will also spur greater interest in these lobsters, with the result that more research will be directed toward a better understanding of larval life history, nisto recruitment, juvenile life history, and adult biology.

This project began with conversations with fellow scyllarid researchers via notices in *The Lobster Newsletter* and with direct one-on-one meetings with such researchers at the Sixth International Workshop on Lobster Biology and Management, held in 2000 in Key West, Florida. We wish to thank all of those who participated in these initial conversations and helped to identify potential contributors. One of us (KLL) has also been lucky enough to have been mentored by many of the pre-eminent researchers on clawed, spiny, and slipper lobsters, and has been significantly influenced by their "feeling for the organism." Therefore, many thanks go to Jelle Atema (for infusing a great interest in the behavior and sensory biology of clawed lobsters), J. Stanley Cobb (for additional enthusiasm for clawed lobsters), Jan Robert Factor (for his love of functional morphology and his role-modeling of what an editor should be), William Herrnkind (for his mastery of teaching techniques, his never-ending enthusiasm for Caribbean spiny lobsters, and his fun and wild way of doing research all day long), and to my co-editor, Ehud Spanier (for his love of the Mediterranean slipper lobster and his enthusiasm during diving in the worst possible conditions). Additional thanks go to my graduate and post-doctoral colleague, Diana Barshaw, who initially piqued my interest in post-larval American lobsters with her own work on them and then convinced me to spend some time overseas working with her and Ehud Spanier on slipper lobsters. Thanks also go to undergraduates Sabrina Hamilton and Betsy Deuster of the College of General Studies at Boston University for their help with initial editing of chapters, and to Megan E. Stover, also of the College of General Studies, for the cover artwork. Both of us would especially like to thank our many collaborators — colleagues, former graduate students, divers, and technicians — who have tirelessly helped us learn more about slipper lobsters.

We thank Ronald Vonk for believing that slipper lobsters would be an appropriate topic for a *Crustacean Issues* volume, and John Sulzycki and Patricia Roberson of Taylor & Francis Group for their valuable instructions, answers to our endless e-mails, and feedback throughout the preparation of this book. We also thank the Research Authority, The University of Haifa, Israel, for its support in the publication of this book. Finally, we are most grateful to the numerous reviewers who gave freely of their time to improve these chapters and to the contributors themselves for their hard work, enthusiasm for the project, and gracious cooperation with our numerous editorial demands.

We dedicate this volume to Lipke B. Holthuis who has worked tirelessly on the taxonomy of slipper lobsters for more than fifty years and who, perhaps single-handedly, has done more than anyone to further our knowledge of the existence of these fascinating creatures.

Editors

Kari L. Lavalli is an assistant professor at the College of General Studies at Boston University. Born in Dearborn, Michigan, she grew up in front of a large forest and was always keenly interested in science. She majored in bio-mathematics at Wells College, a small, private women's college in upstate New York, and in her sophomore year was fortunate enough to be accepted into a short, winter course on animal behavior organized by Dr. Jelle Atema at the Marine Biological Laboratory in Woods Hole, Massachusetts. This course strengthened her interest in animal behavior such that she volunteered to work as an undergraduate at Dr. Melvin Kreithen's homing pigeon laboratory in Pittsburgh, Pennsylvania, in her junior year summer and in Paul Ehrlich's butterfly laboratory in the Colorado Rockies the summer following the earning of her baccalaureate degree. She then entered Boston University's M.A.-Ph.D. program at the Marine Biological Laboratory in Woods Hole, where she worked in Dr. Atema's laboratory on the feeding behavior of post-larval clawed lobsters, earning her Ph.D. in 1992. From there, she continued such work via a post-doctoral position in Dr. Joseph Ayer's laboratory at Northeastern University's marine laboratory in Nahant, Massachusetts, until late 1994, and then was awarded a Fulbright post-doctoral fellowship to work in Israel from 1995 to mid-1998 on slipper lobster behavior with Professor Ehud Spanier and Dr. Diana Barshaw. Together, they investigated comparative defensive morphologies in clawed, spiny, and slipper lobsters, and later investigated the benefits of gregarious grouping behavior in slipper lobsters.

Upon returning to the United States, Dr. Lavalli accepted a position at Southwest Texas State University, where she was in charge of the marine program. She initiated studies of cooperative behavior in the Caribbean spiny lobster with Dr. William Herrnkind of Florida State University and spent a number of summers in the Florida Keys and Panhandle working on this project, as well as working on some slipper lobster morphology projects with graduate students, and participating in an NSF-sponsored program on improving primary and secondary science education via inquiry-based teaching of scientific topics. She also served from 1995 to 2005 as vice-president of The Lobster Conservancy, a non-profit research organization based in Maine that has focused on community-based volunteer monitoring of juvenile lobster populations throughout Maine to Massachusetts. In 2004, she returned to work at her doctoral alma mater in Massachusetts, where she once again studies clawed lobster biology, and continues her work with spiny lobster gregarious behavior and slipper lobster morphology and sensory biology.

She is a member of The Crustacean Society and the Animal Behavior Society, and is active in fundraising for the American Association of University Women.

Ehud Spanier is a full professor at the University of Haifa, Israel, specializing in the field of marine biology. Born in Haifa, he received his B.Sc. in biology and M.Sc. in zoology from Tel Aviv University. His Ph.D. in marine biology and oceanography is from the Rosenstiel School of Marine and Atmospheric Science, University of Miami, Florida.

Throughout his academic career, Professor Spanier was a visiting professor at several American universities including the University of California–Santa Barbara, the University of Rhode Island, and the City University of New York. He was also a visiting scientist at several prominent marine laboratories around the world. Professor Spanier has served on several national and international scientific committees. Currently, he is the director of the Leon Recanati Institute for Maritime Studies and the chairman of the Graduate Department of Maritime Civilizations, University of Haifa, Israel.

For more than thirty years, Professor Spanier has been studying the ecological and behavioral aspects of lobsters from various families — in particular, slipper lobsters and especially the Mediterranean slipper lobster. Many of Professor Spanier's finds, as well as those of his graduate students and colleagues, are represented in this volume.

Contributors

Marco L. Bianchini   Istituto Biologia Agroambientale e Forestale, IBAF/CNR, Monterotondo, Italy

John D. Booth   National Institute of Water and Atmospheric Research, Wellington, New Zealand

Rita Cannas   Dipartimento di Biologia Animale ed Ecologia, Cagliari, Italy

Angelo Cau   Dipartimento di Biologia Animale ed Ecologia, Cagliari, Italy

Elisabetta Coluccia   Dipartimento di Biologia Animale ed Ecologia, Cagliari, Italy

Anna M. Deiana   Dipartimento di Biologia Animale ed Ecologia, Cagliari, Italy

Vinay D. Deshmukh   Central Marine Fisheries Research Institute, Kerala, India

Gerard T. DiNardo   National Oceanic and Atmospheric Administration, National Marine Fisheries Service, Pacific Islands Fisheries Science Center, Honolulu, Hawaii, U.S.A.

Ken J. Graham   Department of Primary Industries, Cronulla Fisheries Research Centre, Cronulla, New South Wales, Australia

Frank Grasso   Department of Psychology, Brooklyn College, City University of New York, Brooklyn, New York, U.S.A.

James A. Haddy   Department of Fisheries and Marine Environment, Australian Maritime College, Beaconsfield, Tasmania, Australia

Alex Hearn   Area de Investigación y Conservación Marina (BIOMAR), Charles Darwin Foundation, Puerto Ayora, Santa Cruz, Galápagos, Ecuador

Francis R. Horne   Department of Biology, Texas State University, San Marcos, Texas, U.S.A.

John H. Hunt   Florida Fish & Wildlife Conservation Commission, Fish & Wildlife Research Institute, Marathon, Florida, U.S.A.

Danielle Johnston   Western Australian Fisheries and Marine Research Laboratories, Hillarys Boat Harbour, Western Australia, Australia

Clive M. Jones   Northern Fisheries Centre, Department of Primary Industries and Fisheries, Cairns, Queensland, Australia

Anna V. Kuballa   Australian Fresh Research & Development Corporation, Department of Primary Industries and Fisheries, Bribie Island Aquaculture Research Centre, Woorim, Bribie Island, Queensland, Australia

Kari L. Lavalli   Division of Natural Science, College of General Studies, Boston University, Boston, Massachusetts, U.S.A.

Angelo Libertini   CNR-Istituto di Scienze Marine, Venezia, Italy

Mary K. Manisseri   Central Marine Fisheries Research Institute, Kerala, India

Camilo Martinez   ECOLAP, University of San Francisco de Quito, Quito, Ecuador

Satoshi Mikami   Australian Fresh Research & Development Corporation, Department of Primary Industries and Fisheries, Bribie Island Aquaculture Research Centre, Woorim, Bribie Island, Queensland, Australia

Robert B. Moffitt   National Oceanic and Atmospheric Administration, National Marine Fisheries Service, Pacific Islands Fisheries Science Center, Honolulu, Hawaii, U.S.A.

Marco Mura   Dipartimento di Biologia Animale ed Ecologia, Cagliari, Italy

Daniela Pessani   Dipartimento di Biologia Animale e dell'Uomo, Torino, Italy

Edakkepravan V. Radhakrishnan   Central Marine Fisheries Research Institute, Kerala, India

Sergio Ragonese   Istituto per l'Ambiente Marino Costiero, IAMC/CNR, Mazara, Italy

Gunther Reck   ECOLAP, University of San Francisco de Quito, Quito, Ecuador

Susanna Salvadori   Dipartimento di Biologia Animale ed Ecologia, Cagliari, Italy

Hideo Sekiguchi   Faculty of Bioresources, Mie University, Mie, Japan

William C. Sharp   Florida Fish & Wildlife Conservation Commission, Division of Marine Fisheries Management, Tallahassee, Florida, U.S.A.

Ehud Spanier   The Leon Recanati Institute for Maritime Studies and the Graduate Department of Maritime Civilizations, University of Haifa, Mount Carmel, Haifa, Israel

John Stewart   Department of Primary Industries, Cronulla Fisheries Research Centre, Cronulla, New South Wales, Australia

Samuel F. Tarsitano   Division of Math, Science, and Engineering, Bristol Community College, Fall River, Massachusetts, U.S.A.

William H. Teehan   Florida Fish & Wildlife Conservation Commission, Division of Marine Fisheries Management, Tallahassee, Florida, U.S.A.

Veronica Toral-Granda   Area de Investigación y Conservación Marina (BIOMAR), Charles Darwin Foundation, Puerto Ayora, Santa Cruz, Galápagos, Ecuador

W. Richard Webber   Museum of New Zealand Te Papa Tongarewa, Wellington, New Zealand

Reviewers

Maria Abate Division of Natural Science, College of General Studies, Boston University, Boston, Massachusetts, U.S.A.

David Aiken Department of Fisheries and Oceans, St. Andrews Biological Station, St. Andrews, New Brunswick, Canada

John D. Booth National Institute of Water and Atmospheric Research, Kilbirnie, Wellington, New Zealand

Mark J. Butler IV Department of Biological Sciences, Old Dominion University, Norfolk, Virginia, U.S.A.

Margarida Castro Centro de Ciências do Mar, Universidade do Algarve, Campus de Gambelas, Faro, Portugal

Michael Clancy Division of Natural Science, College of General Studies, Boston University, Boston, Massachusetts, U.S.A.

Charles Derby Department of Biology, Georgia State University, Atlanta, Georgia, U.S.A.

Jan Robert Factor Natural Sciences, Purchase College, State University of New York, Purchase, New York, U.S.A.

Michael Fogarty National Marine Fisheries Service, Northeast Fisheries Science Center, National Oceanic and Atmospheric Administration, Woods Hole, Massachusetts, U.S.A.

Bella Galil Israel Oceanographic & Limnological Research, Ltd., Tel-Shikmona, Haifa, Israel

Alex Hearn Area de Investgación y Conservación Marina (BIOMAR), Charles Darwin Foundation, Puerto Ayora, Santa Cruz, Galápagos, Ecuador

Lew Incze Research & Development Institute, University of Southern Maine, Portland, Maine, U.S.A.

Clive Jones Northern Fisheries Centre, Department of Primary Industries and Fisheries, Cairns, Queensland, Australia

Alison MacDiarmid Benthic Fisheries & Ecology Group, National Institute of Water & Atmospheric Research (NIWA), Wellington, New Zealand

Gro van der Meeren Institute of Marine Research — Asutevoll, Storebo, Norway

Roy Melville-Smith Rock Lobster & Crab Research, Western Australian Marine Research Laboratories, North Beach, Western Australia, Australia

Sheila Patek Department of Integrative Biology, University of California-Berkeley, Berkeley, California, U.S.A.

Bruce Phillips   Department of Environmental Biology & Applied Biosciences, Muresk Institute, Curtin University of Technology, Perth, Western Australia, Australia

Megan Porter   Department of Biological Sciences, University of Maryland Baltimore County, Baltimore, Maryland, U.S.A.

Deanna Prince   The Lobster Institute, University of Maine-Orono, Orono, Maine, U.S.A.

Dale Tshudy   Geosciences, Edinboro University of Pennsylvania, Edinboro, Pennsylvania, U.S.A.

Susan Waddy   Department of Fisheries and Oceans, St. Andrews Biological Station, St. Andrews, New Brunswick, Canada

Part I

Introduction

1

Introduction to the Biology and Fisheries of Slipper Lobsters

Kari L. Lavalli and Ehud Spanier

CONTENTS

Abstract

The last 30 years have seen a tremendous increase in research on various aspects of lobster biology, but this increase has focused primarily on clawed and spiny lobsters. Research on slipper lobsters remains limited and our current knowledge base, to a certain extent, is less than that for clawed and spiny lobsters 30 years ago. At the same time, slipper lobsters increasingly are becoming targets of commercial fishery operations or are taken as by-product in other fisheries, and it is therefore critical to improve our biological understanding of these animals to effect sustainable management schemes that will ensure the survival of slipper lobster stocks. Furthermore, current physiological studies suggest that slipper lobsters may provide unique models for neurophysiology and ontogeny of gas and ion exchange systems. This volume attempts to pull together the current information on the biology, aquaculture, and fisheries of slipper lobsters to stimulate further interest in these animals, as has been successfully done with other lobster families via similar focused volumes.

1.1 Introduction

Numerous articles and volumes have focused attention on various species of lobsters, particularly those that have sustained long-term fisheries. The most notable of these at the turn of the last century was Francis Hobart Herrick's 1895 monograph on the clawed lobster, *Homarus americanus* H. Milne Edwards, 1837, that summarized the understanding of lobster biology at that time (Herrick 1895). Its publication was partly stimulated by a decline in the fishery for that species, as well as interest in the use of aquaculture to enhance wild stocks (Factor 1995a). Herrick's monograph on clawed lobsters stimulated numerous studies on basic reproductive biology, behavior, habitat preferences, and physiology. By the mid- to late 20th century, crayfish and clawed and spiny lobsters had become significant neurobiological models to understand central pattern generators in the nervous system that exerted control over complex behaviors such as coordinated walking and coordinated mastication of food in the gastric mill (see Evoy & Ayers 1982; Harris-Warrick et al. 1992 for a review). The clawed lobster was also becoming a model for understanding the complex interaction between sensory input and behavior (reviewed in Derby 1984, 2000; Ache & Derby 1985; Blaustein et al. 1987; Derby et al. 1989; Atema & Voigt 1995).

In other parts of the world, spiny lobster fisheries were becoming increasingly important and by the late 1970s, the first workshop on lobster ecology and biology was held in Perth, Australia. The participants of that conference urged the production of a volume that would synthesize all previous and ongoing biological work on all lobsters, as well as various population models that were being used or constructed to help in management of the exploited populations (Cobb & Phillips 1980a, b). In 1980, the two-volume set, *The Biology and Management of Lobsters*, edited by J. Stanley Cobb & Bruce F. Phillips was published by Academic Press. These volumes expanded upon the conference proceedings to provide a comparative approach for the exploration of general lobster biology, growth and control of molting, behavior, and ecology. They also provided the current management schemes and assessment tools for spiny and clawed lobster fisheries and an assessment of aquaculture potential for various species.

These volumes were to remain the main resource for general lobster biology for 14 years and, in that time, many advances were made in terms of improved understanding of lobster physiology, disease, growth, sensory biology, taxonomy and evolution, and the ecologies of all ontogenetic stages of many commercially important species. In the meantime, many general books on natural history of lobsters for the mass public were published, some more scientifically valuable (e.g., *Shrimps, Lobsters and Crabs — Their Fascinating Life Story* by Dorothy Bliss; *Lobsters: Florida, Bahamas, the Caribbean* by Martin Moe, Jr.; *The Western Rock Lobster Panulirus Cygnus Book 1: A Natural History* by Howard Gray; and *Crayfishes, Lobsters and Crabs of Europe — An Illustrated Guide to Common and Traded Species* by Ray Ingle) than others (e.g., *The Compleat Crab and Lobster Book* by Christopher Reaske). In 1988 J. Stanley Cobb and John Pringle started *The Lobster Newsletter* that was a forum through which lobster researchers worldwide could communicate preliminary results, compare catch data, request additional information from fellow researchers, and read about new books or important meetings. What began as a paper version sent to several hundred researchers has now become a web-published newsletter found at http://www.odu.edu/~mbutler/newsletter/index.html and edited by Drs. Mark Butler IV and Peter Lawton. Also in 1988, Austin Williams provided an extremely useful book, entitled *Lobsters of the World — An Illustrated Guide*, which was followed in 1991 by L.B. Holthuis' invaluable *Marine Lobsters of the World. An Annotated and Illustrated Catalogue of the Species of Interest to Fisheries Known to Date. FAO Species Catalogue No. 125, Vol 13*. These two books provided information on commercially exploited species of lobster worldwide and, in the case of Holthuis' synopsis, provided an excellent taxonomic review of the various species in a number of families of lobsters. However, neither volume provided extensive biological information on the species discussed therein.

Other volumes were being published on biological aspects of various crustaceans and often these had important information on aspects of lobster biology — such volumes included *The Biology of Crustacea* (a ten-volume set edited by Dorothy Bliss from 1982 to 1985), *Crustacean Sexual Biology* (edited by Raymond T. Bauer & Joel W. Martin 1991), *Crustacean Issues*, volumes 1, 3, 4, 6, 7, and 14 (series edited by Ronald Vonk 1984 to 2001) on a variety of topics including adult and larval growth, biogeography, functional morphology of feeding appendages, egg production, and decapod larvae, and *The Microscopic Anatomy of Invertebrates, Volume 10: Decapod Crustacea* (edited by Frederick W. Harrison & Arthur G. Humes 1992). However, no update focusing solely on lobster biology was forthcoming until Bruce Phillips, J. Stanley Cobb, and Jiro Kittaka brought together numerous authors working on spiny lobster biology and management to compile the 1994 volume, entitled *Spiny Lobster Management*. The volume was a response to the increase in worldwide demand for spiny lobsters that made obvious the need for better biological information which could be used for improved stock modeling. While it included chapters on spiny lobster biology (general, reproductive biology, functional morphology and digestion, and benthic ecology), its focus was primarily on the worldwide fisheries for numerous species, as well as on biological modeling of stocks, and potential for aquaculture. A year later, Jan R. Factor edited a multiauthored volume, *The Biology of the Lobster, Homarus americanus* that was focused more on basic biology (ecology, physiology, and behavior), with only a few chapters focused on fisheries and management (Factor 1995b). An updated second edition of *Spiny Lobster Management*, entitled *Spiny Lobsters: Fisheries and Culture*, was published in 2000, edited by Bruce Phillips & Jiro Kittaka. This update was more focused on the fisheries around the world, as well as research tools used in those fisheries to assess recruitment and stock numbers, and expanded the aquaculture and marketing sections. Ecology of juveniles, reproductive biology, diseases, and the structure of the digestive system were provided, but only as part of an overall fisheries knowledge base. Most recently, Bruce Phillips has brought together an international group of lobster biologists to once again synthesize the biological knowledge of clawed and spiny lobsters in the 2006 volume entitled, *Lobsters: Biology, Management, Aquaculture and Fisheries*. Phillips' volume is organized into two sections — one focusing solely on biological issues (growth and development, reproduction, behavior and sensory biology, phylogeny and evolution, disease, ecology of various ontogenetic phases) and one focusing on specific commercially important species of lobster (nephropids, palinurids, and scyllarids). Most of the chapters on biology are cowritten by researchers specializing on both clawed and spiny lobsters; thus, these promise to be similar in scope, comparative value, and synthesis to the original 1980 chapters in Phillips & Cobb's 1980 *The Biology and Management of Lobsters*.

In addition to these scholarly volumes, from 1977 to date, seven international conferences and workshops on lobster biology and management have taken place in various locations around the world (Perth, Australia in 1977; St. Andrews, Canada in 1985; Havana, Cuba in 1990; Sanriku, Japan in 1993; Queenstown, New Zealand in 1997; Key West, U.S.A. in 2000; and Hobart, Tasmania in 2004) and have brought together a varied group of researchers involved in research on mostly clawed and spiny lobster, although a few have presented work on slipper lobsters. Each conference has had a particular theme so as to drive additional research in that area: the first Australian conference was primarily directed towards descriptive biology and initial modeling of populations, while that in Canada focused on recruitment issues that were necessary to better understand population dynamics and stock assessment. The Cuban conference focused on biological reference points that would provide important parameters for management and management models. In Japan, the focus was on aquaculture and larval biology, as more and more spiny lobster species were being successfully reared in culture. The New Zealand conference focused on stock assessment, while that in Key West was broadly focused on ecological processes, physiology, behavior, and oceanic processes that impact stocks, along with aquaculture and enhancement techniques (Butler 2001). The conference in Tasmania included a large aquaculture segment, indicating the growing importance of this field in many countries. Other dominant topics were enhancement, environment and ecosystem interactions, and management and assessment with consideration of Marine Protected Areas for lobsters (Frusher & Gardner 2005). Additional presentations dealt with larval transport and recruitment, behavior, and general biology of lobsters. Participation at these conferences has increased from 34 lobster biologists from six countries in 1977 to ~180 participants from 19 countries (Butler 2001). While eight years passed between the first and second conference, the frequency of these international affairs is now every three to four years. The next conference scheduled in 2007 in Prince Edward Island, Canada, promises to continue in the tradition of expanding participation and breadth of focus. Proceedings have evolved from the first conference's edited volume on lobster biology to a series of peer-reviewed articles published in various scientific journals (e.g., *Canadian Journal of Fisheries & Aquatic Sciences*, volume 43; *Crustaceana*, volume 66; *Marine & Freshwater Research*, volume 48(8) and 52(8), and *New Zealand Journal of Marine & Freshwater Research*, volume 39). These volumes, along with the Phillips et al., Phillips, and Factor books have provided invaluable information to anyone interested in lobsters. However, it is immediately obvious that little attention has been paid to species that do not support large fisheries — one such group that has been mostly ignored are the scyllarids, commonly known as slipper, shovel-nosed, or Spanish lobsters.

This volume attempts to address that omission by focusing solely on the biology of these lobsters. Individual chapters summarize our understanding of particular fields of biological research from both an historical and modern perspective. Liberal cross-referencing of chapters helps to underscore the interrelatedness of many biological topics, as well as the obvious interrelationship between genetics, lobster populations, and fisheries management. Given the wide diversity of scyllarid body forms and life-history strategies (addressed in Webber & Booth, Chapter 2 and Sekiguchi et al., Chapter 4), it is difficult to synthesize the biological information available across the subfamilies. However, chapters focusing on genetics (Deiana et al., Chapter 3), factors important in larval growth and molting (Mikami & Kuballa, Chapter 5), the complex process of feeding and digestion (Johnston, Chapter 6), behavior and sensory biology (Lavalli et al., Chapter 7), biomineralization (Horne & Tarsitano, Chapter 8), and growth (Bianchini & Ragonese, Chapter 9) attempt to provide a broad overview that is applicable to many species in a number of subfamilies of scyllarids. Later chapters, focusing on specific species in distinct regions of the world (the Gulf of Mexico, Sharp et al., Chapter 11; the Hawaiian Islands, DiNardo & Moffitt, Chapter 12; the Mediterranean, Pessani & Mura, Chapter 13; the Galápagos Islands, Hearn et al., Chapter 14; India waters, Radhakrishnan et al., Chapter 15; and Australia, Jones, Chapter 16 and Haddy et al., Chapter 17) discuss the various important species in these regions and the fisheries that exist for them. Summary chapters on the biology (Spanier & Lavalli, Chapter 10) and fisheries biology (Spanier & Lavalli, Chapter 18) attempt to synthesize the information provided in the other chapters and to provide directions for future research in these areas. When considered together, it is hoped that the reader will find the chapters in this volume a road map for future directions in scyllarid lobster research, particularly as these lobsters become a greater focus of world fisheries.

1.2   Anatomy of Slipper Lobsters

As with other lobsters, the scyllarid body is segmented and the segments (somites) are fused to form three distinct body regions: the cephalothorax (six somites of the cephalon fused with eight somites of the thorax), the abdomen (six separate somites), and the telson, which is sometimes considered a seventh abdominal segment or a medial appendage of the sixth abdominal segment (Schram 1986; Holthuis 1991). Each body segment bears paired appendages that are highly specialized structures and which are fundamentally biramous, being composed of an endopod and exopod; however, exopodites of pereiopods are lost at the metamorphic molt from final stage phyllosoma to nisto and, in some species of scyllarids, exopods are missing on one or more of the maxillipeds. The head bears the antennules (*aka*, first antennae) on segment 2, the second antennae on segment 3, and the mouthparts (mandibles through second maxilla) on segments four to six. Highly modified thoracic mouthparts are borne on segments seven to nine, while the pereiopods are carried on segments 10 to 14. The abdomen bears paired pleopods on segments 15 to 19 that are specialized for forward swimming in nistos (and possibly small juveniles) and reproduction in adults. The last abdominal segment (20) bears the modified pleopods (uropods) that comprise the tail fan. Figure 1.1, an historical and extremely accurate watercolor of *Scyllarides latus*, shows the body regions of the slipper lobster.

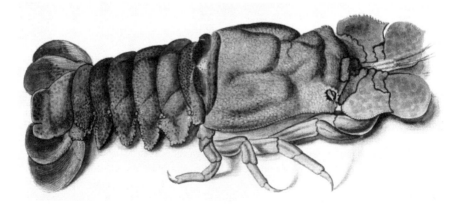

FIGURE 1.1   (See color insert following page 242) Original watercolor of "*Squilla lata*" by Gesner (1558, p. 1097), illustrating *Scyllarides latus*. The detail on this figure is accurate, including the coloration of the body (brown, lighter in some areas, darker in others) and the antennules (blue). (From Holthuis, L.B. 1996. *Zool. Verhand.* 70: 169–196; reprinted with permission.)

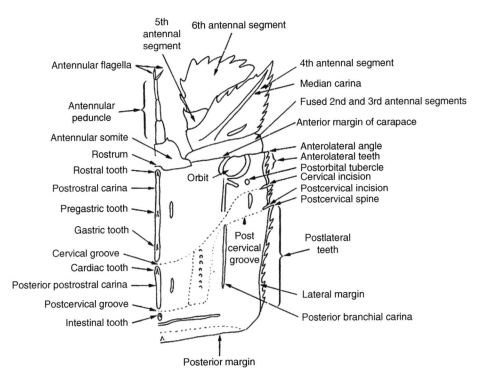

FIGURE 1.2 Dorsal view of the right half of a "general" scyllarid carapace and cephalic appendages with various characters used in taxonomic discrimination. (From Holthuis, L.B. 1991. *Marine Lobsters of the World. An Annotated and Illustrated Catalogue of the Species of Interest to Fisheries Known to Date.* FAO Species Catalogue No. 125, Vol 13: pp. 1–292. Rome: Food and Agriculture Organization of the United Nations; used with permission.)

The shape, sculpturing, pubescence, and spination of the carapace, which overlies the cephalothorax and may project anteriorly to form a rostrum and laterally to the legs to enclose the branchial chamber, generally provides important taxonomic characters used to distinguish species. Likewise, the structure and length of the antennule and the size, shape, and dentition of the second antennae help to distinguish species. In scyllarids, the bifurcated flagellum of the antennule is carried on a long, highly flexible, three-segmented peduncle. In contrast to the whip-like flagellum of the second antennae in nephropids and palinurids, the flagellum of scyllarids is reduced to a single platelike segment and is attached to five additional and highly flattened antennal segments (Holthuis 1991) (Figure 1.2). This character alone distinguishes scyllarids from the other lobster families.

No internal diagrams of scyllarid anatomy exist, but it is presumed that they follow the same internal body plan of other lobsters.

1.3 Life Cycle

As described in Sekiguchi et al. (Chapter 4), the life cycle of scyllarids can be divided into a series of developmental phases that vary in length by species. Embryonic, larval (phyllosomal), postlarval (nisto), juvenile, subadult, and adult phases complete this life cycle.

It is not known for any species of scyllarid whether fertilization of eggs is internal or external, although in those species where the spermatophore is both external and nonpersistent, it is thought to be external (see Radhakrishnan et al., Chapter 15; Jones, Chapter 16; Haddy et al., Chapter 17). In other species, it may be internal (see Lavalli et al., Chapter 7). Newly extruded and fertilized eggs are enclosed in egg envelopes and attached to the female's pleopods where they develop. Females protect the eggs by keeping the abdomen furled such that the tail fan encloses the ventral surface, and presumably groom them with the setal pads on their chelate fifth pereiopods (see Lavalli et al., Chapter 7). At extrusion, eggs are bright orange, small, and spherical; as they develop, they increase in size and develop black eyespots that progressively enlarge as hatching nears. Some scyllarid species apparently hatch as a naupliosoma that has well-developed appendages, except for first maxillipeds and fourth and fifth pereiopods (see Sekiguchi et al, Chapter 4 for description and figures). The naupliosoma is short-lived and molts into the first phyllosomal instar (see Sekiguchi et al., Chapter 4 and Pessani & Mura, Chapter 13 for illustrations of naupliosomas). The number of instars varies among species from 4 to >12 stages (see Sekiguchi et al., Chapter 4 and Pessani & Mura, Chapter 13 for illustrations of a variety of phyllosomas). During successive instars, the appendages enlarge in size, with those not present initially developing into buds and then full appendages, and gain additional setae (see Johnston, Chapter 6 and Lavalli et al., Chapter 7 for a description of these changes). In all phyllosomal instars, the abdomen is extremely small relative to the cephalic shield and thorax and is lacking pleopods (Haond et al. 2001; see Sekiguchi, Chapter 4).

Except for the final-stage phyllosoma, all instars are lacking a branchial cavity and gills. Gill buds appear in the final stage, such that that stage is often called the "gilled" stage (Robertson 1968; Marinovic et al. 1994; Mikami & Greenwood 1997). This situation is unusual for decapods as most (except penaeid shrimp) develop a branchial cavity in which exchange surfaces are present that take the place of gills until metamorphosis (Wolvekamp & Watermann 1960; Vuillemin 1967). In scyllarids, the complete body of the first phyllosoma stains with silver nitrate, indicative of a highly permeable body surface; thus, it is likely that diffusion of gases could occur over the entire body surface, which would be advantageous to a gill-less oceanic form (Haond et al. 2001). In addition, the ventral surface of the cephalic shield has ultrastructural characteristics of an ion-transporting epithelium in that it is thin ($\sim$5 μm) and has an abundance of mitochondria typical of such epithelia (Haond et al. 2001). Both of these types of epithelia would allow phyllosomas to respire over their body surface and osmoregulate via their ventral surface.

The final gilled phyllosoma undergoes a metamorphic molt into the nisto, which can range in size from 3 to 20 mm carapace length depending on species (see Sekiguchi et al., Chapter 4). This stage looks more like an adult, but is typically flatter and similar in form to adult Ibacinae — as such, it was once referred to as a "pseudibacus" (see Pessani & Mura, Chapter 13 for a history of the use of this term). The nisto fulfills the same role as the nephropid postlarva and the palinurid puerulus — that of settlement from the pelagic environment to the benthic environment. In some species, nistos settle over typical adult grounds, while in others, they settle into juvenile grounds and develop there before joining adults in the adult habitat (see Sekiguchi et al., Chapter 4 and Lavalli et al., Chapter 7). Swimming shifts from the thoracic exopodites (which are lost at metamorphosis on all thoracic appendages except the maxillipeds) to the pleopods of the abdomen and is forward directed via the motion of these appendages. Backward swimming (tail flipping) is used as an escape mechanism; however, in adult forms, it can be used also as a long-duration swimming mode (see Lavalli et al., Chapter 7 and Jones, Chapter 16).

Little is currently known of juvenile stages and subadults. In some species, larger juveniles and subadults are trapped with adults; in other species few, if any, juveniles are trapped. Some species grow rapidly and recruit into the adult population within a few years (two to three); other scyllarids grow more slowly and recruit into adult populations after six years (see Hearn et al., Chapter 14; Radhakrishnan et al., Chapter 15, Haddy et al., Chapter 17).

The life history of a "generalized" scyllarid is summarized diagrammatically in Figure 1.3 and includes the naupliosomal stage even though many scyllarid species apparently pass through this stage while still in the egg. In contrast to the building body of knowledge of the life cycles of *Homarus americanus* and several species of spiny lobster, most of the life-history phases of scyllarids are poorly understood at this time. This is an area in great need of investigation because while the emphasis for commercially important species is on the fluctuations of the stocks or populations, populations are ultimately composed of individuals that make decisions about where and when to settle, when and on what to forage, when and with whom to reproduce, and how to interact with others (Butler 1997). Understanding the challenges faced by such individuals at all times of their life cycle can further our understanding of how populations

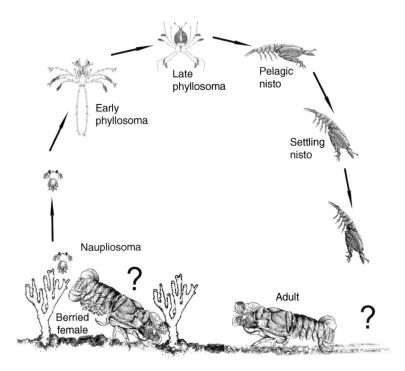

FIGURE 1.3 Life cycle of a "generalized" scyllarid. Question marks indicate life-history phases about which we know almost nothing. (Composite redrawn from Roberston, P.B. 1971. *Bull. Mar. Sci.* 21: 841–865; Crosnier, A. 1972. *Cahiers O.R.S.T.O.M. Oceanogr.* 10: 139–149; Martins, H.R. 1985. *International Council for the Exploration of the Sea C.M.K.*, 52 Shellfish Committee, 13 pp; and Holthuis, L.B. 1996. *Zool. Verhand.* 70: 169–196; used with permission.)

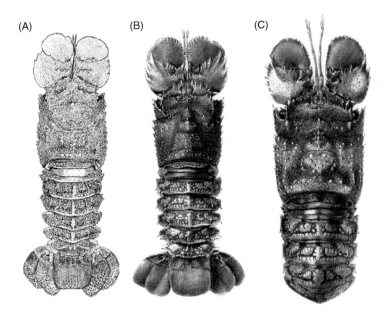

FIGURE 1.4 Lobsters within the *Arctides* genus. (A) *Arctides antipodarum* Holthuis, 1960; (B) *A. guineensis* (Spengler, 1799); (C) *A. regalis* Holthuis, 1963. Lobsters are not to scale. (Modified from Holthuis, L.B. 1991. *Marine Lobsters of the World. An Annotated and Illustrated Catalogue of the Species of Interest to Fisheries Known to Date.* FAO Species Catalogue No. 125, Vol. 13: pp. 1–292. Rome: Food and Agriculture Organization of the United Nations; used with permission.)

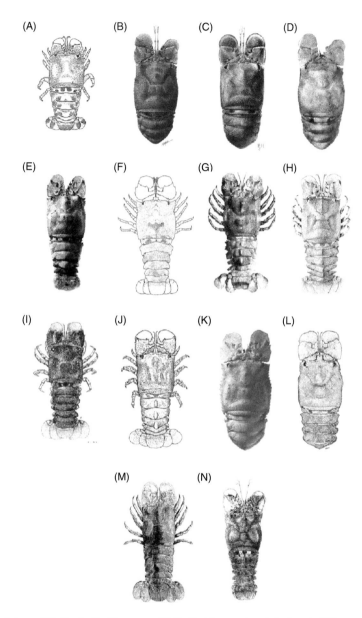

FIGURE 1.5 Lobsters within the *Scyllarides* genus. (A) *Scyllarides aequinoctialis* (Lund, 1793); (B) *S. astori* Holthuis, 1960; (C) *S. brasiliensis* Rathbun, 1906; (D) *S. deceptor* Holthuis, 1963; (E) *S. delfosi* Holthuis, 1960; (F) *S. elisabethae* (Ortmann, 1894); (G) *S. haanii* (De Haan, 1841); (H) *S. herklotsii* (Herklots, 1851); (I) *S. latus* (Latreille, 1802); (J) *S. nodifer* (Stimpson, 1866); (K) *S. obtusus* Holthuis, 1993; (L) *S. roggeveeni* Holthuis, 1967; (M) *S. squammosus* (H. Milne Edwards, 1837); (N) *S. tridacnophaga* Holthuis, 1967. Lobsters are not to scale. (A–J and L–N modified from from Holthuis, L.B. 1991. *Marine Lobsters of the World. An Annotated and Illustrated Catalogue of the Species of Interest to Fisheries Known to Date*. FAO Species Catalogue No. 125, Vol. 13: pp. 1–292. Rome: Food and Agriculture Organization of the United Nations; K modified from Holthuis, L.B. 1993. *Zool. Med. Leiden* 67: 505–515; used with permission.)

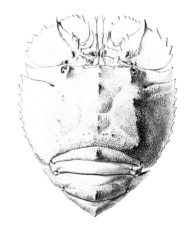

FIGURE 1.6 *Evibacus princeps* S.I. Smith, 1869. (Modified from Holthuis, L.B. 1985. *Zool. Verhand.* 218: 1–130; used with permission.)

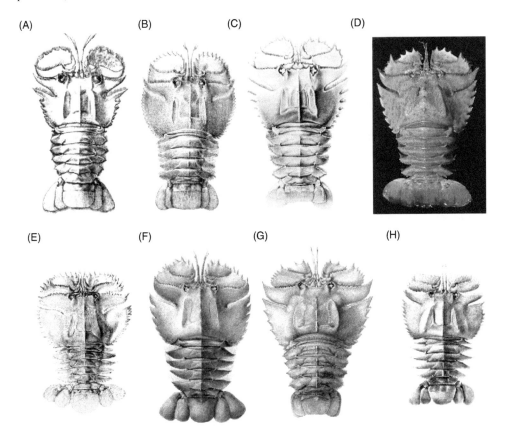

FIGURE 1.7 Lobsters within the *Ibacus* genus. (A) *Ibacus alticrenatus* Bate, 1888; (B) *I. brevipes* Bate, 1888; (C) *I. brucei* Holthuis, 1977; (D) *I. chacei* Brown & Holthuis, 1998; (E) *I. ciliatus* (Von Siebold, 1824); (F) *I. novemdentatus* Gibbes, 1850; (G) *I. peronii* Leach, 1815; (H) *I. pubescens* Holthuis, 1960. Lobsters are not to scale. (A, B, C, E, F, G, H modified from Holthuis, L.B. 1991. *Marine Lobsters of the World. An Annotated and Illustrated Catalogue of the Species of Interest to Fisheries Known to Date.* FAO Species Catalogue No. 125, Vol. 13: pp. 1–292. Rome: Food and Agriculture Organization of the United Nations; D modified from Brown, D.E. & Holthuis, L.B. 1998. *Zool. Med. Leiden* 72: 113–41; used with permission.)

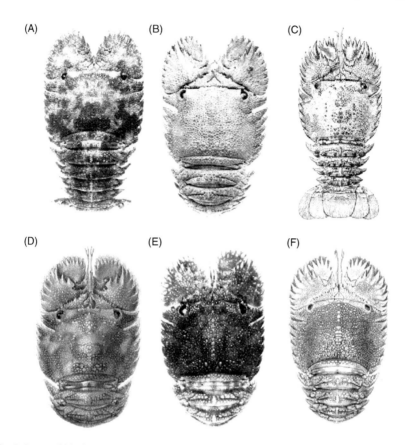

FIGURE 1.8   Lobsters within the *Parribacus* genus. (A) *Parribacus antarcticus* (Lund, 1793); (B) *P. caledonicus* Holthuis, 1960; (C) *P. holthuisi* Forest, 1954; (D) *P. japonicus* Holthuis, 1960; (E) *P. perlatus* Holthuis, 1967; (F) *P. scarlatinus* Holthuis, 1960. (Modified from Holthuis, L.B. 1991. *Marine Lobsters of the World. An Annotated and Illustrated Catalogue of the Species of Interest to Fisheries Known to Date*. FAO Species Catalogue No. 125, Vol. 13: pp. 1–292. Rome: Food and Agriculture Organization of the United Nations; used with permission.)

FIGURE 1.9   *Thenus orientalis* (Lund, 1793). (Modified from Holthuis, 1991. *Marine Lobsters of the World. An Annotated and Illustrated Catalogue of the Species of Interest to Fisheries Known to Date*. FAO Species Catalogue No. 125, Vol. 13: pp. 1–292. Rome: Food and Agriculture Organization of the United Nations; used with permission.)

FIGURE 1.10   Lobsters within the *Acantharctus* genus. (A) *Acantharctus delfini* (Bouvier, 1909); (B) *A. ornatus* (Holthuis, 1960); *Acantharctus posteli* (Forest, 1963) not shown, although an image of this species is currently available on www.ciesm.org/atlas/Scyllarusposteli.htm. Lobsters are not to scale. (A, B modified from Holthuis, L.B. 2002. *Zoosystema* 24: 499–683; used with permission.)

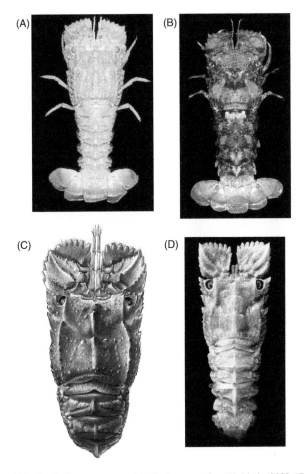

FIGURE 1.11   Lobsters within the *Bathyarctus* genus. (A) *Bathyarctus chani* Holthuis, 2002; (B) *B. formosanus* (Chan & Yu, 1992); (C) *B. rubens* (Alcock & Anderson, 1894); (D) *B. steatopygus* Holthuis, 2002. Lobsters are not to scale. (A, C, D modified from Holthuis, L.B. 2002. *Zoosystema* 24: 499–683; B modified from Chan, T.-Y. & Yu, H.-P. 1992. *Crustaceana* 62: 121–127; used with permission.)

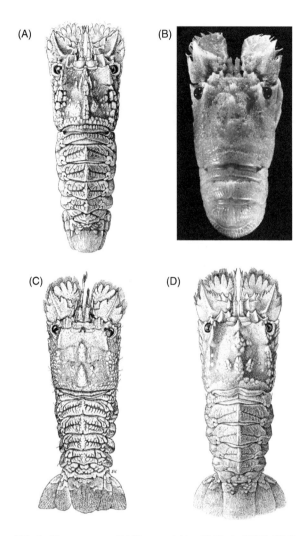

FIGURE 1.12 Lobsters within the *Biarctus* genus. (A) *Biarctus dubius* (Holthuis, 1963); (B) *B. pumilus* (Nobili, 1906); (C) *B. sordidus* (Stimpson, 1860); (D) *B. vitiensis* (Dana 1852). Lobsters are not to scale. (Modified from Holthuis, L.B. 2002. *Zoosystema* 24: 499–683; used with permission.)

respond to ecosystem changes and to exploitation. The chapters in this volume should help the reader in this goal by synthesizing the current knowledge that comprises the biology of scyllarid species.

1.4   Subfamilies of Scyllarids

In 1991, the family Scyllaridae included seven genera distributed in four subfamilies (Arctidinae, Ibacinae, Scyllarinae, and Theninae) (Holthuis, 1991). The Scyllaridae underwent considerable revision from 1991 to 2002 and now has 20 genera distributed within the previous subfamilies; mostly this revision involved species within the Scyllarinae (Holthuis 2002; see Webber & Booth, Chapter 2 for more information on the taxonomic revision of the family). While this chapter does not seek to provide a taxonomic key to each genus, Figure 1.4 to Figure 1.19 illustrate species within the various genera now recognized in each subfamily. Readers interested in knowing the numerous taxonomic distinctions between each species are referred to Holthuis (1991), Brown & Holthuis (1998), and Holthuis (2002); the distinctions are summarized here.

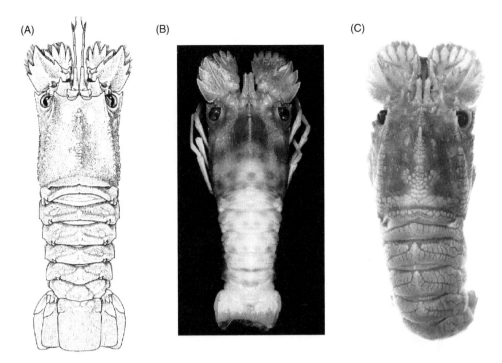

FIGURE 1.13 Lobsters within the *Chelarctus* genus. (A) *Chelarctus aureus* (Whitelegge, 1900); (B) *C. cultrifer* (Ortmann, 1897); (C) *C. crosnieri* Holthuis, 2002. Lobsters are not to scale. (Modified from Holthuis, L.B. 2002. *Zoosystema* 24: 499–683; used with permission.)

FIGURE 1.14 Lobsters within the *Crenarctus* genus. (A) *Crenarctus bicuspidatus* (De Man, 1905); (B) *C. crenatus* (Whitelegge, 1900). Lobsters are not to scale. (Modified from Holthuis, L.B. 2002. *Zoosystema* 24: 499–683; used with permission.)

The subfamily Arctidinae has two genera and 17 species. These are some of the larger scyllarids and thus may be of commercial interest, although Holthuis (1991) states that the *Arctides* species are of no interest (for fisheries information on *Scyllarides* species, see Sharp et al., Chapter 11; DiNardo & Moffitt, Chapter 12; Hearn et al., Chapter 14; and Spanier & Lavalli, Chapter 18). *Arctides* (Figure 1.4) and *Scyllarides* (Figure 1.5) species typically have a highly vaulted carapace, a three-segmented mandibular

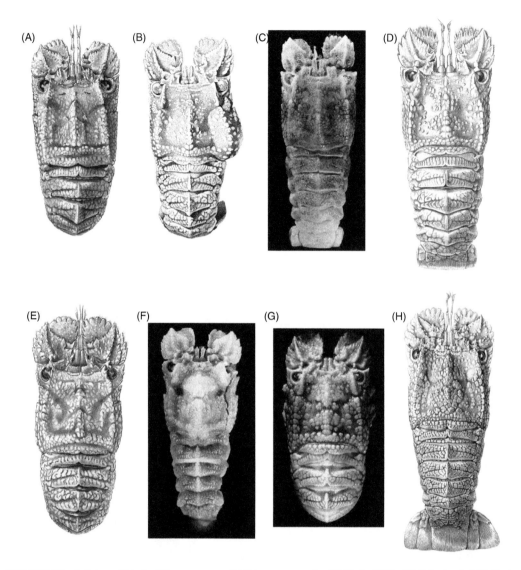

FIGURE 1.15 Lobsters within the *Eduarctus* genus. (A) *Eduarctus aesopius* (Holthuis, 1960); (B) *E. lewinsohni* (Holthuis, 1967); (C) *E. marginatus* (Holthuis, 2002); (D) *Eduarctus martensii* (Pfeffer, 1881); (E) *E. modestus* (Holthuis, 1960); (F) *E. perspicillatus* (Holthuis, 2002); (G) *E. pyrrhonotus* (Holthuis, 2002); (H) *E. reticulatus* (Holthuis, 2002). Lobsters are not to scale. (Modified from Holthuis, L.B. 2002. *Zoosystema* 24: 499–683; used with permission.)

palp, and a shallow cervical incision along the lateral margin of the carapace. They differ in the sculpturation of the dorsal surface of the abdominal somites — *Arctides* spp. are sculptured, whereas *Scyllarides* spp. are not and lack a transverse groove on the first abdominal somite.

The subfamily Ibacinae is comprised of three genera with a total of 15 species. In all species, the carapace is strongly dorsoventrally compressed with a deep cervical incision along the lateral margin of the carapace. The mandibular palp is simple or two segmented, in contrast to the Arctidinae. The single species of *Evibacus* (Figure 1.6) is characterized by entirely closed orbits, a closed cervical incision, the presence of a posteromedian tooth, and smooth, tuberculate body. In contrast, *Ibacus* (Figure 1.7) and *Parribacus* (Figure 1.8) species have orbits that are open anteriorly and possess a distinct, open cervical incision. Distinctions between *Ibacus* and *Parribacus* species include the dorsal body surface (smooth and punctuate with pubescence in *Ibacus*; coarsely squamose-tuberculate in *Parribacus*), the fifth abdominal

FIGURE 1.16 Lobsters within the *Galearctus* genus. (A) *Galearctus aurora* (Holthuis, 1982), (B) *G. rapanus* (Holthuis, 1993); (C) *G. timidus* (Holthuis, 1960); *G. kitanoviriosus* (Harada, 1962) and *G. umbilicatus* (Holthuis, 1977) are not shown. Lobsters are not to scale. (A, B photos by J. Poupin; C modified from Holthuis, L.B. 2002. *Zoosystema* 24: 499–683; used with permission.)

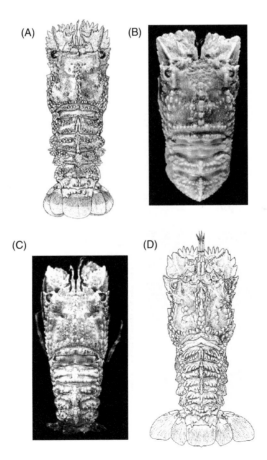

FIGURE 1.17 Lobsters within the *Petrarctus* genus. (A) *Petrarctus brevicornis* (Holthuis, 1946); (B) *P. demani* (Holthuis, 1946); (C) *P. rugosus* (H. Milne Edwards, 1837); (D) *P. veliger* (Holthuis, 2002). Lobsters are not to scale. (A modified from Holthuis, L.B. 1991. *Marine Lobsters of the World. An Annotated and Illustrated Catalogue of the Species of Interest to Fisheries Known to Date*. FAO Species Catalogue No. 125, Vol. 13: pp. 1–292. Rome: Food and Agriculture Organization of the United Nations; B to D modified from Holthuis, L.B. 2002. *Zoosystema* 24: 499–683; used with permission.)

FIGURE 1.18   *Remiarctus bertholdii* (Paulson, 1875). (Modified from Holthuis, L.B. 2002. *Zoosystema* 24: 499–683; used with permission.)

FIGURE 1.19   *Scammarctus batei* (Holthuis, 1946). (Modified from Holthuis, L.B. 2002. *Zoosystema* 24: 499–683; used with permission.)

segment (spined in *Ibacus*; spineless in *Parribacus*), and the number of mandibular palp segments (one in *Ibacus*; two in *Parribacus*).

The subfamily Theninae (Figure 1.9) has been considered monospecific, but there is a move to recognize at least two distinct species (Davie 2002; see Webber & Booth, Chapter 2 and Jones, Chapter 16 for a discussion of this revision and the characters warranting the distinctions between species). This subfamily is similar to the Scyllarinae in that both lack a flagellum on the exopod of the first and third maxillipeds. They differ in the placement of their orbits (to the extreme anterolateral extent of the carapace in Theninae; some distance from the anterolateral edge in Scyllarinae) and in the extent of dorsoventral compression

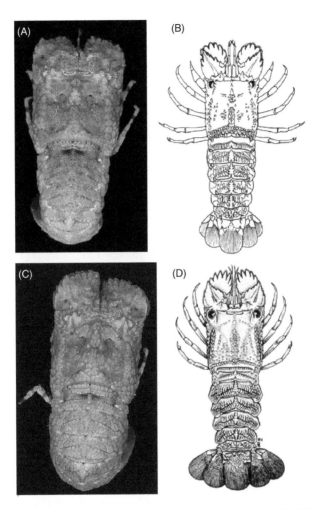

FIGURE 1.20   Lobsters within the *Scyllarus* genus. (A) *Scyllarus americanus* (S.I. Smith, 1869); (B) *S. arctus* (Linnaeus, 1758); (C) *S. chacei* Holthuis, 1960; (D) *S. pygmaeus* (Bate, 1888); *S. depressus* (S.I. Smith, 1881), *S. faxoni* Bouvier, 1917, *S. paradoxus* Miers, 1881, *S. planorbis* Holthuis, 1969, *S. subarctus* Crosnier, 1970, and *S. tutiensis* Srikrishnadhas, Rahman & Anandasekaran, 1991 not shown. *Scyllarus caparti* Holthuis, 1952 also not shown, although an image is currently available at www.ciesm.org/atlas/Scyllaruscaparti.html. Lobsters are not to scale. (A, C from Southeastern Regional Taxonomic Center; B, D modified from Holthuis, 1991. *Marine Lobsters of the World. An Annotated and Illustrated Catalogue of the Species of Interest to Fisheries Known to Date.* FAO Species Catalogue No. 125, Vol. 13: pp. 1–292. Rome: Food and Agriculture Organization of the United Nations.)

of the body (in Theninae the body is highly flattened, whereas in the Scyllarinae, the carapace is vaulted and covered in tubercles).

The subfamily Scyllarinae has 14 genera and 49 named species with two currently unnamed (Holthuis 2002; Radhakrishnan et al., Chapter 15). The species for which illustrations are available are presented in Figure 1.10 to Figure 1.20 and readers are referred to Holthuis (2002) for taxonomic distinctions differentiating the genera. The single species in each of the genera *Antarctus*, *Antipodarctus*, and *Gibbularctus* are not illustrated.

Acknowledgments

We very much appreciate and would like to thank the many authors, journal editors, and organizations that granted permission to use nearly all the available images of scyllarid species so that they could be compiled here in one location to aid identification for interested readers. This generosity on their part makes this introductory chapter more valuable and helps to update Lipke Holthuis' 1991 species catalog.

References

Ache, B.W. & Derby, C.D. 1985. Functional organization of olfaction in crustaceans. *Trends Neurosci.* 8: 356–360.

Atema, J. & Voigt, R. 1995. Behavior and sensory biology. In: Factor, J.R. (ed.), *The Biology of the Lobster Homarus americanus*: pp. 313–344. New York: Academic Press.

Bauer, R.T. & Martin, J.W. 1991. *Crustacean Sexual Biology.* New York: Columbia University Press. 351 pp.

Blaustein, D., Derby, C.D., & Beall, A.C. 1987. The structure of chemosensory centers in the brain of spiny lobsters and crayfish. *Ann. N. Y. Acad. Sci.* 510: 180–183.

Bliss, D.E. 1982–1985. *The Biology of Crustacea*, Vols. 1–10. New York: Academic Press.

Bliss, D.E. 1990. *Shrimps, Lobsters, and Crabs: Their Fascinating Life Story.* Dayton, OH: Morningside Bookshop. 242 pp.

Brown, D.E. & Holthuis, L.B. 1998. The Australian species of the genus *Ibacus* (Crustacea: Decapoda: Scyllaridae), with the description of a new species and addition of new records. *Zool. Med. Leiden* 72: 113–141.

Butler, M.J., IV. 1997. Benthic processes in lobster ecology: report from a workshop. *Mar. Freshw. Res.* 48: 659–662.

Butler, M.J., IV. 2001. The 6th International Conference and Workshop on Lobster Biology and Management: an introduction. *Mar. Freshw. Res.* 52: 1033–1035.

Chan, T.-Y. & Yu, H.-P. 1992. *Scyllarus formosanus*, a new slipper lobster (Decapoda: Scyllaridae) from Taiwan. *Crustaceana* 62: 121–127.

Cobb, J.S. & Phillips, B.F. 1980a. *The Biology and Management of Lobsters, Vol. 1: Physiology and Behavior.* New York: Academic Press. 463 pp.

Cobb, J.S. & Phillips, B.F. 1980b. *The Biology and Management of Lobsters, Vol. 2: Ecology and Management.* New York: Academic Press. 390 pp.

Crosnier, A. 1972. Naupliosoma, phyllosomes et pseudibacus de *Scyllarides herklotsi* (Herklots) (Crustacea, Decapoda, Scyllaridae) recoltes par l'ombango dans le sud du Golfe de Guinee. *Cahiers O.R.S.T.O.M. Oceanogr.* 10: 139–149.

Davie, P.J.F. 2002. Crustacea: Malacostraca, Phyllocarida, Hoplocarida, Eucarida (Part 1). In: Wells, A. & Houston, W.W.K. (eds.), *Zoological Catalogue of Australia*, 19.3A. Melbourne: CSIRO Publishing. 551 pp.

Derby, C.D. 1984. Molecular weight fractions of natural foods that stimulate feeding in crustaceans, with data from the lobster *Homarus americanus. Mar. Behav. Physiol.* 10: 273–282.

Derby, C.D. 2000. Learning from spiny lobsters about chemosensory coding of mixtures. *Physiol. Behav.* 69: 203–209.

Derby, C.D., Girardot, M.-N., Daniel, P.C., & Fine-Levy, J.B. 1989. Olfactory discrimination of mixtures: behavioral, electrophysiological, and theoretical studies using the spiny lobster *Panulirus argus.* In: Laing, D.G., Cain, W.S., McBride, R.L., & Ache, B.W. (eds.), *Perception of Complex Smells and Tastes*: pp. 65–81. Sydney: Academic Press.

Evoy, W. & Ayers, J. 1982. Locomotion and control of limb movements. In: Atwood, D.C. & Sandeman, H. (eds.), *The Biology of Crustacea: Neural Integration and Behavior*, Vol. 4: pp. 62–106. New York: Academic Press.

Factor, J.R. 1995a. Introduction, anatomy, and life history. In: Factor, J.R. (ed.), *The Biology of the Lobster Homarus americanus*: pp. 1–11. New York: Academic Press.

Factor, J.R. (ed.) 1995b. *The Biology of the Lobster Homarus americanus.* New York: Academic Press. 528 pp.

Frusher, S.D. & Gardner, C. 2005. Forward — The 7th International Conferences and Workshops on Lobster Biology and Managements: An introduction. *N. Z. J. Mar. Freshw. Res.* 39: 227–229.

Gray, H. 1992. *The Western Rock Lobster: Panulirus cygnus Book 1: A Natural History.* Geraldton, WA: Westralian Books. 112 pp.

Haond, C., Charmantier, G., Flik, G., & Wendelaar Bonga, S.E. 2001. Identification of respiratory and ion-transporting epithelia in the phyllosoma larvae of the slipper lobster *Scyllarus arctus*. *Cell Tissue Res.* 305: 445–455.

Harris-Warrick, R.M., Marder, E., Selverston, A.I., & Moulins, M. 1992. *Dynamic Biological Networks: The Stomatogastric Nervous System*. Cambridge, MA: The MIT Press.

Harrison, F.W. & Humes, A.G. 1992. *Microscopic Anatomy of the Invertebrates, Volume 10, Decapod Crustacea*. New York: Wiley-Liss, Inc. 474 pp.

Herrick, F.H. 1895. The American lobster: A study of is habits and development. *Bull. U.S. Fish. Comm.* 15: 1–252.

Holthuis, L.B. 1985. A revision of the family Scyllaridae (Crustacea: Decapoda: Macrura). I. Subfamily Ibacinae. *Zool. Verhand.* 218: 1–130.

Holthuis, L.B. 1991. *Marine Lobsters of the World. An Annotated and Illustrated Catalogue of the Species of Interest to Fisheries Known to Date*. FAO Species Catalogue No. 125, Vol. 13: pp. 1–292. Rome: Food and Agriculture Organization of the United Nations.

Holthuis, L.B. 1993. *Scyllarides obtusus* spec. nov., the scyllarid lobster of Saint Helena, Central South Atlantic (Crustacea: Decapoda Reptantia: Scyllaridae). *Zool. Med. Leiden* 67(36): 505–515.

Holthuis, L.B. 1996. Original watercolours donated by Cornelius Sittardus to Conrad Gesner in his (1558–1670) works on aquatic animals. *Zool. Verhand.* 70: 169–196.

Holthuis, LB. 2002. The Indo-Pacific scyllarinid lobsters (Crustacea, Decapoda, Scyllaridae). *Zoosystema* 24: 499–683.

Ingle, R. 1997. *Crayfishes, Lobsters and Crabs of Europe — An Illustrated Guide to Common and Traded Species*. London: Chapman & Hall. 281 pp.

Marinovic, B., Lemmens, J.W.T.J., & Knott, B. 1994. Larval development of *Ibacus peroni* Leach (Decapoda: Scyllaridae) under laboratory conditions. *J. Crust. Biol.* 14: 80–96.

Martins, H.R. 1985. Some observations on the naupliosoma and phyllosoma larvae of the Mediterranean locust lobster, *Scyllarides latus* (Latreille, 1803), from the Azores. *International Council for the Exploration of the Sea C.M.K.*, 52 Shellfish Committee, 13 pp.

Mikami, S. & Greenwood, G. 1997. Complete development and comparative morphology of larval *Thenus orientalis* and *Thenus* sp. (Decapoda: Scyllaridae) reared in the laboratory. *J. Crust. Biol.* 17: 289–308.

Moe, M.A., Jr. 1991. *Lobsters: Florida, Bahamas, the Caribbean*. Plantation, FL: Green Turtle Publications. 510 pp.

Phillips, B.F. 2006. *Lobsters: Biology, Management, Aquaculture and Fisheries*. Oxford, U.K.: Blackwell Publishing Ltd. 506 pp.

Phillips, B.F. & Kittaka, J. 2000. *Spiny Lobsters: Fisheries and Culture*. Oxford, U.K.: Blackwell Publishing Ltd. 704 pp.

Phillips, B.F., Cobb, J.S., & Kittaka, J. 1994. *Spiny Lobster Management*. Oxford, U.K.: Fishing News Books, Ltd. 550 pp.

Reaske, C.R. 1999. *The Compleat Crab and Lobster Book*. Revised Edition. Springfield, NJ: Burford Books. 176 pp.

Robertson, P.B. 1968. The complete larval development of the sand lobster, *Scyllarus americanus* (Smith) (Decapoda, Scyllaridae) in the laboratory with notes on larvae from plankton. *Bull. Mar. Sci.* 18: 294–342.

Robertson, P.B. 1971. The larvae and postlarva of the scyllarid lobster *Scyllarus depressus* (Smith). *Bull. Mar. Sci.* 21: 841–865.

Schram, F.R. 1986. *Crustacea*. New York: Oxford University Press. 620 pp.

Vuillemin, S. 1967. La respiration chez les Crustacés décapodes. *Ann. Biol.* 6: 47–81.

Williams, A.B. 1988. *Lobsters of the World — An Illustrated Guide*. New York: Osprey Books. 186 pp.

Wolvekamp, H.P. & Watermann, T.H. 1960. Respiration. In: Watermann, T.H. (ed.), *The Physiology of Crustacea*, Vol. 1: pp. 35–100. New York: Academic Press.

Part II

Biology of Slipper Lobsters

2

Taxonomy and Evolution

W. Richard Webber and John D. Booth

CONTENTS

Abstract

The Scyllaridae (slipper lobsters) is a highly distinctive family, recognized for millennia. The higher taxonomy of the family is reasonably settled due particularly to the contributions of Dr. Lipke Holthuis. Eighty-five scyllarid species in 20 genera have been identified to date, divided among four subfamilies. Further species will inevitably be described, particularly in the Scyllarinae, which includes well over half the scyllarid species and is distributed throughout a greater range of latitudes and depths than the other three subfamilies. Scyllarids are closely related to the Palinuridae and Synaxidae. The three families (together, the Achelata) share numerous characters, most notably their phyllosoma larvae, that separate the Achelata from all other Decapoda. The plate-like antennal flagellum of slipper lobsters distinguishes them clearly from palinurids and synaxids that possess whiplike antennae. Characters, such as the presence of multiarticulate exopods on the maxillipeds of the subfamilies Ibacinae and Arctidinae and their absence from the Theninae and Scyllarinae, represent well-defined differences between the higher taxa. Relationships within the four subfamilies are less clear and the primitive or derived nature of some morphological characters remains to be resolved. Thirteen new genera have recently been named in the Scyllarinae, by far the largest and least known of the slipper lobster subfamilies. More information is necessary to verify this division. Fossil and morphological evidence, the distribution of extant species, their developmental characteristics, ecology, and biology suggest the Scyllaridae are Tethyan in origin and have evolved with the development of the major oceans. Scyllarids appear to have undergone recent radiations into shallow, shelf, and deep water habitats. The family is pantropical and it is most unlikely that it, or any group within it, is the relict of a formerly more widespread population. Some distinctive genera are, however, characteristic of certain regions. The evolution of scyllarid morphology and behavior appears to have favored cryptic lifestyles. Larval durations in most (but not all) scyllarid species are shorter than in palinurids, and there is evidence of abbreviated development in some continental shelf species, indicating specialization.

2.1   Introduction

The scyllarid lobsters have attracted an interesting variety of common names, compared to their closest relatives, the palinurid and synaxid lobsters. These names include slipper lobsters, shovel-nosed lobsters, squat lobsters, butterfly lobsters, locust lobsters, bugs, and more in various languages. This variety is not because of confusion with other kinds of animals, but rather because slipper lobsters are so distinctive due to their unique platelike antennae — a characteristic that cannot be confused with any other decapod or, for that matter, any other crustacean, and which is common to all 85 species of Scyllaridae named to date. Their distinctiveness has been recognized throughout recorded history, but they have not attracted the level of scientific investigation that palinurid or nephropid lobsters have, despite having more than half again as many species as the Palinuridae and vastly outnumbering nephropid species (see Holthuis 1991a for a compilation of lobster species within various families). The lack of interest is probably due to the fact that they are adept at blending into their background, few are of significant economic interest, and their homogeneity of form and obvious affinity have made it convenient to simply place them in the Scyllaridae and think little more of them.

Scyllarid lobsters have evolved a unique body form, but they are lobsters and clearly related to the Palinuridae and Synaxidae. They share many significant morphological characters particularly their long-lived, teleplanic phyllosoma larvae, a developmental form unique to these three families. The relationship of groups within the reptantian (crawling) decapods has become a hot topic over the past decade or so and the position of these three families within the Reptantia has been very much part of the debate. Despite this, evolution within the Scyllaridae has gained little attention and refinement of the taxonomic classification below family level has been cautious. Preeminent among contributors to the taxonomy of the Scyllaridae has been Dr. Lipke Holthuis of the Netherlands who, over 56 years, formalized much of the present classification of the family. Most recently, he published a revision of Indo-Pacific species of the subfamily Scyllarinae, adding 13 new genera to the seven previously recognized in the Scyllaridae and accounting for more than half the known species in the family (Holthuis 2002). This work emphasizes the need for a better understanding of the evolution of species within the family.

In this chapter, we deal first with the taxonomy of the Scyllaridae — the family's history, its present classification, and a list of currently recognized species — and then we outline characteristics giving rise to the divisions within the family. This is followed by a discussion of the evolution of scyllarids, in which we consider evidence from a number of sources, particularly fossils and the morphology of recent species, but also distributions, developmental characteristics, and aspects of ecology and biology.

2.2   Taxonomy

2.2.1   History

Holthuis (1985), in his revision of the subfamily Ibacinae, provided a synonymy and definition of the family Scyllaridae, a detailed description of scyllarid morphology (including a labeled illustration of carapace features, reproduced in Holthuis 1991a), and a key to the recent genera. In a brief section on the relationships of the Scyllaridae, he observed that they "form a sharply defined natural group, most closely related to the Palinuridae but quite distinct from that family." He also noted that "the short, flat, plate-like antenna of the Scyllaridae distinguishes them immediately from the Palinuridae with cylindrical antennae and long, stiff, multi-articulated flagella." Scyllarids are, therefore, easy to recognize, and the history of the family bears this out.

The pre-Holthuis history of scyllarid taxonomy is well summarized by Holthuis in revisions of the Ibacinae (1985) and Scyllarinae (2002) and in extensive synonymies (Holthuis 1946, 1991a; Brown & Holthuis 1998). Interesting details and illustrations also appear in his descriptions of work by early European naturalists (Holthuis 1991b, 1996a, 1996b; Holthuis & Sakai 1970). The following account gives milestones of this history and more details of post-1946 developments in scyllarid taxonomy.

Holthuis (1985) observed that the distinctiveness of scyllarid lobsters has been recognized for a very long time — Aristotle referred to a crustacean under the name "arctus" which he says is "generally assumed [to have been] a scyllarid." Prior to the system of nomenclature we recognize today, which became established in the tenth edition of Linnaeus's *Systema Naturae* of 1758, various names and naming systems were employed by European naturalists. The five species we now know as *Scyllarus arctus* (Linnaeus, 1758), *Scyllarides latus* (Latreille, 1802), *S. aequinoctialis* (Lund, 1793), *Parribacus antarcticus* (Lund, 1793) and *Thenus orientalis* (Lund, 1793) had been identified and named a number of times in literature, long before 1758. For instance, Holthuis (1985) reported that *S. arctus* and *S. latus* were described and illustrated by Belon (1553) and Rondelet (1554), and Barrelier gave the name "Squilla Ursa minor altera remipes" for *S. arctus* illustrated in a book published in 1714. The quality of some of these early illustrations is such that Holthuis (1991a) was able to select the specimen depicted in a painting of *S. latus* published by Gesner (1558) as lectotype of the species (Holthuis 1996b). See Lavalli & Spanier, Chapter 1 (Figure 1.1) for a depiction of this painting.

Linnaeus (1758) established the first scyllarid species named using the binomial system as *Cancer arctus* and, under this name, included the four other species previously described. The genus *Scyllarus* was erected for *S. arctus* in 1775 by Fabricius to remove it from *Cancer* and, between then and 1815, the four other pre-Linnaean species were moved into the genus *Scyllarus*, along with one new species, *S. guineensis* Spengler, 1799 (now *Arctides guineensis* (Spengler, 1799)). Leach (1815, 1816) erected *Ibacus* for his new species *I. peronii* Leach, 1815 and *Thenus* for *T. indicus* Leach, 1816 (see remark in Table 2.1). *Thenus indicus* was synonymized with *S. orientalis* (as *Thenus orientalis*) by H. Milne Edwards (1837) in a revision of the Crustacea that listed ten species of scyllarid lobsters in the three previously named genera (*Scyllarus, Ibacus* and *Thenus*), including two species of his own (see species list in Table 2.1). He also recognized two groups within the genus *Scyllarus* that correspond to the currently recognized genera *Scyllarus* and *Scyllarides*. De Haan (1849) divided the Scyllaridae into five genera, with the names *Scyllarus* (corresponding to today's *Scyllarides*), *Arctus* (now *Scyllarus*), *Ibacus* A (now *Parribacus*), *Ibacus* B (now *Ibacus*), and *Thenus*. The genus *Parribacus* was erected by Dana (1852) to replace De Haan's *Ibacus* A and *Evibacus* was erected by Smith (1869), thereby establishing six of the seven scyllarid genera recognized until 2002. These six genera and *Arctides*, which Holthuis added in 1960, are still in use today.

Twenty-four species were described between 1841 and 1916 when De Man published a list of 43 species and varieties in the six established genera, as well as five *Pseudibacus* and two *Nisto* species. Bouvier (1915), however, considered *Pseudibacus* and *Nisto* to be the "natant-stage" of species of *Scyllarides* and *Scyllarus*, respectively (see De Man 1916). Holthuis (1991a) confirmed this and recorded three of the *Pseudibacus* species as synonyms of *Scyllarides* species — the other two *Pseudibacus* species have apparently not been identified with named scyllarid species — and the *Nisto* species as synonyms for *Scyllarus arctus*. Holthuis (1946) reviewed De Man's list, made a number of corrections, and recorded seven further names, bringing the number of recognized species and varieties at that time to 50.

From 1917 to 1946, only five additional species of Scyllaridae were described, with Holthuis describing three of these in 1946. Holthuis' 1946 paper on the macruran decapods of the *Snellius* Expedition heralded the beginning of his contribution to the taxonomy of the Scyllaridae — he has since added 35 species to the family, 34 as sole author (Holthuis 1952, 1960, 1963, 1967, 1968, 1977, 1982, 1993a, 1993b, 2002) and one in collaboration (Brown & Holthuis 1998). Six species have also been described by other authors

TABLE 2.1

Classification and Species List of the Family Scyllaridae (Slipper Lobsters)

Order **Decapoda** Latreille, 1802
Suborder **Reptantia** Boas, 1880
Infraorder **Achelata** Scholtz & Richter, 1995
Family **Scyllaridae** Latreille, 1825
Subfamily **Arctidinae** (Holthuis, 1985)
  Arctides Holthuis, 1960
    Arctides antipodarum Holthuis, 1960
    Arctides guineensis (Spengler, 1799)
    Arctides regalis Holthuis, 1963
  Scyllarides Gill, 1898
    Scyllarides aequinoctialis (Lund, 1793)
    Scyllarides astori Holthuis, 1960
    Scyllarides brasiliensis Rathbun, 1906
    Scyllarides deceptor Holthuis, 1963
    Scyllarides delfosi Holthuis, 1960
    Scyllarides elisabethae (Ortmann, 1894)
    Scyllarides haanii (De Haan, 1841)

Scyllarides herklotsii (Herklots, 1851)
Scyllarides latus (Latreille, 1802)
Scyllarides nodifer (Stimpson, 1866)
Scyllarides obtusus Holthuis, 1993
Scyllarides roggeveeni Holthuis, 1967
Scyllarides squammosus (H. Milne Edwards, 1837)
Scyllarides tridacnophaga Holthuis, 1967

Subfamily **Ibacinae** Holthuis, 1985
  Evibacus S.I. Smith, 1869
    Evibacus princeps S.I. Smith, 1869
  Ibacus Leach, 1815
    Ibacus alticrenatus Bate, 1888
    Ibacus brevipes Bate, 1888
    Ibacus brucei Holthuis, 1977
    Ibacus chacei Brown & Holthuis, 1998
  Parribacus Dana, 1852
    Parribacus antarcticus (Lund, 1793)
    Parribacus caledonicus Holthuis, 1960
    Parribacus holthuisi Forest, 1954

Ibacus ciliatus (Von Siebold, 1824)
Ibacus novemdentatus Gibbes, 1850
Ibacus peronii Leach, 1815
Ibacus pubescens Holthuis, 1960

Parribacus japonicus Holthuis, 1960
Parribacus perlatus Holthuis, 1967
Parribacus scarlatinus Holthuis, 1960

Subfamily **Scyllarinae** Latreille, 1825
  Acantharctus Holthuis, 2002
    Acantharctus delfini (Bouvier, 1909)
    Acantharctus ornatus (Holthuis, 1960)
    Acantharctus posteli (Forest, 1963)
  Antarctus Holthuis, 2002
    Antarctus mawsoni (Bage, 1938)
  Antipodarctus Holthuis, 2002
    Antipodarctus aoteanus (Powell, 1949)
  Bathyarctus Holthuis, 2002
    Bathyarctus chani Holthuis, 2002
    Bathyarctus formosanus (Chan & Yu, 1992)
    Bathyarctus rubens (Alcock & Anderson, 1894)
    Bathyarctus steatopygus Holthuis, 2002
  Biarctus Holthuis, 2002
    Biarctus pumilus (Nobili, 1906)
    Biarctus vitiensis (Dana, 1852)
    Biarctus sordidus (Stimpson, 1860)
    Biarctus dubius (Holthuis, 1963)
  Chelarctus Holthuis, 2002
    Chelarctus aureus (Holthuis, 1963)
    Chelarctus cultrifer (Ortmann, 1897)
    Chelarctus crosnieri Holthuis, 2002
  Crenarctus Holthuis, 2002
    Crenarctus crenatus (Whitelegge, 1900)
    Crenarctus bicuspidatus (De Man, 1905)

Eduarctus Holthuis, 2002
  Eduarctus martensii (Pfeffer, 1881)
  Eduarctus aesopius (Holthuis, 1960)
  Eduarctus lewinsohni (Holthuis, 1967)
  Eduarctus marginatus Holthuis, 2002
  Eduarctus modestus (Holthuis, 1960)
  Eduarctus perspicillatus Holthuis, 2002
  Eduarctus pyrrhonotus Holthuis, 2002
  Eduarctus reticulatus Holthuis, 2002
Galearctus Holthuis, 2002
  Galearctus timidus (Holthuis, 1960)
  Galearctus aurora (Holthuis, 1982)
  Galearctus kitanoviriosus (Harada, 1962)
  Galearctus rapanus (Holthuis, 1993)
  Galearctus umbilicatus (Holthuis, 1977)
Gibbularctus Holthuis, 2002
  Gibbularctus gibberosus (De Man, 1905)
Petrarctus Holthuis, 2002
  Petrarctus rugosus (H. Milne Edwards, 1837)
  Petrarctus brevicornis (Holthuis, 1946)
  Petrarctus demani (Holthuis, 1946)
  Petrarctus veliger Holthuis, 2002
Remiarctus Holthuis, 2002
  Remiarctus bertholdii (Paulson, 1875)
Scammarctus Holthuis, 2002
  Scammarctus batei (Holthuis, 1946)

(Continued)

TABLE 2.1

(Continued)

Subfamily **Scyllarinae** Latreille, 1825 (continued)	
Scyllarinae sp 1 (New Zealand)	
Scyllarinae sp 2 (Tasman Sea)	
Scyllarus Fabricius, 1775	
Scyllarus americanus (S.I. Smith, 1869)	*Scyllarus faxoni* Bouvier, 1917
Scyllarus arctus (Linnaeus, 1758)	*Scyllarus paradoxus* Miers, 1881
Scyllarus caparti Holthuis, 1952	*Scyllarus planorbis* Holthuis, 1969
Scyllarus chacei Holthuis, 1960	*Scyllarus pygmaeus* (Bate, 1888)
Scyllarus depressus (S.I. Smith, 1881)	*Scyllarus subarctus* Crosnier, 1970
Subfamily **Theninae** Holthuis, 1985	
Thenus Leach, 1815	*Thenus* sp. (Australia)[a]
Thenus orientalis (Lund, 1793)	*Thenus* sp. (Australia)

[a] Davie (2002) lists *Thenus orientalis* (Lund, 1793) and *T. indicus* Leach, 1816 as presently recognized in Australian waters, but indicates that, in a revision of *Thenus* yet to be published, both will be excluded from the Australian fauna and two new species named in their place. H. Milne Edwards (1837) synonymized *T. indicus* with *T. orientalis* and this synonymy was recognized by Holthuis (1991) and, for the purposes of compiling this list is followed here.

since 1946 and, with the synonymy or rejection of 10 previously recognized species over this time, we have listed 85 currently recognized species (Table 2.1).

Holthuis established the seventh scyllarid genus, *Arctides*, in 1960, but not until 1985 did he set the subfamilies of the Scyllaridae firmly in place. In the latter paper, he provided a definition of the Scyllaridae, a detailed description of scyllarid morphology, and a key to the seven scyllarid genera. Of the four subfamilies, the Ibacinae was defined and revised, while the Arctidinae, Scyllarinae, and Theninae were established without diagnosis; the latter three subfamilies were subsequently defined in an FAO key by Holthuis (1991a).

Holthuis has not been the only provider of systematic information on the family. George & Griffin (1972) described the slipper lobsters of Australia, a work later extended by Griffin & Stoddart (1995). Chan & Yu (1986) published an account of the scyllarid lobsters of Taiwan, and later put out a book on the lobsters of Taiwan with good-quality color photographs of each of the species represented (Chan & Yu 1993). A list of the marine lobsters of the world of interest to U.S. trade, including 63 species of Scyllaridae, was published by Williams (1988). But the best known and least dispensable account of the Scyllaridae, and all other marine lobsters, is the FAO Catalogue *Marine Lobsters of the World* by Holthuis (1991a). This book provides illustrations, distribution maps, and useful biological information on the Ibacinae, Arctidinae, and Theninae, along with seven of the 47 species of the Scyllarinae known at the time. Because they are of little interest to fisheries, the remaining 40 scyllarinid species were simply listed by name, along with superseded names still in use. It would not be until a decade later that this gap was filled.

A paper on Australian species of *Ibacus*, which, in most respects, is a second revision of the genus, was published by Brown and Holthuis in 1998. Seven of the eight known species of the genus have been recorded in Australian waters, and Brown and Holthuis provide accounts of them, along with a key and excellent color photographs of all eight *Ibacus* species.

The most recent major work on the Scyllaridae is that of Holthuis (2002), in which he revised the Indo-Pacific species of the Scyllarinae. A total of 38 species, eight of them new, were assigned to 13 new genera in the subfamily (see the species list in Table 2.1), while all but one of the Atlantic Ocean species (*Acantharctus posteli* (Forest, 1963), previously *S. posteli*) were retained in the genus *Scyllarus*, which had previously encompassed all species in the subfamily.

2.2.2 Classification of Presently Recognized Taxa

Recent interest in the phylogeny of the Decapoda, and particularly the Reptantia (lobsters, lobster-like forms, crabs, crab-like forms), has changed perceptions of taxa above the family Scyllaridae. The usefulness of Linnaean ranks has been questioned, or these ranks have been avoided altogether (e.g., Scholtz & Richter 1995), although the Decapoda and Reptantia have survived recent phylogenetic analyses as recognized taxa. The name "Achelata" (meaning without chelae, although the presence of chelae on the female fifth pereiopod in some groups of achelate lobsters is acknowledged) is the unranked product of one of these analyses, and was coined to incorporate the closely related families Scyllaridae, Synaxidae, and Palinuridae (Scholtz & Richter 1995). Achelata is used here in place of Palinuroidea Latreille, 1802, first applied to these three families by Holthuis (1985) who "corrected" the name from Palinurini Latreille, 1802. Its existence has hardly been acknowledged in recent phylogenetic studies, being replaced by Achelata with the removal of the infraorder Palinura. We make no argument here for or against the disuse of Palinuroidea, but Achelata has gained currency (e.g., Richter & Scholtz 2001; Dixon et al. 2003; Poore 2004) and is a distinctive and probably less confusing name for the group. Achelata is ranked here as an infraorder, in line with the ranks above and below it. This is consistent with the proposal of Ahyong & O'Meally (2004) that requires Achelata to be an infraorder, if their classification of the Reptantia is to be placed in a Linnaean framework.

The list presented in Table 2.1 includes a total of 85 species, four of them unnamed. Two of the unnamed species are of Scyllarinae from the Southwest Pacific; the other two are species of *Thenus* identified from Australia as differing from *Thenus orientalis* and *T. indicus*. *Thenus indicus* was synonymized with *T. orientalis* by H. Milne Edwards (1835), a synonymy that was maintained by Holthuis (1991a), but appears as a separate species in Davie (2002). There are 17 species in the Arctidinae, 15 in the Ibacinae, 50 in the Scyllarinae, and three in the subfamily Theninae (for illustrations of adults, see Lavalli & Spanier, Chapter 1).

With the division of the Scyllarinae into 14 genera (Holthuis 2002), *Scyllarides* has become the largest genus in the family with 14 species, and *Scyllarus* is the next largest genus with 10 species. In the review of Holthuis (2002), *Scyllarus* was included in the key to genera but not otherwise reviewed. A similar review of *Scyllarus* may result in changes and, possibly, additional genera in the subfamily.

2.2.3   Undescribed Species

Offshore exploratory cruises continue to catch undescribed scyllarids, most often species of Scyllarinae. A small scyllarinid from northern New Zealand slope waters awaits description at the Museum of New Zealand and an unnamed species of *Galearctus* was recently collected in deepwater in the northern Tasman Sea (Williams 2003). Further species of Scyllarinae almost certainly remain undetected because all those described in the subfamily (50 to date) are small, include the deepest and rarest scyllarid species yet recorded, and already include 60% of all the scyllarid species known. The greatest potential for more scyllarinid species is probably in the unsampled mid-latitudes of the North and South Pacific, Tasman Sea, and Indian and Atlantic Oceans.

Coastal waters at low latitudes will probably yield further species as well. *Parribacus antarcticus*, for example, was described in 1793 (as *Scyllarus antarcticus*), while two species of this medium-sized and typically shallow-water genus previously confused with *P. antarcticus* (*P. caledonicus* Holthuis, 1960 of the South West Pacific and *P. japonicus* Holthuis, 1960 of southern Japanese waters) and three new species (*P. perlatus* Holthuis, 1967 of Easter Island, *P. holthuisi* Forest, 1954 from the Tuamotu Archipelago and *P. scarlatinus* Holthuis, 1960 of the west central Pacific) were all described relatively recently, between 1954 and 1967. New and improving analytical techniques, notably molecular, sperm morphology, and anatomical studies of existing species, are also likely to lead to revised species definitions. For example, the intention to replace *Thenus orientalis* and *T. indicus* in Australian waters with new names (Davie 2002) was noted earlier, as was a potential revision of *Scyllarus* species in the Atlantic, which may also reveal new taxa.

2.2.4   Definitions of the Taxa

Taxa within the Scyllaridae are primarily distinguished on the basis of adult and, to some extent, larval characters. Figure 2.1 provides a summary and checklist of the adult characters (abbreviated) used to divide the Scyllaridae into its subfamilies and genera and Figure 2.2 illustrates a selection of these characters. The characters are brought together from descriptions and various keys to these taxa in Holthuis (1985, 1991a,

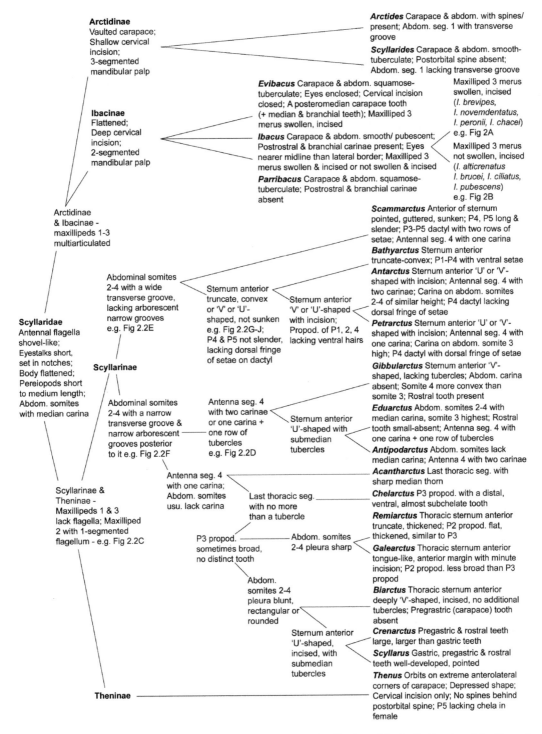

FIGURE 2.1 Diagram summarizing the key characters used in the classification of the Scyllaridae, its subfamilies and genera (based on Holthuis 1985, 1991, 2002). Descriptions of characters are abbreviated and simplified — refer to the papers of Holthuis for more detailed definitions. Note: this diagram does not represent a proposed phylogeny (although the morphological characters used are certainly of significance in indicating evolutionary relationships among the Scyllaridae). Abbreviations: abdom. = abdomen; P2–P5 = pereiopods 2–5; seg. = segment; usu. = usually; propod. = propodite.

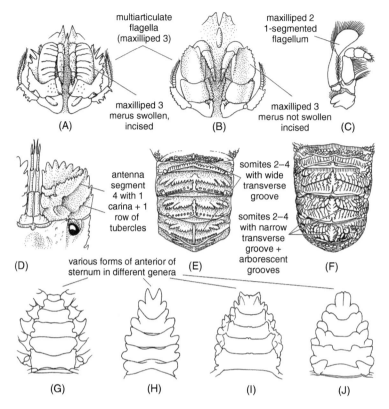

FIGURE 2.2   Illustrations of some of the key characters summarized in Figure 2.1. A and B based on Holthuis (1985) and Poore (2004); D, E, and G–J based on Holthuis (2002); C and F based on unpublished drawings.

2002) and references in those papers, where they are fully defined. In this section, we give an overview of the more distinctive morphological characters of these higher taxa. Prior to 2002, the Scyllarinae was represented by a single genus (*Scyllarus*), the definition of which was at a comparative level of detail to definitions of the other six scyllarid genera. Characters which distinguish genera of the Scyllarinae are more detailed and not discussed fully here, but are included in Figure 2.1.

The family itself is, of course, characterized by its shovel-like, unsegmented antennal flagella (so much so that this is the only character provided by Holthuis 1985 in his family definition), but is also distinguished by having, in combination, short eyestalks set into notches at or immediately behind the front of the carapace, a carapace that is somewhat to extremely flattened with concomitantly blunt to sharp lateral margins, pereiopods of short to medium length with characteristically sharp and hardened dactyl tips, and a median dorsal carina on the abdomen.

Scyllarid phyllosomas are also readily distinguishable from those of the Palinuridae and Synaxidae in having short antennae that become progressively flattened and bilobed during growth. The phyllosoma larvae of the genera *Arctides, Scyllarides, Evibacus, Ibacus, Parribacus* and *Thenus*, and the Scyllarinae are also distinct from each other (see Sekiguchi et al., Chapter 4, Figure 4.4). At this stage, our knowledge of scyllarinid larvae is too fragmentary to reveal larvae typical of genera in the subfamily. A group of some 13 species armed with spines on the abdomen above the origins of pereiopods, and bearing a distinctly forked telson has been recognized (Webber & Booth 2001). However, only five of these species are identified, in four genera that do not match any scyllarinid genus or group of genera, and the unidentified species are only known as phyllosomas. Postlarvae (nistos) of the Scyllaridae are adult-like and quite obviously scyllarid in character but, with few exceptions, cannot be identified to the genus level.

A fundamental division of adult Scyllaridae distinguishes the Arctidinae and Ibacinae from the Scyllarinae and Theninae: the first, second, and third maxillipeds of the Arctidinae and Ibacinae each have a multiarticulate flagellum; the first and third maxillipeds of the Scyllarinae and Theninae lack a flagellum altogether and the second maxilliped has a flagellum of only one segment. Species of the Arctidinae have a vaulted carapace, a shallow cervical incision, and a three-segmented mandibular palp that clearly separate them from those of the Ibacinae. Ibacinae are distinguished by being dorsoventrally compressed with a deep cervical incision laterally in the front half of the carapace, and by having a palp of two segments on the mandible.

The two genera comprising the Arctidinae, *Arctides* and *Scyllarides*, are rather similar, but distinguishable on the basis of upper-surface sculpturing. *Arctides* species are armed with spines, tubercles, and arborescent grooves on the abdominal somites, have a postorbital spine, and have a transverse groove on the first abdominal segment. In *Scyllarides*, upper surfaces are smoothed by flattened, close-set tubercles that tend to cover the abdominal carina and obscure other tubercles; they possess neither a postorbital spine, nor a transverse groove on the first abdominal somite.

The three flattened genera that make up the Ibacinae are more easily distinguished. *Parribacus* has a smoothed-over, squamose-tuberculate carapace and abdomen, armature similar to that of *Scyllarides*, and lacks all evidence of postrostral or branchial carinae. *Evibacus* is also squamose-tuberculate, although more finely than *Parribacus,* and possesses postrostral and branchial carinae armed with rows of blunt teeth. It also has a distinctive posteromedian tubercle on the carapace. The eye becomes uniquely enclosed in *Evibacus*, and the deep cervical groove occluded, and the merus of the third maxilliped is enlarged and swollen with deep transverse incisions on its exposed surface. *Ibacus* species are the most flattened of all scyllarids; the carapace and abdomen are smooth and, in some species pubescent (covered by numerous, fine hairs); postrostral and branchial carinae are clearly evident, although they lack tubercles; and the eyes are positioned nearer to the midline than the lateral borders. A further morphological separation within the Ibacinae can be observed: in *I. brevipes* Bate, 1888*, I. novemdentatus* Gibbes, 1850*, I. peronii* Leach, 1815, and *I. chacei* Brown & Holthuis, 1998 the merus of the third maxilliped is enlarged and swollen, with deep transverse incisions similar to *Evibacus*. In *I. alticrenatus* Bate, 1888, *I. brucei* Holthuis, 1977, *I. ciliatus* (Von Siebold, 1824), and *I. pubescens* Holthuis, 1960, the merus is flat to concave on its posterior (exposed or aboral) surface, as it is in the remaining scyllarid species. This would appear to be a decisive difference between the two ibacinid groups but is, as yet, not formalized in scyllarid taxonomy.

The subfamily Theninae, which is monotypic, is most distinctive in having its orbits at the extreme, anterolateral margin of the carapace with no projections of this margin beneath the eye. In addition, the carapace is dorsoventrally compressed, bears only a cervical incision on the lateral margin, and lacks lateral spines or tubercles posterior to the postcervical spine; in females, the fifth pereiopod is achelate. In the Scyllarinae, the orbits are not at the extreme anterolateral margin of the carapace and this margin projects beneath the eye, the carapace is not strongly flattened but rather more cylindrical, there is both a cervical and a postcervical incision, there are numerous teeth or tubercles on the carapace lateral margin, and the female has a chelate fifth pereiopod.

The genera of Scyllarinae were subdivided by Holthuis (2002) based on the width of the transverse groove on the second to fourth abdominal somites and on the nature of the armature either side of this groove. This division resulted in a group lacking an arborescent pattern of smaller grooves (*Scammarctus, Bathyarctus, Antarctus*, and *Petrarctus*) and a group of the remaining ten genera with arborescent grooves posterior to the transverse groove. Further divisions are based, in part, on the shape, profile, and armature of the anterior margin of the thoracic sternum, and on the presence or absence on the fourth peduncular segment of the antenna of a second carina or row of tubercles lateral to the single carina seen in all scyllarinid genera (see Figure 2.1 and Figure 2.2). In some genera, the median dorsal carina on the abdominal somites is lacking — an interesting development in light of the presence of this carina being a defining character of the family itself.

In summary, the Scyllaridae can be viewed as comprising several distinctive morphological "forms". These forms are clearly adaptive to the various habitats in which they live, described in more detail in a later section on distributions, but touched on here where relevant. *Arctides* and *Scyllarides* are vaulted (subcylindrical), medium to large lobsters of about 20 to 50 cm in total length (TL). The sculpturing of spines and tubercles in *Arctides* and a covering of close-set, flattened tubercles in *Scyllarides* along with dull coloration, clearly enable them to blend with rocky (*Arctides*) and rocky to soft bottom (*Scyllarides*) substrates. The nocturnal habits of *Arctides* further suggest that a cryptic life style is important. *Parribacus* species are small to medium sized (11 to 20 cm TL) and are not unlike *Arctides* in being cryptically sculpted and colored, again to blend with their typically shallow, rocky to sandy habitats. *Ibacus* species are medium sized (12 to 23 cm TL) and have taken the dorsoventrally compressed body form to the extreme. They have also become smooth surfaced, are uniformly colored or lightly patterned, and some are also finely pubescent. These characteristics are no less cryptically adaptive than the roughened to spiny surfaces of other scyllarids: *Ibacus* species inhabit soft bottoms of mud or sand. So, too, does *Evibacus*, another genus that is very flattened and broad, although larger than *Ibacus* species at 33 cm maximum TL (Holthuis 1991a). *Evibacus* also bears low teeth on the carapace and has a finely squamose-tuberculate surface, but there is little doubt that it, too, is well adapted to blending into its habitat. *Thenus* also appears to have adapted to the sedimentary bottoms it inhabits by becoming flattened with low-profile median spines on its carapace, squamose-tuberculate surfaces, and uniform coloration. The genus (long identified as the single species *T. orientalis*), which can reach a total length of 25 cm (Holthuis 1991a), is not as flattened as *Ibacus*, and is very widespread geographically. At this point, the division between the Arctidinae and Ibacinae (with multiarticulate flagella on all three maxillipeds) and the Theninae and Scyllarinae (with a single-segmented flagellum on the second maxilliped only) becomes apparent. The adaptations of the Theninae to soft bottoms imply convergence with the flattened species of the Ibacinae.

While the Scyllarinae include 60% of scyllarid species, occupy the deepest to shallowest depths recorded in the family (e.g., *Eduarctus martensii* (Pfeffer, 1881) ranges from 4 to 887 m depth), and are found on all manner of substrates, they are still distinctive and difficult to confuse with any of the five genera discussed above. Scyllarinids are small species, most being < 10 cm TL. Some, such as *Petrarctus brevicornis* (Holthuis, 1946), reach no more than 5.5 cm TL, but a few, like *Scyllarus arctus*, can attain 16 cm TL. Just as large size may afford some safety to large species of *Scyllarides* that can sometimes be found on open bottoms away from shelter, small size probably confers some advantages to scyllarinids. They are invariably sculpted with grooves, tubercles, spines, and carinae in various combinations. Most also have disruptive color patterns with a few, notably *Chelarctus crosnieri* Holthuis, 2002, very brightly patterned in a variety of hues (Holthuis 2002). Scyllarinids are notoriously difficult to see in their habitats (several species are known from very few specimens indeed), and clearly benefit from this combination of small size and cryptic appearance.

2.3   Evolution

Evolution in the Scyllaridae can be viewed at different levels: evolution of the Achelata within the Reptantia, of the family Scyllaridae within the Achelata, and of species groups and species within the family itself. Data used in assessing evolution include morphology and anatomy of extant species as adults and throughout development, functional morphology, genes and other molecules, fossils, and sperm morphology, along with distribution, habitat, behavior, and other aspects of biology. Such data are used to construct phylogenies that reflect likely relationships, distances between taxa, relative taxa ages, and possible origins. While a variety of characters are used to define taxa and construct classifications, their phylogenetic significance is variable and their use, particularly of morphological structures, is inherently subjective.

2.3.1   Evolution within the Reptantia

Most workers seem to agree that the Decapoda is a monophyletic taxon (see Schram 2001), although this is questioned by some (e.g., Richter & Scholtz 2001). There is also general agreement that the Reptantia Boas, 1880 is monophyletic, and constitutes a more derived group in the Decapoda than the "natant" (shrimp and prawn) group (see Ahyong & O'Meally 2004). It is within the Reptantia, which includes the great majority of decapod species, that the evolutionary relationships of groups has become contentious over the past 20 or so years (see Schram 2001; Richter & Scholtz 2001; Dixon et al. 2003; Ahyong & O'Meally 2004). Schram (1986) observed that the classification of the Reptantia was a "morass" and stated that "[a] single overview using character analysis on all groups is badly needed." Such an overview

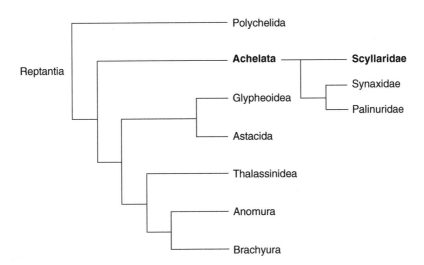

FIGURE 2.3 Diagram of taxa within the Reptantia indicating relationships of Achelata and of families within Achelata (based on Ahyong & O'Meally 2004). Note: lines indicate phylogenetic relationships only and are not intended to indicate distance.

was provided by Scholtz & Richter (1995), who undertook a phylogenetic analysis of the Decapoda using morphological characters.

To date, the most comprehensive phylogeny of the Reptantia to family level is that of Ahyong & O'Meally (2004). Figure 2.3 is based on a cladogram of the Reptantia (Ahyong & O'Meally 2004) that combines 105 scored characters (such as external morphology, anatomy, embryonic growth zones, spermatophore morphology, type of development, and brain morphology) with molecular (rRNA and DNA) data. It indicates that the Achelata evolved early in the Reptantia. While the Polychelida are basal in this scheme, the Achelata is the sister group to all remaining reptants (collectively the Fractosternalia of Sholtz & Richter 1995, although the cohesion of this group was rejected by Ahyong & O'Meally 2004). Ahyong & O'Meally (2004) also indicate unambiguous morphological character states that characterize each node in their cladogram. Those for the Achelata, and thus common to the Scyllaridae, Palinuridae, and Synaxidae, are: thoracic sternal plastron broadening posteriorly; carapace and last thoracic segment connected by a knob-like structure; telson posterolateral spine present mid-laterally; telson posterolateral spine fixed; soft tail fan cuticle; antennular flagellum length similar to that of antennular peduncle segment 2; and antennal basal articles (segments) fused with carapace and epistome. The cladogram (see Figure 2.2) also indicates that the Scyllaridae is a sister group to the Synaxidae plus Palinuridae (although, while *Jasus* and *Palinurus* are sister groups, *Palinurellus* is not sister to them). Add to this the exclusiveness of phyllosoma larvae to the three families of the Achelata and there seems little doubt the infraorder is monophyletic. However, the relationship between these families is not so clear.

2.3.2   Evolution within the Achelata

Evolutionary studies of the three families of the Achelata are few. George & Main (1967) published a detailed study of evolution in the Palinuridae, George (1997) related *Jasus* and *Panulirus* evolution to tectonic plate movements, Baisre (1994) examined the place of phyllosoma larvae in the phylogeny of the Palinuroidea (Achelata), and Davie (1990) discussed the status and relationship of synaxid lobsters to each other and in relation to the Palinuridae. More recently Ptacek et al. (2001) published a molecular phylogeny of the palinurid genus *Panulirus* and Patek & Oakley (2003) compared the evolutionary significance of acoustic systems in palinurids with and without such mechanisms. No similar studies of scyllarids have been published.

While larval and postlarval development and the achelate condition unites the Scyllaridae, Palinuridae, and Synaxidae and differentiates them clearly from other reptantian decapods, the form of the scyllarid antennal flagellum at all stages in the life cycle distinguishes them equally from the other two achelate families. Further characters that distinguish the Scyllaridae are noted above under definitions of the taxa (see Section 2.2.4).

A feature of the Scyllaridae is that they exhibit, as a whole, a cryptic lifestyle. While palinurid and synaxid lobsters are varicolored and sculpted to blend into their backgrounds, the lack of protruding structures (e.g., antennae) in scyllarids, their moderate to extreme flatness, their sculpturing and coloration, and many aspects of their biology, ecology, and behavior point particularly to an evolutionary exploitation of minimized visibility. The tendency to be cryptic is noted in the following discussion of evolution within the family.

2.3.3   Evolution within the Scyllaridae

Scholtz & Richter (1995) considered the Scyllaridae to be a monophyletic group, and it appears that Present-day Scyllaridae do not result from convergence of different lines. Their highly characteristic form is most unlikely to mask such convergence and there is a lack of palaeontological evidence of converging lines. This is not to say, however, that forms within the Scyllaridae have not converged to exploit similar habitats. Therefore, it seems reasonable to conclude that the taxa within the family have evolved from an ancestral form to successfully occupy a variety of habitats at a range of inshore to continental slope depths — from caves through rocky and coral reefs, to shelly and sandy bottoms, to open, soft sedimentary bottoms. Because we do not know of a plesiomorphic sister group to the Scyllaridae or even which genus in the Scyllaridae is basal, we do not know which larval and adult characters are plesiomorphic and which are apomorphic (George & Main 1967; Baisre 1994).

2.3.3.1   Fossil Evidence

Species of Scyllaridae found as fossils are listed in Table 2.2. The list may not be exhaustive as four extant species are included and there are likely to be fossilized specimens of other living species we have not found in the literature. However, we have missed few, if any, extinct species and the list of these in Table 2.2 serves to illustrate both the distribution and approximate ages of fossil scyllarids.

Förster (1984), in describing *Parribacus cristatus* Forster, 1984 and in redescribing *Palibacus praecursor* (Dames, 1886) (previously *Parribacus praecursor*), concluded that all the significant features of the Scyllaridae were fully developed by the mid-Cretaceous (100 to 120 million years ago).

TABLE 2.2

Fossils of the Scyllaridae

Species	Fossil age-range	Locality	Status	Source
Scyllarides punctatus Woods, 1925	≈ 120–110 MYBP	England	Extinct	Woods (1925)
Scyllarides tuberculatus (König, 1825)	≈ 55 MYBP	England	Extinct	Quayle (1987)
Scyllarides aequinoctialis (Lund, 1793)	≈ 110 MYBP	England	Extant	Glaessner (1969)
Scyllarella aspera Rathbun, 1935	≈ 65–56 MYBP	USA (Alabama)	Extinct	Rathbun (1935)
Scyllarella gibbera Rathbun, 1935	≈ 65–56 MYBP	USA (Alabama)	Extinct	Rathbun (1935)
Scyllarella mantelli (Desmarest, 1822)	≈ 65–56 MYBP	England	Extinct	Rathbun (1935)
Scyllarella gardneri (Woods, 1925)	≈ 120–110 MYBP	England	Extinct	Benton (1993)
Ibacus peronii Leach, 1815	≈ 110–25 MYBP	Lebanon-Europe	Extant	Glaessner (1969)
Palibacus praecursor (Dames, 1886)	≈ 100 MYBP	Lebanon	Extinct	Förster (1984)
Parribacus caesius Squires, 2001	≈ 35–55 MYBP	USA (S. California)	Extinct	Squires (2001)
Parribacus antarcticus (Lund, 1793)	≈ 100–65 MYBP	Lebanon	Extant	Förster (1984)
Parribacus cottreauxi (Roger, 1946)	≈ 85 MYBP	Lebanon	Extinct	Förster (1984)
Parribacus cristatus Förster, 1984	≈ 50 MYBP	Italy (Vicenza)	Extinct	Förster (1984)
Parribacus sp. Glaessner, 1965	≈ 35–25 MYBP	Poland	Extinct	Förster (1984)
Biarctus vitiensis (Dana, 1852)	≈ 10,000 YBP	Fiji	Extant	Förster (1984)
Scyllarus junghuhni Böhm, 1922	≈ 20 MYBP	Java	Extinct	Glaessner (1929)

MYBP = million years before present; YBP = years before present.

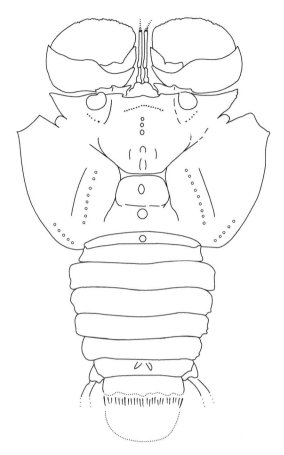

FIGURE 2.4 Reconstructive illustration of fossil of Cretaceous scyllarid *Palibacus praecursor* (Dames, 1886); after Förster (1984).

Fossils of the extant species *Scyllarides aequinoctialis, Ibacus peronii,* and *Parribacus antarcticus* from that time support Förster's observation, as do the oldest known extinct species, *Scyllarides punctatus* Woods, 1925 and *Scyllarella gardneri* (Woods, 1925). Despite living species being among the oldest known, two genera, *Palibacus* and *Scyllarella,* have gone extinct. While some fossils are fragmentary and not easy to compare precisely with living forms, Förster (1984) provided a reconstructive illustration (Figure 2.4) of *P. praecursor,* showing a flattened rather than high-vaulted species, such as the *Scyllarella* species of Rathbun (1935). The width of the carapace in *P. praecursor* is proportionately greater than that in *Ibacus alticrenatus,* which indicates that a strongly dorsoventrally flattened form, suggestive of extant *Ibacus* species, was present during the mid-Cretaceous. In other respects, however, this extinct species was unique in the shape of its lateral extensions of the carapace, anterolateral carapace angle, and cervical groove. The extinct species *Parribacus cristatus* of the lower Eocene in northern Italy is considerably more like Present-day species of *Parribacus.* But here, too, as indicated in Förster's photograph and interpretive drawing (Figure 2.5), a difference from extant species occurs in the apparent lack of spines bordering the sixth and fourth antennal segments and carapace. Only a deep cervical groove breaks the outline of these borders. The outline of the borders along with several other features are, however, unmistakably similar to those of extant *Parribacus* species.

Palibacus praecursor lived around 100 million years ago during the formation of the Tethys Sea, when the marine climate was warmer than it is now. The Tethys and the ocean that was to become the Pacific (Panthalassa) were continuous, as was the pantropical marine biota which spread throughout the Tethys and around the globe (Newman 1991). The distribution of scyllarid fossils between about 25 and 120 million years old from Poland through Europe (including England), to Lebanon, and in Alabama and

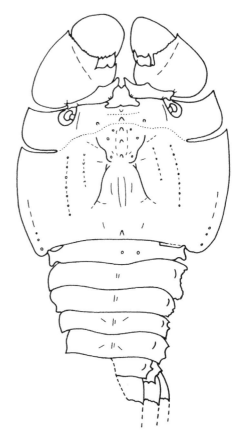

FIGURE 2.5   Photograph and reconstructive illustration of fossil of early Eocene scyllarid *Parribacus cristatus* Förster, 1984; after Förster (1984).

southern California in Present-day United States, would appear to place them all among this late Mesozoic to late Cenozoic Tethyan fauna. It was only much later, with the break up of Gondwana, that the division of the Tethys occurred as the African continent moved northward to meet Eurasia about 18 million years ago (as did India and New Guinea–Australia to meet Asia). Apparently there is no paleontological evidence of extinct genera east of Lebanon that might have given rise to existing Indo-Pacific scyllarid fauna. The gradual widening and deepening of the Atlantic Ocean, which began to emerge 170 to 180 million years ago (Brusca & Brusca 2003), would have separated the Atlantic scyllarids into the eastern and western faunas we see today, but the closure of the isthmus between North and South America, which effectively isolated Indo-Pacific from Atlantic scyllarids, occurred only around three and half million years ago.

The oldest fossils of the subfamily Scyllarinae found so far are the extinct *Scyllarus junghuhni* Böhm, 1922 of about 20 million years of age and the extant *Biarctus vitiensis* (Dana, 1852) of less than 10,000 years of age, both from the Indo-West Pacific region. This is, of course, far too little evidence on which to base a hypothesis, but their age does not conflict with our suggestion provided later that the Scyllarinae may represent a relatively recent radiation.

At 100 to 120 million years, the fossil record of the Scyllaridae is considerably shorter than that of the Palinuridae at about 200 million years, which led Baisre (1994) to suggest the scyllarids evolved more recently than the palinurids. A fossil of possible significance is that of *Cancrinos claviger* Munster, 1839 (Cancrinidae) from the Upper Jurassic (at least 142 MYBP) of southern Germany (Glaessner 1969). Its affinities are not discussed by Glaessner (1969), but he placed it next to the Scyllaridae in his text and it is referred to by George & Main (1967) in relation to palinurid evolution. *Cancrinos*, as illustrated by Glaessner, is certainly an achelate lobster-like creature. Most intriguing are its antennae, which have robust peduncles and club-like flagella, hardly longer than the peduncles, that are thickened about the middle and are comprised of 13 to 19 "rings" (Glaessner 1969), far fewer than found in recent palinurids. Besides being achelate, the first pereiopod of *Cancrinos* is hardly more robust than its other pereiopods; its carapace appears to lack a rostrum and is coarsely granulate, while its abdomen is finely granulate, but without any apparent median dorsal carina. Glaessner's illustration suggests it has lost the uncalcified distal areas of the telson and uropods. The form of the pereiopods and telson is typical of the Achelata and, apart from the lack of a median dorsal carina, the animal could well be ancestral to the Scyllaridae. Since fossil palinurids are found in older formations, *Cancrinos* is more likely to represent a progenitor to the Scyllaridae, if it is not from a separate, extinct branch of the Achelata, than an ancestor of the Palinuridae.

Although no fossil larvae or postlarvae of scyllarids have been found, the well-documented presence of a palinurid-like phyllosoma from Jurassic sediments in Germany (Polz 1996) indicates that this developmental form was established prior to any known fossils of scyllarid lobsters.

2.3.3.2 Evidence from Present-Day Distributions

Present-day geographic distributions are, of course, the result of evolution — of a process of adaptation in the face of competition and habitat change. Based on the time of appearance of scyllarid species in the fossil record, most speciation within the Scyllaridae probably took place during the post-Mesozoic, after the breakup of Gondwana, and during times in which the oceans were divided in much the same way as they are now. In this section, we outline the distribution and habitats of higher taxa of extant Scyllaridae in an attempt to understand the evolutionary processes that resulted in these distributions.

2.3.3.2.1 General Distribution

Some general features of the present distribution of the scyllarid subfamilies and genera follow, with comparisons to the Palinuridae and other Crustacea where relevant.

Scyllarid lobsters are found in coastal, continental shelf, and upper-slope areas across the Equator, at low latitudes, and in temperate latitudes influenced by warm water currents (Figure 2.6). No taxa at subfamilial level or genera of the Arctidinae, Ibacinae, or Theninae are confined to the Northern or Southern Hemispheres; within the Scyllarinae, only two genera, *Antipodarctus* of New Zealand (Figure 2.6D) and *Antarctus* of Australasia (Figure 2.6C), are not pantropical. The only division that can be detected on a broad geographical scale is between the Atlantic and Indo-Pacific regions. The Theninae (Indo-Pacific) is the only subfamily not represented in both regions (Figure 2.6F); only one genus, *Scyllarus*, is exclusive to the Atlantic and Mediterranean (Figure 2.6F), whereas fifteen genera (including *Thenus*) occur in the Indo-Pacific only; and four genera are common to both regions. Scyllarids are not known in northern regions above 45°N (*Scyllarides latus* and *Scyllarus arctus*) or further south than 44°S (*Ibacus alticrenatus*) — this compares with the range of about 60°N and 50°S for the palinurids (Holthuis 1991a). The family has a mainly warm-water distribution with most scyllarid genera and species being found between 30°N and 30°S (Table 2.3). However, the presence of deepwater species of Scyllarinae at low and mid-latitudes, such as *Eduarctus martensii* at 887 m near Fiji (Holthuis 2002) and *Antarctus mawsoni* (Bage, 1938) at 540 m in the southern Tasman Sea (Davie, 2002), does indicate that some scyllarinids inhabit colder water. There is also an indication that juveniles and subadults of *S. latus* may dwell deeper, and hence, in colder water (Spanier & Lavalli 1998; see Lavalli et al., Chapter 7).

Centers of higher species diversity are apparent: the Indo-West Pacific (East and Southeast Asia, Australia) is the region with the greatest number of species, as it is for palinurids and most other crustaceans (Abele 1982). In the Atlantic Ocean, there is a higher diversity of species located around Central America than elsewhere. The eastern Atlantic and, particularly, eastern Pacific Oceans are areas of lower-species diversity — while *Scyllarus* species are distributed fairly evenly on both sides of the Atlantic and at localities between, nonscyllarinid species (notably *Scyllarides*) are focused in the western Atlantic (Figure 2.6A). Only six slipper lobster species occur east of the Tuamotu Archipelago in the Pacific Ocean and only two of these (*Scyllarides astori* Holthus, 1960 and *Evibacus princeps* S.I. Smith, 1869) occur adjacent to the west coast of either North or South America (Holthuis 1991, 2002) (Figures 2.6A, 2.6B).

Several geographical areas are notable for a lack of scyllarids. As in many groups of crustaceans, including the decapods (Abele 1982), scyllarid species do not occur in the eastern South Atlantic south of Angola. Although *Scyllarides obtusus* Holthius, 1993 is present around St. Helena Island at 15°S, no scyllarid species are found at Tristan da Cunha, Vema Seamount, or adjacent islands, unlike the palinurids

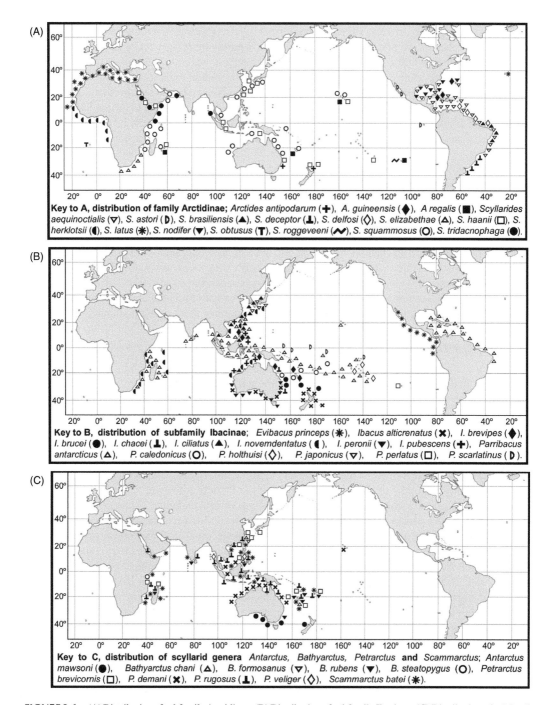

FIGURE 2.6 (A) Distribution of subfamily Arctidinae. (B) Distribution of subfamily Ibacinae. (C) Distribution of subfamily Scyllarinae, genera *Antarctus, Bathyarctus, Petrarctus*, and *Scammarctus*. Distributions are based primarily on Holthuis (1991a, 2002) and Brown & Holthuis (1998). Symbols indicate broad distributions and are not intended to indicate abundance. While distributions of many species are shown in maps by Holthuis (1991a) to be continuous, we have shown symbols only (except for Theninae). Distributions of most species are more continuous than the symbols indicate; refer to Holthuis' papers for more detail.

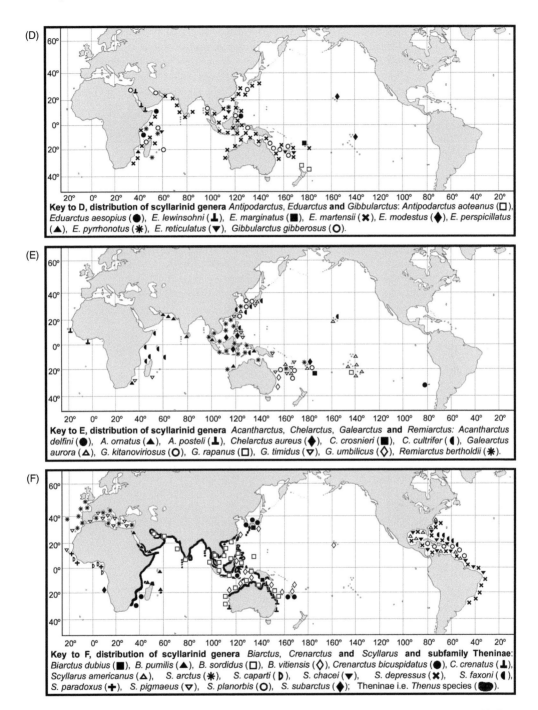

FIGURE 2.6 (continued): (D) Distribution of subfamily Scyllarinae, genera *Antipodarctus, Eduarctus*, and *Gibbularctus*. (E) Distribution of subfamily Scyllarinae, genera *Acantharctus, Chelarctus, Galearctus*, and *Remiarctus*. (F) Distribution of subfamily Scyllarinae, genera *Biarctus, Crenarctus*, and *Scyllarus* and subfamily Theninae. Distributions are based primarily on Holthuis (1991a, 2002) and Brown & Holthuis (1998). Symbols indicate broad distributions and are not intended to indicate abundance. While distributions of many species are shown in maps by Holthuis (1991a) to be continuous, we have shown symbols only (except for Theninae). Distributions of most species are more continuous than the symbols indicate; refer to Holthuis's papers for more detail.

Jasus tristani Holthuis, 1963 and *Projasus parkeri* (Stebbing, 1902) (off Vema Seamount). The eastern part of the South Pacific Ocean (from south of Peru) is also devoid of scyllarids (as it is of palinurids, except the deepwater *Projasus bahamondei* George, 1976), the nearest being at Juan Fernandez, Easter, and Galápagos Islands, and in northern Peru. Saint Paul and Amsterdam Islands in the southern Indian Ocean also lack scyllarids, but *Jasus paulensis* (Heller, 1862) and *Projasus parkeri* are present there and on adjacent seamounts.

2.3.3.2.2 Distribution of Genera

The distribution of scyllarid genera by latitude, depth, and substrate type are summarized in Table 2.3 and Figure 2.6 and briefly outlined here. The six nonscyllarinid genera and the Scyllarinae have distinctive forms (described in Section 2.2.4) indicative of adaptation, to a greater or lesser extent, to certain kinds of habitat. All three species of *Arctides* inhabit shallow, rocky habitats, including underwater caves during the day, and are active nocturnally (Doak 1971; Holthuis 1991). *Arctides* species are sparsely, but fairly evenly spread in tropical to subtropical localities throughout the oceans, with one species (*A. guineensis*) in the western Atlantic and two (*A. regalis* Holthuis, 1963 and *A. antipodarum* Holthuis, 1960) in the Indo-West Pacific (Figure 2.6A). The genus does not occur in the eastern Atlantic.

Scyllarides species are focused very much in the western Atlantic where five species occur, with distributions overlapping considerably (Figure 2.6A). A further three in the central and eastern Atlantic

TABLE 2.3

Latitudinal Ranges, Depth Ranges, and Substrate Types for the Scyllaridae (by genus) Taken Primarily from Holthuis (1991a, 2002) and Brown & Holthuis (1998)

Genus	Latitude	Depths (m)	Substrate Type
Arctidinae			
Arctides	32°N to 36°S	5–146	Rocky
Scyllarides	45°N to 40°S	0–380	Rocky–sedimentary[a]
Ibacinae			
Ibacus	39°N to 44°S	20–750	Sedimentary
Parribacus	36°N to 30°S	0–20	Rocky–sandy
Evibacus	31°N to 4°S	2–90	Sedimentary
Theninae			
Thenus	28°N to 32°S	8–100	Sedimentary
Scyllarinae			
Acantharctus	25°N to 34°S	20–57	Rubble, sand, shell sand
Antarctus	33–44°S	122–440	Unknown
Antipodarctus	34–39°S	≤ 90	Rocky, sedentary invertebrates[b]
Bathyarctus	25°N to 34°S	26–800	Rubble, sedimentary, sedentary invertebrates
Biarctus	32°N to 35°S	0–73	Reefs[c], sedimentary
Chelarctus	36°N to 21°S	100–333	Rock, mud
Crenarctus	39°N to 35°S	0–108	Rock, coral reef, shell sand
Eduarctus	32°N to 30°S	4–112	Rubble, sand, mud
Galearctus	34°N to 37°S	30–>390	Reefs, corals, mud
Gibbularctus	27°N to 15°S	12–57	Rubble, mud
Petrarctus	33°N to 27°S	5–282	Reefs, sand, mud
Remiarctus	25°N to 24°S	18–260	Sedimentary
Scammarctus	25°N to 26°S	152–531	Sedimentary
Scyllarus	45°N to 5°S	0–329	Various including *Posidonia* beds

[a] Sedimentary refers to soft bottoms from shelly sand to mud and mixtures of them.
[b] Some records of scyllarinids indicate they were found on or among sponges, branching corals, and other sedentary invertebrates.
[c] The numerous tropical species referred to as being found on reefs probably frequent coral reefs, although this is not always indicated in records.

brings the total number of *Scyllarides* species in this ocean to eight, whereas only six are present throughout the Indo-Pacific, just two of these in the east Asia–Australasia region, two in the east Pacific, and four in the western Indian Ocean (Figure 2.6A). There is a dearth of *Scyllarides* species in the Indonesia–New Guinea region, which is interesting considering the predominance of other scyllarids in the area. Species of *Scyllarides* are found over a broad range of depths and on a variety of substrates (Holthuis 1991a; Spanier & Lavalli 1998).

A shallow-water, strictly tropical, and hard-substrate dwelling genus is *Parribacus* (Holthuis 1991a). Except for its absence in the Eastern Atlantic and Eastern Pacific, *P. antarcticus* occurs extensively in tropical to subtropical waters in all three major oceans (Figure 2.6B), the only one of the 85 scyllarid species that does. Even so, the genus is focused in the Pacific with five further species occurring mainly around Pacific islands, one of them also in southern Japan. *Evibacus* is confined to a single species on the west coast of Central America (Figure 2.6B). It, too, is a shallow water species, but is found mainly on sedimentary bottoms (Holthuis 1985, 1991a).

The largest genus of the Ibacinae, *Ibacus* (with eight species), is found only in the Indo-West Pacific and is concentrated mainly in the Southeast Asia–Australia region (Figure 2.6B). *Ibacus* species, all flattened and smooth surfaced, live on sedimentary substrates at a wide range of depths, but are particularly common on the extensive continental shelves and slopes that border continents (Holthuis 1985, 1991a; Brown & Holthuis 1998). Seven of the eight species are found off the coasts of Australia, but only one (*I. peronii*) is confined to Australia, while the eighth (*I. ciliatus*) is restricted to East Asia (Chan & Yu 1986, 1993; Brown & Holthuis 1998).

The Theninae are also found only in the Indo-West Pacific, continuously from Mozambique to Japan, and through Southeast Asia to northern Australia (Figure 2.6F). Although they generally live in shallower waters than *Ibacus*, they also inhabit sedimentary substrates (Holthuis 1991a; Davie 2002).

Apart from the genera *Scyllarus* (ten species) and *Acantharctus* (one species in the Atlantic, north of the Congo), the Scyllarinae are very much focused in the Indo-West Pacific (Holthuis 2002) (Figure 2.6C to Figure 2.6F). *Acantharctus* is the only scyllarinid genus recorded in the Atlantic, Pacific, and Indian Oceans, with one species confined to each ocean (Figure 2.6E). Nineteen of the 50 scyllarinid species are found in both the Indian and Pacific Oceans; 13 more have only been found in single localities (some will probably prove to have broader distributions), and the remaining species are confined to the Indian or Pacific Oceans or to waters off Southeast Asia. Thus, despite the apparent similarity of form throughout the subfamily, it is not possible to associate the Scyllarinae, as a whole, with any particular kind of substrate or depth range. Some scyllarinid species depths are as shallow as any in the family, but the group also includes the deepest species on record (Table 2.3). Few scyllarinid genera can be associated with depth ranges and habitat types with any confidence. *Acantharctus* appears to be a shallow water genus (Holthuis 2002), while *Bathyarctus* (Figure 2.6C), despite its name, includes *B. steatopygus* Holthuis, 2002 found as shallow as 26 m (Holthuis 2002). *Galearctus* (Figure 2.6E) and *Petrarctus* (Figure 2.6C) inhabit a wide range of depths and substrates as do *Scyllarus* species (Figure 2.6F), but *Biarctus* (Figure 2.6F), with four species, appears to be confined to shallow, hard substrates (Holthuis 2002). *Eduarctus* (Figure 2.6D), the largest genus outside *Scyllarus*, inhabits rubble to muddy substrates (Holthuis 2002) and includes *E. martensii* found at four to 887 m depth (Holthuis 2002) — the deepest and also one of the shallowest of slipper lobsters on record.

2.3.3.2.3 *Interpretation*

The analysis of palinurid evolution by George & Main (1967) and subsequent works dealing with aspects of palinurid evolution (George 1969, 1997; Pollock 1993) are instructive in considering evolution in the closely related Scyllaridae, and we draw on them in the following discussion. Using present distributions to interpret evolutionary change assumes that: (1) the distributions of taxa are fully known; (2) the taxonomic divisions and relationships on which interpretations are based are correct; (3) similarities between taxa reflect phylogenetic affinity, rather than being the product of convergence; and (4) the present distribution of a given taxon results from radiation, rather than representing the remnant of a formerly more widespread distribution (based on George & Main 1967). In the Scyllaridae there are limits to some of these assumptions.

The distributions of larger scyllarids represented in the Arctidinae, Ibacinae, and Theninae are well-known; it is the Scyllarinae with its numerous small species, some poorly known, for which distribution data are frequently incomplete. We consider the current taxonomy of the Scyllaridae to be reliable since it is the product of careful consideration and, with few exceptions, appears to be widely accepted. Again, most potential for disagreement lies in the recent division of the Scyllarinae into genera based on more subtle character differences than those that distinguish the nonscyllarinid genera. The number of *Thenus* species is uncertain at this stage, but the intended identification of more (Davie 2002) seems unlikely to alter the distinctiveness of this subfamily. While classification does not necessarily reflect affinity, that of the Scyllaridae includes characters that are clearly of phylogenetic significance; however, it is probable that some convergence of forms is reflected in present distributions. The most obvious case in point is, again, *Thenus*, in which the dorsoventrally flattened form is clearly adaptive to its sedimentary habitat as it is for *Ibacus* species, although differences in maxilliped morphology quite clearly place them in different lineages: *Thenus* with the Scyllarinae and *Ibacus* with the remaining Ibacinae and Arctidinae.

Geological and habitat change always has the potential to isolate and divide populations, but there is little palaeontological evidence to suggest that the Scyllaridae has undergone such isolations during its evolution or that existing populations are relicts of former, more widespread taxa. For example, *Scyllarella* and *Palibacus* are extinct, but there is too little evidence to suggest that they might represent the disappearance of major groups; rather, it is more likely that present distributions have resulted from radiation and some specialization to exploit existing habitats.

In Table 2.3 and Figure 2.6 it can be seen that, unlike the palinurids, there is frequent overlap within the bands of occurrence of scyllarid genera. Different palinurid genera rarely inhabit geographic regions that correspond in latitude, depth, or water temperature (George & Main 1967), whereas this appears to occur among scyllarids. Scyllarid species are, however, more numerous and do not penetrate as far into temperate latitudes as palinurids. Instead, they are focused particularly in the East–Southeast Asia–Australia region and, to some extent, the western central Atlantic. Some palinurids also occur at greater depths than any scyllarids and may be better equipped to colonize a wider variety of geographical and depth ranges than scyllarids.

The overlap of scyllarid genera also suggests the presence of mechanisms enabling exploitation of particular habitats. *Thenus* occupies sedimentary substrates, usually in waters of < 100 m depth (according to Davie 2002, *T. indicus* Leach, 1816 is found to 200 m, whereas *T. orientalis* usually occurs at 10 to 50 m depth; see also Jones, Chapter 16), and such substrates are obviously widely available given the continuous distribution of the genus from eastern Africa to eastern Asia and Australia. *Ibacus* species are found on shelf sediments, generally at greater depths than *Thenus*, and may be somewhat limited to wide continental shelves adjacent to continents (although *I. brucei* is also known from the northern central Tasman Sea (NORFANZ Expedition 2003, unpublished data) and northeast of New Zealand (Museum of New Zealand, unpublished record). Both *Arctides* and *Parribacus* are shallow rocky and coral-reef dwellers in all three major oceans (Holthuis 1991). On the other hand, *Scyllarides* species cannot be so easily identified with habitat preferences, since they are found on a variety of substrates and a greater range of depths. The genus does appear to have radiated in the western Atlantic, but has not done so to any extent in the Indo-Pacific, and may not have colonized much of Southeast Asia due to competition or because conditions there are not suited to their development (discussed later). Although occasionally found on muddy bottoms, it may be that *Scyllarides* species are more often associated with coarser sediments and rocky areas (Holthuis 1991a, 1993; Spanier & Lavalli 1998).

Given our knowledge of the Scyllarinae as a whole, it seems that they are particularly unspecialized apart from their size, being found at the greatest range of depths and on all manner of substrates. There is the suggestion that the development of cryptic sculpturing and coloration coupled with small size has enabled a wider radiation into these areas. Some species, such as *Antipodarctus aoteanus* (Powell, 1949) and *Bathyarctus steatopygus*, have been found associated with sponges and other sedentary invertebrates (Holthuis, 2002); *Scyllarus arctus* frequents *Posidonia* beds (Holthuis, 1991a), and further sampling will probably reveal habitat preferences in more species than are apparent at this stage.

Some mechanism has also limited the colonization by scyllarids of higher latitudes and mid-ocean seamounts and islands in the Southern Hemisphere. Whereas *Jasus* has speciated throughout the remote islands of the southern oceans, scyllarids have not — with the exception of Juan Fernandez, where both *Jasus frontalis* (H. Milne Edwards, 1837) and *Acantharctus delfini* (Bouvier, 1909) are found. George (1997) contended that, once the full strength of the Antarctic Circumpolar Current had become established after Africa and Australia had separated from Antarctica, *Jasus* species were able to colonize subantarctic islands and seamounts by taking advantage of their extended larval duration. One explanation for the absence of scyllarids, then, is their generally shorter larval life, preventing them from spanning the necessary distances to colonize remote localities.

Evolution of a longer larval life in scyllarids similar to that of *Jasus* species may, however, have required a similar, high southern latitude distribution of ancestral scyllarids, which seems unlikely. The fossil record suggests an earlier evolution of the palinurids than the Scyllaridae (fossil palinurids are known from the early Mesozoic around 200 MYBP), well before the formation of the Tethys Sea which, as we suggested earlier, existed early in the evolution of the Scyllaridae. Newman (1991) proposed that the presence of *Jasus* species in higher southern latitudes may have resulted from immigration from origins in the Northern Hemisphere with subsequent extinction in the north; however, we can find no similar evidence for such an origin of the Scyllaridae. A Tethyan origin may instead have given rise to the relatively warmer-water scyllarids that have since radiated to produce present distributions.

An absence of scyllarids from the coasts of southwestern Africa (where palinurids are present) and southwestern South America may, likewise, be attributed to shorter larval duration in the scyllarids, but also suggests that the lack of suitable current systems to enable recruitment (to successfully complete the larval phase) has played a part in limiting distribution. Presumably scyllarids with shorter teleplanic larval durations would require smaller or more rapidly cycling current systems, reliable enough to return a critical proportion of their larvae to the benthic (adult) habitat, to maintain a presence. Maintenance of some palinurid populations (for example, *Sagmariasus verreauxi* (H. Milne Edwards, 1851) in northern New Zealand) also requires long-distance migrations during the adult phase. Remembering that many adult scyllarid lobsters are marked by the relative shortness of their pereiopods, it seems likely that such migrations play a less important role in the Scyllaridae than in the Palinuridae.

2.3.3.3 *Implications of Life-History Characters*

Because the life histories of achelate lobsters include characteristically long-lived larvae and a postlarval phase, and studies of their development are numerous (mainly of commercial species), an examination of ontogeny can provide useful information on their evolution. In this section, we attempt to identify characters that might give insight into the phylogeny of the Scyllaridae; these ontogenetic characters are listed in Table 2.4. As is often the case, data are limited, and in several instances we can do no more than suggest that a certain character may be of evolutionary significance.

Ontogeny in scyllarids comprises the embryonic, larval, postlarval, juvenile, and adult phases. Early larval development in the Scyllaridae begins with a "prelarva" (naupliosoma) in at least some species, a larval phase (the phyllosoma), and a postlarval (nisto) phase (see Sekiguchi et al., Chapter 4 for more information on these forms). The naupliosoma is short lived (minutes to hours) and its first three pairs of cephalic appendages (antennules, antennae, and mandibles) are present and setose, whereas appendages posterior to them are absent or rudimentary (Booth et al. 2005). Many characters of potential use in clarifying the phylogeny of scyllarids concern this early life history (Table 2.4).

2.3.3.3.1 *Egg Size*

In general, larger scyllarid species have larger brood sizes (see Table 4.1 of Sekiguchi et al., Chapter 4), but this relationship is modified by variation in egg size, which ranges from 0.4 to 1.2 mm diameter. Adult females of the same size have larger broods when the eggs are smaller and vice versa: larger, fewer eggs result in lower fecundity (see Sekiguchi et al., Chapter 4), but embryos are more advanced in development, are more mobile at hatching, and the length of larval life is abbreviated. The Theninae and *Ibacus* species lay fewer, larger eggs, and hatch more advanced phyllosomas than other scyllarids, whether they hatch as a naupliosoma or not; this advanced development may be viewed as specialization (Baisre 1994).

TABLE 2.4

Characters — Mainly from Early Life History— Considered in Assessing the Phylogeny of the Scyllaridae

Subfamily and Genus	Species	Adult Size	Egg Size	Hatching: Hatch as N or P	First Phyllosoma Advanced	Duration	Size	Phyllosoma: Bilobed Cephalon	Thorax Emarginate	Pereio 5 has Exop	Pereio 5 Reduced	Eyestalk	Offshore Dispersal	Nisto Pleopod: Size	Swim Ability
Ibacinae		Large													
Ibacus			Large		✓	Medium	Large	✓	X	✓	X	Short	Moderate	Large	Well
	I. alticrenatus			N											
	I. ciliatus			N											
	I. peronii			N (?)											
	I. novemdentatus			N (?)											
Parribacus				P		Long	V. large	✓	✓	✓	X	Short	Large	Large	Well (?)
Evibacus							Large	X	✓ (slight)	✓	X	Medium	Moderate	Large	Well (?)
	E. princeps														
Arctidinae		Large													
Arctides						Long	Large	X	✓	✓	X	Long	Large	Large	Well
Scyllarides			Medium		X			X	✓	Variable	Variable	Medium			Well
	S. latus			N											
	S. herklotsii			N											
	S. aequinoctialis			N											
Scyllarinae		Small	Small			Variable	Variable	X	X	X	✓	Short	Variable	Variable	Well
Petrarctus	P. demani			P	X										
Biarctus	B. sordidus			P	X										
Scyllarus	S. americanus			P	X										
Scyllarus	S. depressus			P	X										
Scyllarus	S. planorbis			P	X										
Theninae		Large				Short	Medium	✓	X	X	✓	Medium	Little		Poor
Thenus			Large												
	T. orientalis			P	✓										
	T. sp.			P	✓										

An entry for the subfamily or genus applies to all lower taxa. Small Adult Size = <70 mm carapace length (CL); Small Egg Size = <0.5 mm diameter; Large Egg Size = >1.0 mm diameter; eggs usually hatch as naupliosomas (N) or phyllosomas (P); first-instar phyllosomas may be at an advanced state (where third pereiopods have five segments); P = applies to this taxon; X = does not apply; Short Phyllosoma Duration = <3 months; Long Phyllosoma Duration = >6 months; Large Phyllosoma = 30–60 mm total length; Pereio = pereiopod; Exop = exopod; Little Offshore Dispersal = mainly over continental shelf; Moderate Offshore Dispersal = within, but also beyond shelf break (near 1000 m depth contour); Large Offshore Dispersal = oceanic, mainly well beyond shelf break, including within ocean basins; Large Nisto = >10 mm CL; under Nisto Pleopod, Well = well developed, with appendix interna present, Poor = poorly developed; ? = uncertain; blank = unknown or not applicable. Main data source is Sekiguchi et al., Chapter 4.

2.3.3.3.2   *Presence of a Naupliosoma and the Level of Development of First-stage Phyllosomas*

Species hatching as naupliosomas include *Scyllarides herklotsii* (Herklots, 1851), *S. latus*, *S. aequinoctialis*, *Ibacus alticrenatus*, *I. ciliatus*, and possibly *I. peronii* (Table 2.4). In other scyllarid species, there is either no naupliosoma or one has not been observed. Absence of a naupliosoma is indicative of abbreviated development (Baisre 1994).

2.3.3.3.3   *Phyllosoma Characters*

The principal morphological features distinguishing the Scyllarinae and each of the six nonscyllarinid genera are the shape of the cephalic shield and the shape and size of the thorax (see Sekiguchi et al., Chapter 4). There is an interesting range of shapes from the long, narrow, oval shield and much broader, "emarginated" thorax typical of *Arctides*, through variously oval to round cephalic shields with thoraxes of various widths (but narrower than the shield) in other genera, to *Ibacus* species with the anterolateral borders produced to yield a joined kidney-shaped shield and a narrow thorax that is continuous with the abdomen. Baisre (1994) considered the following features to indicate specialization: (1) the absence of a setose exopod on the fifth pereiopod; (2) the presence of long eyestalks; and (3) the reduction in the size of the fifth pereiopod. The fifth pereiopod exopod is absent in phyllosomas of the Theninae and Scyllarinae (Table 2.4) and, whether this is an apomorphic development or not, it does support a closer relationship of *Thenus* to the Scyllarinae than to other taxa in the family.

The phyllosoma of *Arctides* is notable for the length of its eyestalks, which are far longer than those in the larvae of any other scyllarid genera. This may well be an apomorphy too, although it is accompanied by a long narrow shield and exceptionally wide thorax, giving the impression that the size of each feature may be necessary to maintain trim during swimming. While Baisre (1994) considered eyestalk length to have phylogenetic significance, we think that function may be at least as important. It is interesting to note that a comparable range of shield and thorax shapes occurs in the Palinuridae, suggesting a convergence in form between the two families that results from habitat partitioning between the phyllosomas within each family.

A shorter larval life, with less extensive offshore dispersal, can be viewed as a specialization (Baisre 1994), and is seen particularly in the Theninae, but also among some of the Scyllarinae and Ibacinae. This suggests varying specialization within each subfamily, with the Theninae standing out as being the most specialized. Short larval duration is also accompanied by a very limited swimming ability in the nisto of *Thenus* (Table 2.4).

A tendency toward a shorter larval life is more pronounced in the scyllarids than in the palinurids, possibly because scyllarids evolved more recently in coastal waters. Within the Ibacinae, *Parribacus* species are notable in having a long-lived larval phase (perhaps eight to nine months) that becomes widespread in ocean basins; they are also noteworthy for having phyllosomas that reach a massive 80 mm TL and nistos of up to 21 mm carapace length or more. The three western Atlantic scyllarids, *Scyllarides aequinoctialis*, *S. nodifer*, and *Arctides guineensis*, also have long larval durations as does *S. latus* in the Mediterranean and eastern Atlantic (Martins 1985; Spanier & Lavalli 1998). The similarities in morphology and habitat of adult *Arctides* to tropical palinurids were noted earlier, and perhaps additional similarities are also reflected in their long larval life. Larval durations of eight to nine months in these genera may have evolved to take advantage of larger current-circulation patterns in certain areas (just as *Jasus* species do); the necessity for briefer circulation times for scyllarinid larvae was suggested earlier (see Section 2.3.3.2.3). Currents that take longer to circulate are probably more oceanic than would occur in the Indonesia–New Guinea region, which may explain why *Scyllarides* species are few or absent among these islands.

In his phylogenetic analysis based on larvae of the Achelata (as Palinuroidea), Baisre (1994) identified three groups within the Scyllaridae: a *Scyllarides-Arctides-Parribacus* group, an *Ibacus-Evibacus* group, and a *Thenus-Scyllarus* group. Booth et al. (2005) agreed, in large part, with Baisre's finding, but noted that the subfamily Ibacinae clearly includes some very disparate developmental characteristics, including the relatively advanced stage at hatching in *Ibacus*, and the extremely large and long-lived larvae of *Parribacus* species compared to others in the subfamily. The cluster analysis of larval and adult characters by Baisre (1994), however, indicated that *Parribacus* is closer to *Ibacus* and *Evibacus*, which supports the inclusion of these three genera in the Ibacinae, as was initially done by Holthuis (1985). Also, larval morphology showed more affinities between *Scyllarus* and *Thenus* than are suggested by adult morphology. The results in our Table 2.4 generally support these conclusions.

2.4 Synthesis

Anger (2001) observed that the Achelata are "phylogenetically old reptantians," but, in the case of the Scyllaridae, we appear to have a group that has evolved more recently than the Palinuridae during the formation of Present-day oceans, and that has successfully radiated into warm to temperate habitats. Two genera (*Scyllarella* and *Palibacus*) have become extinct, but the fossil evidence is too scant to suggest that they represent extinctions of major taxa; rather, in the light of their widespread occurrence and high diversity, slipper lobsters are, even now, in their heyday. Baisre (1994) identified primitive and derived characters among scyllarid larvae, and we have attempted to show that the family has evolved forms suited to certain habitat types. However, there does not appear to be an obviously "archetypal" living scyllarid from which we can deduce plesiomorphies and, hence, the likely directions that evolution has taken in the Scyllaridae. The Scyllarinae, because it includes 60% of scyllarid species, is ubiquitous throughout the distribution of the family as whole and frequents all manner of habitats — it may be viewed as the most generalized scyllarid. At the same time, we have suggested that it, too, has specialized, mainly in its small size and cryptic appearance and lifestyle. No fossils of Tethyan origin appear to represent scyllarinids; instead, the only fossil of an extinct species from one subfamily (*Scyllarus junghuhni*) is 20 million years old and from Indonesia, the area of highest scyllarid diversity. Unless the phyllosomas of Scyllarinae species represent a generalized condition in their shape and the lack of an exopod on the fifth pereiopod, there is little evidence of the subfamily representing a plesiomorphic group. In addition, the Scyllarinae share the absence of a fifth pereiopod exopod with *Thenus*, a feature which would appear to be derived.

Thenus has clearly evolved in the Indo-West Pacific region and is most unlikely to result from isolation of an earlier, more widespread form. There is no fossil evidence of an ancestor that might have given rise to *Thenus*, and it has probably evolved since the Lower Miocene (around 18 MYBP) when the African continent came into contact with Eurasia (Newman 1991). There are also no similar genera among the Scyllarinae, although its most obvious affinities are with that subfamily, given the nature of its maxillipeds. *Thenus* is a form that has adapted to an extensive, though more clearly definable, habitat by becoming larger than species of Scyllarinae and in having abbreviated development.

There is little evidence of relict taxa in extant scyllarids, but the closure of the Panama isthmus between 3.1 and 3.6 MYBP (Newman 1991) effectively isolated Atlantic scyllarids from those of the Indo-Pacific region. This closure obviously occurred long after the development of the two branches of scyllarids (those with and without multiarticulate flagella on the maxillipeds) since *Arctides, Scyllarides*, and *Parribacus* (with multiarticulate flagella) and the Scyllarinae (without multiarticulate flagella) are represented in both the Atlantic and Indo-Pacific. The existence of *Evibacus* is, however, more intriguing, as it is distributed along the west coast of Central America in shallow water. Given the very recent closure of Panama, it might be expected that *Evibacus* or similar forms would occur on both sides of the isthmus, but apparently this is not the case. Its similarity to *Ibacus* species (particularly in the form of the merus on its third maxilliped), suggests that it is more closely related to the lobsters of this strictly West Pacific genus. Although well isolated from *Ibacus*, *Evibacus* may have arisen from the settlement of *Ibacus*, or *Ibacus*-like phyllosoma larvae, from the west. Unlike the Atlantic, which developed from the splitting of the Americas from what is now Europe and Africa, the Pacific Ocean originated from the still larger Panthalassa Ocean (Newman 1991). Panthalassa was present long before the development of the Tethys Sea or any fossil evidence of the Scyllaridae. The medium duration of *Ibacus* larvae mitigates against the suggestion that they could have reached the eastern Pacific, but the far more extended larval life of the closely related *Parribacus*, probably far longer than estimated given the immense size of its phyllosomas, implies that long larval life in an ancestor may have enabled the distance to be covered, at least by "island-hopping". Indeed, the origin of *Evibacus* might be closer to *Parribacus*, and the characteristic third maxilliped of *Evibacus* may be a latent character that develops in response to feeding requirements.

The division of *Ibacus* based on the form of the merus of the third maxilliped does not appear to reflect a separation either by geographical distribution or depth. Species distributions overlap and all eight species in the genus overlap, to some extent, in the range of depths they occupy. Thus, it is difficult to relate this morphological difference to habitat difference, and there is not enough information to suggest it relates to diet preference or feeding method, although it seems likely that it does.

A further aspect of scyllarid evolution is their exploitation of camouflage. This appears to be a strong tendency throughout the family, and we have described how the various taxa have adapted to blend in with their backgrounds by evolving smooth or variously sculpted upper surfaces, cryptic coloration, and, in some cases, extreme dorsoventral flattening. Nocturnal feeding and daytime sheltering occur in *Scyllarides latus* (Spanier & Lavalli 1998), *Arctides* species (Doak 1971), and *Thenus* species. (Jones 1988; see Jones, Chapter 16), and other behaviors seem adapted to minimize visibility. Faulkes (2004) established that *Ibacus peronii* and *I. alticrenatus* have lost the neurons that control tail flipping as a response to danger. His experiments indicated that these species do not react to threatening stimuli with such an escape-response, but instead remain motionless. *Ibacus* species inhabit sediments, whereas *S. latus* frequents rocky substrates and retains the escape-response ability (Spanier et al. 1991; see Lavalli et al., Chapter 7), but is also able to cling strongly to its preferred rocky habitat if a predator attempts to pry it off (Spanier & Lavalli 1998). During daytime, *S. latus* hides in lairs in rocks or underwater caves (as does *Arctides antipodarum*). The relative shortness of slipper lobster pereiopods is a characteristic of the family as a whole and most, if not all, species can conceal the legs beneath the body at rest. The possible consequence of short pereiopods is that contranatant migration of lobsters to correct for displacement following nisto settlement downstream of adult lobster habitats may be limited, as opposed to the case in the longer-legged palinurids where such contranatant migrations occur. Some scyllarid lobsters have, however, developed the ability to swim in a controlled manner (Jones 1988; Spanier et al., 1991; Faulkes, 2004). There may be advantages in controlled bursts of swimming over continuous walking migrations, in allowing scyllarid lobsters to lie low between bursts and in daytime, taking advantage of their camouflage, and placing fewer requirements on them to aggregate for migrations as the palinurids do.

2.5 Conclusions

The Scyllaridae are a highly characteristic family of lobsters, recognized for millennia. Slipper lobsters are clearly closely related to the Palinuridae and Synaxidae (together defined as the Achelata), through the many characters they share, particularly their distinctive larvae that are like no others in the Decapoda. The higher taxonomy of the family is also reasonably settled due, in large part, to the contributions of Dr. Lipke Holthuis. There are, however, some relationships that still need to be settled, particularly those involving primitive and derived characters, such as the form of the maxillipeds in species of *Ibacus*, and the question of the number of species constituting the Theninae. The recent establishment of 13 new genera in the Scyllarinae, by far the largest and least known group of slipper lobsters, has yet to stand the test of time and the accumulation of much needed additional information. Further species in this subfamily will inevitably be discovered.

The slipper lobsters appear to have evolved more recently than the Palinuridae, along with the development of the major oceans to their present configuration, and the present diversity of species (85 known to date) does not suggest the family, or any group within it, represents a remnant of a formerly more widespread population. The presence of multiarticulate exopods on the maxillipeds of the Ibacinae and Arctidinae and their absence from the Theninae and Scyllarinae represents an unequivocal difference between these two groups of taxa. The similarity of these characters within each of the two pairs of subfamilies also indicates their close relationships, aligning the two groups in an evolutionary sense. Shared larval characters, including the absence of a naupliosoma and lack of an exopod on the fifth pereiopod, support the relationship of the Scyllarinae and Theninae. At the same time, some *Ibacus* species and particularly the Theninae have abbreviated development with shorter larval durations, resulting in less-extensive offshore migrations. This, no doubt, relates to their adaptation to sedimentary coastal and shelf habitats where prevailing current systems may not otherwise recycle larvae for successful recruitment to adult habitats.

In conclusion it appears that in accordance with the theory that there is directional dispersal from the largest and most favorable niches into smaller and less favorable, more specialized ones (Abele 1982) the scyllarids appear to have undergone recent radiations into the various shallow, shelf and deepwater habitats in which we find them today.

Acknowledgments

Our thanks to the two referees for their helpful comments. Their suggestions were constructive and even generous, which was of real help in our completing this review of a sometimes speculative subject. Grateful thanks also to Raymond Coory and Sam Webber for their help with the electronic production of our illustrations.

References

Abele, L.G. 1982. Biogeography. In: Abele, L.G. (ed.), *The Biology of the Crustacea, Vol. 1, Systematics, the Fossil Record, and Biogeography*: pp. 241–304. New York: Academic Press.

Ahyong, S.T. & O'Meally, D. 2004. Phylogeny of the Decapoda Reptantia: Resolution using three molecular loci and morphology. *Raffles Bull. Zool.* 52: 673–693.

Anger, K. 2001. The Biology of Decapod Crustacean Larvae. *Crustacean Issues, Vol. 14*. Lisse, The Netherlands: A.A. Balkema. 420 pp.

Baisre, J.A. 1994. Phyllosoma larvae and the phylogeny of Palinuroidea (Crustacea: Decapoda): A review. *Aust. J. Mar. Freshw. Res.* 45: 925–944.

Booth, J.D., Webber, W.R., Sekiguchi, H. & Coutures, E. 2005. Diverse recruitment strategies within the Scyllaridae. *N. Z. J. Mar. Freshw. Res.* 39: 581–592.

Bouvier, E.L. 1915. Sur les formes adaptives du *Scyllarus arctus* L. et sur le développement post-larvaire des scyllares. *Compte Rendu Sci. Paris* 160: 288–291.

Brown, D.E. & Holthuis, L.B. 1998. The Australian species of the genus *Ibacus* (Crustacea: Decapoda: Scyllaridae), with the description of a new species and addition of new records. *Zool. Med.* 72: 113–141.

Brusca, R.C. & Brusca, G.J. 2003. *Invertebrates,* 2nd Edition. Sunderland, MA: Sinauer Associates, Inc. 936 pp.

Chan T.-Y. & Yu, H.-P. 1986. A report on the *Scyllarus* lobsters (Crustacea: Decapoda: Scyllaridae) from Taiwan. *J. Taiwan Mus.* 39: 147–174.

Chan T.-Y. & Yu, H.-P. 1993. *The Illustrated Lobsters of Taiwan*. Taipei: SMC Publishing Inc. 251 pp.

Dana, J.D. 1852. Conspectus crustaceorum quae in orbis terrerum circumnavigatione, Carolo Wilkes e classe reipublicae foederatae duce, lexit et descripsit. *Proc. Acad. Natl. Sci. Philadelphia* 6: 6–28.

Davie, P.J.F. 1990. A new genus and species of marine crayfish, *Palibythus magnificus*, and new records of *Palinurellus* (Decapoda: Palinuridae) from the Pacific Ocean. *Invert. Taxon.* 4: 685–695.

Davie, P.J.F. 2002. Crustacea: Malacostraca, Phyllocarida, Hoplocarida, Eucarida (Part 1). In: Wells, A. & Houston, W.W.K. (eds.), *Zoological Catalogue of Australia* 19.3A, Melbourne: CSIRO Publishing. 551 pp.

Dixon, C.J., Ahyong, S.T., & Schram, F.R. 2003. A new hypothesis of decapod phylogeny. *Crustaceana* 76: 935–975.

Doak, W. 1971. *The Cliff Dwellers: An Undersea Community*. Auckland: Hodder & Stoughton. 80 pp.

Fabricius, J.C. 1775. Systema entomologiae, sistens insectorum classes, ordines, genera, species, adiectis synonymis, locis, descriptionibus, observationibus. Flensburg et Lipsiae, in Offic. Libr. Kortii: 1–832.

Faulkes, Z. 2004. Loss of escape response and giant neurons in the tailflipping circuits of slipper lobsters, *Ibacus* spp. (Decapoda, Palinura, Scyllaridae). *Arthrop. Struc. Dev.* 33: 113–123.

Forest, J. 1954. Scyllaridae. Crustacés Décapodes Marcheurs des îles de Tahiti et des Tuamotu II. *Bull. Mus. d'Hist. naturelle Paris* 26: 345–352.

Förster, R. 1984. Bärenkrebse (Crustacea, Decapoda) aus dem Cenoman des Libanon und dem Eozän Italiens. *Mitteilungen. Bayerische, Straatsam. für Paläont. Hist. Geol.* 24: 57–66.

George, R.W. 1969. Natural distribution and speciation of marine animals. *J. R. Soc. West. Aust.* 52: 33–40.

George, R.W. 1997. Tectonic plate movements and the evolution of *Jasus* and *Panulirus* spiny lobsters (Palinuridae). *Mar. Freshw. Res.* 48: 1121–1130.

George, R.W. & Main, A.R. 1967. The evolution of spiny lobsters (Palinuridae): A study of evolution in the marine environment. *Evolution* 21: 803–820.

George, R.W. & Griffin, D.J.G. 1972. The shovel nosed lobsters of Australia. *Aust. Nat. Hist.* 17: 227–231.

Glaessner, M.F. 1929. Crustacea Decapoda. In: Pompeckj J.F. (ed.), *Fossilium Catalogus* 1: *Animalia, 41.* Berlin: W. Junk. 464 pp.

Glaessner, M.F. 1969. Decapoda. In: Moore, R.C. (ed.), *Treatise on Invertebrate Paleontology, Pt. R, Arthropoda* 4(2): pp. R400–R533. Lawrence, KS: University of Kansas & Geology Society of America.

Griffin, D.J.G. & Stoddart, H.E. 1995. Deep-water decapod Crustacea from eastern Australia: Lobsters of the families Nephropidae, Palinuridae, Polychelidae and Scyllaridae. *Rec. Aust. Mus.* 47: 231–263.

Haan, W. de. 1833–1850. Crustacea. In: de Siebold, P.F. (ed.), *Fauna Japonica sive Description Animalium, Quae in Itinere per Japoniam, Jussu et Auspiciis Uperiorum, qui Summum in India Batava Imperium Tentent, Suscepto, Annis 1823–1830 Collegit, Notis, Observationibus et Adumbrationibus Illustravit.* Leiden. 243 pp.

Holthuis, L.B. 1946. The Stenopodidae, Nephropsidae, Scyllaridae and Palinuridae. The Decapoda Macrura of the *Snellius* Expedition. I. Biological results of the *Snellius* Expedition. XIV. *Temminckia* 7: 1–178, Pls. I–XI.

Holthuis, L.B. 1952. The Crustacea Decapoda Macrura of Chile. *Reports of the Lund University Chile Expedition 1948–49. 5.* Con resumen en español. *Lunds Univ. Arsskrift (n. s.)* (2) 47: 1–110.

Holthuis, L.B. 1960. Preliminary descriptions of one new genus, twelve new species and three new subspecies of Scyllarid lobsters (Crustacea Decapoda Macrura). *Proc. Biol. Soc. Washington* 73: 147–154.

Holthuis, L.B. 1963. Preliminary descriptions of some new species of Palinuridea (Crustacea Decapoda, Macrura, Reptantia). *Proc. Koninklijke Nederlandse Akad. Wetenschappen Series C, Biol. Med. Sci.* 66: 54–60.

Holthuis, L.B. 1967. Some new species of Scyllaridae. *Proc. Koninklijke Nederlandse Akad. Wetenschappen Series C, Biol. Med. Sci.* 70: 305–308.

Holthuis, L.B. 1968. The Palinuridae and Scyllaridae of the Red Sea. The second Israel South Red Sea Expedition, 1965, Report No.7. *Zool. Med.* 42: 281–301.

Holthuis, L.B. 1977. Two new species of scyllarid lobsters (Crustacea Decapoda, Palinuridea) from Australia and the Kermadec Islands, New Zealand. *Zool. Med.* 52: 191–200.

Holthuis, L.B. 1982. A new species of *Scyllarus* (Crustacea Decapoda Palinuridea) from the Pacific Ocean. *Bull. Mus. Nat. d'Hist. Natur. Paris* 4e ser., 3, section A no. 3: 847–853.

Holthuis, L.B., 1985. A revision of the family Scyllaridae (Crustacea: Decapoda: Macrura). I. Subfamily Ibacinae. *Zool. Verhand.* 218: 1–130.

Holthuis, L.B. 1991a. *Marine Lobsters of the World. An Annotated and Illustrated Catalogue of the Species of Interest to Fisheries Known to Date.* FAO Species Catalogue No. 125, Vol 13: pp. 1–292. Rome: Food and Agriculture Organization of the United Nations.

Holthuis, L.B. 1991b. Marcgraf's (1648) Brazilian Crustacea. *Zool. Verhand.* 268: 1–123.

Holthuis, L.B. 1993a. *Scyllarus rapanus*, a new species of locust lobster from the South Pacific (Crustacea, Decapoda, Scyllaridae). *Bull. Mus. Nat. d'Hist. Natur. Paris* 4e ser. 15, section A, no. 1–4: 179–186.

Holthuis, L.B. 1993b. *Scyllarides obtusus* spec. nov., the scyllarid lobster of Saint Helena, Central South Atlantic (Crustacea: Decapoda Reptantia: Scyllaridae). *Zool. Med.* 67: 505–515.

Holthuis L.B. 1996a. The scyllarid lobsters (Crustacea: Decapoda: Palinuridea) collected by F. Péron and C.A. Leseur during the 1800–1804 expedition to Australia. *Zool. Med.* 70: 261–270.

Holthuis, L.B. 1996b. Original watercolours donated by Cornelius Sittardus to Conrad Gesner in his (1558–1670) works on aquatic animals. *Zool. Verhand.* 70: 169–196.

Holthuis, LB. 2002. The Indo-Pacific scyllarinid lobsters (Crustacea, Decapoda, Scyllaridae). *Zoosystema* 24: 499–683.

Holthuis, L.B. & Sakai, T. 1970. *Ph. F. von Siebold and Fauna Japonica. A History of Early Japanese Zoology.* Tokyo: Academic Press of Japan. 323 pp.

Jones, C.M. 1988. *The Biology and Behaviour of Bay Lobsters, Thenus* spp. (*Decapoda: Scyllaridae) in Northern Queensland, Australia.* Ph.D. thesis: University of Queensland, Brisbane, Australia, 190 pp.

Leach, W.E. 1816. A tabular view of the external characters of four classes of animals, which Linné arranged under Insecta: With the distribution of the genera composing three of these classes into orders, &c. and descriptions of several new genera and species. *Trans. Linnaean Soc. Lond.* 11: 306–400.

Linnaeus, C. 1758. *Systema Naturae per Regna Tria Naturae, Secundum Classes, Ordines, Genera, Species, cum Characteribus, Differentiis, Synonymis, Locis*. 10th edition, vol. 1. London: British Museum (N.H.). 824 pp.

Man, J.G. de. 1916. The Decapoda of the Siboga Expedition. Part III. Families Eryonidae, Palinuridae, Scyllaridae and Nephropidae. *Siboga Expeditie* 39: 1–122.

Martins, H.R. 1985. Biological studies of the exploited stock of the Mediterranean locust lobster *Scyllarus latus* (Latreille, 1802) (Decapoda: Scyllaridae) in the Azores. *J. Crust. Bio.* 5: 294–305.

Milne Edwards, H. 1837. Histoire Naturelle des Crustacés, Comprenant L'anatomie, La Physiologie et La Classification de ces Animaux. Paris: Libraire Encyclopedique de Roret T.II. 468 pp.

Newman, W.A. 1991. Origins of Southern Hemisphere endemism, especially among marine Crustacea. *Mem. Queensland Mus.* 31: 51–76.

Patek, S.N. & Oakley, T.H. 2003. Comparative tests of evolutionary trade-offs in a palinurid lobster acoustic system. *Evolution* 57: 2082–2100.

Pollock, D.E. 1993. Speciation in spiny lobsters — Clues to climatically-induced changes in ocean circulation patterns. *Bull. Mar. Sci.* 53: 937–944.

Polz, H. 1996. Eine Form-C-Krebslarve mit erhaltenem Kopfschild (Crustacea, Decapoda, Palinuriodea) aus den Solnhofener Plattenkalken. *Archaeopteryx* 14: 43–50.

Poore, G.C.B. 2004. *Marine decapod Crustacea of Southern Australia: A Guide to Identification with a Chapter on Stomatopoda by S.T. Ahyong*. Collingwood, Victoria: CSIRO Publishing. 574 pp.

Ptacek, M.B., Sarver, S.K., Childress, M.J., & Herrnkind, W.F. 2001. Molecular phylogeny of the spiny lobster genus *Panulirus* (Decapoda; Palinuridae). *Mar. Freshw. Res.* 52: 1037–1047.

Quayle, W.J. 1987. English Eocene Crustacea (lobsters and stomatopod). *Palaeontology* 30: 581–612.

Rathbun, M.J. 1935. Fossil Crustacea of the Atlantic and Gulf Coastal Plain. *Geo. Soc. Am. Spec. Pap.* 2: 160 pp.

Richter, S. & Scholtz, G. 2001. Phylogenetic analysis of the Malacostraca (Crustacea). *J. Zool. Syst. Evol. Res.* 39: 113–136.

Scholtz, G. & Richter, S. 1995. Phylogenetic systematics of the reptantian Decapoda (Crustacea, Malacostraca). *Zool. J. Linn. Soc.* 113: 289–328.

Schram, F.R. 1986. *Crustacea*. New York: Oxford University Press, Inc. 606 pp.

Schram, F.R. 2001. Phylogeny of decapods: moving towards a consensus. *Hydrobiologia* 449: 1–20.

Smith, S.I. 1869. Descriptions of a genus and two new species of Scyllaridae, and of a new species of *Aethra* from North America. *Am. J. Sci.* 48: 118–121.

Spanier, E. & Lavalli K.L. 1998. Natural history of *Scyllarides latus* (Crustacea: Decapoda): a review of the contemporary biological knowledge of the Mediterranean slipper lobster. *J. Nat. Hist.* 32: 1769–1786.

Spanier, E., Weihs, D. & Almog-Shtayer, G. 1991. Swimming of the Mediterranean slipper lobster. *J. Exp. Biol.* 145: 15–31.

Squires, R.L. 2001. Additions to the Eocene invertebrate fauna of the Llajas Formation, Simi Valley, Southern California. *Nat. Hist. Mus. Los Angeles County, Cont. Sci.* 480: 1–40.

Stewart, J. & Kennelly, S.L. 1997. Fecundity and egg-size of the Balmain bug *Ibacus peronii* (Leach, 1815) (Decapoda, Scyllaridae) off the east coast of Australia. *Crustaceana* 70: 191–197.

Webber, W.R. & Booth, J.D. 2001. Larval stages, developmental ecology, and distribution of *Scyllarus* sp. Z (probably *Scyllarus aoteanus* Powell, 1949) (Decapoda: Scyllaridae). *N. Z. J. Mar. Freshw. Res.* 35: 1025–1056.

Williams, A. 2003 (compiled). *Biodiversity Survey of Seamounts & Slopes of the Norfolk Ridge and Lord Howe Rise*. Progress Report 1 to Australia's National Oceans Office. CSIRO Marine Research: 1–50, Appendixes 1–5.

Williams, A.B. 1988. *Lobsters of the World — An Illustrated Guide*. New York: Osprey Books. 186 pp.

Woods, H. 1925. *A Monograph on the Fossil Macrurous Crustacea of England*. London: Adlard & Son Ltd. 122 pp.

3

Genetics of Slipper Lobsters

Anna M. Deiana, Angelo Libertini, Angelo Cau, Rita Cannas, Elisabetta Coluccia, and Susanna Salvadori

CONTENTS

Abstract

Aware of the paucity of genetic data available for slipper lobsters, in the following paragraphs, we review the cytogenetic and molecular data known for this family. Moreover, we integrate the scyllarid data with some examples of how genetic data have helped in evolutionary studies and have been applied to practical issues, especially those important to fisheries, for the closely related family Palinuridae.

3.1 Introduction

The family Scyllaridae consists of four extant subfamilies and 85 species, four of them unnamed (Webber & Booth, Chapter 2). Information on scyllarid evolution is fragmentary; since few scyllarid species form significant fisheries, fewer studies have been published on them than on spiny lobsters (Booth et al. 2005). Slipper lobster phylogeny has been reconstructed using almost exclusively morphological characters of larvae and adults and only recently has scyllarid taxonomy undergone important revisions (Holthuis 2002; see Webber & Booth, Chapter 2 for additional details); thus, many issues remain open in the taxonomic status of this family.

Above the family level, the Scyllaridae has been included in the newly coined infraorder "Achelata" together with the Palinuridae and Synaxidae families (Scholtz & Richter 1995). There seems to be little doubt that the infraorder Achelata is monophyletic; however, the relationships among the three Achelata families are not so clear (see Webber & Booth, Chapter 2). Ahyong & O'Meally (2004) indicate that the Scyllaridae are a sister taxon to the Synaxidae-plus-Palinuridae.

At the family level, Holthuis (2002), analyzing the Indo-Pacific species of the subfamily Scyllarinae, divided the species previously attributed to the single genus *Scyllarus* into 14 genera. Further revisions may result by the reanalysis of the Atlantic *Scyllarus* species (Webber & Booth, Chapter 2). Moreover, in a recent review on scyllarid larvae, Booth et al. (2005) point out some discrepancies in the subfamily Ibacinae (genera *Ibacus*, *Evibacus* and *Parribacus*), which suggests the need of its future revision.

At the species level, new species continue to be reported, especially from deeper waters (>300 m), mostly among the Scyllarinae, a subfamily that includes over half the known scyllarids. According to Webber & Booth (2001) there are descriptions of at least 57 scyllarinid (formerly *Scyllarus*) phyllosoma, although no more than 20 of them were attributed to known species. Most scyllarid larvae collected, even those recently described, remain unclassified beyond the genus level. Further new species and taxonomic revisions can be expected (Booth et al. 2005). New molecular techniques, the combination of different types of data (molecular and morphological), and the application of new phylogenetic methods can help in scyllarid evolutionary studies, offering new suites of characters for analyzing relationships among species, genera, families, etc. Furthermore, genetic markers — including chromosomes, genome sizes and composition, microsatellites, and sequences of nuclear and mitochondrial DNA genes — can allow a solid discrimination of species and the identification of reproductive stocks in commercial species, both of which are very relevant to fishery-management strategies. Cytogenetics and molecular genetics, therefore, can be powerful tools in clarifying and diagnosing systematic and phylogenetic issues in Scyllaridae.

In this chapter we gather all the genetic data available for slipper lobsters, focusing on molecular genetics and cytogenetics. Although the available data are few and refer mainly to Mediterranean species, they allow the formation of some hypotheses on the evolution of genome organization within the family. Moreover, the genetic information on scyllarids has been compared to that available on the better studied Palinuridae.

3.2   Importance of Genetic Data

Taxonomic uncertainties result predominantly from a lack of data. As discussed in the previous chapters of this book, sometimes fossil or extant morphological characters are not definitive in addressing problematic issues, especially those concerning recognition of new species, discrimination of cryptic forms, and reconstruction of evolutionary pathways.

Since genomes are huge in information content, DNA, contained in any biological material, can be used to get an enormous amount of data, that is both stable and heritable. Apart from phylogenetic studies, genetic analyses can contribute by providing essential information on little-known aspects of species biology, in recognizing and characterizing populations, in defining management units within species, in detecting hybridization, in describing speciation mechanisms, etc.

3.2.1   Importance of Molecular Genetics

The use of molecular genetic techniques has increased greatly over the past years, together with the awareness of the value of genetic data. Many current molecular genetic methods rely on the polymerase chain reaction (PCR) (Saiki et al. 1988; Sambrook & Russel 2001a), which allows amplification of specific DNA sequences from tiny amounts of starting tissue. PCR-based assays are proliferating rapidly and have become the main approach in molecular genetics research due to their speed of sample processing, possibility of nondestructive sampling, and range of specificity. Direct sequencing of PCR-amplified fragments is the most common tool in molecular phylogenetic studies (see Sambrook & Russell (2001b) for a comprehensive description of the techniques). It is a relatively time-consuming and expensive step; but sequencing gives the highest level of resolution.

At the intraspecific level, mitochondrial DNA (mtDNA) is one of the most frequently used markers in molecular population genetics because of its maternal inheritance, haploid nature, smaller effective size, sensitivity to genetic drift phenomena, and rapid rate of evolution (Moritz et al. 1987; Boore 1999). Mitochondrial DNA sequence analyses found substantial genetic homogeneity in several spiny lobster species: *Jasus edwardsii* (Hutton, 1875) (Ovenden et al. 1992), *Panulirus argus* (Latreille, 1804) (Silberman et al. 1994), and *Panulirus interruptus* (Randall, 1840) (Garcia-Rodriguez & Perez-Enriquez 2005). Similarly, in *Palinurus gilchristi* Stebbing, 1900, despite the existence of two distinct, phenotypically distinguishable stocks, sequences of a portion of the mtDNA control region highlighted a panmictic genetic structure for the species (Tolley et al. 2005). Furthermore, mitochondrial sequence analyses have been used to recognize subspecies and new species. Sarver et al. (1998) found high levels of sequence divergence between Caribbean and Brazilian *Panulirus argus* and suggested the recognition of two subspecies: *P. argus argus* in the Caribbean and *P. argus westonii* in Brazil. In *Panulirus longipes* (H. Milne Edwards, 1868), Ravago & Juinio-Menez (2002) provided molecular evidence for the recognition of *Panulirus femoristriga* (Von Martens, 1872) as a distinct species from *Panulirus longipes bispinosus* Borradaile, 1899.

Recently, microsatellites have become the marker of choice for population studies. Microsatellite loci are tandem repeats of short DNA motifs, typically one to five bases in length. The number of microsatellite repeats is highly variable due to slippage during DNA replication. Since the microsatellites are highly variable, individual genotypes can be directly inferred. They have shown to be very useful in studying the genetic diversity and spatial population structure of crustacean species. In the American lobster, *Homarus americanus* H. Milne Edwards, 1837, microsatellite analyses have been applied to study the mating system, revealing female promiscuity and multiple paternities (Gosselin et al. 2005). Microsatellite loci are described for a growing number of decapods: several shrimps species (Maggioni & Rogers 2002; Robainas et al. 2002; Ball & Chapman 2003; Maggioni et al. 2003; Meadows et al. 2003; Meehan et al. 2003; Pan et al. 2004; Chand et al. 2005), a few crab species (Gopurenko et al. 2002; Yap et al. 2002; Hänfling & Weetman 2003; Jensen & Bentzen 2004; Stevens et al. 2005), and lobsters (Streiff et al. 2001; Jones et al. 2003; Diniz et al. 2004; Jørstad et al. 2004; Gosselin et al. 2005).

3.2.2   Importance of Cytogenetics

Chromosomes are the morphological manifestation of the genome, and the analysis of their size, shape, and number, as well as the structural organization of chromatin is the fundamental step for the karyotype characterization of a species. In several taxa, in addition to classical cytogenetic techniques such as C- and fluorochrome banding, molecular techniques such as fluorescence *in situ* hybridization (FISH) have been applied to locate genes or specific sequences on chromosomes. Beside the species-specific karyotype characterization, comparative cytogenetics is now widely used to study karyotype evolution among species and represents a very useful approach in phylogenetic analysis (Almeida-Toledo 1996; Pisano & Ozouf-Costaz 2000). Cytogenetic information may represent a useful phylogenetic tool when morphological and molecular data cannot resolve evolutionary relationships, as in the case of sibling species (Volobouev et al. 2002; Dobigny et al. 2004).

Data on cytogenetics are scarce for decapods, as well as for most crustaceans. Among Decapoda, the Nephropidae and Palinuridae families, which include the most economically important species, are also the most cytogenetically studied (Nakamura et al. 1988; Lècher et al. 1995; Coluccia et al. 2004 and references therein; Coluccia et al. 2005). In these two families, classical bandings and FISH of specific sequences have been used, allowing the characterization of the chromosome complement and comparison among species.

The overall paucity of cytogenetic data on Decapoda is principally due to the technical constraints in obtaining good chromosomal preparations. Decapods have high chromosome numbers, ranging from $2n = 54$ in the swimming crab, *Liocarcinus vernalis* Risso, 1816 (Trentini et al. 1989) to 254 in the hermit crab, *Eupagurus ochotensis* Brandt, 1851 (Niiyama 1959; for a review see Lécher et al. 1995). Polymorphism of the chromosome number is reported for several species. Variability in chromosome numbers could be explained by the presence, in many species, of a large number of small-sized chromosomes rendering an accurate count difficult or by possible chromosome loss during cytological preparation.

Moreover, in some species of Palinuridae and Nephropidae, the presence of supernumerary (B) chromosomes was hypothesized and demonstrated (Coluccia et al. 2004). B chromosomes are supernumerary to the standard chromosome complement, variable in number within species and populations, mainly heterochromatic, and in meiosis they show non-Mendelian segregation. B chromosomes appear to be widespread in animals, plants, and fungi; among Crustacea, the presence of B chromosomes have been reported in 20 species (Camacho 2005). Among Decapoda, B chromosomes have been demonstrated in the clawed lobsters *Nephrops norvegicus* (Linnaeus, 1758) and *Homarus americanus* (Nephropidae), in the spiny lobsters *Palinurus elephas* (Fabricius, 1787), *P. mauritanicus* Gruvel, 1911, and *P. gilchristi* (Palinuridae), but not in the slipper lobsters *Scyllarides latus* (Latreille, 1802), *Scyllarus arctus* (Linnaeus, 1758) (for a review see Coluccia et al. 2004) and *S. pygmaeus* (Bate, 1888) (Deiana et al., unpublished data).

3.2.3   Importance of Genome Size and DNA Base Composition

Genome size (GS) is a fundamental parameter of a given species and the study of its variation among living organisms is important for practical and theoretical perspectives. Among many other characters, GS has been shown to correlate directly with cellular and nuclear sizes in animals and plants; duration of mitosis and meiosis in plants; body mass in mammals, annelids, flatworms, insects, and copepods; and duration of development in amphibians, beetles, aphids, and copepods. Likewise, GS has been inversely correlated with the rate of basal metabolism in mammals and passerine birds and with morphological complexity in the brains of amphibians. In such cases, the observed variation in genome size reflects different adaptive needs or the efficacy of natural selection in different organisms (references in: Vinogradov 1998; Petrov 2001; Gregory 2005a).

The simultaneous determination of GS and DNA base composition (AT- or GC-percent) is more informative than the mere determination of GS. G + C content was found to be significantly positively correlated with genome size among vertebrates in general, albeit in a nonlinear way. Higher GC-percent of larger genomes suggests the existence of selection pressure for GC-enrichment of repetitive DNA, conferring to the genomes a higher physical stability and providing a certain protection of genes against chemical damage (Vinogradov 1998). DNA base composition has been related to the regulation capability of body temperature in vertebrates and to resistance to air exposure in terrestrial and semiterrestrial molluscs (Vinogradov 1998; Gregory 2005a; Vitturi et al. 2005 and references therein).

Data on genome size and DNA base composition are required for studies on genome structure and evolution, and on biodiversity at the genetic level (Vinogradov 1998). DNA contents have been also used for determination of weighting factors in the estimation of ecological exergy, a quality indicator for ecosystems (Fonseca et al. 2000). Moreover, in biotechnology applied to fisheries, knowledge of DNA content may be basic in the choice of model species for genome mapping and sequencing, or in planning experiments of hybridization and chromosome set manipulation.

3.3   Molecular Genetics of Slipper Lobsters

To solve some existing taxonomic and phylogenetic uncertainties in Scyllaridae, briefly summarized in the previous sections of this chapter and fully described in Chapter 2 of this book, genetics could be a significantly helpful tool. Unfortunately, genetic methods have been rarely applied to scyllarid species, despite the fact that these methodologies have assumed an increasing importance in evolutionary studies.

Protein analysis was used to elucidate the genetic structure of *Scyllarides latus* from different Sicilian areas, and showed low levels of variability (Bianchini et al. 2003). However, little DNA sequencing has been done on scyllarids. Some DNA sequences have been included as outgroups in phylogenetic studies on palinurids and nephropids: the partial sequence of 16S rDNA of *Scyllarides nodifer* (Stimpson, 1866) (Tam & Kornfield 1998), the 28S, 18S, and the 16S rDNA sequences of *Scyllarus arctus* (Patek & Oakley, 2003), and the partial sequence of the mitochondrial gene for cytochrome oxydase subunit I of *Parribacus antarcticus* (Lund, 1793) (Ravago & Juinio-Menez 2002). Other sequences, cited in Patek & Oakley (2003), are deposited in GenBank (http://www.ncbi.nlm.nih.gov/). Partial DNA sequences for the 28S and 16S rRNA for three Mediterranean scyllarid species have been obtained (Cannas et al., unpublished data: *Scyllarus pygmaeus* GenBank accession number DQ377984, DQ377973; *Scyllarus arctus* DQ377985, and *Scyllarides latus* DQ377983, DQ377974) and aligned with Clustal X (Thompson et al. 1997). The 28S rDNA sequences are more conserved than the 16S sequences, but both regions clearly differentiate the species of the subfamily Scyllarinae from the Arctidinae.

Furthermore, the alignment of the 18S rDNA sequences recovered from GenBank for species belonging to different genera (*S. arctus*: GeneBank Accession number AF498677, *S. latus*: AF498669, and *Parribacus antarcticus*: AF498676), points out a closer affinity between *Parribacus* and *Scyllarides* (Cannas et al., unpublished data), confirming the traditional morphological scyllarid phylogeny (Baisre 1994). Ribosomal DNA sequences have been demonstrated to be valid phylogenetic markers, but sequencing more regions and species, belonging to all the four subfamilies, could provide further insights for Scyllaridae systematics.

3.4 Cytogenetics of Slipper Lobsters

3.4.1 Karyology

Only three species of Scyllaridae have been studied cytogenetically: *Scyllarides latus* (Deiana et al. 1997), *Scyllarus arctus* (Deiana et al. 1992), and *Scyllarus pygmaeus* (Deiana et al., unpublished data). Because of the lack of availability of specimens and technical constraints in obtaining good karyological preparations, the karyological data obtained by Giemsa staining and C- and fluorochrome bandings support only general information on the chromosome complement in term of chromosomal size, shape, and chromosome number.

In these species, chromosome preparations were obtained by direct methods from gonads (mainly testis) and hepatopancreas. Cell culture techniques for decapods have not been widely successful; thus far, only *in vitro* maintenance of lymphoid organ cells of *Penaeus japonicus* (Bate, 1888) was obtained, and growth of these cells did not occur (Itami et al. 1989). As a result, direct methods are the only ones available for obtaining chromosome preparations from crustaceans. In particular, the air-drying technique is based on tissue dissociation, followed by hypotonization and fixation of cells in suspension (Doussau De Bazignan & Ozouf-Costaz 1985). Different tissues can be used, but the most suitable in terms of mitotic activity are: gonads, hepatopancreas, antennal gland, and gills. The most used tissue is testis since a good number of metaphases may be obtained, but its activity varies with season and maturity stage. The disadvantage of noncultured cells is the very low number of metaphases and the presence of cellular debris. Hayashi & Fujiwara (1988) obtained satisfactory numbers of mitoses using regeneration blastema in *P. japonicus*. An alternative air-drying method has been used, especially for embryos or small pieces of tissue (Libertini et al. 2000).

Cytogenetic data on the Scyllaridae studied have been obtained using different techniques: (1) Giemsa staining that uniformly stains chromatin and allows for the study of size, shape, and number of mitotic and meiotic chromosomes, and (2) heterochromatic chromosome bandings, that allow identification of specific chromosomal regions. The heterochromatic regions, composed of moderately and highly repetitive DNA, have been studied by C- (Deiana et al. 1996) and fluorochrome bandings. C-banding is a technique used for the identification and localization of heterochromatin in chromosomes; these regions may represent very useful markers in cytogenetics. AT- or GC-specific fluorochromes identify heterochromatic regions with different base composition. In particular, in the scyllarids studied, quinacrine (Casperson et al. 1968) and DAPI (4, 6-diamidino-2 phenylindole) (Schweizer 1976) bandings, specific for AT-rich DNA, have been applied.

The chromosome complements of *S. latus*, *S. arctus*, and *S. pygmaeus* are similarly composed of some large pairs of meta-submetacentric chromosomes and a large number of small chromosomes, mostly metacentric in shape, even if the morphology of many of them is difficult to identify since they are very small. Heterochromatic regions, studied by C- and fluorescence bandings, have been localized mainly in centromeric regions in all of the species; furthermore, some species-specific bands have been found. By labeling the centromeres, these banding techniques have been useful in understanding the chromosomal morphology and in identifying individual chromosomes, characterized by specific bands.

The chromosomal numbers (diploid and haploid) have been difficult to determine because of the difficulty in counting a large number of small chromosomes and of the low numbers of specimens analyzed, especially in the case of *S. latus* which has more than 100 chromosomes. For this reason, a range of values has been reported for each species. The presence of supernumerary B chromosomes, generally causing numerical variability among individuals and populations within a species, was not reported in the scyllarids studied (Coluccia et al. 2004).

In *S. latus* the few data available come from a preliminary study carried out on two male specimens (Deiana et al. 1997). The diploid value, obtained from the count of spermatogonial metaphases, ranged from 126 to 138; the haploid value, obtained from the count of meiotic metaphases I and II, ranged from 65 to 70. The chromosome complement is primarily made up of metacentric and submetacentric chromosomes. Six chromosome pairs are clearly recognizable because of their large size: four submetacentric and two metacentric; the remaining complement is made up of smaller chromosomes of decreasing length (Figure 3.1A). Banding with quinacrine fluorochrome produced a Q-banding pattern that localized AT-rich heterochromatin on the long arm of some large chromosomes (Figure 3.1B).

In *S. arctus* the reported data refer to a preliminary study (Deiana et al. 1992) and to our unpublished data (Deiana et al., unpublished data). The chromosome complement of *S. arctus* is comprised mainly of meta-submetacentric chromosomes (Figure 3.1C). Counts of mitotic and male meiotic metaphases, performed on 20 male specimens, gave a diploid number of 70 to 72 and a haploid number of 35 to 36 (Deiana et al. 1992). Banding with DAPI fluorochrome identified some AT-rich heterochromatic regions, mainly located in the centromeric region of several chromosomes (Figure 3.1D) (Deiana et al., unpublished data).

Scyllarus pygmaeus specimens are very difficult to find and data available come from personal observations of the authors (Deiana et al., unpublished data).The chromosome complement of *S. pygmaeus* is mainly composed of meta-submetacentric chromosomes (Figure 3.1E). The diploid number, drawn from the counting of four somatic mitoses from three specimens, ranged from 70 to 74. C-bands were localized in all centromeric regions; furthermore, some specific large bands were localized in some chromosomes (Figure 3.1F).

In addition, the analysis of meiotic metaphases has been performed in the three species. In late prophase and metaphase I of all of these species, pairs of homologous chromosomes are visible as tetrads of different morphology: dumbbell (Figure 3.2A), cross- (Figure 3.2B), ring-shaped (Figure 3.2C), or multiple rings (Figure 3.2D). The dumbbell figure is typical of crustaceans (Niiyama 1959) and derives from a chromosomal end-to-end pairing, the cross figure results from a single chiasma, the ring figure is determined by two chiasmata, and the multiple rings by three chiasmata. In *S. latus*, tetrads of all types of meiotic figures are present (Figure 3.2E) (Deiana et al. 1997). In *S. arctus* (Deiana et al. 1992) and *S. pygmaeus* (Deiana et al., unpublished data), multiple rings are absent (Figure 3.2F); in *S. arctus* metaphases I, C-banding has been a useful technique for a better comprehension of meiotic tetrad shape revealing constitutive heterochromatin in centromeric regions. Furthermore, in some tetrads, large bands were found (Figure 3.2F). In all studied species, the presence of asynaptic or abnormally paired chromosomes was not detected.

3.4.2 Comparison of the Karyological Data within Achelata

Cytogenetics on Scyllaridae is still at preliminary stage. The reported data, referring to few specimens, are not sufficient to describe the karyotype of the three studied species; furthermore, the small number of species studied does not allow the drawing of a general trend in chromosome evolution within the family. Altogether, comparing these preliminary data with those of the better known Palinuridae species some remarks can be made. Among Achelata, the diploid number ranges from 70 to 158; in this context, the diploid number of Palinuridae species is always above 100: from 111 to 158 (Nakamura et al. 1988; Salvadori et al. 1995; Coluccia et al. 2003, 2005; Cannas et al. 2001) with some species having the highest $2n$ value. Among Scyllaridae, in the subfamily Scyllarinae (with the two species *S. arctus* and *S. pygmaeus*), the diploid number ranges from 70 to 74 (the lowest number among Achelata) while the diploid number of *S. latus*, the only studied species of the subfamily Arctidinae, is 126 to 138, or, about twice the diploid number in the subfamily Scyllarinae. For almost all the Achelata species, polymorphism of the diploid number is reported, while the presence of B chromosomes has been demonstrated in three

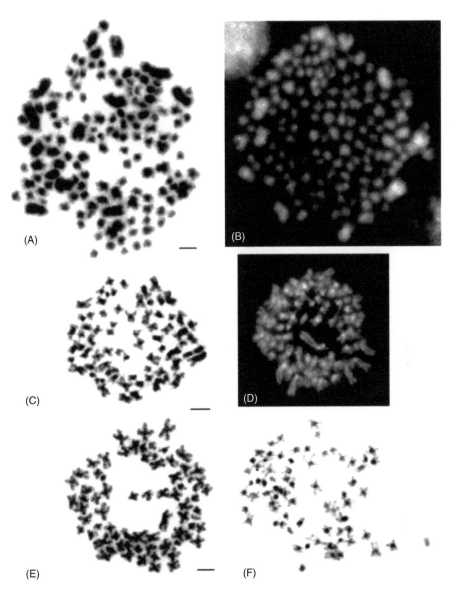

FIGURE 3.1 (A) Mitotic chromosomes of *Scyllarides latus* after Giemsa staining and (B) banding with the fluorochrome quinacrine; in (B) brightly fluorescent bands are present on the long arm of some large chromosomes. (C) Mitotic chromosomes of *Scyllarus arctus* after Giemsa staining and (D) banding with the fluorochrome DAPI; in (D) fluorescent bands are located mainly in the centromeric region of several chromosomes. (E) Mitotic chromosomes of *Scyllarus pygmaeus* after Giemsa staining and (F) C-banding; C-bands are present in the centromeric regions and large pericentromeric bands are present in some chromosomes. Scale bars = 5 μm.

species, *Palinurus elephas*, *P. mauritanicus*, and *P. gilchristi* all belonging to Palinuridae family (Salvadori et al. 1995; Coluccia et al. 2003, 2005). The presence or absence and the number of B chromosomes can represent a useful marker to identify and characterize different populations of a species, since B chromosomes can be found in different numbers among conspecific populations. The absence of B chromosomes in the Mediterranean specimens of *S. latus*, *S. arctus*, and *S. pygmaeus* could represent a species-specific character even if more specimens, from different geographical sites, should be examined. Furthermore, since these three species are the only Scyllaridae cytogenetically studied, the presence of B chrmosomes within the family could not be excluded.

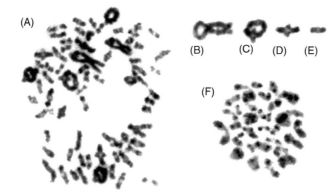

FIGURE 3.2 (A) to (D) Meiotic configurations of *Scyllarides latus*: the main meiotic figures visible in metaphases I are: (A) dumbbell, (B) cross, (C) ring, (D) multiple rings. (E) *Scyllarides latus* metaphase I. (F) *Scyllarus arctus* metaphase I after C-banding; C-bands are present in centromeric and in some pericentromeric regions.

TABLE 3.1

Summary of the data on Nuclear DNA Contents in decapod crustaceans. For the taxa higher than genus, data are reported as a range, with the number of considered species in brackets. For slipper lobster species, data are reported in detail and Genome Size and AT-DNA are expressed as Mean ±2 (standard error).

Taxon	Genome Size (C-value, pg)	Haploid AT-DNA (pg)	AT-percent	GC-percent
Order Decapoda	1.07–38.00 (84)		34.92–47.46 (12)	52.54–65.08 (12)
Suborder Dendrobranchiata	2.37–5.23 (8)		41.77 (1)	58.23 (1)
Suborder Pleocyemata	1.07–38.00 (76)		34.92–47.46 (11)	52.54–65.08 (11)
Infraorder Anomura	2.10–8.40 (12)		N/A	N/A
Infraorder Astacidea	3.91–5.64 (7)		34.92–46.00 (3)	54.00–65.08 (3)
Infraorder Brachyura	1.07–4.55 (30)		N/A	N/A
Infraorder Caridea	3.30–38.00 (18)		N/A	N/A
Infraorder Achelata (Palinura)	1.94–6.99 (9)		37.49–47.46 (8)	52.54–62.51 (8)
Family Palinuridae	3.15–5.33 (5)		37.49–47.46 (5)	52.54–62.51 (5)
Family Scyllaridae	1.94–6.99 (4)		41.90–42.64 (3)	57.36–58.10 (3)
Subfamily Arctidinae	6.82–6.99 (2)		42.64 (1)	57.36 (1)
Scyllarides herklotsii	6.82±0.13	N/A	N/A	N/A
Scyllarides latus	6.99±0.21	2.98±0.17	42.64	57.36
Subfamily Scyllarinae	1.94–2.01 (2)		41.90–42.27 (2)	51.73–58.10 (2)
Scyllarus arctus	2.01±0.02	0.84±0.03	41.90	58.10
Scyllarus pygmaeus	1.94±0.03	0.82±0.01	42.27	51.73

Source: Deiana, A.M., Cau, A., Coluccia, E., Cannas, R., Milia, A., Salvadori, S., & Libertini, A. 1999. In: Schram, F.R. & von Vaupel Klein, J.C. (eds.), *Crustaceans and the Biodiversity Crisis*: pp. 981–985. Leiden, The Netherlands: Brill Academic Publisher; Gregory, T.R. 2005b. Animal Genome Size Database. http://www.genomesize.com.

3.4.3 Genome Size and DNA Base Composition

With regard to slipper lobsters, GS was determined by means of flow cytometry for *Scyllarides herklotsii* (Herklots, 1851), *S. latus*, *S. arctus*, and *S. pygmaeus*, while DNA base composition, expressed as AT-percent, was acquired with the same method for the last three species (Deiana et al. 1999). Table 3.1 reports detailed data on haploid GS (C-value), AT-DNA, calculated AT-percent, and calculated GC-percent in the slipper lobsters studied so far, as well as the range of variation of these parameters along the taxonomic hierarchy up to the order Crustacea.

Although scanty, data on GS and DNA base composition in slipper lobsters allow the drawing of some preliminary inferences that future studies should be able to confirm. The family Scyllaridae seems characterized by a wide range of GS and almost constant DNA base composition. By considering Scyllaridae as a monophyletic group according to Scholtz & Richter (1995), GS changes during the evolution of this family seem to have occurred without any preferential enrichment in GC- or AT-DNA. At present, in the subfamily Arctidinae (with two studied species both belonging to the genus *Scyllarides*), GS values are more than three times those in the subfamily Scyllarinae (two species both belonging to the genus *Scyllarus*). Body size in the two *Scyllarides* species is larger than that in the two *Scyllarus* species (Holthuis 1991) and its ranking among the analyzed species is the same as GS; therefore, a positive relationship between GS and body size seems to characterize the Scyllaridae. Similarly, since Arctidinae shows a longer larval life history than Scyllarinae (Baisre 1994; Booth et al. 2005), this biological parameter also appears to be positively related to GS.

The other family within the infraorder Achelata for which data on GS and DNA base composition are available is Palinuridae (Table 3.1). In comparison with Scyllaridae, C-values within the Palinuridae are intermediate in a narrower range, while the range of AT- or GC-percent is wider. Thus, the two families appear to have undergone different pathways in genome size evolution. Within Palinuridae, available data on life history further support the hypothetical relationship between GS and length of larval life while, in contrast to the Scyllaridae, body size and GS seem to be unrelated (Holthuis 1991; Baisre 1994; Deiana et al. 1999; George 2005).

Nuclear DNA content has been evaluated for 84 species of Decapoda (Gregory 2005b and references therein). In this order, GS values range from 1.07 to 38.00 pg, average 5.10 ± 5.16 (mean $\pm$ standard deviation), and are widely scattered and skewed toward lower values, with the highest values generally pertaining to the infraorder Caridea (Table 3.1). A similar distribution of GS values emerges also for the suborder Pleocyemata (mean 5.33 ± 5.37 pg). Among Pleocyemata, the infraorder Achelata shows a GS range as wide as the other infraorders with the exception of Caridea (Table 3.1): the GS values in the genus *Scyllarus* are among the lowest, while they are among the highest in the genus *Scyllarides*. In 12 decapod species studied so far (Deiana et al. 1999), AT-percent ranges from 34.9 to 47.5% (Table 3.1) with slipper lobsters (all sharing about 42% of AT-DNA) being characterized by intermediate values.

3.4.4 Methods for Evaluation of Genome Size and DNA Base Pair Composition

With regards to decapod crustaceans, data on genome size have been acquired for about 80 species by Feulgen microdensitometry (79% of analyzed species) or flow cytometry (21%) (Gregory 2005b). AT-percent was determined only for 12 species by flow cytometry (Deiana et al. 1999). In the past, the main technique for nuclear DNA content determination was absorption cytometry using Feulgen staining; this technique was recently improved by the application of computerized image analysis, namely "Feulgen image analysis densitometry" (Hardie et al. 2002). With a method similar to absorption cytometry, but using AT- or GC-specific dyes instead of Feulgen stain, and measuring fluorescence emission by a microphotometer or with a computerized image analysis system, DNA base pair composition may be evaluated. Alternatively, DNA base composition may be assessed by density caesium chloride gradient centrifugation (Hudson et al. 1980) or by ultraviolet absorbance analysis during DNA thermal denaturation (Mandel & Marmur 1968).

Since flow cytometry was successfully applied to evaluate both genome size and AT-percent in four species of slipper lobsters (Deiana et al. 1999), this technique will be described herein; for other techniques to measure nuclear DNA contents see cited references. Flow cytometry provides a much higher precision and reproducibility of measurements (Muirhead et al. 1985; Vinogradov 1998). This technique is rapid and accurate, but it has some important limitations related to the need to place the cells in suspension and to the large number of nuclei required for analysis (Hardie et al. 2002). These limitations can be easily overcome by choosing suitable tissues as a cell source and by the application of microdissection techniques in sample preparation (Libertini et al. 2000; Libertini & Bertotto 2004). Flow cytometry can be applied to tissue samples that are ready as a suspension of single cells or whose disgregation can easily provide free suspended cells. For slipper lobsters, the antennal gland and gills proved to be a good material whose dissociation provided suspensions of single cells suitable for flow cytometric analysis (Deiana et al. 1999).

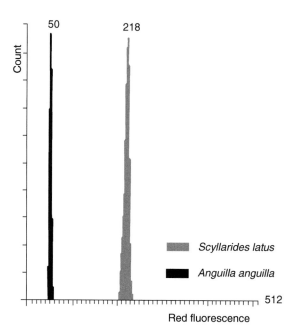

FIGURE 3.3 Histogram of fluorescence in a flow cytometry assay of a mixed cell sample from a slipper lobster and an eel (reference species). DNA content is calculated by comparison of target and reference cell peaks.

Haemolymph and embryos were successfully utilized with other crustaceans (Chow et al. 1990; Libertini et al. 1990, 2000) and may be potentially used for scyllarids. Sampled tissues can be stored for long periods of time at −80°C by applying the method proposed by Gold et al. (1991) for fish. Fixation with cold 70% ethanol and storage at −20°C can be also performed, but it is usually applied to cells already suspended in a saline buffer solution. Debris-free cell suspensions are obtained by mincing soft organs and embryos with scissors, followed by filtration through a 30 μm nylon filter, and by two consecutive steps of centrifugation–resuspension in buffer saline.

The DNA content of target cells is quantified relative to a standard DNA content in cells from a reference species. In DNA content flow cytometric evaluation, samples are usually prepared with an internal reference (Figure 3.3). Erythrocytes from chicken, *Gallus domesticus* (Linnaeus, 1758), eel, *Anguilla anguilla* (Linnaeus, 1758), and rainbow trout, *Oncorhynchus mykiss* (Walbaum, 1792), have been more frequently used as standards with crustaceans (Gregory 2005b). Other reference species may be chosen, taking into account that it is crucial that the reference cells and the target cells have a similar DNA content to minimize possible zero shift error (Vindelov et al. 1983) and, in the case of an internal reference, that the values for the reference and target cells do not overlap. For surveys on GS and DNA base composition in potential reference species see Tiersch et al. (1989) and Vinogradov (1998).

According to the type of DNA content that is going to be evaluated (GS, AT-DNA, GC-DNA), cell nuclei have to be stained with a solution containing a specific fluorescent dye, buffering salts, and a small amount of detergent. Propidium iodide is the most commonly used dye to quantitatively assess the whole genome size, although there are several different, available dyes that bind stoichiometrically to DNA. Among the UV-light excited dyes, Hoechst 33342, Hoechst 33258, or DAPI are suitable for AT-DNA evaluation, and olivomycin or chromomycin A_3 is suitable for GC-DNA. An important factor in the choice of the stain is the type of flow cytometer used and its light source. AT- and GC-specific fluorochromes require UV-light excitation at a specific wavelength and this kind of light is generally not available with commercial argon laser flow cytometers (with emission in the blue at 488 nm and in the green at 514 nm). Arc-lamp flow cytometers are more versatile and can provide either UV- or visible-light excitation by means of specific optical filter sets.

3.5 Conclusions

The infraorder Achelata and the family Scyllaridae have been considered as monophyletic (Scholtz & Richter 1995; Webber & Booth, Chapter 2). Within the Achelata, the fossil record suggests an earlier evolution of the Palinuridae than the Scyllaridae, and an earlier evolution of the genus *Scyllarides* than the genus *Scyllarus*, within the Scyllaridae (Baisre 1994; Webber & Booth, Chapter 2).

All the genetic data available, both cytogenetic and molecular, show a clear separation between Arctidinae and Scyllarinae. From the cytogenetical point of view, *Scyllarides* species are endowed by approximately threefold the GS of *Scyllarus* species and *Scyllarides latus* has about twice the chromosome number of *Scyllarus arctus* and *S. pygmaeus*. These data indicate that a drastic reduction of genome size and chromosome number ought to have accompanied evolutionary radiation of Scyllarinae; such a reduction might have played an adaptive role favoring shorter larval lives and a reduction in body size. The hypothesis of a trend toward reduction of karyotypical parameters within Scyllaridae is further supported by the finding of medium to high GSs and high chromosome numbers (Deiana et al. 1999; Coluccia et al. 2005), in the more ancient and related family Palinuridae (Anger 2001; Webber & Booth, Chapter 2). Within the order Decapoda, as well as in the infraorder Achelata, high GSs and chromosome numbers are frequent (Nakamura et al. 1988; Deiana et al. 1999; Coluccia et al. 2005) and, thus, could be considered as primitive karyotypical features. The previously reported hypothesis about derivation of the *Scyllarides* ancestor from *Scyllarus* by polyploidization, although supported by an apparent constant DNA base composition among scyllarids (Deiana et al. 1999), seems weaker than the reductional hypothesis. More complete cytogenetic analyses, also extended to other subfamilies presently unstudied, will provide a better understanding of the pathways of the evolution of genome organization in the family Scyllaridae.

Similarly, additional molecular genetic data could provide important insights in Scyllaridae systematics, clarifying the relationships among subfamilies, genera, and species. For instance, molecular markers could help in resolving the discrepancies in the subfamily Ibacinae (Booth et al. 2005) and in examining the phylogenetic relationships within the Scyllarinae — in particular, extending the analyses not only to the recently revised Indo-Pacific but also to the Atlantic species. In general, more molecular sequencing could provide important data for describing speciation mechanisms and divergence times among species, as has been done successfully for the Palinuridae genera *Jasus* (Ovenden et al. 1997) and *Panulirus* (Ptacek et al. 2001). Furthermore, considering the hard task of classifying scyllarid larvae (Webber & Booth 2001), molecular markers could be useful in the discrimination of species in any development stage, including the larval stage (see Silberman & Walsh 1992; Cannas et al. 2001 for a similar application of molecular data). In conclusion, filling the wide gap of genetic knowledge on Scyllaridae will help in solving many taxonomic and phylogenetic issues, as well as aid in planning fishery-management strategies for various stocks.

Acknowledgment

The authors wish to thank Prof. Ugo Laudani for very useful comments.

References

Ahyong, S.T. & O'Meally, D. 2004. Phylogeny of the Decapoda reptantia: Resolution using three molecular loci and morphology. *Raffles Bull. Zool.* 52: 673–693.

Almeida-Toledo, L.F. 1996. Molecular and immunocytogenetics of Brazilian fishes. *Ciénc. Cult. J. Braz. Assoc. Adv. Sci.* 48: 377–382.

Anger, K. 2001. *The Biology of Decapod Crustacean Larvae*. In: Vonk, R. (ed.), *Crustacean Issues,* Vol. 14: pp. 1–405. Lisse, The Netherlands: A.A. Balkema Publishers.

Baisre, J.A. 1994. Phyllosoma larvae and the phylogeny of Palinuroidea (Crustacea, Decapoda). A review. *Aust. J. Mar. Freshw. Res.* 45: 925–944.

Ball, A.O. & Chapman, R.W. 2003. Population genetic analysis of white shrimp, *Litopenaeus setiferus*, using microsatellite genetic markers. *Mol. Ecol.* 12: 2319–2330.

Bianchini, M.L., Spanier, E., & Ragonese, S. 2003. Enzymatic variability of Mediterranean slipper lobsters, *Scyllarides latus*, from Sicilian waters. *Ann. (Ann. Istrian & Med. Stud.), Series Hist. Nat.* 13: 43–50.

Boore, J.L. 1999. Animal mitochondrial genomes. *Nucleic Acids Res.* 27: 1767–1780.

Booth, J.D., Sekiguchi, H., Webber, W.R., & Coutures, E. 2005. Diverse larval recruitment strategies within the Scyllaridae. *N. Z. J. Mar. Freshw. Res.* 39: 581–592.

Camacho, J.P.M. 2005. B chromosomes. In: Gregory, T.R. (ed.), *The Evolution of the Genome*: pp. 223–285. London, U.K.: Elsevier Academic Press.

Cannas, R., Tagliavini, J., Relini, M., Torchia, G., & Cau, A. 2001. Marcatori molecolari per l'identificazione delle aragoste mediterranee *Palinurus elephas* (Fabricius, 1787) e *P. mauritanicus* Gruvel, 1911. *Biol. Mar. Medit.* 8: 814–816.

Caspersson, T., Farber, S., Foley, G.E., Kudynowski, J., Modest, E.J., Simonsson, E., Wagh, U., & Zech, L. 1968. Chemical differentiation along metaphase chromosomes. *Exp. Cell Res.* 49: 219–222.

Chand, V., De Bruyn, M., & Mather, P.B. 2005. Microsatellite loci in the eastern form of the giant freshwater prawn (*Macrobrachium rosenbergii*). *Mol. Ecol. Notes* 5: 308–310.

Chow, S., Dougherty, W.J., & Sandifer, P.A. 1990. Meiotic chromosome complements and nuclear DNA contents of four species of shrimps of the genus *Penaeus*. *J. Crust. Biol.* 10: 29–36.

Coluccia, E., Salvadori, S., Cannas, R., & Deiana, A.M. 2003. Study of mitotic and meiotic chromosomes of *Palinurus mauritanicus* (Crustacea, Decapoda). *Biol. Mar. Medit.* 10: 1069–1071.

Coluccia, E., Cannas, R., Cau, A., Deiana, A.M., & Salvadori, S. 2004. B chromosomes in Crustacea Decapoda. *Cytogenet. Genome Res.* 106: 215–221.

Coluccia, E., Salvadori, S., Cannas, R., Milia, A., & Deiana, A.M. 2005. First data on the karyology of *Palinurus gilchristi* (Crustacea, Decapoda). *Biol. Mar. Medit.* 12: 661–663.

Deiana, A.M., Coluccia, E., Milia, A., & Salvadori, S. 1992. Mitosis and meiosis of two Decapoda species. *Oebalia* Suppl. 17: 571–572.

Deiana, A.M., Coluccia, E., Milia, A., & Salvadori, S. 1996. Supernumerary chromosomes in *Nephrops norvegicus* L. (Crustacea, Decapoda). *Heredity* 76: 92–99.

Deiana, A.M., Bianchini, M.L., Coluccia, E., Milia, A., Cannas, R., Serra, D., & Salvadori, S. 1997. Dati preliminari sulla cariologia della magnosa (*Scyllarides latus*, Crustacea, Decapoda). *Biol. Mar. Medit.* 4: 640–642.

Deiana, A.M., Cau, A., Coluccia, E., Cannas, R., Milia, A., Salvadori, S., & Libertini, A. 1999. Genome size and AT-DNA content in thirteen species of Decapoda. In: Schram, F.R. & von Vaupel Klein, J.C. (eds.), *Crustaceans and the Biodiversity Crisis*: pp. 981–985. Leiden, The Netherlands: Brill Academic Publisher.

Diniz, F.M., Maclean, N., Paterson, I.G., & Bentzen, P. 2004. Polymorphic tetranucleotide microsatellite markers in the Caribbean spiny lobster, *Panulirus argus*. *Mol. Ecol. Notes* 4: 327–329.

Dobigny, G., Ducroz, J.F., Robinson, T.J., & Volobouev, V. 2004. Cytogenetics and cladistics. *Syst. Biol.* 53: 470–484.

Doussau De Bazignan, M. & Ozouf-Costaz, C. 1985. Une tecnique rapid d'analyse chromosomique appliquée a sept especies de poissons antarticques. *Cybium* 9: 57–74.

Fonseca, J.C., Marques, J.C., Paiva, A.A., Freitas, A.M., Madeira, V.M.C., & Jørgensen, S.E. 2000. Nuclear DNA in the determination of weighting factors to estimate energy from organisms biomass. *Ecol. Modell.* 126: 179–189.

Garcia-Rodriguez, F.J. & Perez-Enriquez, R. 2005. Genetic differentiation of the California spiny lobster *Panulirus interruptus* (Randall, 1840) along the west coast of the Baja California Peninsula, Mexico. *Mar. Biol.* 148: 621–629.

George, R.W. 2005. Evolution of spiny lobster life cycles. *N. Z. J. Mar. Freshw. Res.* 39: 503–514.

Gold, J.R., Ragland, C.J., Birkner, M.C., & Garrett, G.P. 1991. A simple procedure for long-term storage and preparation of fish cells for DNA content analysis using flow cytometry. *Prog. Fish-Cult.* 53: 108–110.

Gopurenko, D., Hughes, J.M., & Ma, J. 2002. Identification of polymorphic microsatellite loci in the mud crab *Scylla serrata* (Brachyura: Portunidae). *Mol. Ecol. Notes* 2: 481–483.

Gosselin, T., Sainte-Marie, B., & Bernatchez, L. 2005. Geographic variation of multiple paternity in the American lobster, *Homarus americanus*. *Mol. Ecol.* 14: 1517–1525.

Gregory, T.R. 2005a. Genome size evolution in animals. In: Gregory, T.R. (ed.), *The Evolution of the Genome*: pp. 3–87. London, U.K.: Elsevier Academic Press.

Gregory, T.R. 2005b. Animal Genome Size Database. http://www.genomesize.com (5 January 2006).

Hänfling, B. & Weetman, D. 2003. Characterization of microsatellite loci for the Chinese mitten crab, *Eriocheir sinensis*. *Mol. Ecol. Notes* 3: 15–17.

Hardie, D.C., Gregory, T.R., & Hebert, P.D.N. 2002. From pixels to picograms: A beginners' guide to genome quantification by Feulgen image analysis densitometry. *J. Histochem. Cytochem.* 50: 735–749.

Hayashi, K.I. & Fujiwara, Y. 1988. A new method for obtaining metaphase chromosomes from the regeneration blastema of *Penaeus (Marsupenaeus) japonicus*. *Nippon Suisan Gakk.* 54: 1563–1565.

Holthuis, L.B. 1991. *Marine Lobsters of the World. An Annotated and Illustrated Catalogue of the Species of Interest to Fisheries Known to Date*. FAO Species Catalogue No. 125, Vol. 13: pp. 1–292. Rome: Food and Agriculture Organization of the United Nations.

Holthuis, L.B. 2002. The Indo-Pacific scyllarine lobsters (Crustacea, Decapoda, Scyllaridae). *Zoosystema* 24: 488–529.

Hudson, A.P., Cuny, G., Cortadas, J., Haschemeyer, A.E., & Bernardi, G. 1980. An analysis of fish genomes by density gradient centrifugation. *Eur. J. Biochem.* 112: 203–210.

Itami, T., Aoki, J., Hayashi, K.I., Yu, Y., & Takayashi, Y. 1989. *In vitro* maintenance of cells of lymphoid organ in kuruma shrimp *Penaeus japonicus*. *Nippon Suisan Gakk.* 55: 2205.

Jensen, P.C. & Bentzen, P. 2004. Isolation and inheritance of microsatellite loci in the Dungeness crab (Brachyura: Cancridae: *Cancer magister*). *Genome* 47: 325–331.

Jones, M.W., O'Reilly, P.T., McPherson, A.A., McParland, T.L., Armstrong, D.E., Cox, A.J., Spence, K.R., Kenchington, E.L., Taggart, C.T., & Bentzen, P. 2003. Development, characterisation, inheritance, and cross-species utility of American lobster (*Homarus americanus*) microsatellite and mtDNA PCR-RFLP markers. *Genome* 46: 59–69.

Jørstad, K.E., Prodöhl, P.A., Agnalt, A.-L., Hughes, M., Apostolidis, A.P., Triantafyllidis, A., Farestveit, E., Kristiansen, T.S., Mercer, J., & Svåsand, T. 2004. Sub-arctic populations of European lobster, *Homarus gammarus*, in Northern Norway. *Environ. Biol. Fish.* 69: 223–231.

Lécher, P., Defaye, D., & Noel, P. 1995. Chromosome and nuclear DNA of Crustacea. *Invertebr. Reprod. Dev.* 27: 85–114.

Libertini, A. & Bertotto, D. 2004. Una microtecnica per allestire campioni da singole larve di pesci teleostei per la diagnosi della ploidia tramite citometria a flusso. *Lett. GIC* 13: 9–12.

Libertini, A., Panozzo, M., & Scovacricchi, T. 1990. Nuclear DNA content in *Penaeus kerathurus* (Forskål) and *P. japonicus* Bate, 1888. pp. 102. *Abstracts of the 25th E.M.B.S.,* Ferrara, Italy, 10–15 Sep. 1990.

Libertini, A., Colomba, M.S., & Vitturi, R. 2000. Cytogenetics of the amphipod *Jassa marmorata* Holmes, 1903 (Corophioidea: Ischyroceridae): Karyotype morphology, chromosome banding, fluorescent *in situ* hybridization and nuclear DNA content. *J. Crust. Biol.* 20: 350–356.

Maggioni, R. & Rogers, A.D. 2002. Microsatellite primers for three Western Atlantic *Farfantepenaeus* prawn species. *Mol. Ecol. Notes* 2: 51–53.

Maggioni, R., Rogers, A.D., & Maclean, N. 2003. Population structure of *Litopenaeus schmitti* (Decapoda: Penaeidae) from the Brazilian coast identified using six polymorphic microsatellite loci. *Mol. Ecol.* 12: 3213–3217.

Mandel, C.M. & Marmur, J. 1968. Use of ultraviolet absorbance temperature profile for determining the guanine plus cytosine content of DNA. In: Grossman, L. & Moldave, K. (eds.), *Methods in Enzymology*, Vol. SII: pp. 195–206. New York: Academic Press, Inc.

Meadows, J.R.S., Ward, R.D., Grewe, P.M., Dierens, L.M., & Lehnert, S.A. 2003. Characterization of 23 Tri- and tetranucleotide microsatellite loci in the brown tiger prawn, *Penaeus esculentus*. *Mol. Ecol. Notes* 3: 454–456.

Meehan, D., Xu, Z., Zuniga, G., & Alcivar-Warren, A. 2003. High frequency and large number of polymorphic microsatellites in cultured shrimp, *Penaeus (Litopenaeus) vannamei* [Crustacea: Decapoda]. *Mar. Biotechnol.* 5: 311–330.

Moritz, C., Dowling, T.E., & Brown, W.M. 1987. Evolution of animal mitochondrial DNA: Relevance for population biology and systematics. *Ann. Rev. Ecol. Syst.* 18: 269–292.

Muirhead, K.A., Horan, P.K., & Poste, G. 1985. Flow cytometry: Past and present. *Biotechnology* 3: 337–356.

Nakamura, H.K., Makii, A., Wada, K.T., Awaji, M., & Townsley, S.J. 1988. A check list of Decapod chromosomes (Crustacea). *Bull. Natl. Res. Inst. Aquacult.* 13: 1–9.

Niiyama, H. 1959. A comparative study of the chromosomes in decapods, isopods and amphipods, with some remarks on cytotaxonomy and sex-determination in the Crustacea. *Mem. Fac. Fish. Hokkaido Univ.* 7: 1–59.

Ovenden, J.R., Brasher, D.J., & White, R.W.G. 1992. Mitochondrial DNA analyses of the red rock lobster *Jasus edwardsii* supports an apparent absence of population subdivision throughout Australasia. *Mar. Biol.* 112: 319–326.

Ovenden, J.R., Booth, J.D., & Smolenski, A.J. 1997. Mitochondrial DNA phylogeny of red and green rock lobsters (genus *Jasus*). *Mar. Freshw. Res.* 48: 1131–1136.

Pan, Y.-W., Chou, H.-H., You, E.-M., & Yu, H.-T. 2004. Isolation and characterization of 23 polymorphic microsatellite markers for diversity and stock analysis in tiger shrimp (*Penaeus monodon*). *Mol. Ecol. Notes* 4: 345–347.

Patek, S.N. & Oakley, T.H. 2003. Comparative tests of evolutionary trade-offs in a palinurid lobster acoustic system. *Evolution* 57: 2082–2100.

Petrov, D.A. 2001. Evolution of genome size: New approaches to an old problem. *Trends Genet.* 17: 23–28.

Pisano, E. & Ozouf-Costaz, C. 2000. Chromosome change and the evolution in the antarctic fish suborder Notothenioidei. *Antarct. Sci.* 12: 334–342.

Ptacek, M.B., Sarver, S.K., Childress, M.J., & Herrnkind, W.F. 2001. Molecular phylogeny of the spiny lobster genus *Panulirus* (Decapoda: Palinuridae). *Mar. Freshw. Res.* 52: 1037–1047.

Ravago, R.G. & Juinio-Menez, M.A. 2002. Phylogenetic position of the striped-legged forms of *Panulirus longipes* (A. Milne-Edwards 1868) (Decapoda, Palinuridae) inferred from mitochondrial DNA sequences. *Crustaceana* 75: 1047–1059.

Robainas, A., Monnerot, M., Solignac, M., Dennebouy, N., Espinosa, G., & García-Machado, E. 2002. Microsatellite loci from the pink shrimp *Farfantenaeus notialis* (Crustacea, Decapoda). *Mol. Ecol. Notes* 2: 344–345.

Saiki, R.K., Gelfand, D.H., Stoffel, S., Scharf, S.J., Higuchi, R., Horn, G.T., Mullis, K.B., & Erlich, H.A. 1988. Primer-directed enzymatic amplification of DNA with a thermostable DNA polymerase. *Science* 239: 487–491.

Salvadori, S., Coluccia, E., Milia, A., Davini, M.A., & Deiana, A.M. 1995. The karyology of *Palinurus elephas*. *Biol. Mar. Medit.* 2: 581–583.

Sambrook, J. & Russell, D.W. 2001a. *In vitro* amplification of DNA by the polymerase chain reaction. In: *Molecular Cloning: A Laboratory Manual*, 3rd Edition, Vol. 2: pp. 8.1. Cold Spring Harbor, NY: Cold Spring Harbor Laboratory Press.

Sambrook, J. & Russell, D.W. 2001b. DNA sequencing. In: *Molecular Cloning: A Laboratory Manual*, 3rd Edition, Vol. 2: pp. 12.1. Cold Spring Harbor, NY: Cold Spring Harbor Laboratory Press.

Sarver, S.K., Silberman, J.D., & Walsh, P.J. 1998. Mitochondrial DNA sequence evidence supporting the recognition of two subspecies or species of the Florida spiny lobster *Panulirus argus*. *J. Crust. Biol.* 18: 177–186.

Scholtz, G. & Richter, S. 1995. Phylogenetic systematics of the reptantian Decapoda (Crustacea, Malacostraca). *Zool. J. Linn. Soc. Lond.* 113: 289–328.

Schweizer, D. 1976. Reverse fluorescent chromosome banding with chromomycin and DAPI. *Chromosoma* 58: 307–320.

Silberman, J.D. & Walsh, P.J. 1992. Species identification of spiny lobster phyllosome larvae via ribosomal DNA analysis. *Mol. Mar. Biol. Biotechnol.* 1: 195–205.

Silberman, J.D., Sarver, S.K., & Walsh, P.J. 1994. Mitochondrial DNA variation and population structure in the spiny lobster *Panulirus argus*. *Mar. Biol.* 120: 601–608.

Steven, C.R., Hill, J., Masters, B., & Place, A.R. 2005. Genetic markers in blue crabs (*Callinectes sapidus*): isolation and characterization of microsatellite markers. *J. Exp. Mar. Biol. Ecol.* 319: 3–14.

Streiff, R., Guillemaud, T., Alberto, F., Magalhães, J., Castro, M., & Cancela, M.L. 2001. Isolation and characterization of microsatellite loci in the Norway lobster (*Nephrops norvegicus*). *Mol. Ecol. Notes* 1: 71–72.

Tam, Y.K. & Kornfield, I. 1998. Phylogenetic relationships of clawed lobster genera (Decapoda, Nephropidae) based on mitochondrial 16S rRNA gene sequences. *J. Crust. Biol.* 18: 138–146.

Thompson, J.D., Gibson, T.J., Plewniak, F., Jeanmougin, F., & Higgins, D.G. 1997. The ClustalX windows interface: Flexible strategies for multiple sequence alignment aided by quality analysis tools. *Nucleic Acids Res.* 24: 4876–4882.

Tiersch, T.R., Chandler, R.W., Wachtel, S.S., & Elias, S. 1989. Reference standards for flow cytometry and application in comparative studies of nuclear DNA content. *Cytometry* 10: 706–710.

Trentini, M., Corni, M.G., & Froglia, C. 1989. The chromosomes of *Liocarcinus vernalis* (Risso 1816) and *Liocarcinus depurator* (L., 1758) (Decapoda, Brachyura, Portunidae). *Biol. Zbl.* 108: 163–166.

Tolley, K.A., Groeneveld, J.C., Gopal, K., & Matthee, C.A. 2005. Mitochondrial DNA panmixia in spiny lobster *Palinurus gilchristi* suggests a population expansion. *Mar. Ecol. Prog. Ser.* 297: 225–231.

Vindelov, L.L., Christensen, I.J., & Nissen, N.I. 1983. Standardization of high-resolution flow cytometric DNA analysis by the simultaneous use of chicken and trout red blood cells as internal reference standards. *Cytometry* 3: 328–331.

Vinogradov, A.E. 1998. Genome size and GC-percent in vertebrates as determined by flow cytometry: the triangular relationship. *Cytometry* 31: 100–109.

Vitturi, R., Libertini, A., Sineo, L., Sparacio, I., Lannino, A., & Colomba, M.S. 2005. Cytogenetics in the land snails *Cantareus aspersus* and *C. mazzullii* (Mollusca: Gastropoda: Pulmonata). *Micron* 36: 351–357.

Volobouev, V., Aniskin, V., Lecompte, E., & Ducroz, J.F. 2002. Patterns of karyotype evolution in complexes of sibling species within three genera of African murid rodents inferred from the comparison of cytogenetic and molecular data. *Cytogenet. Genome Res.* 96: 261–275.

Webber, W.R. & Booth, J.D. 2001. Larval stages, developmental ecology, and distribution of *Scyllarus* sp. Z (probably *Scyllarus aoteanus* Powell, 1949) (Decapoda: Scyllaridae). *N. Z. J. Mar. Freshw. Res.* 35: 1025–1056.

Yap, E.S, Sezmis, E., Chaplin, J.A., Potter, I.C., &. Spencer, P.B.S. 2002. Isolation and characterization of microsatellite loci in *Portunus pelagicus* (Crustacea: Portunidae). *Mol. Ecol. Notes* 2: 30–32.

4

Early Life Histories of Slipper Lobsters

Hideo Sekiguchi, John D. Booth, and W. Richard Webber

CONTENTS

Abstract

The small size of adult scyllarids means that few are commercially important, which, in turn, means that there has been relatively little work directed toward their early life histories. Some slipper lobsters hatch as a naupliosoma larva and all (together with the spiny and the coral lobsters) have a unique, long-lived, planktonic larva — the phyllosoma — that metamorphoses into a nektonic/benthic postlarva known in the Scyllaridae as a "nisto." Phyllosomas of genera with large eggs (e.g., *Ibacus* and *Thenus* species) have more developed stage I phyllosomas than those with small eggs (e.g., *Scyllarides* and *Scyllarus* species), but the evolutionary trade-off for producing larger eggs is lower fecundity. The mid- and late-stage phyllosomas of scyllarids with long larval durations (several months, e.g., *Arctides*, *Scyllarides*, and *Parribacus* species) are rarely found close to the coast. However, short larval durations have evolved among some scyllarids, such that with the development of particular behavioral patterns, larvae are not transported far from the coast. Between these two extremes there appear to be a number of intermediate larval recruitment strategies. Scyllarid phyllosomas appear to feed mainly on soft, fleshy foods; in turn, they are eaten by a diverse range of mainly pelagic fishes. Swimming abilities appear to vary widely among genera, with *Thenus* species being particularly weak swimmers. Many genera in their late stages of development can be found associated with gelatinous zooplankton, particularly medusae.

The nisto — which is probably nonfeeding — in at least some species, appears to be more similar to its first juvenile instar than is the puerulus of palinurids to their first juvenile instar. But the nistos of different species show considerable variability in swimming, burrowing, and other behaviors, suggesting that the precise role of the nisto may vary among genera.

4.1 Introduction

Slipper lobsters (family Scyllaridae) are very closely related to the spiny or rock lobsters (Palinuridae) and the coral or furry lobsters (Synaxidae). All three families have a unique, long-lived, planktonic larva — the phyllosoma — that metamorphoses into a nektonic/benthic postlarva known as the "nisto" in scyllarids and the "puerulus" in palinurids and synaxids. Due to the small size of the adults of most scyllarid species, few are important commercially compared with many palinurids, such as *Jasus* and *Panulirus* species, that form important, fully exploited fisheries. Therefore, there have been fewer detailed studies of the biology and ecology of scyllarids — particularly of their early life history — compared with the palinurids (for palinurids, see the reviews of Phillips & Sastry 1980 and Booth & Phillips 1994). Little is known about the synaxids, which are of little to no fishery importance (see Holthuis 1991).

Scyllarids are widespread in temperate and tropical seas, from tropical lagoons to continental shelves and slopes, and on deep-sea ridges and seamounts. They are a very diverse group in terms of size and habitat. They range in total length from less than 50 mm to half a meter, and live from shallow coastal depths down to habitats beyond natural light. The highly flattened form of many scyllarids is more suited to homogeneous, often soft substrates than the more vaulted shape of the palinurids, but some scyllarids are also vaulted and live in similar habitats to the palinurids (Holthuis 1985, 1991, 2002). This diversity among the scyllarids suggests that there could be a variety of larval recruitment strategies. Our knowledge of the biology and behavior of the early life-history stages in the Scyllaridae, although generally poor, largely descriptive, and often speculative (e.g., see Sekiguchi & Inoue 2002; Booth et al. 2005), does indeed suggest that there are different larval recruitment strategies. At the same time, however, all scyllarid species known so far share many early life-history features, and indeed there are many similarities in larval and postlarval development and behavior between the scyllarids and the palinurids.

This chapter reviews the early life histories of scyllarid lobsters, focusing on morphological features of the phyllosomas and nistos, their spatial distributions, how these distributions may relate to larval-recruitment strategies, and their biology and ecology. Previous useful reviews include: (1) Phillips & Sastry (1980) who dealt with larval and postlarval development and behavior, as well as factors affecting growth and survival of larvae; (2) Baisre (1994) who reviewed larval morphology to derive phylogenies; (3) Holthuis (1985, 2002) who reviewed the taxonomic status of the subfamilies Ibacinae and Scyllarinae, at the same time referencing data on larvae and postlarvae; and (4) Booth et al. (2005) who focused on the diverse larval recruitment strategies among the scyllarids.

4.2 Egg Size and Fecundity

In general, species of larger scyllarids have greater egg numbers per brood or clutch (i.e., have higher fecundity) than smaller scyllarids, but this relationship is modified by variation in egg size among species, which ranges from about 0.4 to 1.2 mm in diameter (Table 4.1). Adult females of the same size have larger broods when the eggs are smaller and vice versa; the evolutionary trade-off for producing larger eggs with more developed, possibly more mobile offspring at hatching and a shorter time spent in the plankton is lower fecundity. As examples of species with small eggs, the brood size of *Scyllarides squammosus* (H. Milne Edwards 1837) (females up to about 130 mm carapace length [CL]; egg diameter about 0.7 mm) reaches around 227,000 eggs (DeMartini & Williams 2001), whereas in the much smaller *Scyllarus americanus* (S.I. Smith 1869) (females <30 mm CL; egg diameter about 0.6 mm), the brood

TABLE 4.1

Adult Size, Fecundity, and Egg Size of Scyllarid Lobsters

Species	Female Size (mm CL)	Fecundity	Mean Egg Diameter (mm)	Reference
Scyllarides latus	120	100,000–356,000	0.7	Martins (1985); Almog-Shtayer (1988); Spanier & Lavalli (1998); Bianchini & Ragonese (2003)
Scyllarides squammosus	130	54,000–227,000	0.7	DeMartini & Williams (2001)
Scyllarides aequinoctialis	110		0.6–0.7	Robertson (1969a)
Parribacus caledonicus	70	156,000–211,000		Coutures (2003)
Ibacus novemdentatus (as *I. ciliatus*)	80		1.0	Harada (1958)
Ibacus peronii	70	5500–37,000	1.2	Stewart & Kennelly (1997)
Ibacus alticrenatus	50	2000–15,000	0.9–1.3	Haddy et al. (2005)
Ibacus brucei	65	2000–61,000	0.7–1.0	Haddy et al. (2005)
Ibacus chacei	70	2000–29,000	1.0–1.4	Haddy et al. (2005)
Petrarctus demani	28	12,000–17,000	0.5	Ritz (1977); Ito & Lucas (1990)
Biarctus sordidus	24		0.5	Ritz (1977)
Scyllarus depressus	26		0.4	Robertson (1971)
Scyllarus americanus	25	8000–9000	0.6	Robertson (1968)
Thenus orientalis	80	5000–53,000	1.1	Jones (1988) in Stewart & Kennelly (1997), Kagwade & Kabli (1996)

CL: carapace length. Measurements given under female size are toward the maximum size of mature females, based on Lyons (1970), Holthuis (1991, 2002), and Haddy et al. (2005). Fecundity is brood/clutch size, irrespective of the number of broods per year; for most species there is a range of individual fecundities because there was more than one female and more than one study involved.

is up to about 9000 eggs (Robertson 1968). Species with large eggs that hatch at an advanced point in their development include *Ibacus peronii* Leach 1815 and *Thenus orientalis* (Lund 1793) (females 70 and 80 mm CL and egg diameter 1.2 and 1.1 mm, respectively), with broods of up to about 37,000 and 53,000, respectively (Jones 1988; Stewart & Kennelly 1997) (Table 4.1). In species with small eggs, the stage I phyllosomas typically have fully developed first and second pereiopods and a partially developed third pereiopod, with any other pereiopods being bud-like or absent (e.g., *Petrarctus demani* (Holthuis 1946) — Ito & Lucas 1990; *Scyllarus depressus* (S.I. Smith 1881) — Robertson 1971). In contrast, in species such as *I. peronii* and *T. orientalis* with larger eggs, the first three pereiopods are complete, the fourth pereiopod lacks only a fully developed exopod, and the fifth pereiopod is often partially developed (Marinovic et al. 1994; Mikami & Greenwood 1997a). The pereiopods are primarily swimming appendages (Williamson 1969), so it seems likely that the phyllosomas with additional pereiopods, hatching from larger eggs, are stronger swimmers.

4.3 Postembryonic Development

4.3.1 Developmental Characteristics

Postembryonic development in the Scyllaridae consists of a series of free-living instars separated by molts. Some of these molts involve a major change in body form, and the series of instars between these major molts are termed "phases" in development (Williamson 1969): naupliosoma (in some species), phyllosoma, nisto, juvenile lobster, and adult lobster phases. The first three phases are broadly equivalent to the nauplius, zoea, and megalopa defined by Williamson (1969) for decapods as a whole. The change from the phyllosoma to the nisto is a "metamorphosis" (Williamson 1969). Whereas there is more than

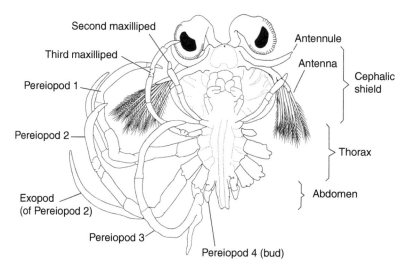

FIGURE 4.1 Naupliosoma of *Ibacus alticrenatus* Bate, 1888 labeled to indicate main external morphological features. (Based on Lesser J.H.R. 1974. *Crustaceana* 27: 259–277; used with permission.)

one instar in the phyllosoma phase, only one is known in the naupliosoma (e.g., Lesser 1974) and the nisto (e.g., Barnett et al. 1986).

There may be significant morphological change among phyllosomal instars (e.g., development of appendix internae on the pleopods), sometimes coinciding with a marked change in behavior (e.g., stronger swimming). Such morphological changes have led to the designation of a new "stage" in development, clearly recognizable by the level of segmentation and development of appendages. In species with few instars (e.g., *Thenus orientalis* — Mikami & Greenwood 1997a), each stage is usually associated with a single instar (i.e., each molt is accompanied by morphological changes interpreted as a change in stage). On the other hand, in those species with long larval lives such as *Scyllarides* species, there will probably be more than one instar (i.e., more than one molt) in the later stages, as is seen in the palinurids (e.g., Kittaka 2000). Therefore, whereas the stage and instar number are obvious for larvae reared in the laboratory, only the stages of those species with long larval lives can be defined with certainty for field-caught larvae.

Some scyllarids hatch as a naupliosoma, but all develop through a series of planktonic phyllosomal instars. Species sometimes (but not necessarily always) reported to have hatched as a naupliosoma — and apparently known only from tank and culture studies — include *Ibacus alticrenatus* Bate, 1888 (Lesser 1974), *I. ciliatus* (Von Siebold 1824) (Harada (1958), but possibly *I. novemdentatus* Gibbes, 1850 — see Holthuis (1985)), and possibly *I. peronii* (Stewart et al. 1997); and *Scyllarides latus* (Latreille 1802) (Martins 1985) and its congeners *S. herklotsii* (Herklots 1851) (Crosnier 1972) and *S. aequinoctialis* (Lund, 1793) (Robertson 1969a). The naupliosoma (Figure 4.1) is short-lived (minutes to a few hours) and bears large stalked eyes, uniramous antennules, biramous antennae (with natatory sensae on both rami), well-developed second and third maxillipeds, and quite well-developed first through third pairs of thoracic appendages. For other scyllarids, either there is no naupliosoma or apparently one has not been observed.

Phyllosomas (Figure 4.2) are adapted to planktonic life by being dorsoventrally flattened and transparent. They bear swimming exopods on some or all thoracic appendages, depending on the species and stage of development, with pleopods being absent or rudimentary until the later stages. The final-stage phyllosoma metamorphoses into the nisto. The nisto links the planktonic larval and the benthic juvenile phases of the life cycle, presumably by swimming to the settlement areas as does the puerulus stage of palinurids (Booth & Phillips 1994).

Morphologically, the nisto (Figure 4.3) is similar to the juvenile, with the particularly distinctive second antennae being formed into flat plates. However, nistos are easily distinguished from juveniles by being virtually transparent during the earlier part of their lives and in having setose swimming pleopods usually (e.g., *Petrarctus demani* — Ito & Lucas 1990), but not always (e.g., *Thenus orientalis* — Barnett et al.

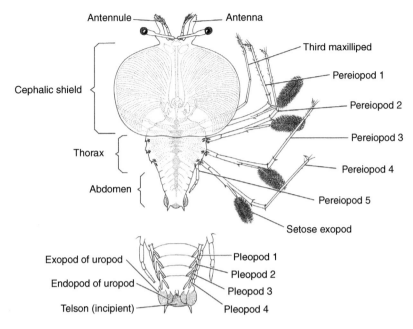

FIGURE 4.2 Final-stage phyllosoma of *Antipodarctus aoteanus* (Powell, 1949) labeled to indicate main external morphological features. (Based on Webber, W.R. & Booth, J.D. 2001. *N. Z. J. Mar. Freshw. Res.* 35: 1025–1056; used with permission.)

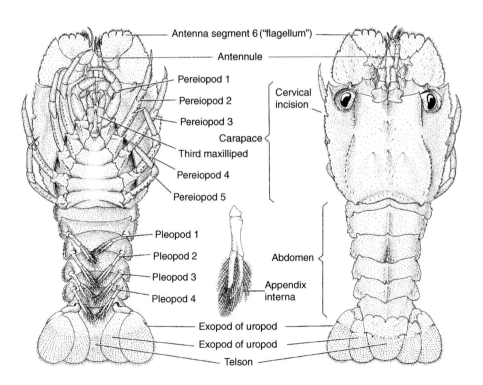

FIGURE 4.3 Nisto of Scyllarinae sp. labeled to indicate main external morphological features. (Reproduced from Coutures, E. & Webber, R. 2005. *Zootaxa* 873: 1–20; used with permission.)

1984), with well-developed appendix internae on abdominal somites two through five; the appendix internae couple the pleopod pairs to effect stronger swimming.

4.3.2   Identification of Phyllosomas and Nistos

Scyllarid phyllosomas (Figure 4.2) usually are readily distinguishable from those of palinurids. Their third maxillipeds do not have a setose exopod, a characteristic that can be used for identification at all stages of development. This exopod is present in all palinurid phyllosomas known except: (1) *Jasus* species (Baisre 1994), and (2) *Sagmariasus verreauxi* (H. Milne Edwards, 1851) in which it does not appear until the late stages, and then only intermittently (Kittaka et al. 1997). (It is probably also absent in the phyllosomas of the very closely related *Projasus*.) In addition, mid- and late-stage scyllarid phyllosomas have second antennae that are clearly dorsoventrally compressed, with a lateral process, whereas in palinurids, the second antennae are cylindrical and unbranched.

Numerous descriptions of scyllarid larval morphology and development exist, but few deal with all stages and not many give the numbers of instars or stages (Table 4.2). Not only are most identities not formally confirmed, but many scyllarid larvae collected in the wild, even those described recently (e.g., McWilliam et al. 1995; Coutures & Webber 2005; Inoue & Sekiguchi 2005), remain unidentified beyond the level of genus. (See Pessani & Mura, Chapter 13 for an example of problems in identifying scyllarid larvae.)

Phyllosoma larvae of at least 57 of the 85 species of scyllarids known from the adult phase have been described, although only about 30 have been reliably identified with adults of scyllarid species (Table 4.2). Scyllarid genera for which some or all phyllosoma stages have been described include *Arctides* (e.g., *A. guineensis* (Spengler 1799) — Robertson 1969b), *Scyllarides* (e.g., *S. herklotsii* — Crosnier 1972), *Evibacus* (*E. princeps* S.I. Smith 1869 — Johnson 1968, 1971a), *Ibacus* (e.g., *I. peronii* — Marinovic et al. 1994), *Parribacus* (e.g., *P. antarcticus* (Lund 1793) — Johnson 1971b), *Antipodarctus* (*A. aoteanus* (Powell 1949) — Webber & Booth 2001), *Biarctus* (e.g., *B. sordidus* (Stimpson 1860) — Ritz 1977), *Chelarctus* (e.g., *C. cultrifer* (Ortmann 1897) — Higa & Shokita 2004; Inoue & Sekiguchi 2006), *Crenarctus* (e.g., *C. bicuspidatus* (De Man 1905) — Phillips et al. 1981, but note that Higa & Shokita 2004 and Inoue & Sekiguchi 2006 suggest that the identification may not be correct), *Eduarctus* (e.g., *E. martensii* (Pfeffer 1881) — Phillips & McWilliam 1986a), *Galearctus* (e.g., *G. kitanoviriosus* (Harada 1962) — Higa & Saisho 1983), *Petrarctus* (e.g., *P. demani* — Ito & Lucas 1990), *Scammarctus* (*S. batei* (Holthuis 1946) — Prasad et al. 1975), *Scyllarus* (e.g., *S. americanus* — Robertson 1968), and *Thenus* (e.g., *T. orientalis* — Mikami & Greenwood 1997a).

These phyllosoma identifications have been made using both direct and indirect methods. In the direct approach, eggs hatched in captivity, or larvae caught at sea are cultured, sometimes to the nisto phase. Species cultured from egg to nisto include *Ibacus ciliatus* (Takahashi & Saisho 1978), *I. novemdentatus* (Takahashi & Saisho 1978), *I. peronii* (Marinovic et al. 1994), *P. demani* (Ito & Lucas 1990), *S. americanus* (Robertson 1968), *T. orientalis* (Mikami & Greenwood 1997a), and *Thenus* sp. (Mikami & Greenwood 1997a). The indirect method — which is much less reliable (e.g., see Higa & Shokita 2004) — involves knowing the identity of scyllarid species present as adults in the region where the larvae have been caught and deducing the probable phyllosoma identity based on adult distribution and abundance, while at the same time comparing the morphology of the larvae with published accounts (e.g., Michel 1968; Prasad et al. 1975; Webber & Booth 2001). These morphological features include whether or not the cephalic shield is bilobed, whether or not the fifth pereiopod bears a setose exopod, whether or not the abdomen is a tapered continuation of the thorax, and whether the dorsal surface of the abdomen is with or without spines (Key; Figure 4.2). In mid- and late-stage phyllosomas, the form of the cephalic shield and abdomen can be a good guide to genus; Figure 4.4 schematically illustrates these for the final-stage phyllosomas of the nonscyllarinid genera and the Scyllarinae.

Nistos of at least 25 species have been described and identified with an adult scyllarid (Table 4.2); others have been described, but identified only to subfamily (e.g., Coutures & Webber 2005). Some nistos have been identified to species directly, the nisto having been cultured to an identifiable juvenile (e.g., *I. alticrenatus* — Atkinson & Boustead 1982); others have been indirectly identified by being associated with mid- to late-stage phyllosomas whose specific identity is known (e.g., *T. orientalis* — Barnett et al.

TABLE 4.2

Scyllarid Reproductive and Larval Parameters

Subfamily/Species	Female Size (mm CL)	No. Instars	No. Stages	Estimated Length of Larval Life	Size of Final Phyllosoma (mm TL)	Extent of Offshore Distribution of Late Stages	Nisto Duration	Nisto Size (mm CL)
Arctidinae								
Arctides guineensis	50[2]		13[a,34]	~8–9mo[a,34]	59[a,34]	Oceanic[34]		13[a,22]
Scyllarides aequinoctialis	110[2,22]		11[a,b,13]	~8–9mo[a,b,13]	36–48[a,13]	Oceanic[13,37]		9–11[a,22]
Scyllarides nodifer	100[2,22]		12[13]	9 mo[13]	37[a,13]	Oceanic[13]		13–15[a,23,44]
Scyllarides herklotsii	130[2]		11[a,23]		~25[a,23]	Oceanic[23]		
Scyllarides astori	100[2]				44[a,8]	Oceanic[8]		15[a,10]
Scyllarides squammosus	130[2]				48[a,10]	Oceanic[29,30,39,43]		13[a,45]
Scyllarides elisabethae	90[2]							
Ibacinae								
Parribacus antarcticus	80[2]		>11[a,30]	~9mo[a,30]	75–83[a,14,30]	Oceanic[29,30]		20[a,17,30]
Parribacus scarlatinus	70[2]							21[a,9]
Evibacus princeps	130[2]		11[a,24]		32[a,8,24]	Coastal & intermediate[8,24]		11–12[a,17]
Ibacus alticrenatus	43[17]		7[a,38]	4–6 mo[a,38]	36–44[a,38]	Intermediate[38]		12–15[a,38]
Ibacus ciliatus	80[2]	7–8[b,15,19]		54–76 d[b,15,19]	40–46[a,b,15,19,26]	Coastal & intermediate[4]		16–20[a,b,15,19,26]
Ibacus novemdentatus	70[2]	7[b,19]		65 d[b,19]	21–33[a,b,4,19,26]	Coastal & intermediate[4]		13–15[a,b,4,17,19,26]
Ibacus peronii	70[2]	6[b,36]	7[a,25]	71–97 d[b,36]	39[a,25] 25[b,36]	Intermediate & oceanic[25,35,39]	17–24 d[a,b,25,36]	11–12[a,b,25,36]
Scyllarinae								
Antipodarctus aoteanus	30[1]		10[a,3]	Several mo[a,3]	19–31[a,3]	Intermediate & oceanic[3]		6.5[a,3]
Biarctus sordidus	24[16]		8[a,41]		14[a,41]	Coastal & intermediate[41]		3.5[a,16]
Chelarctus cultrifer	31[16]				20[b,29]	Coastal & intermediate[29,40,42]	10–11 d[b,29]	5[b,29]
Crenarctus bicuspidatus	25[16]			4 mo[a,35]	21[a,35]	Coastal, intermediate & oceanic[35,39,40,42]		

(Continued)

TABLE 4.2

(Continued)

Subfamily/Species	Female Size (mm CL)	No. Instars	No. Stages	Estimated Length of Larval Life	Size of Final Phyllosoma (mm TL)	Extent of Offshore Distribution of Late Stages	Nisto Duration	Nisto Size (mm CL)
Eduarctus martensii	14[16]		8–10[a,6,18,28]	2–3 mo[a,18]	11–15[a,6,28,29]	Coastal & intermediate[18,28,40,43]	7–9 d[a,32]	3[a,6]
Eduarctus modestus	14[16]		8[a,30]		13[a,30]	? Intermediate & oceanic[30]		
Galearctus kitanoviriosus	36[16]				17–21[a,27]	Coastal & intermediate[40,42]	2 wk[a,27]	4–5[a,27]
Galearctus timidus	27[16]		9[a,30]		23[a,30]	? Intermediate & oceanic[30]		8[a,16]
Petrarctus demani	28[16]		6[a,b,41]	42–53 d[b,20]	10–11[a,b,20,41]	Coastal & intermediate[41]	5–6 d[b,20]	4–6[a,b,20,32]
Petrarctus rugosus	22[16]	8[b,20]	12[a,28,29]		15–19[a,28,29]	Coastal & intermediate[28]		
Scammarctus batei	33[16]		10[a,28]		26–29[a,28]	Coastal & intermediate[28]		
Scyllarus americanus	25[22]	6–7[a,b,5]	6–7[a,b,5,33]	32–40 d[b,5]	8–13[a,b,5]	Coastal & intermediate[5,33,37]		3[a,b,5,22]
Scyllarus chacei	25[22]					Coastal & intermediate[37]		3–4[a,22]
Scyllarus depressus	26[22]		9–10[a,b,7]	2.5 mo[a,b,7]	24–27[a,7]	Coastal & intermediate[37]		6–7[a,7,22]
Scyllarus planorbis		8[b,12]	8[b,12]	54 d[b,12]	8[b,12]	Coastal, intermediate & oceanic[7]		
Theninae								
Thenus orientalis	80[2]	4[b,11,21]	4[a,b,11,21,31]	27–45 d[b,11,21]	20[a,31] 16–20[b,11,21]	Coastal[31]	7 d[b,11]	8[a,31] 7–8[b,11,21]
Thenus sp.	80[2]	4[b,21]	4[b,21]	~27–45 d[b,21]	13–19[b,21]			7–9[b,21]

[a]denotes data based on wild lobsters; [b]denotes data based on cultured lobsters; ? = the literature is unclear on this point. Number of instars given only if an estimate is available (otherwise it is at least the number of stages). Female sizes are toward the maximum size of mature females, obtained from the literature. For the extent of offshore distribution: coastal, mainly over continental shelf; intermediate, within but also beyond shelf break (the shelf break being at about the 1000 m depth contour); oceanic, mainly well beyond shelf break, including within ocean basins. Note that there has been confusion between the identities of *Ibacus ciliatus* and *I. novemdentatus* (see Holthuis 1985).

[1]Powell (1949), [2]Holthuis (1991), [3]Webber & Booth (2001), [4]Shojima (1973), [5]Robertson (1968), [6]Phillips & McWilliam (1986a), [7]Robertson (1971), [8]Johnson (1971a), [9]Coutures et al. 2002, [10]Michel (1968), [11]Mikami & Greenwood (1997b), [12]Robertson (1979), [13]Robertson (1969a), [14]Baisre (1994), [15]Mikami & Takashima (1993), [16]Holthuis (2002), [17]Holthuis (1985), [18]Rothlisberg et al. (1994), [19]Takahashi & Saisho (1978), [20]Ito & Lucas (1990), [21]Mikami & Greenwood (1997a), [22]Lyons (1970), [23]Crosnier (1972), [24]Johnson (1968), [25]Ritz & Thomas (1973) (but Marinovic et al. (1994) questioned the number of stages), [26]Dotsu et al. (1966), [27]Higa & Saisho (1983), [28]Prasad et al. (1975), [29]Higa & Shokita (2004), [30]Johnson (1971b), [31]Barnett et al. (1984), [32]Barnett et al. (1986), [33]Olvera Limas & Ordonez Alcala (1988), [34]Robertson (1969b), [35]Phillips et al. (1981), [36]Marinovic et al. (1994), [37]Yeung & McGowan (1991), [38]Atkinson & Boustead (1982), [39]McWilliam & Phillips (1983), [40]Inoue et al. (2001), [41]Ritz (1977), [42]Sekiguchi & Inoue (2002), [43]Coutures (2000), [44]Chace (1966), [45]Barnard (1950).

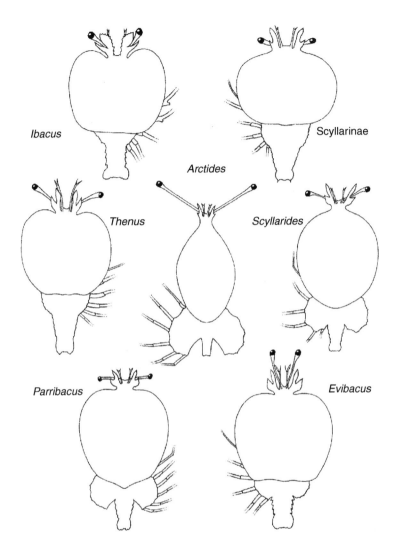

FIGURE 4.4 Outline illustrations of the form of the cephalic shield and abdomen of final-stage scyllarid phyllosomas of the nonscyllarinid genera and the Scyllarinae. The phyllosomas are not drawn to scale: *Ibacus* typically reach about 40 mm total length (TL), the Scyllarinae about 25 mm TL, *Thenus* about 20 mm TL, *Arctides* about 55 mm TL, *Scyllarides* about 40 mm TL, *Parribacus* about 80 mm TL, and *Evibacus* about 30 mm TL. Illustration by Richard Webber, based on phyllosomas referred to in Table 4.2.

1986), or with adults in their area (e.g., *Antipodarctus aoteanus* — Webber & Booth 2001). Distinguishing field-caught nistos within genera is usually very difficult because of similarities in shape. Because the nistos of so few genera have been described (Table 4.2), it is not possible to generalize about the form of the nisto by genus at this time. But there are some useful distinguishing characters. For example, (1) as in the adult, the eyes of the nisto of *Thenus* species are near the lateral margin of the carapace (e.g., Mikami & Greenwood 1997a), and (2) there are strong cervical incisions (lateral notches) on the carapace of the nisto of *Ibacus* species (e.g., Atkinson & Boustead 1982) not so far seen in the nistos of *Thenus*, *Scyllarides*, or the Scyllarinae. (See Pessani & Mura, Chapter 13, for further information about confusion in identifying the nisto stages in the Mediterranean.)

Molecular methods, which have been used to unambiguously differentiate larvae of some palinurid species (e.g., Silberman & Walsh 1992), suggest a way to distinguish scyllarid phyllosomas and nistos of different species that are similar in appearance.

4.4   Spatial Distributions of Phyllosomas and Nistos

4.4.1   Horizontal Distributions

The distribution and ecology of scyllarid phyllosomas have been less well studied than those of palinurids but, nevertheless, scyllarid phyllosomas are represented in widely distributed plankton samples. Early reports on the distribution of scyllarid phyllosomas (e.g., Gurney 1936; Prasad & Tampi 1960; Saisho 1966; Johnson 1968, 1971a, 1971b; Prasad et al. 1975) came mainly from wide-ranging plankton sampling in which catches were often small (<10 individuals of each species). Incorrect or unresolved identifications are a common problem with field-caught larvae of marine species and are particularly evident in this early scyllarid literature (see Holthuis 1985). Gurney (1936), for example, reported the distributions of a number of *Scyllarus* species extending into ocean basins, but in many instances, the species he described remain unidentified or unconfirmed. Most of the detailed distributional data for scyllarid larvae have come from more recent collections — and mainly as by-catch where other crustaceans (particularly palinurids) or fishes were the primary focus. Examples of this are Phillips et al. (1981), in which 9721 larvae were identified as *Crenarctus bicuspidatus* (De Man 1905); Rothlisberg et al. (1994), which included 751 *Eduarctus martensii* (Pfeffer 1881) phyllosomas; Inoue et al. (2000) with 166 scyllarid phyllosomas belonging to four species; Webber & Booth (2001) with 878 *Antipodarctus aoteanus* phyllosomas; Inoue et al. (2001) with 198 scyllarid phyllosomas belonging to 10 species; Minami et al. (2001) with 62 scyllarid phyllosomas belonging to four species; Inoue et al. (2004) with 362 scyllarid phyllosomas belonging to seven species; and Inoue & Sekiguchi (2005) with 122 scyllarid phyllosomas belonging to nine species. However, scyllarids have been a primary focus of some of the recent studies, including those of Barnett et al. (1984, 1986) with 815 and 51 mainly *Thenus* phyllosomas, respectively; Yeung & McGowan (1991) with 243 scyllarid phyllosomas from two genera; and Coutures (2001) with 433 scyllarid phyllosomas from three genera. Detailed geographic distributional data have come mainly from Western Australian waters (Phillips et al. 1981), the Gulf of Carpentaria in northern Australia (Rothlisberg et al. 1994), around New Zealand (Webber & Booth 2001), in the western North Pacific (Inoue et al. 2000, 2001, 2004; Minami et al. 2001; Inoue & Sekiguchi 2005), and off Florida (Yeung & McGowan 1991).

The estimated larval duration among scyllarids varies from several weeks to many months (Table 4.2), giving phyllosomas varying potential among species for long-distance dispersal to oceanic regions. (The term "coastal" refers to waters over the continental shelf, the "shelf break" is near the 1000 m depth contour, and "oceanic" refers to waters well beyond the shelf break, including within ocean basins.) Knowing the spatial distribution and abundance of scyllarid larvae in the field allows hypotheses to be developed concerning larval dispersal and retention that take into account both larval behavior and the oceanographic conditions. With this approach, an understanding can grow concerning the mechanisms by which variations in recruitment are driven, and the extent to which particular benthic populations are maintained by their own larval production vs. input from larval production elsewhere. Although scyllarids are generally highly fecund (many thousands of eggs in each brood — Table 4.1), a long larval life means that only a small percentage of hatched larvae can be expected to survive and contribute to the adult phase.

The extent of dispersal varies among species (Table 4.2). For example, the larvae of *Parribacus*, *Scyllarides*, and *Arctides* species are distributed much further from shore (Johnson 1968, 1971a, 1971b; Robertson 1969a; Crosnier 1972; Berry 1974; McWilliam & Phillips 1983; Yeung & McGowan 1991; Baisre 1994; Coutures 2000) than those of *Evibacus* and such scyllarinid species as *Petrarctus demani*, *Scammarctus batei*, and *Scyllarus americanus* (Johnson 1968, 1971a; Robertson 1968; Prasad et al. 1975; Ritz 1977; Yeung & McGowan 1991). Indeed, it has recently been shown that some scyllarinids (referred to as unidentified *Scyllarus* species) complete their entire larval development within the lagoons formed by coral island barrier reefs (Coutures 2000). Even more spatially restricted recruitment mechanisms may exist: Coutures (University of New Caledonia, personal communication) has observed that some unidentified New Caledonian scyllarinid phyllosomas (regardless of their stage of development) and nistos are commonly caught in barrier-reef crest nets, but are seldom taken within or beyond adjacent lagoon waters. However, even among the scyllarinids, the phyllosomas of some species can be found hundreds to thousands of kilometers from shore. Examples of this include *Antipodarctus aoteanus* (Webber & Booth 2001),

Crenarctus bicuspidatus (Phillips et al. 1981), *Galearctus timidus* (Holthuis, 1960) (Johnson 1971b), and *Scyllarus depressus* (Robertson 1971), all taken well beyond the known depth distribution of the adults.

In those species where phyllosomas appear not to become widely dispersed (such as *Evibacus princeps*, *Ibacus ciliatus* and *I. novemdentatus*, *Biarctus sordidus* (Stimpson, 1860), *Chelarctus cultrifer* (Ortmann 1897), *Eduarctus martensii*, *Petrarctus demani* and *P. rugosus* (H. Milne Edwards 1837), *Scammarctus batei*, *Scyllarus americanus* and *S. chacei* Holthuis 1960, and particularly *Thenus* species; Table 4.2), mechanisms must exist to enable larvae to remain close to parent grounds because even those species with the shortest larval development period (about four weeks) have the potential to be transported large distances from shore. Recent work has helped clarify the role of ocean processes in limiting the dispersal of palinurids, and indicates that local eddies, gyres, counter currents, and wind-induced currents are particularly important (reviewed, for example, by Phillips & McWilliam (1986b), Booth & Phillips (1994), and Sekiguchi & Inoue (2002)). Such mechanisms have also been widely invoked as being important in scyllarid larval dispersal and retention, but, reflecting the sampling difficulties, seldom with any strong supporting evidence. As examples, Johnson (1971b) suggested entrapment of scyllarid larvae in eddies and counterflows that form in the wake of the Hawaiian Islands, while Berry (1974) suggested that short-term reversals of waters inshore of the Agulhas Current prevented a significant proportion of the scyllarid larvae from being carried away by the Agulhas Current. A similar explanation probably applies to *Crenarctus bicuspidatus* and other scyllarid larvae being retained close to shore off Japan (Sekiguchi 1986b; Sekiguchi & Inoue 2002). Lee & Williams (1999) concluded that the combined influences of the downstream flow of the Florida Current, onshore Ekman transports in the upper layer, coastal countercurrents, and cyclonic circulation in the Tortugas Gyre aid retention of scyllarid larvae off Florida, and mesoscale eddies were thought important by Fiedler & Spanier (1999) in the retention of *Scyllarus arctus* (Linnaeus, 1758) larvae in the Mediterranean Sea.

A variety of scenarios, therefore, may explain how recruitment to existing adult stocks eventually comes about, including: (1) a proportion of hatched larvae are retained near adult populations, and later recruit to them, with those larvae dispersing farther afield being destined to perish because they end up in waters too deep or otherwise unsuitable for settlement and survival; (2) larvae disperse widely offshore into oceanic waters, some returning shoreward by means of water circulation systems coupled with larval behavior that makes use of such systems; or (3) larvae drift large distances before settling as nistos and return to the adult grounds as migrating juveniles.

An example of the first scenario is illustrated in Figure 4.5: Sekiguchi (1986b) and Inoue & Sekiguchi (2005) found that *Chelarctus cultrifer* larvae that hatched along the coast of Japan and remained in coastal waters north of the Kuroshio Current contributed to recruitment to local benthic populations. In contrast, *C. cultrifer* larvae that dispersed far offshore (south of the Kuroshio Current) — a long-distance transport within the Kuroshio subgyre circulation that also carries *Panulirus japonicus* (Von Siebold 1824) larvae (Inoue & Sekiguchi 2001) — may be destined to perish because their larval duration is too short to allow them to return to coastal waters and there are currently no known populations of juvenile or adult *C. cultrifer* in the deep, offshore waters. However, the longer larval life span of *P. japonicus* allows such wide dispersal and subsequent return to near the shelf break to metamorphose and recruit.

The second scenario appears to have examples in the long-lived larvae of *Scyllarides* and *Arctides* (Table 4.2), which in late stages are found exclusively in oceanic waters. A species that displays the third scenario is *Ibacus chacei* Brown & Holthuis 1998 off New South Wales and south Queensland, Australia, where significant proportions of lobsters >30 mm CL migrate northward along the coast, moving from settlement areas to breeding areas (Stewart & Kennelly 1998; Haddy et al. 2005). It is thought that this is a contranatant migration to counter larval drift: unless some animals migrate upstream against the current, the center of distribution of the species may be continually displaced in the direction of the currents which transport the larvae.

Two further general observations emerge from examination of the literature on scyllarid phyllosoma distributions. First, although both shallow water scyllarinid and palinurid species have long-lived phyllosomas that begin this phase of development near the shore, most scyllarinid phyllosomas do not seem to disperse as far offshore as palinurid phyllosomas in the same area. An example of this is the distribution further offshore of *Panulirus cygnus* George 1962 phyllosomas compared with those of *Crenarctus bicuspidatus* (Phillips et al. 1979, 1981). The main reasons for scyllarid larvae remaining closer to shore

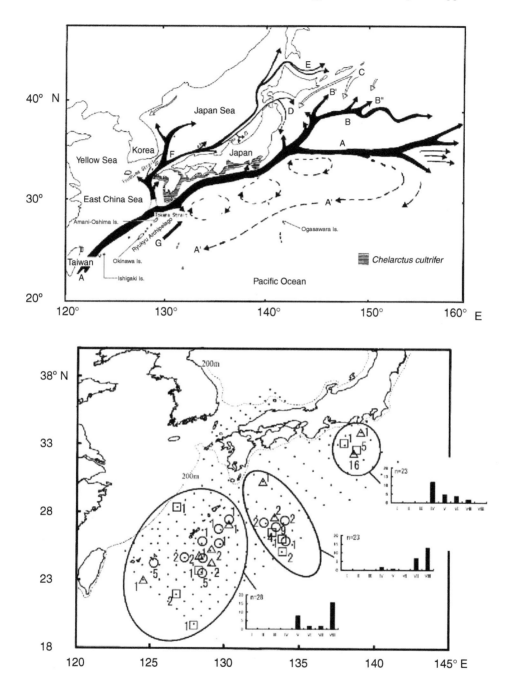

FIGURE 4.5 An example of a scyllarid for which a proportion of hatched larvae are retained near adult populations, and later recruit to them, with those larvae dispersing farther afield probably being destined to perish because they end up in waters too deep or otherwise unsuitable for settlement and survival; **Upper panel** Schematic illustration of current systems in the Kuroshio Subgyre. (From Sekiguchi, H. & Inoue, N. 2002. *J. Oceanogr.* 58: 747–757; used with permission.) A, warm Kuroshio Current; A′, Kuroshio-Counter Current; B, B′, B″, branches of the Kuroshio; C, cool Oyashio Current; D, warm Tsushima Current; E, warm Soya Current; F, warm Tsushima Current; G, Ryukyu Current System; hatched areas, geographical distribution of *Chelarctus cultrifer* (Ortmann, 1897). **Lower panel** Oceanic distribution of *C. cultrifer* phyllosomas which are unlikely to survive. (From Inoue, N. & Sekiguchi, H. 2005. *J. Oceanogr.* 61: 389–398; used with permission.) Dots, sampling stations; number with open triangle, open square, and open circles are the numbers of stage IV/V, stage VI/VII, and stage VIII (final) phyllosomas, respectively; hatched areas, geographical distribution of *C. cultrifer*; dotted lines, 200 m isobath; numerals in vertical and horizontal axes of inserted graphs, individual numbers collected with large plankton nets (2 m-diameter ring net, oblique tow from 80 m depth to the surface) and development stage, respectively.

are probably: (1) that although the larval life in both families is long, it is nevertheless shorter among the scyllarinids; and (2) vertical migration behavior differs between the pairs of species (see Section 4.4.2).

Second, scyllarinid species often are not only the numerically dominant scyllarid species among larvae collected from coastal waters, but frequently the most abundant of all phyllosomas in those waters (e.g., McWilliam & Phillips 1983; Rothlisberg et al. 1994; Sekiguchi & Inoue 2002). Their abundance is due, in part, to their prolonged egg-hatching periods and repetitive breeding in a single year (Baisre 1994; Booth et al. 2005). In addition, scyllarinid phyllosomas generally remain concentrated closer to shore as a result of their shorter larval lives and their behavior (e.g., vertical migrations), where they are more readily collected.

The distribution and ecology of the nisto are not as well-known as they are for the phyllosomas: nistos are less commonly taken, in part, because their duration is much shorter (several species around one week, with some up to three weeks; Table 4.2). Although nistos appear to have a similar nektonic/benthic function as palinurid pueruli, which swim shoreward from or beyond the continental shelf break (Booth & Phillips 1994), it appears that the final-stage phyllosomas of some scyllarids are found much closer to the coast before metamorphosis than final-stage palinurid phyllosomas. Such scyllarids include *Thenus* species, the late-stage larvae and the nistos of which rarely appear to be taken beyond the shelf break (Barnett et al. 1984, 1986). Precisely where metamorphosis takes place in those species with long larval lives and wide offshore dispersal is unknown.

4.4.2 Vertical Distributions

Knowing the vertical distributional patterns of phyllosomas and nistos is important for unraveling and modeling dispersal and recruitment processes. The extended larval period gives potential for widespread dispersal in wind-driven surface currents and in subsurface flows. Since metamorphosis needs to take place within the swimming range of the nisto for it to reach its settlement grounds, there must be some mechanism that prevents phyllosomas dispersing too far from these grounds or, alternatively, that returns them there after having dispersed offshore. (Note that the only evidence for settlement grounds being significantly separate from the adult populations is for *Ibacus chacei*, as discussed earlier.)

One mechanism with the potential to place late-stage phyllosomas close enough to settlement grounds is their exploitation by vertical migration of different subsurface currents. The most common form of such movements — seen among the palinurids (Booth & Phillips 1994) — is diel vertical migration, in which the larvae rise near the surface at night and descend to deeper layers during the day. Diel vertical migration has been recorded in the phyllosomas of several scyllarid species: *Crenarctus bicuspidatus* in Western Australian waters (Phillips et al. 1981), *Scyllarus* and *Scyllarides* species off the Florida Keys, USA (Yeung & McGowan 1991), and *Chelarctus cultrifer* in the western North Pacific (Minami et al. 2001). There is some evidence that the vertical range of distribution may be determined not just by negative phototaxis, but also by the bottom depth of the mixed layer (Minami et al. 2001).

There are very limited data on the extent of vertical migration of scyllarid phyllosomas in relation to developmental stage, but available information indicates that changes in depth can occur during development, the extent probably depending on the species and the local oceanography. For example, Phillips et al. (1981) found that early-stage phyllosomas of *Chelarctus bicuspidatus* off Western Australia exhibited vertical distribution patterns different from those of the mid- and late stages. All stages of *C. bicuspidatus* undergo diel vertical migration, with peak densities of early stages occurring at shallower depths than those of mid- and late-stages. During the day, the greatest density of early stages was at 40 to 80 m depth, whereas mid- and late-stage densities peaked at 80 to 100 m depth. At night, early-stage phyllosomas were found at the surface during both new and full moons, but numbers of mid- and late-stages peaked at the surface only near the new moon. On nights near full moon, the densities of mid- and late-stages peaked at depths of 15 to 25 m. Presumably the different depths of the larvae at different stages of development influence the seaward dispersal of larvae, and then their transport back toward shore. The greater offshore distribution of *Panulirus cygnus* phyllosomas, compared with that of *C. bicuspidatus* (Phillips et al. 1979, 1981), is likely due, at least in part, to different vertical distribution patterns between the two species.

The nisto is the last planktonic phase, which settles to the seafloor. Its vertical distribution is unknown, except that nistos are often taken in plankton tows (e.g., *Antipodarctus aoteanus* — Webber & Booth 2001) and have been observed swimming in the water column under lights (e.g., *Scyllarides astori* Holthuis 1960 — Holthuis & Loesch 1967) (see Section 4.5.2). This indicates that at times the nistos of some species are pelagic. However, the extent and frequency with which they move up into the water column probably varies among species (see Section 4.5.2).

4.5   General Biology

4.5.1   Phyllosomas

Many aspects of phyllosoma biology essential to understanding larval recruitment are poorly known. These include food and feeding, behavior (including swimming ability), predation, and sensory systems. Tank studies suggest that phyllosomas of most, if not all, scyllarids are primarily predators and use their pereiopods to fix and hold food items (e.g., Robertson 1968; Ito & Lucas 1990; Marinovic et al. 1994; Mikami & Greenwood 1997a). The mouth and foregut structures also suggest consumption of soft fleshy foods (Mikami et al. 1994; Mikami & Takashima 2000; also see Johnston, Chapter 6 for information on the digestive system and its functional morphology). Therefore, foods such as *Artemia* species, fish larvae, bivalve gonad and flesh, and jellyfish have been useful in phyllosoma culture, particularly for the early stages (e.g., Ritz 1977; Takahashi & Saisho 1978; Marinovic et al. 1994; Mikami & Greenwood 1997a; also see Mikami & Kuballa, Chapter 5). In nature, similar types of food (larvae, jellyfish, small zooplankton) are more abundant in coastal waters, at upwelling sites, and in areas along water-mass boundaries, than in offshore, more oligotrophic waters.

Scyllarid phyllosomas, at least in their late stages, are consumed by a wide variety of fish. For example, *Ibacus alticrenatus* phyllosomas off the coast of New Zealand were found in the stomachs of pilot fish (*Naucrates ductor* (Linnaeus 1758)), albacore (*Thunnus alalunga* (Bonnaterre 1788)), rudderfish (*Centrolophus niger* (Gmelin 1789)), and sunfish (*Mola mola* (Linnaeus 1758)) out of 27 fish species taken incidentally in the purse-seine skipjack fishery (Bailey & Habib 1982). Many such pelagic fish are most often found in areas of high productivity — including coastal waters — meaning that scyllarid phyllosomas that remain close to shore during their development are particularly prone to predation.

There have been many reports of scyllarid phyllosomas, usually late and final stages, being closely associated with — and sometimes "riding" on — medusae (and possibly salps and ctenophores) (reviewed in Booth et al. 2005). Phyllosomas of several scyllarid genera have been observed associating with gelatinous zooplankton in this way: *Ibacus* (Shojima 1963, 1973; Thomas 1963; Booth & Mathews 1994), *Scyllarides* (Shojima 1963), *Thenus* (Barnett et al. 1986), *Scyllarus* (Herrnkind et al. 1976; Barnett et al. 1986), and *Petrarctus* and *Eduarctus* (Barnett et al. 1986). Medusa riding does not appear to be accidental or entirely intermittent: for example, among 402 individual *Aurelia aurita* (Linnaeus, 1758) examined, Herrnkind et al. (1976) found 80 to have an associated *Scyllarus* sp. larva, and Barnett et al. (1986) found 49 out of 51 collected larvae (mostly final-stage phyllosomas of *Thenus orientalis*, *Petrarctus demani*, *Eduarctus martensii*, and *Scyllarus* sp.) to be closely associated with medusae. Interestingly, similar observations have not been reported for palinurid phyllosomas.

It is not known if the phyllosoma–medusa/salp relationship is for food, protection, transport, or a combination of these. Nematocysts in the feces of a giant phyllosoma (probably a *Parribacus* spp. — Holthuis 1985) and of other (unidentified) phyllosomas suggested a feeding association to Sims & Brown (1968), and Shojima (1963) observed *Scyllarus* species and *Scyllarides* species larvae carrying hydromedusae "as if for food." In support of a possible food source, Kittaka (2000) concluded that medusae may be nutritionally satisfactory for palinurid phyllosomas in laboratory culture. The nematocysts of medusae, which appear not to affect the phyllosomas, may confer some protection from predation on the phyllosomas. Gelatinous zooplankton is more common within the shelf break than beyond it, and is found in areas of high primary productivity. Association of phyllosomas with such zooplankton may, therefore, play a role in reducing the extent of larval dispersal in some species.

The swimming behavior of scyllarid phyllosomas at similar stages of development appears to differ among genera. Some, such as *Thenus orientalis*, appear to be weak swimmers, even in the final stage

(Barnett et al. 1984). Other scyllarid phyllosomas, such as *Crenarctus bicuspidatus* and *Chelarctus cultrifer*, appear to display swimming ability similar to that of the palinurids found in the same water masses (Phillips et al. 1981; Minami et al. 2001). Any horizontal swimming is, on its own, insufficient to effect a return to the coast of species that become widely distributed. It is more likely that shoreward transport of phyllosomas is achieved more passively by larvae being in particular strata in the water column.

4.5.2 Nistos

Recruitment is completed at the settlement of the nisto, which provides the link between the planktonic and benthic life phases in scyllarids, as does the puerulus for palinurids. However, the nisto of at least some scyllarids may be a more transient phase and may be more similar to the first juvenile instar than is the puerulus (Barnett et al. 1986).

Almost all nistos described to date are at first virtually transparent (like pueruli), a characteristic undoubtedly evolved to help avoid predation as they move inshore. The onset of pigmentation seems to vary: Barnett et al. (1986) found that the nisto of *Eduarctus martensii* remained transparent throughout the duration of the phase while that of *Thenus orientalis* completely lost its pigmentation at night but reestablished it during the day. In contrast, *Petrarctus demani* showed a slight loss of pigmentation at night. The changes in pigmentation observed in the latter two species presumably assist the nisto in retaining its cryptic coloration against changing backgrounds associated with swimming in the water column and then resting on the substrate.

Nistos are active mainly at night. The nocturnal activity of *Eduarctus martensii*, *Petrarctus demani*, and *Thenus orientalis*, for example, included backward swimming — short bursts of oblique swimming controlled by flexing of the abdomen and telson (Barnett et al. 1986), similar to that seen in the adults of some scyllarids (Jacklyn & Ritz 1986). This was succeeded by more passive sinking in which the pereiopods were slightly splayed. In *P. demani*, backward swimming was also achieved by rapid beating of the setose pleopods. Forward swimming by vigorous beating of the pleopods, which is typically associated with time spent in the water column, has also been observed in some scyllarid species, for example *Scyllarus americanus* (Robertson 1968) and *Galearctus kitanoviriosus* (Higa & Saisho 1983).

Differences in forward-swimming ability are presumably correlated with the marked differences in the size and development of the pleopods among different scyllarids. The nisto of *Thenus orientalis* for example, has poorly developed pleopods and is probably poorly equipped for forward swimming and directed shoreward progression (Barnett et al. 1984). Weak-swimming ability in such nistos suggests that the phyllosomas do not stray far from adult grounds. In contrast, nistos of several other scyllarinid species have well-developed, setose pleopods, each bearing a prominent appendix interna which links pleopod pairs. The nistos of such species have been found relatively far from shore (e.g., *Antipodarctus aoteanus*, Webber & Booth 2001) and presumably are stronger swimmers.

Sand-burying during the day, a behavior that is becoming increasingly evident among palinurid pueruli (e.g., see Booth 2001), may also be quite widespread in the nisto stage. The nistos of *Thenus orientalis* and *Eduarctus martensii* were found by Barnett et al. (1986) to bury in sandy substrates during the day, the effectiveness of burrowing governed by the coarseness of the substrate (for more information on *Thenus* burying behavior, see Jones, Chapter 16). It is likely that nistos alternate between being active at night and taking refuge in or on the substrate during the day as they move toward their settlement areas. Thus the nisto is probably neither completely nektonic nor entirely benthic.

The nisto appears to be a nonfeeding stage, relying on energy reserves accumulated in the final phyllosoma stage, and so is similar to the puerulus in this regard. For example, attempts to feed the nistos of *Eduarctus martensii*, *Petrarctus demani*, and *Thenus orientalis* were unsuccessful (Barnett et al. 1986). If nistos do feed, their diet is probably confined to small, soft materials, judging by observations of those cultured in the laboratory, histological examination of the proventriculus, and examination of the mouthparts (e.g., Mikami & Takashima 1993, 2000; Mikami & Greenwood 1997a; see also Johnston, Chapter 6). The absence of the need to feed allows the animal to concentrate its activities on finding a suitable settlement place.

Nistos have only occasionally been reported on collectors deployed for pueruli (e.g., *Ibacus peronii* and *Antipodarctus aoteanus* — Phillips & Booth 1994), suggesting that in most scyllarids settlement requirements (and possibly settlement depths) are different from those of palinurids. Palinurids appear to prefer holes and crevices in hard substrates or structurally complex growths of algae with sedentary invertebrates in shallow waters (e.g., see Booth & Phillips 1994). In contrast, the flattened form of scyllarids — including the nisto — suggests adaptation to living on or burrowing into softer, more homogeneous substrates than those inhabited by most spiny lobsters.

The duration of the nisto stage is broadly known for only six species, mainly small Scyllarinae and *Thenus orientalis* (Table 4.2). For most of these, the nisto is very short lived (under culture, about a week), much shorter than that of the puerulus (several weeks — Booth & Phillips 1994). This has led some workers, such as Barnett et al. (1986), to suggest that the nisto may not have precisely the same function as the puerulus. The impression is that the nisto, at least of the scyllarinids and theninids, is more like the first juvenile instar than is the puerulus: it is generally shorter lived than the puerulus; shows less propensity to swim forward; more often swims backward, as does the adult (and not just in escape); and appears to have a greater propensity to burrow. However, given the variation in early life-history characteristics among scyllarids discussed earlier, the great range in nisto size (3 to 21 mm carapace length — Table 4.2), and the fact that at least some scyllarids have quite long nisto durations (e.g., about three weeks in *Ibacus peronii* — Table 4.2), it is likely that the precise role of the nisto varies considerably among species.

4.5.3   Subfamily Characteristics

Drawing on the material presented in Table 4.1 and Table 4.2, recruitment strategies among the subfamilies can be examined from a phylogenetic perspective. In the Arctidinae, adult females are large (mostly >100 mm CL). The larval development period is long (at least eight months) with many stages (at least 11), and results in large phyllosomas (most 35 mm total length [TL] or larger) and large nistos (at least 10 mm CL). The long larval duration (and large phyllosomas) means that arctidinid larvae are capable of dispersing great distances offshore, well beyond the shelf break.

In the Ibacinae, adult females are usually relatively large (most are >70 mm CL). *Ibacus* eggs are large and the larvae hatch at an advanced stage. Larval development in *Ibacus* is of medium duration (usually two to four months), consists of at least seven stages, and results in large phyllosomas (>30 mm TL) and large nistos (at least 12 mm CL). Because of the shortened planktonic existence, there is relatively limited offshore dispersal of the phyllosomas. *Parribacus* spp. differ in that the phyllosomas are much larger with extensive offshore dispersal.

In the Scyllarinae, adult females are small (mostly <30 mm CL) and, although there are usually numerous larval stages (at least 6 to 12), other parameters are highly variable: the larval development period ranges from one to four months or more, the final phyllosoma ranges from about 10 to 30 mm TL, and the nisto is a corresponding 3 to 8 mm CL. Scyllarinids exhibit a wide range of late-stage larval distributions, from being confined close to shore to becoming widely dispersed beyond the shelf break.

In the Theninae, adult females are relatively large (about 80 mm CL), and their larvae hatch at an advanced stage of development, leading to an abbreviated planktonic existence similar to that of *Ibacus* larvae. There are few instars/stages (about four) and a short larval-development period (about one month); the final phyllosoma averages about 16 mm TL and the nisto is moderately large (seven to nine mm CL). Theninid larvae do not disperse far from shore.

Baisre (1994) assessed the significance of phyllosoma larvae in the phylogeny of the Palinuroidea using a number of larval characters and, based on these alone, identified three groups within the Scyllaridae: a *Scyllarides–Arctides–Parribacus* group, an *Ibacus–Evibacus* group, and a *Thenus–Scyllarus* group. Booth et al. (2005) agreed, in large part, with Baisre's finding, but noted that the subfamily Ibacinae clearly includes some very disparate developmental characteristics, including the relatively advanced stage at hatching in *Ibacus* species, and the extremely large and long-lived larvae of *Parribacus* species (up to 80 mm TL attained over at least eight to nine months) compared to others in the subfamily. The Scyllarinae also display considerable variability in developmental characteristics, with patterns among the genera expected to become clearer as more information becomes available.

Key to Identification to Generic/Subfamily Level of Final-Stage Scyllarid Phyllosomas

This key distinguishes among genera within the Scyllaridae and also distinguishes scyllaridae from the Palinuridae and Synaxidae. See Figure 4.4 for scyllarid larval shapes. There are insufficient data to allow the scyllarinid genera to be distinguished.

1 (a) 3rd maxilliped with setose exopod . . . Synaxidae, Palinuridae except *Jasus* & *Sagmariasus**
  (b) 3rd maxilliped without setose exopod . . . . (2)
2 (a) Antenna cylindrical . . . . *Jasus, Sagmariasus*
  (b) Antenna dorsoventrally compressed . . . . **Scyllaridae** (3)
3 (a) Cephalic shield bilobed, similar to two joined kidneys . . . . (4)
  (b) Cephalic shield round, elongate, oval to somewhat rectangular . . . . (5)
4 (a) 5th pereiopod with setose exopod; posterior margin of thorax not emarginate[†] . . . . . . *Ibacus*
  (b) 5th pereiopod with setose exopod; posterior margin of thorax emarginate . . . . . . *Parribacus*
  (c) 5th pereiopod without setose exopod . . . . *Thenus*
5 (a) Abdomen tapered posteriorly; posterior margin of thorax not emarginate; 5th pereiopod without setose exopod and underdeveloped compared to other pereiopods . . . . . . Scyllarinae
  (b) Abdomen tapered posteriorly; posterior margin of thorax slightly emarginate; 5th pereiopod with setose exopod, similar to other pereiopods . . . . *Evibacus*
  (c) Abdomen slightly tapered posteriorly; posterior margin of thorax emarginate . . . . (6)
6 (a) Cephalic shield narrower than thorax . . . . *Arctides*
  (b) Cephalic shield marginally or notably wider than thorax . . . . *Scyllarides*

*The palinurid *Sagmariasus verreauxi* develops a setose exopod on the third maxilliped in late stages, and then apparently only intermittently (Kittaka et al. 1997). The third maxilliped probably lacks a setose exopod also in the palinurid *Projasus*, but this remains unconfirmed.
[†] Thorax emarginate, margin of thorax not continuous with that of abdomen.

4.6 Conclusions

Most scyllarids have eggs that are relatively small among the Decapoda, and provide little food for the growth of an embryo that hatches as a larva at an early stage of development and then pursues a teleplanic lifestyle (a protracted planktonic existence). Generally, larger scyllarids have larger broods, but this relationship is modified by variation in egg size, which ranges from about 0.4 to 1.2 mm diameter (Table 4.1). The trade-off for producing larger eggs is lower fecundity. Based on this, and on information presented in Table 4.2, there is a clear relationship among species between female size (and therefore fecundity) and larval recruitment strategy. *Ibacus* species and *Thenus* species have larger eggs and more advanced larval development at hatching; thus, their larvae are more likely to survive and be retained until settlement within coastal waters than are *Scyllarides* species (and probably also *Parribacus* species). The higher fecundity of *Scyllarides* species is presumably an adaptation to offset high larval loss in oceanic waters.

It seems likely that short larval durations have evolved among many scyllarid species so that fewer larvae are carried far from the coast by currents, allowing them to recruit to their benthic inshore populations. It appears that scyllarids with longer larval durations lasting several months, such as *Scyllarides*, *Parribacus*, and *Arctides* species, are similar to palinurids in their dispersal because mid- and late-stage phyllosomas are rarely found close to the coast. This gives greater opportunity for dispersal, but nisto recruitment of these species may be less regular from year to year than it is for the more nearshore genera such as *Thenus* species.

Scyllarid phyllosomas appear to feed mainly on soft, fleshy foods; in turn, they are eaten by a diverse range of pelagic fishes. Swimming abilities appear to vary widely among genera. The phyllosomas of many genera, in their later stages of development, can be found associated with gelatinous zooplankton, particularly medusae. The nisto — which is probably nonfeeding — appears in some scyllarid species to be more similar to its first juvenile instar than is the puerulus of palinurids to theirs. But the nistos of different species show considerable variability in swimming, burrowing, and other behavior, suggesting that the precise role of the nisto may vary among genera.

Although close to two thirds of all known scyllarids have had one or more phyllosoma instars described, only about half of these have been reliably identified with adults of scyllarid species; about 25 nistos have been described. Only seven species have been cultured through their full larval development. Addressing these gaps in our knowledge of the early life-history stages of scyllarids is an obvious area of endeavor for the future; very importantly, this will allow field samples to be identified with greater certainty. Also, much less is known of the ecology and behavior of scyllarid larvae and postlarvae than for palinurids, reflecting fewer investigations into scyllarids, mainly because of their lower commercial value. Again these gaps suggest further fruitful grounds of investigation.

Acknowledgments

We gratefully acknowledge the critical reviews by two unknown referees, one in particular who offered many very good suggestions for improvements to this chapter.

References

Almog-Shtayer, G. 1988. Behavioural-ecological aspects of Mediterranean slipper lobsters in the past and of the slipper lobster *Scyllarides latus* in the present. M.A. thesis: University of Haifa, Israel. 165 pp.

Atkinson, J.M. & Boustead, N.C. 1982. The complete larval development of the scyllarid lobster *Ibacus alticrenatus* Bate, 1888 in New Zealand waters. *Crustaceana* 42: 275–287.

Bailey, K.N. & Habib, G. 1982. Food of incidental fish species taken in the purse-seine skipjack fishery, 1976–1981. *Fish. Res. Div. Occ. Publ. Data Ser.* 6: 1–24.

Baisre, J.A. 1994. Phyllosoma larvae and the phylogeny of Palinuroidea (Crustacea: Decapoda): A review. *Aust. J. Mar. Freshw. Res.* 45: 925–944.

Barnard, K.H. 1950. Descriptive catalogue of South African decapod Crustacea. *Ann. S. Afr. Mus.* 38: 1–837.

Barnett, B.M., Hartwick, R.F., & Milward, N.E. 1984. Phyllosoma and nisto stage of the Moreton Bay bug, *Thenus orientalis* (Lund) (Crustacea: Decapoda: Scyllaridae), from shelf waters of the Great Barrier Reef. *Aust. J. Mar. Freshw. Res.* 35: 143–152.

Barnett, B.M., Hartwick, R.F., & Milward, N.E. 1986. Descriptions of the nisto stage of *Scyllarus demani* Holthuis, two unidentified *Scyllarus* species, and the juvenile of *Scyllarus martensii* Pfeffer (Crustacea: Decapoda: Scyllaridae), reared in the laboratory; and behavioural observations of the nistos of *S. demani*, *S. martensii* and *Thenus orientalis* (Lund). *Aust. J. Mar. Freshw. Res.* 37: 595–608.

Berry, P.F. 1974. Palinurid and scyllarid lobster larvae of the Natal coast, South Africa. *Invest. Rep. S. Afr.* 34: 1–44.

Bianchini, M.L. & Ragonese, S. 2003. *In ovo* embryonic development of the Mediterranean slipper lobster, *Scyllarides latus*. *Lobster Newslett.* 10(1): 10–12.

Booth, J.D. 2001. Habitat preferences and behaviour of newly settled *Jasus edwardsii* (Palinuridae). *Mar. Freshw. Res.* 52: 1055–1065.

Booth, J.D. & Matthews, R. 1994. Phyllosomas riding jellyfish. *Lobster Newslett.* 7(1): 12.

Booth, J.D. & Phillips, B.F. 1994. Early life history of spiny lobster. *Crustaceana* 66: 271–294.

Booth, J.D., Webber, W.R., Sekiguchi, H., & Coutures, E. 2005. Diverse larval recruitment strategies within the Scyllaridae. *N. Z. J. Mar. Freshw. Res.* 39: 581–592.

Chace, F.A. 1966. Decapod crustaceans from St. Helena Island, South Atlantic. *Proc. U.S. Nat. Mus.* 118: 623–661.

Coutures, E. 2000. Distribution of phyllosoma larvae of Scyllaridae and Palinuridae (Decapoda: Palinuridea) in the south-western lagoon of New Caledonia. *Mar. Freshw. Res.* 51: 363–369.

Coutures, E. 2003. The biology and artisanal fishery of lobsters of American Samoa. *Depart. Mar. Wildl. Resources Biol. Rep.* 103: 1–18.

Coutures, E. & Webber, R. 2005. Phyllosoma and nisto stages of Scyllarinae sp. D (Crustacea: Decapoda: Scyllaridae) from the south-west lagoon of New Caledonia. *Zootaxa* 873: 1–20.

Coutures, E., Hebert, P., Wantiez, L., & Chauvet, C. 2002. Lobster catches using crest nets in Uvea (Wallis and Futuna). *Lobster Newslett.* 15(1): 3–5.

Crosnier, A. 1972. Naupliosoma, phyllosomes et pseudibacus de *Scyllarides herklotsi* (Herklots) (Crustacea, Decapoda, Scyllaridae) recoltes par l'ombango dans le sud Golfe de Guinee. *Cahiers ORSTOM Oceanogr.* 10: 139–149.

DeMartini, E.E. & Williams, H.A. 2001. Fecundity and egg size of *Scyllarides squammosus* (Decapoda: Scyllaridae) at Maro Reef, Northwestern Hawaiian Islands. *J. Crust. Biol.* 21: 891–896.

Dotsu, Y., Tanaka, O., Shojima, Y., & Seno, K. 1966. Metamorphosis of the phyllosomas of *Ibacus ciliatus* (Von Siebold) and *I. novemdentatus* Gibbes (Crustacea: Reptantia) to the reptant larvae. *Bull. Fac. Fish. Nagasaki Univ.* 21: 195–221.

Fiedler, U. & Spanier, E. 1999. Occurrence of *Scyllarus arctus* (Crustacea, Decapoda, Scyllaridae) in the eastern Mediterranean — preliminary results. *Ann. Istrian. Med. Stud.* 17: 153–158.

Gurney, R. 1936. Larvae of decapod Crustacea: Part III Phyllosoma. *Discovery Rep.* 12: 379–440.

Haddy, J.A., Courtney, A.J., & Roy, D.P. 2005. Aspects of the reproductive biology and growth of Balmain bugs (*Ibacus* spp.) (Scyllaridae). *J. Crust. Biol.* 25: 263–273.

Harada, E. 1958. Notes on the naupliosoma and newly hatched phyllosoma of *Ibacus ciliatus* (Von Siebold). *Publ. Seto Mar. Biol. Lab.* 7: 173–179.

Herrnkind, W.F., Halusky, J., & Kanciruk, P. 1976. A further note on phyllosoma larvae associated with medusae. *Bull. Mar. Sci.* 26: 110–112.

Higa, T. & Saisho, T. 1983. Metamorphosis and growth of the late-stage phyllosoma of *Scyllarus kitanoviriosus* Harada (Decapoda, Scyllaridae). *Mem. Kagoshima Univ. Res. Center South Pacific* 3: 86–98.

Higa, T. & Shokita, S. 2004. Late-stage phyllosoma larvae and metamorphosis of a scyllarid lobster, *Chelarctus cultrifer* (Crustacea: Decapoda: Scyllaridae), from the northwestern Pacific. *Spec. Diversity* 9: 221–249.

Holthuis, L.B. 1985. A revision of the family Scyllaridae (Crustacea: Decapoda: Macrura). 1. Subfamily Ibacinae. *Zool. Verhand.* 218: 1–130.

Holthuis, L.B. 1991. Marine Lobsters of the World. *An Annotated and Illustrated Catalogue of the Species of Interest to Fisheries Known to Date.* FAO Species Catalogue No. 125, Vol 13: pp. 1–292. Rome: Food and Agriculture Organization of the United Nations.

Holthuis, L.B. 2002. The Indo-Pacific scyllarine lobsters (Crustacea, Decapoda, Scyllaridae). *Zoosystema* 24: 499–683.

Holthuis, L.B. & Loesch, H. 1967. The lobsters of the Galapagos Islands (Decapoda, Palinuridea). *Crustaceana* 12: 214–222.

Inoue, N. & Sekiguchi, H. 2001. Distribution of late-stage phyllosoma larvae of *Panulirus japonicus* in the Kuroshio Subgyre. *Mar. Freshw. Res.* 52: 1201–1209.

Inoue, N. & Sekiguchi, H. 2006. Descriptions of phyllosoma larvae of *Scyllarus bicuspidatus* and *S. cultrifer* (Decapoda, Scyllaridae) collected in Japanese waters. *Plankton Benthos Res.* 1: 26–41.

Inoue, N. & Sekiguchi, H. 2005. Distribution of scyllarid phyllosoma larvae (Crustacea: Decapoda: Scyllaridae) in the Kuroshio Subgyre. *J. Oceanogr.* 61: 389–398.

Inoue, N., Sekiguchi, H., & Nagasawa, T. 2000. Distribution and identification of phyllosoma larvae in the Tsushima Current region. *Bull. Japan. Soc. Fish. Oceanogr.* 64: 129–137.

Inoue, N., Sekiguchi, H., & Shinn-Pyng, Y. 2001. Spatial distributions of phyllosoma larvae (Crustacea: Decapoda: Palinuridae and Scyllaridae) in Taiwanese waters. *J. Oceanogr.* 57: 535–548.

Inoue, N., Minami, H., & Sekiguchi, H. 2004. Distribution of phyllosoma larvae (Crustacea: Decapoda: Palinuridae, Scyllaridae and Synaxidae) in the western North Pacific. *J. Oceanogr.* 60: 963–976.

Ito, M. & Lucas, J.S. 1990. The complete larval development of the scyllarid lobster, *Scyllarus demani* Holthuis, 1946 (Decapoda, Scyllaridae), in the laboratory. *Crustaceana* 58: 144–167.

Jacklyn, P.M. & Ritz, D.A. 1986. Hydrodynamics of swimming in scyllarid lobsters. *J. Exp. Mar. Biol. Ecol.* 101: 85–99.

Johnson, M.W. 1968. The phyllosoma larvae of scyllarid lobsters in the Gulf of California and off Central America with special reference to *Evibacus princeps* (Palinuridea). *Crustaceana Suppl.* 2: 98–116.

Johnson, M.W. 1971a. The palinurid and scyllarid lobster larvae of the tropical eastern Pacific and their distribution as related to the prevailing oceanography. *Bull. Scripps Inst. Oceanogr. Univ. California* 19: 1–36.

Johnson, M.W. 1971b. The phyllosoma larvae of slipper lobsters from the Hawaiian Islands and adjacent areas (Decapoda, Scyllaridae). *Crustaceana* 20: 77–103.

Jones, C. 1988. The biology and behaviour of bay lobsters, *Thenus* spp. (Decapoda: Scyllaridae) in northern Queensland, Australia. Ph.D. dissertation: University of Queensland, Brisbane, Australia.

Kagwade, P.V. & Kabli, L.M. 1996. Reproductive biology of the sand lobster *Thenus orientalis* (Lund) from Bombay waters. *Indian J. Fish.* 43: 13–25.

Kittaka, J. 2000. Culture of larval spiny lobsters. In: Phillips, B.F. & Kittaka, J. (eds.), *Spiny Lobster: Fisheries and Culture*: pp. 508–532. Oxford, U.K.: Fishing News Books.

Kittaka, J., Ono, K., & Booth, J.D. 1997. Complete development of the green rock lobster, *Jasus verreauxi* from egg to juvenile. *Bull. Mar. Sci.* 61: 57–71.

Lee, T.N. & Williams, E. 1999. Mean distribution and seasonal variability of coastal currents and temperature in the Florida Keys with implications for larval recruitment. *Bull. Mar. Sci.* 64: 35–56.

Lesser, J.H.R. 1974. Identification of early larvae of New Zealand spiny and shovel-nosed lobsters (Decapoda, Palinuridae and Scyllaridae). *Crustaceana* 27: 259–277.

Lyons, W.G. 1970. Scyllarid lobsters (Crustacea, Decapoda). *Mem. Hourglass Cruises* 1: 1–74.

Marinovic, B., Lemmens, J.W.T.J., & Knott, B. 1994. Larval development of *Ibacus peronii* Leach (Decapoda: Scyllaridae) under laboratory conditions. *J. Crust. Biol.* 14: 80–96.

Martins, H.R. 1985. Biological studies of the exploited stock of the Mediterranean locust lobster *Scyllarides latus* (Latreille, 1802) (Decapoda: Scyllaridae) in the Azores. *J. Crust. Biol.* 5: 294–305.

McWilliam, P.S. & Phillips, B.F. 1983. Phyllosoma larvae and other crustacean macrozooplankton associated with Eddy J, a warm-core eddy off south-eastern Australia. *Aust. J. Mar. Freshw. Res.* 34: 653–663.

McWilliam, P.S., Phillips, B.F., & Kelly, S. 1995. Phyllosoma larvae of *Scyllarus* species (Decapoda, Scyllaridae) from the shelf waters of Australia. *Crustaceana* 68: 537–565.

Michel, A. 1968. Les larves phyllosomes et la post-larve de *Scyllarides squamosus* (H. Milne Edwards) — Scyllaridae (Crustaces Decapodes). *Cahier ORSTOM Oceanogr.* 6: 47–53.

Mikami, S. & Greenwood, J.G. 1997a. Complete development and comparative morphology of larval *Thenus orientalis* and *Thenus* sp. (Decapoda: Scyllaridae) reared in the laboratory. *J. Crust. Biol.* 17: 289–308.

Mikami, S. & Greenwood, J.G. 1997b. Influence of light regimes on phyllosomal growth and timing of moulting in *Thenus orientalis* (Lund) (Decapoda: Scyllaridae). *Mar. Freshw. Res.* 48: 777–782.

Mikami, S. & Takashima, F. 1993. Development of the proventriculus in larvae of the slipper lobster, *Ibacus ciliatus* (Decapoda: Scyllaridae). *Aquaculture* 116: 199–217.

Mikami, S. & Takashima, F. 2000. Functional morphology of the digestive system. In: Phillips, B.F. & Kittaka, J. (eds.), *Spiny Lobsters: Fisheries and Culture*: pp. 601–610. Oxford, U.K.: Blackwell Science.

Mikami, S., Greenwood, J.G., & Takashima, F. 1994. Functional morphology and cytology of the phyllosomal digestive system of *Ibacus ciliatus* and *Panulirus japonicus* (Decapoda, Scyllaridae and Palinuridae). *Crustaceana* 67: 212–225.

Minami, H., Inoue, N., & Sekiguchi, H. 2001. Vertical distributions of phyllosoma larvae of palinurid and scyllarid lobsters in the western North Pacific. *J. Oceanogr.* 57: 743–748.

Olvera Limas, R. Ma., & Ordonez Alcala, L. 1988. Distribution, relative abundance and larval development of the lobsters *Panulirus argus* and *Scyllarus americanus* in the Exclusive Economic Zone of the Gulf of Mexico and the Caribbean Sea. *Ciencia Pesquera* 6: 7–31.

Phillips, B.F. & Booth, J.D. 1994. Design, use, and effectiveness of collectors for catching the puerulus stage of spiny lobsters. *Rev. Fish. Sci.* 2: 255–289.

Phillips, B.F. & McWilliam, P.S. 1986a. Phyllosoma stages and nisto of *Scyllarus martensii* Pfeffer (Decapoda, Scyllaridae) from the Gulf of Carpentaria, Australia. *Crustaceana* 51: 133–154.

Phillips, B.F. & McWilliam, P.S. 1986b. The pelagic phase of rock lobster development. *Can. J. Fish. Aquat. Sci.* 43: 2153–2163.

Phillips, B.F. & Sastry, A.N. 1980. Larval ecology. In: Cobb, J.S. & Phillips, B.F. (eds.), *The Biology and Management of Lobsters, Vol. 2: Ecology and Management*: pp. 11–57. New York: Academic Press.

Phillips, B.F., Brown, P.A., Rimmer, D.W., & Reid, D.D. 1979. Distribution and dispersal of the phyllosoma larvae of the western rock lobster *Panulirus cygnus* in the southeastern Indian ocean. *Aust. J. Mar. Freshw. Res.* 30: 773–783.

Phillips, B.F., Brown, P.A., Rimmer, D.W., & Braine, S.J. 1981. Description, distribution and abundance of late larval stages of the Scyllaridae (slipper lobsters) in the south-eastern Indian Ocean. *Aust. J. Mar. Freshw. Res.* 32: 417–437.

Powell, A.W.B. 1949. New species of Crustacea from New Zealand of the genera *Scyllarus* and *Ctenocheles* with notes on *Lyreidus tridentatus*. *Rec. Auckland Mus. New Zealand* 3: 368–371.

Prasad, R.R. & Tampi, P.R.S. 1960. Phyllosomas of scyllarid lobsters from the Arabian Sea. *J. Mar. Biol. Ass. India* 2: 241–249.

Prasad, R.R., Tampi, P.R.S., & George, M.J. 1975. Phyllosoma larvae from the Indian Ocean collected by Dana Expedition 1928–1930. *J. Mar. Biol. Assoc. India* 17: 56–107.

Ritz, D.A. 1977. The larval stages of *Scyllarus demani* Holthuis, with notes on the larvae of *S. sordidus* (Stimpson) and *S. timidus* Holthuis (Decapoda, Palinuridea). *Crustaceana* 32: 229–240.

Ritz, D.A. & Thomas, L.R. 1973. The larval and postlarval stages of *Ibacus peronii* Leach (Decapoda, Reptantia, Scyllaridae). *Crustaceana* 24: 5–16.

Robertson, P.B. 1968. The complete larval development of the sand lobster, *Scyllarus americanus* (Smith), (Decapoda, Scyllaridae) in the laboratory, with notes on larvae from the plankton. *Bull. Mar. Sci.* 18: 294–342.

Robertson, P.B. 1969a. The early larval development of the scyllarid lobster *Scyllarides aequinoctialis* (Lund) in the laboratory, with a revision of the larval characters of the genus. *Deep-Sea Res.* 16: 557–586.

Robertson, P.B. 1969b. Biological investigations of the deep sea. No. 48. Phyllosoma larvae of a scyllarid lobster, *Arctides guineensis*, from the western Atlantic. *Mar. Biol.* 4: 143–151.

Robertson, P.B. 1971. The larvae and postlarva of the scyllarid lobster *Scyllarus depressus* (Smith). *Bull. Mar. Sci.* 21: 841–865.

Robertson, P.B. 1979. Biological results of the University of Miami Deep-Sea Expeditions. 131. Larval development of the scyllarid lobster *Scyllarus planorbis* Holthuis reared in the laboratory. *Bull. Mar. Sci.* 29: 320–328.

Rothlisberg, P.C., Jackson, C.J., Phillips, B.F., & McWilliam, P.S. 1994. Distribution and abundance of scyllarid and palinurid lobster larvae in the Gulf of Carpentaria, Australia. *Aust. J. Mar. Freshw. Res.* 45: 337–349.

Saisho, T. 1966. Studies on the phyllosoma larvae with reference to the oceanographical conditions. *Mem. Fac. Fish. Kagoshima Univ.* 15: 177–239.

Sekiguchi, H. 1986a. Identification of late-stage phyllosoma larvae of the scyllarid and palinurid lobsters in the Japanese waters. *Bull. Japan. Soc. Sci. Fish.* 52: 1289–1294.

Sekiguchi, H. 1986b. Spatial distribution and abundance of phyllosoma larvae in the Kumano- and Enshu-nada seas north of the Kuroshio Current. *Bull. Japan Soc. Fish. Oceanogr.* 50: 289–297.

Sekiguchi, H. & Inoue, N. 2002. Recent advances in larval recruitment processes of scyllarid and palinurid lobsters in Japanese waters. *J. Oceanogr.* 58: 747–757.

Shojima, Y. 1963. Scyllarid phyllosomas' habit of accompanying the jelly-fish. *Bull. Japan Soc. Sci. Fish.* 29: 349–353.

Shojima, Y. 1973. The phyllosoma larvae of Palinura in the East China Sea and adjacent waters 1. *Ibacus novemdentatus*. *Bull. Seikai Reg. Fish. Res. Lab.* 43: 105–115.

Silberman, J.D. & Walsh, P.J. 1992. Species identification of spiny lobster phyllosome larvae via ribosomal DNA analysis. *Mol. Mar. Biol. Biotechnol.* 1: 195–205.

Sims, H.W. & Brown, C.L. 1968. A giant scyllarid phyllosoma larva taken north of Bermuda (Palinuridea). *Crustaceana Suppl.* 2: 80–82.

Spanier, E. & Lavalli, K.L. 1998. Natural history of *Scyllarides latus* (Crustacea: Decapoda): A review of the contemporary biological knowledge of the Mediterranean slipper lobster. *J. Nat. Hist.* 32: 1769–1786.

Stewart, J. & Kennelly, S.L. 1997. Fecundity and egg-size of the Balmain bug *Ibacus peronii* (Leach, 1815) (Decapoda, Scyllaridae) off the east coast of Australia. *Crustaceana* 70: 191–197.

Stewart, J. & Kennelly, S.J. 1998. Contrasting movements of two exploited scyllarid lobsters of the genus *Ibacus* off the east coast of Australia. *Fish. Res.* 36: 127–132.

Stewart, J., Kennelly, S.L., & Hoegh-Guldberg, O. 1997. Size at sexual maturity and the reproductive biology of two species of scyllarid lobster from New South Wales and Victoria, Australia. *Crustaceana* 70: 344–367.

Takahashi, M. & Saisho, T. 1978. The complete larval development of the scyllarid lobster, *Ibacus ciliatus* (Von Siebold) and *Ibacus novemdentatus* Gibbes in the laboratory. *Mem. Fac. Fish. Kagoshima Univ.* 27: 305–353.

Thomas, L.R. 1963. Phyllosoma larvae associated with medusae. *Nature* 198 (4876): 208.

Webber, W.R. & Booth, J.D. 2001. Larval stages, developmental ecology, and distribution of *Scyllarus* sp. Z (probably *Scyllarus aoteanus* Powell, 1949) (Decapoda: Scyllaridae). *N. Z. J. Mar. Freshw. Res.* 35: 1025–1056.

Williamson, D.I. 1969. Names of larvae in the Decapoda and Euphausiacea. *Crustaceana* 16: 210–213.

Yeung, C. & McGowan, M.F. 1991. Differences in inshore-offshore and vertical distribution of phyllosoma larvae of *Panulirus*, *Scyllarus* and *Scyllarides* in the Florida Keys in May–June, 1989. *Bull. Mar. Sci.* 49: 699–714.

5

Factors Important in Larval and Postlarval Molting, Growth, and Rearing

Satoshi Mikami and Anna V. Kuballa

CONTENTS

Abstract

Scyllarid lobsters are a diverse group in terms of size, habitat, and distribution. The life history of scyllarid lobsters consists of an initial, delicate, planktonic phyllosoma; a non-feeding, intermediate, nisto stage; and a final, benthic, juvenile/adult stage. Each life stage has specific nutritional/environmental requirements, which must be comprehensively understood in order to successfully rear scyllarid lobsters in captivity. Maintenance of the fragile phyllosomas in culture conditions is the most difficult aspect of developing scyllarid lobster aquaculture techniques. Until recently, the main focus of larval rearing in the laboratory has been on larval morphology, but this has led to limited numbers of juveniles being produced. Significant advances, however, have been made with *Thenus* phyllosoma culture, allowing for the mass production of juveniles. The key to this success has been to first comprehensively understand larval nutritional, environmental, and physiological requirements and then to optimize rearing conditions to suit commercial scale culture. This chapter summarizes the important aspects of larval physiology and details the past and present attempts at larval rearing.

5.1   Introduction

The families of Scyllaridae and Palinuridae exhibit a unique larval phase called the "phyllosoma" (derived from the Greek *phyllo*, meaning leaf, and *soma*, meaning body). Phyllosomas are planktonic larvae, with flattened, transparent bodies, and long appendages. Unfortunately, the understanding of how phyllosoma biology relates to oceanic currents during this typically wide dispersal phase is still fragmentary. In fact, larval recruitment has mainly been studied in commercially important palinurid lobsters (e.g., *Panulirus cygnus* (George, 1962), *Panulirus argus* (Latreille, 1804), and *Jasus edwardsii* (Hutton, 1875)), and the nephropid lobsters (*Homarus americanus* H. Milne-Edwards, 1837 and *Homarus gammarus* (Linneaus, 1758); and see Phillips & Sastry 1980; Ennis 1995; Factor 1995; Lawton & Lavalli 1995 for reviews. Except for nephropid lobsters, little attention has been paid to important aspects of the biology of the lobster larval stages.

After a number of molt stages in the ocean, planktonic scyllarid phyllosomas metamorphose to the brief, free-swimming, postlarval phase called the "nisto" (equivalent to the palinurid "puerulus"). The nisto is the transition stage between the planktonic phyllosoma and benthic juvenile stages. As with phyllosomas, the biology of the nisto is still unexplored. Although the scyllarid nisto appears to have the same transitional function as the palinurid puerulus stage, which swims long distances from offshore to inshore habitats, the role of some scyllarid nistos may differ from that of palinurid puerulii because of the diverse range of adult scyllarid habitats. For example, adult distributions of *Thenus* spp. and *Ibacus* spp. are offshore, suggesting that the nistos in these species may not be required to swim long distances, and therefore, these species may have a compromised swimming ability compared to the palinurid puerulus (Barnett et al. 1986; Booth et al. 2005; see Sekiguchi et al., Chapter 4 for a discussion of the nisto role).

Research into larval rearing should provide vital information on scyllarid larval biology. The major problem in attempting larval rearing lies in the successful maintenance of the long-lived phyllosoma stage. There is a great advantage for scyllarid phyllosoma rearing in that some species of this family (e.g., *Thenus*, *Ibacus*, and *Scyllarus* species) have a much reduced larval duration when compared to palinurid species and others within the Scyllaridae (Booth et al. 2005). Although some scyllarid species have been successfully cultured through to the juvenile stage, the aim of these studies often focused only on describing larval morphology, without advancing our understanding of larval biology (Harada 1958; Saisho & Nakahara 1960; Robertson 1968, 1969, 1971, 1979; Sandifer 1971; Saisho & Sone 1971; Crosnier 1972; Ritz & Thomas 1973; Lesser 1974; Ritz 1977; Takahashi & Saisho 1978; Barnett et al. 1986; Ito & Lucas 1990; Marinovic et al. 1994; Mikami & Greenwood 1997a; Pessani et al. 1999). A comprehensive understanding of larval biology and larval physiological requirements is important for the development of commercial-scale larval culture (aquaculture); hence, previous attempts at larval rearing have failed to provide adequate information to produce large numbers of juveniles. A comprehensive study of optimal rearing conditions has been made recently, showing that the duration of the complete larval life cycle of *Thenus* sp. lasts only 25 to 30 days with high survival (>80%) (Mikami 1995). This study demonstrates that there is a great potential for the commercialization of scyllarid lobster species.

This chapter reviews past and current research on larval rearing of scyllarid lobster species, and provides knowledge of larval biology gleaned from such aquaculture efforts, including information on growth, molting physiology, and nutritional requirements. It also examines the potential to apply information and techniques obtained from the larval rearing of *Thenus* sp. to other scyllarid lobsters.

5.2   Previous Attempts to Rear Phyllosomas of Scyllarid Lobsters

The life cycle of the Scyllaridae consists of a short-lived "prelarval" stage, called the naupliosoma that lasts only minutes (Harada 1958; Crosnier 1972; Lesser 1974; Martins 1985), the larval or phyllosomal stage lasting weeks to months (Booth et al. 2005), and a postlarval or nisto stage lasting days to weeks (Booth et al. 2005). Several studies have reported not observing a naupliosomal stage (see review by Phillips & Sastry 1980 and Sekiguchi et al., Chapter 4); however, given the evolutionary history of the families Scyllaridae and Palinuridae (see Webber & Booth, Chapter 2 for a discussion of evolutionary relationships) and the briefness of this naupliosomal stage, it may simply have been overlooked. The phyllosomal stages of the diverse scyllarid species vary greatly in terms of both duration and dispersal strategies (Phillips et al. 1981; Baisre 1994). This variation is not surprising given the ecological diversity of the Scyllaridae, members of which differ in habitat preference from rocky shores to deep sandy waters. Likewise, while the nisto stage is typically short, the dispersal strategies vary among the diverse species (Booth et al. 2005; see Sekiguchi et al., Chapter 4 for more information).

Understanding the larval life cycle based on wild-caught phyllosomas can be particularly frustrating as this information is often incomplete. Ignorance of all aspects of the larval life cycle makes the culture of phyllosomas problematic. Only a limited number of scyllarid species have been cultured though all of the larval stages, including *Ibacus ciliatus* (Von Siebold, 1824), *Ibacus novemdentatus* Gibbes, 1850, *Ibacus peronii* Leach, 1815, *Scyllarus americanus* (S.I. Smith, 1869), *Crenarctus bicuspidatus* (De Man, 1905), *Chelarctus cultrifer* (Ortmann, 1897), *Petrarctus demani* (Holthuis, 1946), *Scyllarus arctus* (Linneaus, 1758), *Thenus orientalis* (Lund, 1793), and *Thenus indicus* (Leach, 1816). Table 5.1 is a chronological summary of all the previous attempts to rear scyllarid species.

Traditionally, the main focus for larval rearing has been the study of morphology and the desire to accurately stage the phyllosoma of cultured vs. wild-caught specimens (Booth et al. 2005). Unfortunately, this focus on larval morphology has resulted in minimal optimization of larval rearing conditions, and has led to a limited understanding of the actual biology of the phyllosoma. Nutritional inadequacies are thought to be a major contributor of the inability to culture phyllosomas through all stages. A study by Mikami (1995), in which four different food regimes were tested, found that superfluous/excessive molts, atypical morphological development, and poor survival are directly related to dietary deficiencies. Similar dietary work with nephropids has also shown that extra molts, incomplete metamorphoses, and morphological anomalies are common with poor larval nutrition or other unfavorable conditions (Templeman 1936; Charmantier & Aiken 1987).

Establishing the nutritional requirements of the phyllosoma throughout the larval stages is an essential, yet difficult aspect of larval rearing, as often different types and sizes of food are required by different stages. This can only be determined by trial-and-error because intact food particles are rarely observed in the gut system of wild-caught phyllosomas. Observing the morphology of the maxillipeds (mouthparts) and thoracic appendages can often aid in the assessment of feeding habits across species (Phillips & Sastry 1980). To date, the majority of successful larval rearing trials have supplemented the use of *Artemia* sp. nauplii as the only food source, with the chopped flesh of bivalves (Takahashi & Saisho 1978; Ito & Lukas 1990; Mikami & Takashima 1993; Marinovic et al. 1994; Lemmens & Knott 1994; Mikami & Greenwood 1997a) or whisked fish and beef (Pessani et al. 1999) being given to later stages. Chopped bivalves and whisked fish/beef are not necessarily readily available in the wild; however, mimicking exactly what is found in nature may not be essential or even suitable for the laboratory or mass culture of phyllosomas.

5.3   Rearing Factors Affecting Phyllosoma Growth

The optimal rearing conditions needed for phyllosoma growth and survival may vary among species. There have been several successful attempts at rearing scyllarid phyllosomas in the laboratory under different environmental conditions (e.g., Takahashi & Saisho 1978; Ito & Lucas 1990; Marinovic et al.

TABLE 5.1

Previous Attempts to Rear Scyllarid Phyllosomas

Species	No. Stages Attained through Culture	Length (days) of Larval Life Attained through Culture	Feed	Temperature (°C)	Total Length of Final Phyllosoma Attained through Culture (mm)	Author
Arctidinae						
Scyllarides aequinoctialis (Lund, 1793)	8 (incomplete)	94	*Artemia* sp. nauplii	24	5.7	Robertson (1969)
Scyllarides haanii (De Haan, 1841)	8 (16 instars) no final gilled stage (incomplete)	197	Initially *Artemia* sp. nauplii; at stage 5 supplemented with minced muscle of the oyster *Crassostrea virginica*	24	29	Kazama (pers. com.)
Scyllarides squammosus (H. Milne Edwards, 1837)	6 (incomplete)	35	*Artemia* sp. nauplii	No info	3.8	Saisho & Sone (1971)
Ibacinae						
Ibacus alticrenatus Bate, 1888	Wild caught at stage 6, molted through to 7, then nisto, then juvenile		Finely chopped mussel *Mytilus edulis aeteanus*		39.7	Atkinson & Boustead (1982)
Ibacus ciliatus (Von Siebold, 1824)	1 (incomplete)	15	Boiled egg yolk	No info	2.6–2.9	Harada (1958)
Ibacus ciliatus	4 (incomplete)	28–34	*Artemia* sp. nauplii	23.5–28.4	6	Saisho & Nakahara (1960)
Ibacus ciliatus	4 (incomplete)	32	*Artemia* sp. nauplii	20–29	6	Saisho (1966)
Ibacus ciliatus	7–8 + nisto (complete)	54–76	Initially fed on *Artemia* nauplii; later stages supplemented with the meat of the short necked clam *Tapes philippinarum*	25	32–41	Takahashi & Saisho (1978)
Ibacus ciliatus	7–8 + nisto (complete)	17	*Artemia* sp. nauplii; later chopped *Mytilus edulis* flesh	24	30	Mikami & Takashima (1993)
Ibacus novemdentatus Gibbes, 1850	7 + nisto (complete)	65	Initially fed on *Artemia* nauplii; later stages supplemented with the meat of the short necked clam *Tapes philippinarum*	23–25	23	Takahashi & Saisho (1978)

Species	Stages	Duration	Food	Temperature (°C)		Reference
Ibacus peronii Leach, 1815	6 + nisto + juveniles (complete)	79	Initially *Artemia* sp. nauplii; after the 3rd instar fed the ovaries of mussel *Mytilus edulis*	20.5 and 23.5	24.9	Marinovic et al. (1994)
Parribacus antarcticus (Lund, 1793)	3 (incomplete)	22	*Artemia* sp. nauplii	20–29	2.6	Saisho (1966)
Scyllarinae						
Chelarctus cultrifer (Ortmann, 1897)	Stage 10 + nisto (complete)	159	Initially *Artemia* sp. nauplii; after 30 days supplemented with chopped gonads of the mussel *Mytilus galloprovincialis*	24.3	22	Matsuda & Mikami (unpublished)
Crenarctus bicuspidatus (De Man, 1905)	9 (incomplete)	78	*Artemia* sp. nauplii	20–29	4.8	Saisho (1966)
Crenarctus bicuspidatus	Stage 11 + nisto + juvenile (complete)	51–62	*Artemia* nauplii with chopped *Mytilus edulis*	24	15.6	Matsuda (unpublished)
Scyllarus americanus (S.I. Smith, 1869)	6–7 Postlarvae (complete)	32–40	*Artemia* sp. nauplii	25	8.9	Robertson (1968)
Scyllarus arctus (Linnaeus, 1758)	16 Postlarvae (complete)	192 at 20 ± 1	*Artemia salina* for ~80 days then whisked fish and beef	20 ± 1°C		Pessani et al. (1999)
Scyllarus depressus (S.I. Smith, 1881)	7 (incomplete)	74 at 20 54 at 25	*Artemia* sp. nauplii	20 and 25	6.5	Robertson (1971)
Scyllarus depressus	2 (incomplete)	10	*Artemia* sp. nauplii	21–25	1.5	Sandifer (1971)
Galearctus timidus (Holthuis, 1960) (formerly, *Scyllarus timidus* Holthuis, 1960)	4 (incomplete)	two months	Initially fed on *Artemia* sp. nauplii then supplemented with *Dunaliella* sp.	No info	2.7–3.4	Ritz (1977)
Scyllarus planorbis Holthuis, 1969	8 (final gilled stage); no nisto	54	*Artemia* sp. nauplii	24	8.4	Robertson (1979)
Petractus demani (Holthuis, 1946)	8 + nisto + juvenile lobster instar 1 (complete)	46 to nisto	*Artemia* sp. nauplii supplemented after the 5th instar with chopped *Gafrarium* sp.	25.5	10.5	Ito & Lucas (1990)
Theninae						
Thenus orientalis (Lund, 1793)	4 + nisto + juvenile (complete)	28	Fresh flesh of *Donax brazieri*	27	18	Mikami & Greenwood (1997a)
Thenus indicus (Leach, 1816)	4 + nisto + juvenile (complete)	28	Fresh flesh of *Donax brazieri*	27	16	Mikami & Greenwood (1997a)

1994); however, survival in these studies (except the recent study of *Thenus* sp. by Mikami & Greenwood (1997a)) is always extremely low, with only a few phyllosomas successfully metamorphosing to the nisto stage. Suboptimal laboratory conditions may provide inadequate nutrition and decreased feeding frequency of phyllosomas, which, in turn, may cause slow growth and molt stage variability (Minagawa 1993; Mikami 1995). Four key factors are addressed here to gain a better understanding of the behavioral responses of oceanic phyllosomas under laboratory conditions: environmental (temperature, salinity, light); physical (tank shape, hydrodynamics, larval densities); biological (bacteria, viruses, fungi); and nutritional factors. Because these factors may act synergistically, comprehensive studies addressed at understanding the interactions between these factors are necessary to obtain optimal rearing conditions for scyllarid phyllosomal growth.

5.3.1   Environmental Factors

5.3.1.1   *Temperature*

Perhaps the most influential environmental factor affecting phyllosoma growth is temperature. Phyllosomal growth (molt increments and durations) can be accelerated with increased temperature due to the elevation of the animal's metabolic rate. However, when the optimal temperature is exceeded, reduced molt increment occurs, often resulting in reduced overall phyllosomal survival (Mikami 1995). Hence, there is an upper limit, observed through high mortalities, of the maximum desirable rearing temperature and this should be determined for each species. *Scyllarides aequinoctialis* (Lund, 1793) phyllosomas, for example, survived best at 24°C when tested within the range 10 to 30°C (Robertson 1979). *Scyllarus americanus* phyllosomas were reared under five temperature regimes between 10 and 30°C. The shortest duration of the phyllosoma stages was obtained at 25°C, although the highest survival rate through to the nisto stage was obtained at 20°C (Robertson 1968). *Scyllarus depressus* (S.I. Smith, 1881) phyllosomas showed the fastest growth at temperatures near 25°C (Robertson 1971), while in *Scyllarus arctus*, a thermal regime of 20 to 21°C provided full development to the nisto stage after 192 days (Pessani et al. 1999). *Ibacus peronii* phyllosomas reared at 20.7 and 23.3°C had higher survival at lower temperatures, despite a decreased molt frequency (Marinovic et al. 1994). It is, however, difficult to pinpoint the cause of high mortalities under higher temperature regimes; they could be caused by the limitation of temperature tolerance or by some other factor, such as poor nutritional condition and high bacterial growth in the system (Mikami 1995). Elevation of current optimal rearing temperatures may be possible if other rearing conditions are improved (Mikami 1995).

5.3.1.2   *Salinity*

Salinity is also an important environmental factor affecting phyllosomal growth. Since phyllosomas are widely distributed in offshore waters, the general oceanic salinity level (~35 ppt) is thought to be the optimal salinity level for growth. An interesting finding was made by Robertson (1968) when *Scyllarus americanus* phyllosomas were reared under three salinity regimes (25.8, 29.3, and 32.7 ppt); phyllosomas reared under the two lower salinity regimes passed through an extra molt stage before metamorphosing to the nisto stage. Mikami (1995) also observed that *Thenus orientalis* phyllosomas had significantly reduced growth and a prolonged intermolt period at lower salinity levels (20 and 25 ppt). These studies indicate that suboptimal salinities, especially lower salinity levels, result in slow phyllosomal growth, even though phyllosomas have a certain level of tolerance for salinity changes. Reduced growth at lower salinity levels is thought to be the result of increased osmoregulatory demands and lower food consumption. In general, marine crustacean larvae maintain their hemolymph hyperosmotic to seawater by a weak hyper-regulation, a mechanism considered to provide the osmotic gradient required to accomplish endosmosis of water (Mantel & Farmer 1983). Under lower salinity levels, the mechanisms of hyper-regulation may require more energy to maintain body fluids within the phyllosomas at the required osmolarity, and this energy drain may result in slower growth. Lower salinity levels have also been shown to cause lower food consumption rates in spanner crab zoeas, *Ranina ranina* (Linneaus, 1758), resulting in a prolonged intermolt period and decreased molt increment (Minagawa 1992).

5.3.1.3   Light Intensity

Light intensity is probably the single most important factor influencing the vertical migration of phyllosomas in the wild (Phillips & Sastry 1980). There are several reports that early stage phyllosomas show a strong positive phototaxis to dim light (Saisho 1966; Mikami 1995). Mikami & Greenwood (1997b) concluded that this phototactic reaction to the light source may affect feeding opportunities of phyllosomas, resulting in slower growth. However, under conditions of total darkness, there was no effect on the growth of either *Thenus orientalis* phyllosomas or nistos (which are nonfeeding stage), suggesting that light may modify their behavior, but does not necessarily affect their metabolic rate (Mikami & Greenwood 1997b).

5.3.2   Physical Factors

Phyllosomas of many scyllarid species are long-lived members of the plankton (Booth et al. 2005). For example, in the case of *Scyllarus* spp., the phyllosomal period extends for a number of months with many molting stages, and total body length changes from just over 1 mm to more than 30 mm (Robertson 1979). Optimal tank design with a consideration of hydrodynamics for water circulation is crucial for the maintenance of these fragile, long-lived phyllosomas. Because of their unique morphology (flat body and long appendages), strong aeration can damage body parts (Kittaka & Booth 2000). No water movement, or only gentle water movement, should be used in the tank. Water movement also needs to take into consideration the behavior of the phyllosomas (swimming, feeding, and phototaxis), food distribution, and accessibility to larvae in the tank system. Rearing methods previously reported can be divided into three categories, depending on water movement: static (Matsuda & Takenouchi 2005), vertical (Illingworth et al. 1997), and horizontal (Mikami 1995) (Table 5.2). Static water movement refers to traditional bowl culture, which involves daily (or twice daily) transferal of individual larva from one bowl to a freshly prepared bowl to accommodate water exchange. This method minimizes physical damage to phyllosoma limbs due to a lack of water current; however, the disadvantages are the time consuming nature of this technique and a difficulty in upscaling for commercial ventures. Vertical water current describes an upwelling column of water which accommodates the vertical movement of larvae. Upwelling systems may minimize the floor space required; however, accessibility to the larvae is greatly reduced. The raceway and donut-shaped tank belong to the horizontal water movement category, in which larvae travel in a horizontal direction propelled by small water inlets. Although floor space requirements are greatly increased, accessibility to larvae is also greatly increased, as is space available for rearing large numbers of larvae, making this the only system currently able to produce large numbers of juveniles (Mikami & Kuballa 2004).

TABLE 5.2

Rearing Methods for Phyllosomas

	Water movement		
	Static	**Vertical**	**Horizontal**
Rearing vessels	Glass bowl (beaker), Oval shape tank	Upwelling tank Parabolic tank	Donut-shape tank Raceway
Advantages	No physical damage for larvae	Compact, less floor space needed	Even flow, easy maintenance
Disadvantages	Upscaling to a larger water volume, maintenance	Upscaling to a larger water volume	Large floor space needed
Reared species	*S. americanus* (Robertson 1968), *I. ciliatus* (Takahashi & Saisho 1978), *I. novemdentatus* (Takahashi & Saisho 1978), *P. demani* (Ito & Lucas 1990), *I. peronii* (Marinovic et al. 1994), *T. orientalis* (Mikami & Greenwood 1997a)	*J. edwardsii* (Illingworth et al. 1997)	*P. japonicus* (Inoue 1981), *J. lalandii* (Kittaka 1988), *P. elephas* (Kittaka & Ikegami 1988), *T. orientalis* (Mikami & Kuballa 2004)

5.3.3   Biological Factors

5.3.3.1   *Population Density*

Larval rearing density affects the growth of phyllosomas (Mikami 1995). Solitary rearing usually yields maximum growth rates (Mikami 1995). When *Ibacus ciliatus* phyllosomas were reared individually, maximum molt increments and short intermolt periods were achieved (Mikami & Takashima 1993). Mikami (1995) investigated rearing densities for each phyllosoma instar of *Thenus orientalis*, and observed high levels of cannibalism only in cases where insufficient food was provided. This suggests that phyllosomas can be reared at high densities (>10 phyllosomas per liter) if adequate food is available.

5.3.3.2   *Microbe Presence*

Microbial control is another key issue for the rearing of long-lived oceanic larvae under controlled environments. The natural habitat of the phyllosoma is the open ocean where the water is low in nutrients and virtually pathogen free (Phillips & Sastry 1980); therefore, phyllosomas are very susceptible to microorganisms in the water column. Traditionally, a number of antibiotics (e.g., oxytetracycline, streptomycin, chloramphenicol), as well as chemicals (e.g., formalin), have been heavily used for controlling bacterial colonies in the rearing water. Although antibiotics and chemicals are strong agents for minimizing bacterial growth in the rearing water, they are not considered a suitable, long-term solution for the commercial scale rearing of phyllosomas due to microbial resistance (Kittaka & Booth 2000). Alternative sterilization methods, via the use of ultraviolet light (UV) and ozone (O_3) sterilizers, have been introduced in modern rearing techniques, and have improved survival dramatically (Mikami & Kuballa 2004). UV light can prevent microbial contamination from incoming water and maintain low microbial levels in recirculated rearing water by irradiating microorganisms upon direct contact. Ozone also can prevent microbial growth in the rearing water by the oxidization of organic matter. Ozone, however, is toxic to all organisms including phyllosomas and, therefore, ozone should either be completely removed before entering the rearing water, or must be maintained at low levels (<0.05 ppm).

Microalgae may also control nutrients and bacterial growth in the rearing water during long-term phyllosoma culture (Igarashi et al. 1991; Kittaka & Booth 2000). However, very little information on controlling bacterial growth using microalgae is available, and the potential benefits of microalgae remain largely unexplored.

5.3.4   Nutritional Factors

Nutrition is an essential factor for the successful rearing of phyllosomas. Mikami et al. (1995) directly demonstrated the importance of nutrition on molting and growth by showing that starvation and feeding duration affected the survival, intermolt period, and growth of phyllosomas of *Thenus indicus*. The results indicate that molting was delayed for a period proportionate to the first starvation, up to a period of 50% level of point-of-no-return (PNR_{50}, the initial starvation period at which 50% of larvae molted to the next instar; 1.7 days for *T. indicus* phyllosomas at 28°C). If phyllosomas are fed within the PNR period, there are no consequential effects of the initial starvation on the next instar phyllosomas in terms of larval survival, intermolt period, and growth. On the other hand, if phyllosomas are fed later than the 50% level of point-of-reserve-saturation (PRS_{50}, the initial feeding period at which 50% of larvae molted to the next instar; 4.6 days for *T. indicus* phyllosomas at 28°C), effects of subsequent starvation periods are evident on larval growth, but there are no consequential effects on the next instar phyllosomas in terms of intermolt period and survival. These results indicate that *Thenus* phyllosomas first assimilate an energy reserve for molting, with any additional reserves being available for growth.

Most attempts to rear scyllarid phyllosomas have involved the use of live *Artemia* nauplii (see Table 5.1). Similarly, zooplankton such as *Daphnia* sp. and *Sagitta* sp., fish larvae, and hydromedusa have been fed to phyllosoma, but none of the studies were able to complete the larval life cycle (Saisho 1966; Ritz & Thomas 1973). Several studies have found that successful rearing of phyllosomas was possible when fresh marine bivalves were fed to the phyllosomas (Takahashi & Saisho 1978; Ito & Lukas 1990; Mikami & Takashima 1993; Marinovic et al. 1994; Mikami & Greenwood 1997a). Takahashi & Saisho (1978) successfully reared *Ibacus ciliatus* and *Ibacus novemdentatus* phyllosomas using the short neck clam, *Tapes philippinarum* (Adams & Reeve, 1850). Ito & Lucas (1990) reported complete larval rearing of *Panulirus demani* Borradale, 1899 using the Venus clam, *Gafrarium* sp. The blue mussel, *Mytilus edulis* Linneaus, 1758, has been widely introduced in the successful rearing of several palinurid species (Kittaka & Booth 2000) and *Ibacus peronii* (Marinovic et al. 1994). The introduction of molluskan flesh seems to be a key factor contributing to the complete rearing of these phyllosomas, although Pessani et al. (1999) achieved full development to the nisto stage in *Scyllarus arctus* using whipped fish and beef instead of molluskan flesh. Mikami & Greenwood (1997a) also found that fresh bivalve flesh contributed to high survival of *Thenus orientalis* phyllosomas, with larger molt increments and shorter intermolt periods being achieved. However, this was not the case when phyllosomas were fed frozen mollusks, indicating a reduction in nutritional value during the freezing process.

Although there have been a number of biochemical analyses of *Artemia* sp. (see review by Kanazawa 2000) used to assess the nutritional value for early stage phyllosoma growth, research into the nutritional requirements of any phyllosoma species are virtually nonexistent.

Scyllarid phyllosomas are associated with various species of hydromedusae and may be feeding on the tissues of those medusae (Shojima 1963; Thomas 1963; Herrnkind et al. 1976; Phillips & Sastry 1980). Evidence lending support to the idea that the phyllosomas may be eating the tissues of medusae is provided by Sims & Brown (1968), who observed a specimen of a scyllarid phyllosoma that contained undigested nematocysts in its fecal material. Hence, such tissue may also be useful in rearing specific species of scyllarids.

5.4   Molting and Growth of Phyllosomas

Molting is a process common to all crustaceans and is essential for the growth and development of phyllosomas. During molting, the outer exoskeleton is shed to expose a soft new exoskeleton that expands and then hardens to allow for growth of the tissues within (Skinner 1985; see Horne & Tarsitano, Chapter 8 for additional information). Molting also allows for morphological changes between each stage of larval development. These changes result in appendage development and increased sensory function via increased setation on appendages (see Johnston, Chapter 6 for descriptions of changes in larval, postlarval, and adult feeding appendages and Lavalli et al., Chapter 7 for a discussion of sensory biology of scyllarids). Very few studies have been conducted on molting in scyllarid phyllosomas (Mikami & Greenwood 1997b; Mikami 2005). However, studies in adult crustaceans have revealed that molting is a complex process that is affected by a range of environmental cues and regulated by a cascade of hormonal signals (Skinner 1985). In nephropid lobsters, endocrine control of the molting cycle is considered similar in larval and adult lobsters because the physiological responses of larvae to eyestalk ablation and ecdysteroid treatment are similar (Charmantier et al. 1988; Snyder & Chang 1986). No studies have yet investigated these processes in scyllarids.

The molt cycle refers to the period between two successive molts, and has been subdivided into stages in adult crustaceans. Initially, Herrick (1909) recognized three distinct stages: molt preparation, shedding, and recently molted. Drach (1939) further developed this scheme, taking into account not only morphological changes that occurred, but also physiological changes. He recognized four stages: molt (ecdysis), postmolt (metecdysis), intermolt (anecdysis), and premolt (proecdysis). The molt itself involves the shedding of the old exoskeleton through a rapid uptake of water from the environment. This water uptake causes the old exoskeleton to lift above the new, underlying exoskeleton, and eventually to rupture at decalcified suture lines (Mykles 1980). During postmolt, further water uptake expands the new, still soft, exoskeleton; this expansion is essential for the growth of the animal, as space is made for tissue growth within the exoskeleton (Drach 1944, and see Bianchini & Ragonese, Chapter 9 for details on growth in scyllarids). Exoskeleton mineralization and hardening then occur (see Horne & Tarsitano, Chapter 8, for additional details of the mineralization process). The intermolt period is the so-called period of nonactivity, and by far is the longest stage of the molt cycle. During this time, muscle regeneration occurs, tissue is added, energy reserves, such as glycogen and lipids, are accumulated in the hemolymph and midgut for the succeeding molt, ovaries mature, and spawning generally occurs (Passano 1960). Premolt sees the atrophy of somatic muscle, the resorption of the old exoskeleton, and the formation of a new exoskeleton in preparation for the onset of ecdysis (Drach 1939, 1944).

The process of molting in crustaceans is controlled by the release of hormones produced by specialized endocrine systems. Two types of endocrine organs are associated with the molting process: traditional androgenic glands and neurosecretory cells, which are specialized neurons that form part of the central nervous system (CNS). The principle endocrine glands in crustaceans are the Y-organs, which secrete ecdysteroids, such as ecdysone that induce molting, and the mandibular organs, that secrete methyl farnesoate (MF), important for metamorphosis (Skinner 1985) and implicated in the regulation of the Y-organs (Laufer et al. 1987). Whereas ecdysone initiates the cascade of physiological and behavioral events that precede the molt, MF appears to be a status-quo hormone that maintains the phyllosoma in its immature form and prevents its metamorphosis (Abdu et al. 1998). High levels of MF retard larval growth and development, and produce "intermediate" individuals, which exhibit both larval and postlarval morphology and behavior (Abdu et al. 1998).

The neurosecretory cells in crustaceans are located in the eyestalk and referred to as the sinus gland/X-organ complex. The sinus gland is responsible for the storage and secretion of a number of neuroendocrine hormones involved in molt regulation, including molt-inhibiting hormone (MIH), mandibular-organ-inhibiting hormone (MOIH), and crustacean hyperglycemic hormone (CHH) (Fingerman 1992). The X-organ consists of a group of neurosecretory neurons responsible for the production of these hormones (Skinner 1985). MIH is considered the primary hormone responsible for the inhibition of molting in crustaceans and functions by negatively regulating ecdysteroid production by the Y-organs (Skinner 1985). The sinus gland/X-organ complex has been observed, via fluorescence microscopy, in the phyllosoma of *Panulirus japonicus* (Von Siebold, 1824) as early as the first stage (Mikami, unpublished data 1990; see Figure 5.1). In a separate study, MIH gene expression was detected in sand crab larvae (Kuballa, unpublished data 2004). These data indicate that at least some of these hormones are functional even at the earliest stages of larval development.

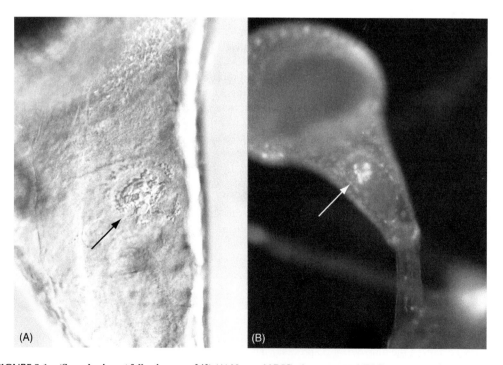

FIGURE 5.1   (See color insert following page 242) (A) Nomarski DIC microscopy and (B) fluorescence microscopy image of a first-stage phyllosoma of *Panulirus japonicus* in which the sinus gland/X-organ complex is clearly visible (arrows).

CHH is a multifunctional hormone that plays a central role in carbohydrate metabolism (Van Herp 1998); it has also been shown to be involved in molt regulation by inhibiting ecdysteroid synthesis (Sedlimeier 1985). Another important function for CHH is the facilitation of water and ion uptake at the time of molting. A dramatic surge of CHH (100-fold increase) occurs immediately prior to, and continues throughout, the molt itself, causing an influx of water that results in the swelling necessary for ecdysis to occur (Chung et al. 1999).

The control over the timing and frequency of molting resides in the physiological mechanisms that control the secretion of ecdysteroids. Within the natural environment, factors such as photoperiod and temperature clearly affect the duration and timing of the molt cycle (Skinner 1985). Research into the effect of photoperiod on molting synchrony and phyllosomal development indicates that a natural dark:light period is necessary for synchronized molting (which typically occurs around dawn) in *Thenus orientalis* and that both continuous light or dark periods result in irregular, unsynchronized molting behavior (Mikami & Greenwood 1997a). Studies have also been conducted on the effects of temperature on phyllosomal growth and molting increments (Inoue 1981; Marinovic et al. 1994; Mikami 1995). In *T. orientalis*, a rise in temperature (up to 27°C) generally facilitates a shorter molting period, while a drop (down to 18°C) in temperature extends the molting period (Mikami 1995). Survival rates of *T. orientalis* phyllosomas decrease dramatically when larvae are exposed to temperatures above 30°C or below 16°C (Mikami 1995).

Molting is required for phyllosomal growth and metamorphosis. Poor molt increment, prolonged duration of instars, and the addition of extra instars during the life cycle of phyllosomas stress the importance of developing a greater understanding of the molting process and its regulation. As the hormonal control of molting in larval scyllarids is further explored, it may be possible to manipulate endocrine pathways to facilitate the culture of scyllarid species that have a particularly long larval life cycle, such as *Scyllarides* and *Parribacus*, either by reducing instar duration or by reducing the number of stages required for metamorphosis (induction of metamorphosis).

5.5   Physiology and Rearing of Nistos

After the final planktonic stage, phyllosomas metamorphose into the transitional postlarval stage called the nisto. Both the nisto of the scyllarid lobster and the puerulus of the palinurid lobster are adult-like, nonfeeding stages that function to settle in the final habitat of the benthic juvenile/adult. The nistos of rocky-shore species, such as *Scyllarides* sp. and *Parribacus* sp., are required to swim long distances, whereas the nistos of the offshore species, such as *Thenus* sp. and *Ibacus* sp., merely settle to the bottom (see Sekiguchi et al, Chapter 4 for more information about recruitment strategies).

It is thought that the swimming capability of the nisto/puerulus correlates with the habitat of the adult stage (see Sekiguchi et al., Chapter 4 for a detailed description of nisto swimming ability). The nistos of "flat" offshore species, such as *Ibacus*, *Thenus*, and *Evibacus* species, that dwell in a sandy-bottom habitat, simply bury themselves in the sand and are not required to swim long distances. The phyllosomas of rocky-shore species, however, hatch and drift in offshore waters, then metamorphose into the nisto that swims back to the coast over days to weeks to settle in the shallow benthos (Phillips & Sastry 1980; Booth et al. 2005). The morphology of the uropods reflects the distinction between those species that are good swimmers and those that are not in the number of filaments that are attached to the uropods and the size of the uropods themselves (Barnett et al. 1986). Large uropods with many filaments result in a greater swimming potential (e.g., *Scyllarides*, *Panulirus*, *Jasus* species) than small uropods without filaments (e.g., *Thenus* and *Ibacus* species).

During the culture of *Thenus* sp. it was observed that the nistos settle immediately after metamorphosis and bury themselves in the sand (Mikami 1995). In this state, high densities of nistos can be held in small tanks (Figure 5.2). Good water movement is essential in such conditions, as static water has been observed to cause death (Mikami 1995).

There have been several studies which show that both scyllarid nistos (Mikami & Greenwood 1997a) and palinurid pueruli (Kittaka 1988; Kittaka & Ikegami 1988) do not feed. In fact, during the successful

FIGURE 5.2   *Thenus orientalis* nistos at high densities in a holding tank.

culture of *Thenus* spp., nistos were not fed throughout their entire seven-day stage (Mikami & Greenwood 1997a). The structure of the mouthparts and foregut of the nisto and puerulus support this observation of nonfeeding (see Johnston, Chapter 6 for a detailed discussion of feeding appendages and digestive capabilities of nistos), as they show poor morphological development when compared to the phyllosomas or juveniles (Nishida et al. 1990; Mikami & Takshima 1994), and are ill equipped to both ingest and digest food. This suggests that the postlarval stage relies solely on energy reserves accumulated throughout the phyllosomal stages. The nonfeeding status of nistos is supported by the presence of fat bodies in the cephalic hemocoel of pueruli of *Jasus edwardsii*, which gradually decrease in size as the pueruli age (Takahashi et al. 1994). As reduction of the fat bodies occurs, an increase of lipid droplets in the hepatopancreas is observed (Takahashi et al. 1994), suggesting that the puerulus utilizes the fat bodies as a source of metabolic energy for settlement, developmental change, and molting. Further work by Lemmens (1994) examined the change in energy reserves in phyllosomas, pueruli, and juveniles of *Panulirus cygnus* in order to demonstrate biochemically that the puerulus is, in fact, nonfeeding. Energy reserves were accumulated during the final-stage phyllosoma and were gradually depleted during the puerulus stage.

5.6   Prospects for Scyllarid Aquaculture

Recently, considerable advances in phyllosomal rearing of palinurid lobsters have been made by Japanese researchers, with the production of a few hundred juveniles per year (see summary by Mikami & Kuballa 2004). Despite these recent advances, there are still a number of technical difficulties to be overcome in developing large-scale production (>1000 juveniles); as a result, commercialization of lobster aquaculture is not yet possible. The major problems of phyllosomal rearing, in comparison with other commercially available crustacean species such as prawns and crabs, arise from a prolonged larval life, the unique and fragile phyllosoma morphology, and the fundamental lack of information on most aspects of larval biology. It is very difficult to maintain hygienic conditions in a large-scale hatchery operation during the

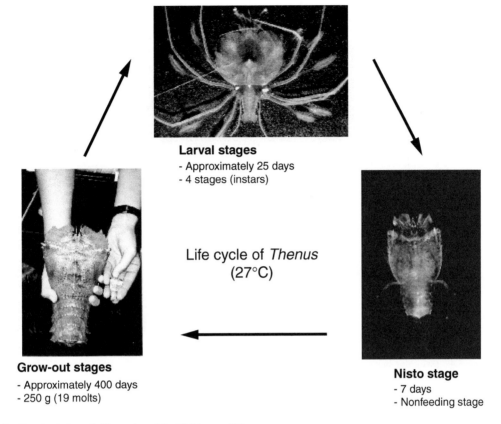

Larval stages
- Approximately 25 days
- 4 stages (instars)

Life cycle of *Thenus*
(27°C)

Grow-out stages
- Approximately 400 days
- 250 g (19 molts)

Nisto stage
- 7 days
- Nonfeeding stage

FIGURE 5.3   Schematic illustration of the life history of *Thenus* sp.

lengthy larval phase, and current operations are labor intensive. The long appendages of the phyllosomas are easily damaged in confined conditions, by aeration, or strong water movement, requiring detailed studies into the hydrodynamics associated with tank design.

In the Scyllaridae, there is a wide range in the duration of the larval stages: some *Scyllarus*, *Ibacus*, and *Thenus* species hatch at a more advanced stage than palinurids, and thus have shorter larval stages (one to two months) (see Table 5.1). This short larval life and the hardy nature of the larvae, compared to the phyllosomas of palinurids, have aided in the development of large-scale culture of *Thenus* sp. in recent years. A pilot system for growing *Thenus* sp. phyllosomas has been operating successfully at the Australian Fresh Research & Development Corporation Pty, Ltd (AFR&DC), and the establishment of a commercial-scale operation is now underway. The key for success is perhaps the relatively short phyllosoma period and fast juvenile growth (Figure 5.3). The following is a summary of *Thenus* culture protocol at AFR&DC.

5.6.1   Broodstock Management and Hatching

Berried females are caught from the wild throughout the year from Australian waters. Females usually spawn twice per year; in some cases, the second spawning occurs within a few days of hatching. Holding temperature for berried females is between 24 and 28°C. Different stages of egg development are characterized by egg coloration (Figure 5.4), while an accurate estimation of the date of larval hatching can be determined by microscopic observation (Figure 5.5).

Hatching occurs just before sunrise. Larvae exhibit a short nauplisoma stage and then molt into the first instar within a few minutes. Larvae are phototactic, so harvesting is made easy with the use of a light

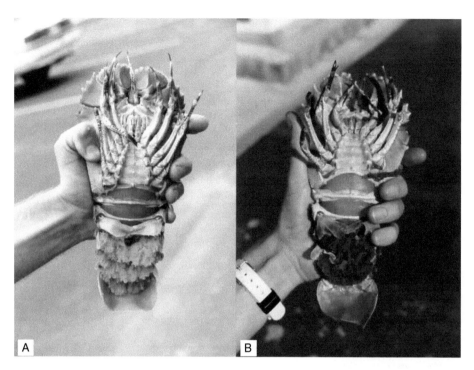

FIGURE 5.4 (**See color insert following page 242**) Different stages of egg development. (A) Newly spawned–orange-colored egg mass indicates early embryonic development, (B) prehatching–brown colored egg mass indicates appearance of eye pigment of embryo and reduction of yolk.

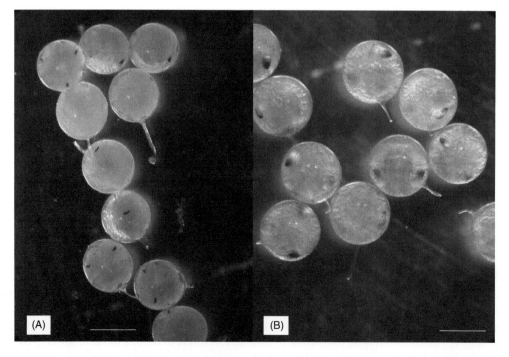

FIGURE 5.5 (**See color insert following page 242**) Microscopic images of egg development. (A) Newly spawned eggs, (B) pre-hatching eggs. Scale bars indicate 1 mm.

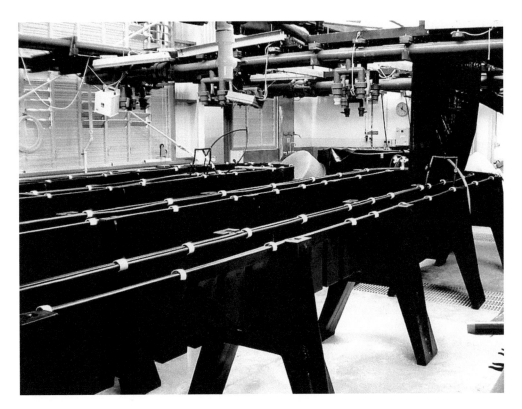

FIGURE 5.6 Photo of prototype-300 mm wide raceways for larval rearing.

source to which they are attracted. Larvae are carefully harvested from the hatching tank using a small container (not via a net), and transferred to the larval rearing system. Larvae are stocked at a density of approximately 10 larvae per liter of water.

5.6.2 Larval Rearing

Newly hatched phyllosomas pass through only four molt stages within four weeks before metamorphosing to the settling nisto stage. Sufficient exchange of good quality water is essential for the naturally oceanic phyllosomas. At high-rearing densities, the delicate phyllosomas are easily damaged by physical abrasion caused by water circulation, aeration, and structure of the rearing tank. Therefore, the rearing tank design must provide high-quality water while, at the same time, allowing the phyllosomas to remain in suspension without causing physical damage (Figure 5.6). A constant supply of an adequate quantity of quality food is also crucial to the success of a large-scale operation of phyllosoma rearing.

In the case of *Thenus* sp., feeding of mollusk flesh results in high survival and optimal growth of phyllosomas. The optimal rearing temperature for *Thenus* phyllosomas is 27°C. Molting into the next phyllosoma instar occurs just before sunrise. Phyllosomas of *Thenus* sp. are hardy compared to any other palinurid or scyllarid species, but they are still vulnerable at the time of molting; this is when the majority of mortalities are likely to occur. Understanding and controlling the molting process is, therefore, the key to obtaining high survival throughout the operation (see Section 5.4).

Just as understanding the physiological requirements of phyllosomas is essential for larval rearing, the further culture of scyllarids through to the juvenile stage requires knowledge of the basic biology of the postlarval and juvenile stages. Metamorphosis into the nonfeeding nisto stage occurs just after sunset. The nistos of scyllarid lobsters require the least maintenance of all of the life stages. Their

FIGURE 5.7   Photo showing high rearing density of *Thenus* juveniles.

lack of nutritional requirements and their inclination to settle immediately after metamorphosis make this stage ideally suited for high-density stocking. This is very important in any large-scale aquaculture production facility, as tank size and husbandry costs are principle considerations for any commercial facility.

5.6.3   Juvenile Grow-Out

Grow-out duration is relatively short, with juveniles reaching a typical market size of about 250 g within 400 days after hatching (Mikami 1995), although the time from hatch to market size can be altered by temperature and food supply. *Thenus* sp. juveniles also can be reared at high density due to their placid nature (Figure 5.7 and Figure 5.8), even though they are widely dispersed on the sandy bottom of the ocean (Kailola et al. 1993). Juveniles of *Thenus* sp. prefer mollusks and, in the case of grow-out trials, chopped frozen squid flesh was fed to obtain optimal growth.

5.7   Conclusions

Key issues concerning the biological and technical aspects of scyllarid larval rearing have been addressed within this chapter. Future successful aquaculture (both in the laboratory and large-scale commercial operations) of scyllarids relies heavily on filling in the "biological gaps" that currently exist in our understanding of the larval life cycle, molting, and nutritional needs. The high survival through to nisto and the mass larval-rearing capabilities demonstrated by recent research into *Thenus* sp., provide insight into the direction of further research, and provide a basis for the potential large-scale rearing of scyllarid lobsters.

In order to successfully develop the aquaculture of scyllarid species, biological, technical, and economic issues should be addressed. The critical biological needs include the identification of factors affecting phyllosoma growth and the ways in which larval and juvenile nutrition can be optimized. Technical aspects include the development of productive, large-scale rearing systems with appropriate hydrodynamics and control of microbial conditions within the system. Economics drive the development of large-scale aquaculture for any species and include the evaluation of market price and the evaluation of costs for optimal operations.

FIGURE 5.8 Photo of *Thenus* juveniles, demonstrating their suitability for high-density culture.

References

Abdu, U, Takac, P., Laufer, H., & Sagi, A. 1998. Effect of methyl farnesoate on late larval development and metamorphosis in the prawn *Macrobrachium rosenbergii* (Decapoda, Palaemonidae): A juvenoid-like effect? *Biol. Bull. (Woods Hole, Mass.)* 195: 112–119.

Atkinson, J. & Boustead, N.C. 1982. The complete larval development of the scyllarid lobster *Ibacus alticrenatus* Bate, 1888 in New Zealand waters. *Crustaceana* 42: 275–287.

Baisre, J.A. 1994. Phyllosoma larvae and the phylogeny of the Palinuridae (Crustacea: Decapda): A review. *Aust. J. Mar. Freshw. Res.* 45: 925–944.

Barnett, B.M., Hartwick, R.F., & Milward, N.W. 1986. Descriptions of the nisto stage of *Scyllarus demani* Holthuis, two unidentified *Scyllarus* species, and the juvenile of *Scyllarus martensii* Pfeffer (Crustacea: Decapoda: Scyllaridae), reared in the laboratory; and behavioural observation of the nistos of *S. demani*, *S. martensii*, and *Thenus orientalis. Aust. J. Mar. Freshw. Res.* 37: 595–608.

Booth, D.J., Webber., W.R., Sekiguchi, H., & Coutures, E. 2005. Diverse larval recruitment strategies within the Scyllaridae. *N. Z. J. Mar. Freshw. Res.* 39: 581–592.

Charmantier, G. & Aiken, D.E. 1987. Intermediate larval and postlarval stages of *Homarus americanus* H. Milne Edwards, 1837 (Crustacea: Decapoda). *J. Crust. Biol.* 7: 525–535.

Charmantier, G., Charmantier-Daures, M., & Aiken, D.E. 1988. Larval development and metamorphosis of the American lobster *Homarus americanus* (Crustacea: Decapoda): Effect of eyestalk ablation and juvenile hormone injection. *Gen. Comp. Endocrinol.* 70: 319–333.

Chung, J.S., Dircksen, H., & Webster, S.G. 1999. A remarkable, precisely timed release of hyperglycemic hormone from endocrine cells in the gut is associated with ecdysis in the crab *Carcinus maenas*. *Proc. Natl. Acad. Sci. U.S.A.* 96: 13103–13107.

Crosnier, A. 1972. Naupliosoma, phyllosomes et pseudibacus de *Scyllarides herklotsi* (Herklots) (Crustacea, Decapoda, Scyllaridae) recoltes par l'ombango dans le sud Golfe de Guinee. *Cahiers ORSTOM Oceanogr.* 10: 139–149.

Drach, P. 1939. Mue et cycle d'intermue chez les Crustacé Décapodes. *Ann. Inst. Océanog. (Paris)* 18: 103–391.

Drach, P. 1944. Etude préliminaire sur le cycle d'intermue et son conditionnement hormonal chez *Leander serratus* (Pennant). *Bull. Biol. Fr. Belg.* 78: 40–62.

Ennis, G.P. 1995. Larval and postlarval ecology. In: Factor, J.R. (ed.), *Biology of the Lobster Homarus americanus*: pp. 23–46. New York: Academic Press.

Factor, J.R. 1995. *Biology of the Lobster Homarus americanus*. New York: Academic Press.

Fingerman, M. 1992. Glands and secretion. In: Frederick, W. & Humes, A.G. (eds.), *Microscopic Anatomy of Invertebrates, Vol. 10, Decapod Crustacea*: pp. 345–394. New York: Wiley-Liss.

Harada, E. 1958. Notes on the nauplisoma and newly-hatched phyllosoma of *Ibacus ciliatus* (von Siebold). *Pub. Seto Mar. Biol. Lab.* 7: 173–180.

Herrick, F.H. 1909. Natural history of the American lobster. *Bull. U.S. Bur. Fish.* 29: 149–408.

Herrnkind, W.F., Halusky, J., & Kanciruk, P. 1976. A further note on phyllosoma larvae associated with medusa. *Bull. Mar. Sci.* 26: 110–112.

Igarashi, M.A., Romero, S.F., & Kittaka, J. 1991. Bacteriological character in the culture water of penaeid, homarid and palinurid larvae. *Nippon Suisan Gakkaishi* 57: 2255–2260.

Illingworth, J., Tong, L.J., Moss, G.A., & Pickering, T.D. 1997. Upwelling tank for culturing rock lobster (*Jasus edwardsii*) phyllosomas. *Mar. Freshw. Res.* 48: 911–914.

Inoue, M. 1981. Studies on the cultured phyllosoma larvae of the Japanese spiny lobster, *Panulirus japonicus* (V. Siebold). *Special Rep. Kanagawa Pref. Fish. Exp. Stn.* 1: 1–91.

Ito, M. & Lucas, J.S. 1990. The complete larval development of the scyllarid lobster, *Scyllarus demani* Holthuis, 1946 (Decapoda, Scyllaridae), in the laboratory. *Crustaceana* 58: 144–167.

Kailola, P.J., William, M.J., Stewart, P.C., Reichelt, R.E., McNee, A., & Grieve, C. 1993. *Australian Fisheries Resources*: pp. 168–170. Canberra: Bureau of Resource Sciences, Department of Primary Industries and Energy/Fisheries Research and Development Corporation.

Kanazawa, A. 2000. Nutrition and food. In : Phillips, B.F. & Kittaka, J. (eds.), *Spiny Lobsters Fisheries and Culture*, 2nd Edition: pp. 611–624. Oxford: Blackwell Scientific.

Kittaka, J. 1988. Culture of the palinurid *Jasus lalandii* from egg stage to puerulus. *Nippon Suisan Gakkaishi* 54: 87–93.

Kittaka, J. & Booth, J.D. 2000. Prospectus for aquaculture. In: Phillips, B.F. & Kittaka, J. (eds.), *Spiny Lobsters Fisheries and Culture*, 2nd Edition: pp. 465–473. Oxford: Blackwell Scientific.

Kittaka, J. & Ikegami, E. 1988. Culture of the palinurid *Palinurus elephas* from egg stage to puerulus. *Nippon Suisan Gakkaishi* 54: 1149–1154.

Laufer, H., Landau, M., Homola, E., & Borst, D.W. 1987. Methyl Farnesoate: Its site of synthesis and regulation of secretion in a juvenile crustacean. *Insect Biochem.* 17: 1129–1131.

Lawton, P. & Lavalli, K.L. 1995. Postlarval, juvenile, adolescent, and adult ecology. In: Factor, J.R. (ed.), *Biology of the Lobster Homarus americanus*: pp. 47–88. New York: Academic Press.

Lemmens, J.W.T.J. 1994. The western rock lobster *Panulirus cygnus* (George, 1962) (Decapoda: Palinuridae): The effect of temperature and developmental stage on energy requirements of pueruli. *J. Exp. Mar. Biol. Ecol.* 180: 221–234.

Lemmens, J.W.T.J. & Knott, B. 1994. Morphological changes in external and internal feeding structures during the transition phyllosoma-puerulus-juvenile in the western rock lobster (*Panulirus cygnus*, Decapoda: Palinuridae). *J. Morph.* 220: 271–280.

Lesser, J.H.R. 1974. Identification of early larvae of New Zealand spiny and shovel-nosed lobsters (Decapoda, Palinuridae and Scyllaridae). *Crustaceana* 27: 259–277.

Mantel, L.H. & Farmer, L.L. 1983. Osmotic and ionic regulation. In: Mantel, L.H. (ed.), *The Biology of Crustacea: Internal Anatomy and Physiological Regulation*, Vol. 5: pp. 53–161. New York: Academic Press.

Marinovic, B., Lemmens, J.W.T.J., & Knott, B. 1994. Larval development of *Ibacus peronii* Leach (Decapoda: Scyllaridae) under laboratory conditions. *J. Crust. Biol.* 14: 80–96.

Martins, H.R. 1985. Biological studies on the exploited stock of the Mediterranean locust lobster *Scyllarides latus* (Latreille, 1803) (Decapoda: Scyllaridae) in the Azores. *J. Crust. Biol.* 5: 294–305.

Matsuda, H. & Takenouchi, T. 2005. New tank design for larval culture of Japanese spiny lobster, *Panurlirus japonicus*. *N. Z. J. Mar. Freshw. Res.* 39: 279–285.

Mikami, S. 1995. *Larviculture of Thenus (Decapoda, Scyllaridae), the Moreton Bay Bugs.* Ph.D. thesis: University of Queensland, Brisbane, Australia.

Mikami, S. 2005. Moulting behaviour responses of Bay lobster, *Thenus orientalis*, to environmental manipulation. *N. Z. J. Mar. Freshw. Res.* 39: 297–292.

Mikami, S. & Greenwood, J.G. 1997a. Complete development and comparative morphology of larva *Thenus orientalis* and *Thenus* sp. (Decapoda: Scyllaridae) reared in the laboratory. *J. Crust. Biol.* 17: 289–308.

Mikami, S. & Greenwood, J.G. 1997b. Influence of light regimes on phyllosomal growth and timing of moulting in *Thenus orientalis* (Lund) (Decapoda: Scyllaridae). *Mar. Freshw. Res.* 48: 777–782.

Mikami, S. & Kuballa, A. 2004. Overview of lobster aquaculture research. In: Kolkovski, S., Heine, J., & Clarke, S. (eds.), *Proceedings of the Second Hatchery Feeds and Technology Workshop*: pp. 132–135. Sydney: Western Australian Department of Fisheries — Research Division.

Mikami, S. & Takashima, F. 1993. Development of the proventriculus in larvae of the slipper lobster, *Ibacus ciliatus* (Decapoda: Scyllaridae). *Aquaculture* 116: 199–217.

Mikami, S. & Takashima, F. 1994. Functional morphology and cytology of the phyllosomal digestive system of *Ibacus ciliatus* and *Panulirus japonicus* (Decapoda, Scyllaridae and Palinuridae). *Crustaceana* 67: 212–225.

Mikami, S., Greenwood, J.G., & Gillespie, N.C. 1995. The effect of starvation and feeding regimes on survival, intermoult period and growth of cultured *Panulirus japonicus* and *Thenus* sp. phyllosomas (Decapoda: Palinuridae and Scyllaridae). *Crustaceana* 68: 160–169.

Minagawa, M. 1992. Effects of salinity on survival, feeding, and development of larvae of the red frog crab *Ranina ranina*. *Nippon Suisan Gakkaishi* 58: 1855–1860.

Minagawa, M. 1993. *Studies on Larval Rearing of the Red Frog Crab Ranina ranina (Crustacea, Decapoda, Raninidae).* Ph.D. thesis: Tokyo University of Fisheries, Tokyo, Japan.

Mykles, D.L. 1980. The mechanism of fluid absorption at ecdysis in the American lobster, *Homarus americanus*. *J. Exp. Biol.* 84: 89–101.

Nishida, S., Quigley, B.D., Booth, J.D., Nemoto, T., & Kittaka, J. 1990. Comparative morphology of the mouthparts and foregut of the final-stage phyllosoma, puerulus, and postpuerulus of the rock lobster *Jasus edwardsii* (Decapoda: Palinuridae). *J. Crust. Biol.* 10: 293–305.

Passano, L.M. 1960. Molting and its control. In: Waterman, T.H. (ed.), *The Physiology of Crustacea*, Vol. 1: pp. 473–536. New York: Academic Press.

Pessani, D., Pisa, G., & Gattelli, R. 1999. The complete larval development of *Scyllarus arctus* (Decapoda, Scyllaridae) in the laboratory. *Abstracts of the 7th Colloquium Crustacea Decapoda Mediterranea*, Lisbon, September 6–9, 1999. pp. 143–144.

Phillips, B.F. & Sastry, A.N. 1980. Larval ecology. In: Cobb, J.S. & Phillips, B.F. (eds.), *The Biology and Management of Lobsters*, Vol. 2: pp. 11–57. New York: Academic Press.

Phillips, B.F., Brown, P.A., Rimmer, D.W., & Braine, J.S. 1981. Description, distribution and abundance of late larval stages of the Scyllaridae (slipper lobster) in the south-eastern Indian Ocean. *Aust. J. Mar. Freshw. Res.* 32: 417–437.

Ritz, D.A. 1977. The larval stage of *Scyllarus demani* Holthuis with notes on the larvae of *S. sordidus* (Stimpson) and *S. timidus* Holthuis (Decapoda, Palinuridae). *Crustaceana* 32: 229–240.

Ritz, D.A. & Thomas, L.R. 1973. The larval and postlarval stages of *Ibacus peronii* Leach (Decapoda, Reptantia, Scyllaridae). *Crustaceana* 24: 5–17.

Robertson, P.B. 1968. The complete larval development of the sand lobster, *Scyllarus americanus* (Smith), (Decapoda, Scyllaridae) in the laboratory with notes on larvae from the plankton. *Bull. Mar. Sci. Gulf Caribb.* 18: 294–342.

Robertson, P.B. 1969. The early larval development of the sand lobster, *Scyllarides aequinoctalis* (Lund) in the laboratory, with a revision of the larval characters of the genus. *Deep Sea Res.* 16: 557–586.

Robertson, P.B. 1971. The larvae and postlarva of the scyllarid lobster *Scyllarus depressus* (Smith). *Bull. Mar. Sci.* 29: 841–865.

Robertson, P.B. 1979. Larval development of the scyllarid lobster *Scyllarus planorbis* Holthuis reared in the laboratory. *Bull. Mar. Sci.* 29: 320–328.

Saisho, T. 1966. Studies on the phyllosoma larvae with reference to the oceanographical conditions. *Mem. Fac. Fish. Kagoshima Univ.* 15: 177–239.

Saisho, T. & Nakahara, K. 1960. On the development of *Ibacus ciliatus* and *Panulirus longipes*. *Mem. Fac. Fish. Kagoshima Univ.* 9: 84–90.

Saisho, T. & Sone, M. 1971. Notes on the early development of a scyllarid lobster, *Scyllarides squamosus* (H. Milne-Edwards). *Mem. Fac. Fish. Kagoshima Univ.* 20: 191–226.

Sandifer, P.A. 1971. The first two phyllosomas of the sand lobster, *Scyllarus depressus* (Smith) (Decapoda, Scyllaridae). *J. Elisha Mitchell Sci. Soc.* Winter: 183.

Sedlmeier, D. 1985. Mode of action of the crustacean hyperglycemic hormone. *Am. Zool.* 25: 223–232.

Shojima, Y. 1963. Scyllarid phyllosomas' habit of accompanying the jelly-fish (preliminary report). *Bull. Japan Soc. Sci. Fish.* 29: 349–353.

Sims, H.W. & Brown, C.L. 1968. A giant scyllarid phyllosoma larvae taken north of Bermuda (Palinuridae). *Crustaceana Suppl.* 2: 80–82.

Skinner, D.M. 1985. Moulting and regeneration. In: Bliss, D.E. (ed.), *The Biology of Crustacea: Integument, Pigments, and Hormonal Processes*, Vol. 9: pp. 44–128. New York: Academic Press.

Snyder, M.J. & Chang, E.S. 1986. Effects of eyestalk ablation on larval molting rates and morphological development of the American lobster, *Homarus americanus*. *Biol. Bull. (Woods Hole, Mass.)* 170: 232–243.

Takahashi, M. & Saisho, T. 1978. The complete larval development of the scyllarid lobster, *Ibacus ciliatus* (von Siebold) and *Ibacus novemdentatus* Gibbes in the laboratory. *Mem. Fac. Fish. Kagoshima Univ.* 27: 305–353.

Takahashi, Y., Nishida, S., & Kittaka, J. 1994. Histological characteristics of fat bodies in the puerulus of the rock lobster *Jasus edwardsii* (Hutton, 1875) (Decapoda, Palinuridae). *Crustaceana* 66: 318–325.

Templeman, W. 1936. Fourth stage larvae of *Homarus americanus* intermediate in form between normal third and fourth stages. *J. Biol. Board Can.* 2: 249–354.

Thomas, L.R. 1963. Phyllosoma larvae associated with medusae. *Nature* 198: 208.

Van Herp, F. 1998. Molecular, cytological and physiological aspects of the crustacean hyperglycemic hormone family. In: Coast, G.M. & Webster, S.G. (eds.), *Recent Advances in Arthropod Endocrinology*: pp. 53–70. Cambridge: Cambridge University Press.

6

Feeding Morphology and Digestive System of Slipper Lobsters

Danielle Johnston

CONTENTS

Abstract

Slipper lobsters are a unique group characterized by their distinctive, dorsoventrally flattened body. The mouthpart and digestive system morphology of larval, juvenile, and adult slipper lobsters are summarized in this chapter, and dietary preferences, feeding behavior, ingestion mechanisms, and digestive system function are inferred from the morphology. Bivalve mollusks are the preferred diet of scyllarids and are wedged open using sharp pereiopod dactyls. Adult mouthparts are not heavily calcified, and a shearing action of the asymmetrical mandibles masticates flesh. A large, prominent, membranous lobe (modified metastome) is characteristic of scyllarids and plays an important role in ingestion to lubricate food and allow quick and efficient swallowing of large items by retraction of its anterior lip. Compared with palinurid and nephropid lobsters, the cardiac stomach of adult slipper lobsters is simple with reduced ossicles and simple ventral filtration channels. The gastric mill teeth are well-developed with large incisor processes and chitinous plates. It is a simplified cutting apparatus and the major masticatory action is a cutting action by the lateral teeth, with only a minor role for the medial tooth. Digestive gland structure and function is similar to other lobsters with high concentrations of proteolytic enzymes consistent with a carnivorous diet. Dietary preferences, feeding, and digestion in phyllosoma remain poorly understood. Phyllosoma mouthparts are well-developed from stage I, with sharp incisor processes on the mandibles and well-developed setation on the first maxillae and second maxillipeds. Mastication mostly occurs through the action of well-developed incisor and molar mandibular processes. In contrast to nephropid lobster larvae, which lack the teeth of the gastric mill until stage III, the foregut of the scyllarid phyllosoma lacks a gastric mill in all stages and does not grind food, but functions mainly to sort and filter food particles previously masticated by the mandibles. These characteristics are suggestive of a gelatinous diet that requires little additional mastication by internal teeth.

6.1   Introduction

Decapod lobsters include some of the world's most important commercial species. Rapidly expanding demand and concern about the sustainability of any increase in exploitation of existing wild fisheries has sparked considerable interest in culture and expansion of fisheries into unexploited species. Scyllarid or slipper lobsters are a relatively unexploited group of lobsters that may be a candidate for culture or that may provide a new fishery with scope for expansion. Unfortunately, much of the basic biological information that we take for granted when considering other species has not been collected for the scyllarid lobsters. This severely limits our capacity to culture the animals and manage their fisheries effectively.

The natural diet of slipper lobsters remains one of these significant gaps in our understanding. An improved understanding of the functional morphology and ontogeny of slipper lobster mouthparts and digestive system provides an indication of digestive capacity and possibly an insight into dietary preferences. This knowledge can be used to assess natural prey species and preferred habitat, and may help to determine the suitability of particular food types for larval and grow-out diets. Understanding digestive system function can also be used to indicate whether commercial diets that are presently available for members of other lobster families are likely to be properly utilized by phyllosomas and juveniles of scyllarids.

There are many publications reviewing the feeding morphology and digestive system of decapods in general (Dall & Moriarty 1983; Factor 1989; Dall et al. 1990; Felgenhauer 1992; Icely & Nott 1992). The majority of research has focused on the adult stages, although a recent comprehensive account has been published on larval decapods (Anger 2001). Of the lobsters, most is known about the diet, mouthparts, and digestive system of nephropid (clawed) lobsters (Factor 1995) and, more recently, palinurid (spiny) lobsters (general reviews by Mikami & Takashima 2000; Cox & Johnston 2003a), but little attention has focused on the scyllarid lobsters.

This chapter summarizes our current understanding of the mouthpart and digestive system morphology of slipper lobsters, including larval, juvenile, and adult stages of development, where possible. The chapter begins by outlining what is currently known of the dietary preferences and feeding behavior of slipper lobsters as these relate closely to morphological and functional aspects of the digestive system. Next, it details the current knowledge of mouthpart structure and ingestion mechanisms, and then details digestive system structure and function in slipper lobster adults and phyllosomas. Finally, the chapter concludes by summarizing the major findings in each area and, in doing so, highlights gaps in our knowledge of scyllarid digestive system functional morphology. Although the focus is entirely on scyllarids, comparisons with palinurid lobsters have often been made in the literature and will be included where appropriate. While similarities do exist among genera, care has been taken not to extrapolate information from other species-specific studies to scyllarids in general.

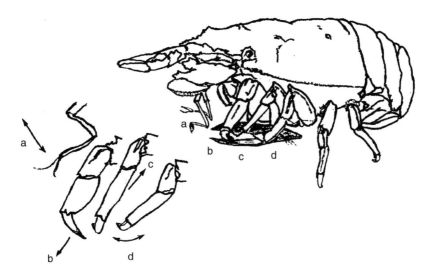

FIGURE 6.1   The orientation and direction of movement of the pereiopods during wedging of the bivalve *Isognomon incisum* (T.A. Conrad, 1837) by the slipper lobster, *Scyllarides squammosus* (H. Milne Edwards, 1837). (a) Antennae are continually flicked up and down; (b) the dactyls of the first pereiopods create a wedge that is forced in the lip of the shell, and the shell is pushed outwards; (c) the second pereiopods pull the opposite valve toward the lobster; and (d) the third pereiopods scrape the adductor muscle. From Lau 1987; used with permission.

6.2   Diet and Foraging Behavior

To exploit a wide range of habitats, decapod crustaceans have developed an extensive repertoire of feeding strategies that generally involves a highly proficient system of food acquisition and ingestion, which they exercise selectively, depending on the food available. These strategies include filter and suspension feeding, scavenging, predation, and parasitism. Scyllarid juveniles and adults are primarily carnivorous scavengers with a preference for small benthic invertebrates, including mollusks, polycheates, and crustaceans (Suthers & Anderson 1981; Lau 1988; Johnston & Yellowlees 1998). *Scyllarides* spp., *Parribacus* spp., and *Arctides* spp. specialize in mollusks, and the bivalves *Glycymeris pilosus* (Linnaeus, 1767), *Ostrea sandvicensis* (Sowerby, 1871), *Isognomon incisum* (T.A. Conrad, 1837), and *Venus verrucosa* Linnaeus, 1758 are preferentially selected (Lau 1987, 1988; Spanier 1987). *Thenus orientalis* (Lund, 1793) preferentially selected scallops (*Amusium* sp.) when presented with a variety of species from the most abundant macrofaunal groups associated with this scyllarid (Jones 1988; Johnston & Yellowlees 1998). *Scyllarides latus* (Latreille, 1802) appears similarly specialized for feeding on mollusks with the bivalves *V. verrucosa*, *G. pilosus*, and *Pinctada radiata* (Leach, 1814) found near dens and readily eaten in the laboratory (Spanier 1987). Limpet radulae have also been found in the gut of *S. latus*, and limpets were readily ingested when fed in the laboratory (Martins 1985).

Decapods that preferentially feed on mollusks, principally bivalves, have developed complex attack behaviors and specialized structures to capture, open, and manipulate their prey to remove edible flesh. These attack behaviors are detailed elsewhere in this chapter and book (see Section 6.3.4 and Lavalli et al., Chapter 7). Nonchelate decapods, such as slipper lobsters, have adopted specialized tactics for opening the shell of bivalves. A common technique adopted by many species of slipper lobster, including *T. orientalis* and *Scyllarides squammosus* (H. Milne Edwards, 1837), is "wedging." The edible flesh is removed by a progressive prying apart of the shell valves using the sharp, chitinized, pereiopod dactyls (Lau 1987; Spanier et al. 1988; Johnston & Yellowlees 1998) (Figure 6.1).

The natural diet of slipper lobster phyllosoma is currently unknown, although phyllosomas of *Ibacus* spp. have been associated with scyphozoan medusae, *Aurelia aurita* (Linnaeus, 1758), revealing that they may prefer soft, fleshy, gelatinous prey (Herrnkind et al. 1976). Slipper lobster phyllosomas have been successfully grown in culture on a mixture of pippi (a bivalve), scallop flesh, and *Artemia* nauplii (Mikami & Greenwood 1997), or fresh mussel flesh (Marinovic et al. 1994) (see Mikami & Kuballa, Chapter 5 for more information on phyllosomal culture). However, in these cases, preference and suitability was not determined. Preference studies aimed at determining the diet of spiny (palinurid) lobster phyllosomas have shown a preference for medusae, ctenophores, salps, chaetognaths, and other soft-bodied foods (Mitchell 1971), and a number of recent studies on palinurid phyllosoma speculate that soft-bodied zooplankton, as well as krill and fish larvae, may be potentially ingested (Johnston & Ritar 2001; Cox & Johnston 2003a; Johnston et al. 2004). Relative success in culture has also been achieved for palinurid phyllosomas using *Artemia* nauplii and chopped mussel gonad (Kittaka 1994, 1997; Tong et al. 1997).

6.3   Mouthpart Structure and Ingestion Mechanisms

Slipper lobsters have evolved specialized feeding appendages to maximize their functional efficiency. Hence, the overall morphology of mouthparts is closely related to diet and can be used as a general indicator of dietary preference. One of the most important adaptations is setation, the morphology, position, and innervation of which provides an indication of its role in feeding and grooming (see related lobster species literature in Derby 1982; Lavalli & Factor 1992, 1995).

There are very few studies describing mouthpart structure and ingestion of adult scyllarids, the most notable being on *Ibacus peronii* Leach, 1815 (Suthers & Anderson 1981) and *Thenus orientalis* (Johnston 1994; Johnston & Alexander 1999). There are, however, several studies describing the mouthpart structure of larval stages, the majority of which are incorporated with general morphological descriptions of larval development (Atkinson & Boustead 1982; Barnett et al. 1984; Phillips & McWilliam 1986; Ito & Lucas 1990; Marinovic et al. 1994; Mikami & Greenwood 1997; Coutures 2001; Webber & Booth 2001). The only detailed study on the mouthparts of slipper lobster phyllosoma was conducted by Mikami et al. (1994) on *Ibacus ciliatus* (Von Siebold, 1824).

6.3.1   Phyllosoma Mouthpart Structure

Slipper lobster phyllosomas have six pairs of mouthparts: mandibles, first and second maxillae, and first, second, and third maxillipeds, although, as in palinurid phyllosomas, the first maxilliped is rudimentary (Barnett et al. 1986; Ito & Lucas 1990; Mikami & Greenwood 1997). All mouthparts are present from stage I and generally increase in the coverage and complexity of setation during development. The large labrum and paragnaths form a well-developed, semienclosed chamber covering the mandibles (Mikami et al. 1994) (Figure 6.2A). The mandibles are not well described, possibly because they are difficult to examine as they are covered by maxilla I. They are divided into three portions: the upper portion has a relatively blunt, multipronged, canine-like process, the middle portion has multiple sharp incisor teeth along the inner medial edge, and the lower portion has a flattened molar process with small tubercles on its surface (Mikami et al. 1994) (Figure 6.2B). Although not stated, it is likely that the mandibles are fully developed in stage I larvae, as is the case for palinurid phyllosomas (Wolfe & Felgenhauer 1991; Johnston & Ritar 2001; Cox & Johnston 2003b). The paragnaths have small denticles and pores covering the medial surface, the latter of which are connected to tegumental glands from which mucus is secreted (Mikami & Takashima 1993; Mikami et al. 1994).

Maxilla I is biramous with the basal endite of stage I larvae bearing two, coarse, elongated, terminal spines covered in spinules, and the coxal endite bearing one long and one short terminal spine (Ito & Lucas 1990). The number of setae on either the basal and coxal endite differs among species; for example, the basal endite of *Thenus* sp. bears three spines and the coxal endite bears two spines (Barnett et al. 1986; Mikami & Greenwood 1997), while the basal endite of *Scyllarus* sp. bears two spines and the coxal endite bears three spines (Webber & Booth 2001). During development, the number of terminal spines increases to four on the basal endite and up to three on the coxal endite, with many smaller setae present by stage IV, but remains relatively constant thereafter (Figure 6.2A).

Maxilla II is slightly removed from the rest of the inner mouthparts and is a paddle-shaped lobe bearing between two and four long plumose setae, although the latter number is most common (Barnett et al. 1984;

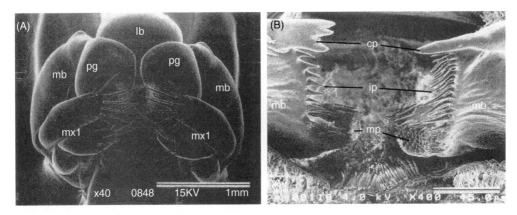

FIGURE 6.2 (A) Scanning electron micrograph of the inner mouthparts of the stage VI phyllosoma of the slipper lobster, *Ibacus ciliatus*, aboral view. Scale 1 mm. lb, labrum; mb, mandibles; mx1, first maxillae; pg, paragnaths. (B) Scanning electron micrograph of the mandibles of the stage VI phyllosoma of the slipper lobster, *Ibacus ciliatus*. Scale, 45 μm. cp, canine-like process; ip, incisor process; mb, mandibles; mp, molar process. From Mikami et al. 1994; used with permission.

Phillips & McWilliam 1986; Ito & Lucas 1990; Mikami & Greenwood 1997). By stage VI, the second maxilla begins to expand distally and the plumose setae appear to disappear (although closer examination with SEM may be needed to confirm this) (Ito & Lucas 1990). By stage X, maxilla II is larger and flatter with anterior and posterior lobes (Ito & Lucas 1990; Webber & Booth 2001), and more closely resembles the scaphognathite of adult scyllarids.

Maxilliped II has no exopod and consists of five segments, with a number of spines and setae on the distal two to three segments (Barnett et al. 1984; Webber & Booth 2001). The number and robustness of these setae and spines increase during development, although their structural detail has not been described. Maxilliped III is very elongated, with several setae distributed on each of the five segments, but most on the distal segments. A ventral coxal spine is present from stage I, with comb-like setae on the distal segment (Barnett et al. 1984). By stage IV, an exopod bud has formed that gets progressively larger during development (Webber & Booth 2001). Scanning electron microscopy has not yet been employed as a tool for describing slipper lobster phyllosoma mouthparts (with the exception of two images in Mikami et al. 1994), which prevents a more detailed description of setae, such as those that are available for palinurid larvae.

6.3.2 Phyllosoma Ingestion Mechanisms

The ingestion process has not been described in detail for scyllarid phyllosoma. The main mastication occurs through the action of the well-developed incisor and molar processes of the mandibles, with ingestion assisted by mucus secretions from the paragnaths (Mikami & Takashima 1993). More detailed descriptions are available for palinurid phyllosomas (Cox & Bruce 2002; Nelson et al. 2002; Cox & Johnston 2003b) and, based on similarities in mouthpart structure, it seems likely that ingestion processes in scyllarid and palinurid phyllosoma are relatively similar.

6.3.3 Nisto/Adult Mouthpart Structure

There are a number of significant structural changes in the mouthparts between the phyllosoma and nisto stages, after which there are only minor developmental changes through to adult. For this reason, this section will concentrate on describing the adult form. However, it must be noted that the following descriptions are based on two species, as detailed descriptions are only available for *Ibacus peronii* (Suthers & Anderson 1981) and *Thenus orientalis* (Johnston & Alexander 1999). The mouthparts of *I. peronii* and *T. orientalis* are generally similar and appear well adapted to ingest soft flesh, which is consistent with the diet of slipper lobsters (Johnston & Alexander 1999). Adult slipper lobsters have six pairs of mouthparts, although, unlike the phyllosoma, the adult first maxilliped is well formed (Suthers & Anderson 1981;

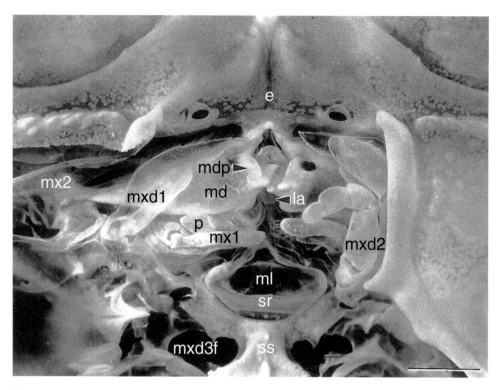

FIGURE 6.3 Photograph of *in situ* position of the mouthparts within the preoral cavity of *Thenus orientalis*, ventral view. The third maxillipeds and left second maxillipeds have been removed to reveal the mouthparts within the branchial chamber. Note, the distal and proximal endite of the second maxilla are not visible. Scale 4 mm. e, epistome; ml, membranous lobe; la, labrum; md, mandible; mdp, mandibular palp; mx1, first maxilla; mx2, second maxilla scaphognathite; mxd1, first maxilliped; mxd2, second maxilliped; mxd3f, foramen of third maxilliped; p, paragnath; ss, sternal skeleton; sr, strengthening rod. From Johnston 1994; used with permission.

Johnston & Alexander 1999). The mouthparts have simple setation and fusion of endites, and are generally less complex than some palinurid lobsters (Figure 6.3).

The mandibles are large, asymmetric structures with a well-developed left mandible, the inner medial margin of which has a calcified molar process that extends over the smaller right mandible at rest (Suthers & Anderson 1981; Johnston & Alexander 1999). Both the left and right mandibles have an incisor process ventromedially with tooth-like serrations. The mandibular palps are noncalcified and simple, lying along the dorsomedial surface of the mandible. Each palp has densely arranged pappose setae along its upper margins (Suthers & Anderson 1981; Johnston & Alexander 1999) (Figure 6.4).

A large, broad, fleshy protuberance — the membranous lobe — arises from the sternal sclerite between the second maxillipeds of *T. orientalis* (Johnston 1994) (Figure 6.3). The lobe extends anteriorly toward the mandibles and forms the roof of the preoral cavity and posterior boundary of the mouth. Its anterior lip is continuous with the esophageal wall and paragnaths, which are, in turn, connected anteriorly to the labrum (Johnston 1994). The lower part of the labrum lies between the mandibular incisor processes and has numerous, ventrally-directed microspines across its surface.

The membranous lobe of *T. orientalis* appears to be a modified metastome, similar to the metastome illustrated in *Scyllarus americanus* (S.I. Smith, 1869) (Snodgrass 1951), and is most likely the muscular protuberance illustrated in *I. peronii* (Suthers & Anderson 1981). The lobe has a flattened, calcified, strengthening rod extending the width of its aboral surface that is covered by three types of anteriorly-directed annulate setae and densely arranged clusters of microscales and microspines (Johnston 1994) (Figure 6.3). Beneath the surface are tegumental glands that secrete acid mucopolysaccharides to lubricate food during ingestion. The role of the lobe in ingestion is described later.

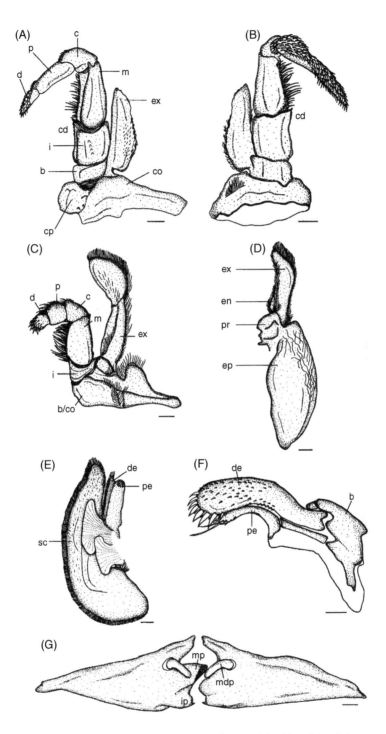

FIGURE 6.4 Illustrations of *Thenus orientalis* mouthparts; all are from the right side and aboral view, except for (B) which is an oral view. (A) Third maxilliped, scale 1.5 mm; (B) third maxilliped, scale 1.4 mm; (C) second maxilliped, scale 1.4 mm; (D) first maxilliped, scale 1.6 mm; (E) second maxilla, scale 1.2 mm; (F) first maxilla; note large hook-like simple stout setae on the distal endite, scale 0.6 mm; (G) mandibles, scale 0.6 mm. Pappose setae on mandibular palps are not illustrated. b, basis; c, carpus; cd, crista dentata; co, coxopodite; cp, calcified process, dactylus; de, distal endite; en, endite; ep, epipod; ex, exopod; i, ischium; ip, incisor process; m, merus; mdp, mandibular palp; mp, molar process; p, propodus; pe, proximal endite; pr, protopod; sc, scaphognathite. From Johnston & Alexander 1999; used with permission.

The paragnaths of slipper lobsters are fleshy protuberances that extend along the lower margin of the mandibles (Suthers & Anderson 1981; Johnston & Alexander 1999). Pores are densely distributed over the oral and aboral surfaces and each pore is covered by a V-shaped lip. The paragnaths contain two types of tegumental rosette glands that are distinguishable from one another by the large metachromatic apical vacuole in the secretory cells of type 2 rosette glands (Johnston & Alexander 1999). Both secrete acid mucopolysaccharides for lubrication during ingestion.

Maxilla I has a small proximal and large distal endite. Numerous, simple, stout setae on the aboral surface of the distal endite are extremely large and hook-like along the distal margin (Johnston & Alexander 1999). Two elongate, multidenticulate, cuneate setae project terminally from the proximal endite. The large scaphognathite of the second maxilla is fringed with plumose setae, and simple stout and tapered setae cover the aboral and oral surfaces, respectively (Johnston & Alexander 1999). The distal and proximal endites are fused; the inner margin of the former is covered with pappose setae and the latter is covered with simple, tapered setae (Johnston & Alexander 1999) (Figure 6.4 and Figure 6.5).

The distal and proximal endites of maxilliped I are reduced, with the endopod fused to the elongate, membranous exopod (Suthers & Anderson 1981; Johnston & Alexander 1999). The endopod and exopod are lined with simple, pappose, and plumose setae, and in *I. peronii* the exopod forms a terminal flagellum that is not present in *T. orientalis* (Suthers & Anderson 1981; Johnston & Alexander 1999). The membranous, flattened epipod forms the floor of the exhalent respiratory channel (Suthers & Anderson 1981) (Figure 6.4 and Figure 6.5).

The basis and coxa of maxilliped II are fused and the ischium is reduced. Short, simple, stout setae extend from the upper margins of the propodus and carpus, with larger, hook-like, simple, stout setae along the periphery of the dactylus (Suthers & Anderson 1981; Johnston & Alexander 1999). The exopod is dorsoventrally flattened and membranous distally, with no flagellum.

Maxilliped III is large and relatively robust, although it is more robust in *I. peronii* than *T. orientalis*. The exopod of *T. orientalis* is reduced, dorsoventrally flattened, and lacks a flagellum, but is elongate with a flagellum in *I. peronii* (Suthers & Anderson 1981; Johnston & Alexander 1999). In contrast to many decapod crustaceans, the ischium has only small, blunt, crista dentata. The inner medial margin of the merus has multidenticulate, spinose setae and a range of other setal types covering the oral surface of the carpus and merus, with simple, stout setae covering the entire oral surface and part of the aboral surface of the dactylus (Suthers & Anderson 1981; Johnston & Alexander 1999).

6.3.4   Nisto/Adult Ingestion Mechanisms

Slipper lobsters adopt both microphagous and macrophagous modes of feeding. In contrast to other decapods, the first maxillipeds and second maxillae are not directly involved in ingestion (Suthers & Anderson 1981; Johnston & Alexander 1999). At rest, the third maxilliped endopods are completely opposed. They flex at the merus–carpus articulation, with the dactyls resting on the inner edge of the ischium and the exopods lying flat against the carapace (Johnston 1994) (Figure 6.6). The third maxillipeds have little masticatory function, except at the crista dentata. Food received from the periopods is grasped between the crista dentata, and the simple stout and elongate setae on the dactyl and propus aid in this gripping. The dactyls are swung downward and outward to manipulate the food mass and pull it toward the crista dentata (Suthers & Anderson 1981; Johnston & Alexander 1999) (Figure 6.6). Food is then inserted between the mandibles, which close in a shearing action. By pushing out ventroposteriorly, the third maxillipeds pull on the food mass, which is stretched between them and the mandibles in a tearing action (Suthers & Anderson 1981; Johnston & Alexander 1999). The stretched food is threshed and macerated by simple stout and elongate setae on the dactyl and propus via alternate strokes of the second maxillipeds. When food is cut, the innermost end is pushed into the esophagus by the first maxillae. The third maxillipeds then open and the dactyls reposition the remainder of food between the crista dentata for insertion between the mandibles and further mastication (Figure 6.6). Using this macrophagous mode of feeding, a piece of scallop flesh 10 mm^3 in size can be ingested by *T. orientalis* in 4.8 sec (Johnston & Alexander 1999). This appears to be faster than *I. peronii*, which ingested a 5 to 10 mm^3 cube of prawn meat in one to two min (Suthers & Anderson 1981).

FIGURE 6.5 Scanning electron micrographs of setae on the mouthparts of *Thenus orientalis*. (A) Simple elongate setae interspersed with multidenticulate spinose setae along the inner aboral margin of the merus, scale 250 μm. Inset, multidenticulate spinose seta, scale 50 μm. (B) Multidenticulate comb setae on the carpus, scale 50 μm. Inset, fine structure of comb seta, scale 20 μm. (C) Fine structure of multidenticulate serrate setae; note the small cuneate denticules along the setal ridge, scale 50 μm. (D) Simple stout setae, scale 100 μm. (E) Large and hook-like simple stout setae at the tip of the dactyl, scale 200 μm. (F) Fine structure of plumose setae on the distal margin of the exopod, scale 50 μm. (G) Fine structure of a multidenticulate digitate seta; these setae are dispersed on the aboral surface of the epipod, scale 10 μm. msp, multidenticulate spinous setae. From Johnston & Alexander 1999; used with permission.

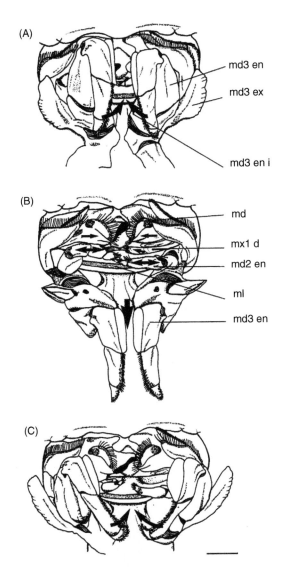

FIGURE 6.6 Illustration of sequential mouthpart movements of *Thenus orientalis* during large and hard food ingestion (macrophagous). Refer to text for detailed explanation. Arrows indicate: (A) inward and dorsal movement of the third maxillipeds to grasp food and present it to the mandibles; (B) the mandibles close and the third maxillipeds move ventrally as the second maxillipeds move lateromedially, followed by the dorsomedial movement of the right first maxilla; (C) after opening, the third maxillipeds move inward and dorsally to insert another section of food between the mandibles. Note, black outline on left mandible indicates pigmented calcification of the molar process. Scale 3 mm. md, mandible; md2 en, second maxilliped endopod; md3 en, third maxilliped endopod; md3 en i, third maxilliped endopod ischium; md3 ex, third maxilliped exopod; ml, membranous lobe; mx1d, first maxilla distal endite. From Johnston & Alexander 1999; used with permission. See also Johnston 1994 for illustrations of microphagous ingestion.

During ingestion, food is pushed over the membranous lobe (metastoma, muscular protuberance) and between the mandibles by the second maxillipeds and second maxillae (Johnston 1994). At this time, the lobe lip is retracted ventroposteriorly and the labrum rotates anterodorsally to enlarge the preoral cavity and allow for quick and efficient ingestion of lubricated food. In addition to this role, setae and microscales on the lobe help grip food during ingestion and secrete mucus for lubrication (Johnston 1994). During ingestion, the relatively reduced exopods of the third and second maxillipeds of slipper lobsters do not display the vigorous beating used for cleaning and rejection of debris by larger decapods, such as is seen in palinurid lobsters (Suthers & Anderson 1981).

Microphagous ingestion is essentially similar to macrophagous ingestion, except that the third maxillipeds do not pull the food between the mandibles and crista dentata. Instead, small pieces of food are passed by the dactyls of the third maxillipeds to the second maxillipeds that then alternate quickly to insert the food between the mandibles. Food is further pushed into the mouth by the second maxillae with little or no chewing by the mandibles. A detailed description of microphagous ingestion by *T. orientalis* is given by Johnston (1994).

6.4   Digestive System

Compared with other lobsters, there is little information on the digestive system of scyllarids. The entire alimentary tract has been described in detail for *Thenus orientalis* adults (Johnston & Alexander 1999), whereas specific studies have been conducted on the proventriculus and gastric mill function of *Ibacus peronii* (Suthers & Anderson 1981). In addition, the structure and function of the digestive glands (Johnston et al. 1998) and the tegumental glands (including the characterization of associated secretions) (Shyamasundari & Hanamantha Rao 1978) have been investigated in *T. orientalis*. The digestive enzymes of adult *T. orientalis* were also detailed and correlated with the diet of this species (Johnston & Yellowlees 1998). Fewer studies have been conducted on phyllosomas. The structure of the proventriculus was described, and possible pathways for particle flow through the proventriculus have been postulated for *Ibacus ciliatus* (Mikami & Takashima 1993). A more general description of the proventriculus, midgut, digestive gland, and hindgut has also been recorded for the same species (Mikami et al. 1994). Some comparisons have been drawn between spiny and slipper lobster phyllosomas in the review of the spiny lobster digestive system by Mikami & Takashima (2000).

6.4.1   Adult Digestive System Structure

The alimentary tract of adult slipper lobsters is divided into the foregut, midgut, and hindgut, representing approximately 25, 5, and 70% of total tract length, respectively. The ectodermal foregut and hindgut are lined with chitinous cuticle and the endodermal midgut is without cuticle. The foregut has a short esophagus that leads into the proventriculus. In adults, the proventriculus is divided into two chambers: an anterior cardiac stomach and a posterior pyloric stomach, separated by the cardiopyloric valve. This division is absent in phyllosomas, which have a simple single chamber or tube. The midgut proper is reduced to a short tube from which extends a small dorsal cecum and large bilobed digestive gland. The hindgut is a simple tube that terminates at the anus on the telson.

6.4.1.1   Foregut

The short esophagus is distensible to allow rapid ingestion of large food particles. It has four longitudinal folds, the dorsal and ventral of which are continuous with the labrum and membranous lobe, respectively (Johnston & Alexander 1999). Beneath the cuticle are columnar epithelium and rosette glands containing acid mucopolysaccharides (Shyamasundari & Hanamantha Rao 1978). Compared with other lobsters, the proventricular ossicles of slipper lobsters are reduced and less calcified, although in some species, they are rudimentary or entirely absent (Johnston & Alexander 1999) (Figure 6.7A). Consequently, the cardiac stomach is essentially a voluminous membranous sac with considerable food storage capacity via extension of its membranous walls and reduced ossicles (Suthers & Anderson 1981; Johnston & Alexander 1999). Slipper lobsters also lack the complex arrangement of ventral filtration channels of other decapods (Figure 6.7B). Two simple ventral filtration channels extend to the gastric mill where they diverge to either side of the cardiopyloric valve and enter the ventral pyloric chamber. Above the ventral channels is a pair of ventrolateral channels and above these a pair of lateral channels that merge into two dorsal fluid channels running along the length of the dorsal pyloric chamber (Figure 6.7B). Setae line the channels, forming a series of filters (Johnston & Alexander 1999).

At the posterior of the cardiac stomach is the gastric mill, which is less calcified than those of other carnivorous decapods and lacks a well-developed molar process. The median tooth has a bifurcate, ventroposteriorly-directed process and four raised molar cusps (Figure 6.7C). The paired lateral teeth have

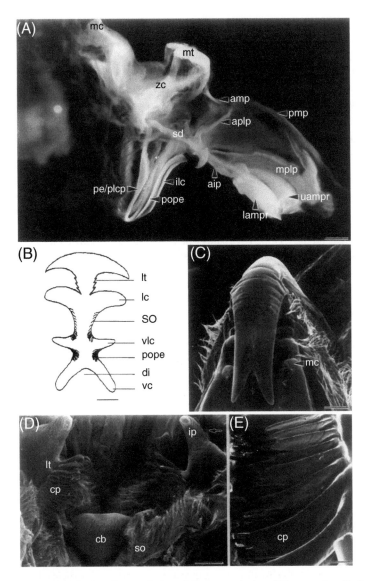

FIGURE 6.7   (A) Photograph of the lateral external view of the proventriculus of *Thenus orientalis* showing the position of ossicles, scale 1 mm. (B) Diagrammatic transverse section through the cardiac stomach just anterior to the gastric mill, scale 0.5 mm. Scanning electron micrographs: (C) gastric mill median tooth, scale 300 μm; (D) dorsal view of the lateral teeth of the gastric mill and cardiopyloric valve, *in situ*, scale 500 μm. Top is anterior. Note arrangement of chitinous plates and subsidiary teeth on incisor process of lateral teeth (arrow); (E) masticating edges of the chitinous plates, scale 150 μm. Ossicles: aip, anterior inferior pyloric; amp, anterior mesopyloric; aplp, anterior pleuropyloric; ilc, inferior lateral cardiac; lampr, lower ampullary roof; mc, mesocardiac; mplp, middle pleuropyloric; mt, median tooth; pe/plcp, pectineal/posterior lateral cardiac plate; pmp, posterior mesopyloric; pope, postpectineal; sd, subdentary; uampr, upper ampullary roof; zc, zygocardiac; cb, central bulb of cardiopyloric valve; cp, chitinous plates; di, dorsal infolding; ip, incisor process; lc, lateral channel; lt, lateral tooth; mc, molar cusp; mt median tooth; so, setose outfolding; vc, ventral channel; vlc, ventrolateral channel. From Johnston & Alexander 1999; used with permission.

a large, medially-directed incisor process and rows of vertically-directed chitinous plates (Figure 6.7D and Figure 6.7E). The masticating surfaces are sharp and serrated, and adapted to macerate and tear flesh when the teeth are opposed (Johnston & Alexander 1999). Slipper lobsters do not possess the spines or accessory teeth on the cardiopyloric valve that are common in other carnivorous decapods, and this reflects the generally soft, fleshy diet of scyllarids (Johnston & Alexander 1999).

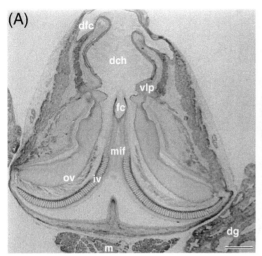 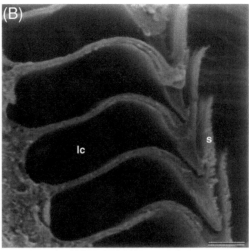

FIGURE 6.8 (A) Transverse section (6 μm) through the pyloric stomach of *Thenus orientalis*, scale 500 μm. (B) Scanning electron micrograph of the inner valve filtration setae showing the extent of overlap and longitudinal channels, scale 20 μm. Labels: dch, dorsal chamber; dfc, dorsal fluid channel; dg, digestive gland; fc, filter crest; iv, inner valve; lc, longitudinal channel; m, muscle; mif, medial infolding; ov, outer valve; s, setae; vlp, ventrolateral partition. From Johnston & Alexander 1999; used with permission.

The structure of the pyloric stomach is similar to other lobsters, with a dorsal and ventral chamber separated by a setose ventrolateral partition (Johnston & Alexander 1999) (Figure 6.8A). Above the dorsal chamber are two dorsal fluid channels that are devoid of setae, except for the small, ventral outfolding separating them from the dorsal chamber (Figure 6.8A). The outer valves of the pyloric filter press are covered in a dense mat of elongate, simple setae. The longitudinal channels of the inner valves bear rows of dorsally-directed filtration setae, each row overlapping the next (Johnston & Alexander 1999) (Figure 6.8B). Three types of sheaths extend from the pyloric stomach into the midgut: a rigid dorsal sheath, two setose lateral sheaths, and a flattened ventral sheath that overlies the entrance to the primary ducts of the digestive gland.

6.4.1.2 Midgut and Digestive Gland

The midgut of adult slipper lobsters is very short and comprises a short tube 3 to 6 mm in length, a single dorsal cecum, and the large, bilobed digestive gland (Johnston et al. 1998; Johnston & Alexander 1999). The dorsal cecum arises from the midgut tube via a single duct in the region of the primary ducts of the digestive gland to rest anteriorly on the roof of the pyloric stomach (Johnston & Alexander 1999). At this point, the cecum is a small, semicircular chamber in cross-section. Posteriorly, the lateral walls of the dorsal cecum extend ventrally down either side of the midgut tube and a short portion of the hindgut. The cecum has columnar epithelium containing apical accumulations of acid mucin granules and protein below the nuclei (Johnston & Alexander 1999). The epithelial cells are lined by microvilli, and bands of circular muscle lie beneath the basal lamina.

The digestive gland primary ducts are composed of one type of columnar epithelial cell that stains positive for protein (Johnston et al. 1998). The ducts contain extensive musculature, which suggests that they actively move fluid into and out of the glands. Four epithelial cells — E, R, F, and B — are present in the gland tubules, with R- and F-cells the most and least abundant, respectively (Figure 6.9A). Cytoplasmic characteristics of each type of cell are similar to other decapods, with the F-cells staining positive for protein and containing considerable rough endoplasmic reticulum and mitochondria (Figure 6.9B). Immunolocalization of trypsin of *Thenus orientalis* (Johnston et al. 1995) has shown that F-cells are the only sites of trypsin synthesis and secretion in the digestive gland of this species (Johnston et al. 1998) (Figure 6.9D and Figure 6.9E). Trypsin is usually concentrated in the apical cytoplasm and is secreted

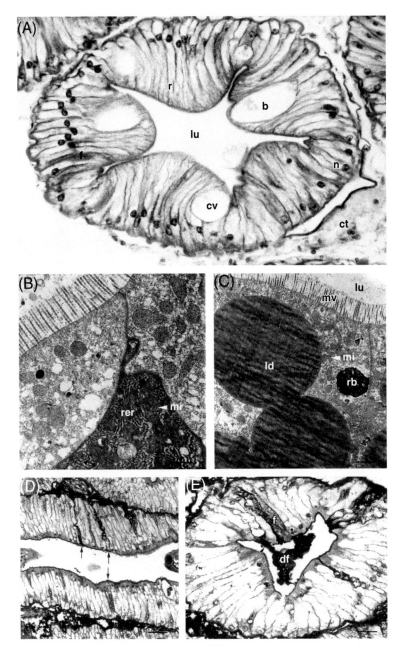

FIGURE 6.9 (A) Transverse section through a single digestive gland tubule of *Thenus orientalis*, showing the types of epithelial cells (R, F, B), scale 50 μm. Note the obvious central vacuole in the B-cells. (B) Transmission electron micrograph of two R cells and an F-cell, scale, 0.8 μm. In the F-cell note the extensive rRER and numerous mitochondria. (C) Transmission electron micrograph of an R-cell with lipid deposits and residual body in the apical cytoplasm, scale 1.8 μm. (D), (E) Immunohistochemistry of digestive gland tubules of *Thenus orientalis*, confirming that trypsin is localized only in F-cells. (D) Longitudinal serial section (6 μm) of a digestive gland tubule, scale, 70 μm and (E) Transverse serial section (6 μm) of a digestive gland tubule, scale, 40 μm. The positive reaction of F-cells to trypsin with diaminobenzidine is indicated by dark staining in the F-cell cytoplasm (arrows in D) and secretory products in the lumen (arrow in E). b, B-cell with large central vacuole; ct, connective tissue; cv, central vacuole within a B-cell; df, digestive fluid; f, F-cell; ld, lipid deposit; lu, lumen; mi, mitochondria; mv, microvilli; n, nucleus; r, R-cell; rb, residual body; rer, rough endoplasmic reticulum. From Johnston et al. 1998; used with permission.

into the lumen. Trypsin was not present in the oral region, or in epithelial cells of the cardiac and pyloric stomach and hindgut, confirming that the digestive glands are the only site of digestive enzyme synthesis in the digestive system (Johnston et al. 1998). Substantial accumulations of lipid in the R-cells indicate a role in the absorption and storage of digestion products (Figure 6.9C). Residual bodies in the apical cytoplasm are similar to the metal-storing granular inclusions and supranuclear vacuoles of other decapods. The presence of an apical complex of pinocytotic vesicles and substantial vacuoles suggest an absorptive role for B-cells (Johnston et al. 1998).

6.4.1.3 Hindgut

The anterior walls of the slipper lobster hindgut are covered in fine spines, are thickly cuticularized, and have well-developed, striated, longitudinal muscle. Circular muscle bundles extend the length of the hindgut, becoming enlarged toward the anus (Johnston & Alexander 1999).

6.4.2 Adult Digestive Function and Physiology

The reduced ossicle framework and extensive folding of the cardiac stomach of slipper lobsters allows considerable storage of food. Food storage is beneficial for scavengers, like slipper lobsters, that feed opportunistically and may have to survive periods of food deprivation. Suthers & Anderson (1981) observed that food was kept in the stomach for up to 6 h before being moved posteriorly by peristaltic action of the cardiac wall musculature to the gastric mill. The gastric mill teeth are considerably less calcified than other macrophagous carnivores, although their masticatory edges are sharp and serrated, particularly the chitinized plates of the lateral teeth, which are adapted to tear and macerate flesh when the teeth are opposed (Johnston & Alexander 1999). In general, the gastric mill is a simplified cutting apparatus compared with the powerful grinding mill of the palinurid and nephropid lobsters. The major masticatory action is a cutting action by the lateral teeth, with only a minor role for the medial tooth (Suthers & Anderson 1981; Johnston & Alexander 1999). Each masticatory cycle is between 3 and 10 sec long, depending on the extent of contraction of the lateral teeth, and each cycle is repeated for long periods. The absence of spines and accessory teeth on the cardiopyloric valve, a common feature of other carnivorous decapods, also reflects the soft, fleshy diet of scyllarids, because additional mastication by the valve is not required.

The circulation of food particles and digestive gland fluids through the proventriculus of slipper lobsters has not been documented. Only food circulation and masticatory action arising from the movements of the gastric mill have been described for the scyllarid *Ibacus peronii* (Suthers & Anderson 1981). Structural characteristics similar to other lobsters and decapods suggest that it is likely that fluid and particle circulation is similar to that of other species that have been described (Powell 1974; Dall & Moriarty 1983; King & Alexander 1994; Mikami & Takashima 2000).

The microvillus epithelium of the dorsal cecum suggests an absorptive role, although its precise nature is not known. Granules of acid mucin in the apical cytoplasm indicate that the cells may be involved in mucus production to lubricate the indigestible food bolus as it passes into the hindgut (Johnston & Alexander 1999). Mucus may also aid in binding this material prior to enclosure in the peritrophic membrane. It is unlikely that protein found in the proximal cytoplasm of the dorsal cecum epithelium is digestive enzyme (Johnston et al. 1998; Johnston & Alexander 1999). Instead, the protein secretions may contribute to the protein-mucopolysaccharide matrix of the peritrophic membrane in this species (Johnston et al. 1998; Johnston & Alexander 1999).

The cuticle-lined hindgut of slipper lobsters is involved in defecation, which is facilitated by the peristaltic contractions of the circular and longitudinal muscles of the hindgut wall (Johnston & Alexander 1999). These peristaltic contractions allow fine spines on the hindgut wall to grasp the fecal pellet and maintain traction on the peritrophic membrane as it moves posteriorly toward the anus. Circular muscle around the anus constricts the hindgut lumen and forces the fecal pellet through the anus to the exterior (Johnston & Alexander 1999).

Very little is known of the digestive physiology of slipper lobsters. The digestive enzyme profiles of *Thenus orientalis* are consistent with the carnivorous diet of slipper lobsters (Johnston & Yellowlees 1998). The proteases trypsin and chymotrypsin are produced in high concentrations, and trypsin represents a considerable proportion (up to 35%) of digestive gland protein. This trend is similar to other crusta-ceans investigated where trypsin plays a significant role in proteolysis (Johnston et al. 1995; Johnston & Yellowlees 1998). Chymotrypsin activity, however, was significantly higher than in most other decapods documented, irrespective of their dietary preferences. Purified trypsin characterized from *T. orientalis* is a glycoprotein with a molecular mass of 35 kDa and has a strong homology with other crustacean trypsins (Johnston et al. 1995).

Carbohydrase activity was relatively low compared with the proteases, reflecting the smaller propor-tion of carbohydrate compared to protein ingested by slipper lobsters (Johnston & Yellowlees 1998). Nevertheless, the wide range of carbohydrases produced, including polysaccharases, glucosidases, and galactosidases, indicates that most dietary carbohydrates can be digested into their component monosac-charides. The presence of α-amylase and α-glucosidase indicate that *T. orientalis* is capable of completely hydrolyzing carbohydrates like glycogen, the main storage product found in bivalves that are a common dietary item of slipper lobsters (Johnston & Yellowlees 1998). Significant N-acetyl β-D-glucosaminidase activity revealed an ability to hydrolyze chitooligosaccharides, which indicates that slipper lobsters can digest chitin from crustacean exoskeletons. Negligible quantities of cellobiase and laminarinase are con-sistent with the small proportion of plant and algal tissue ingested by scyllarids. The digestive fluid had an acidic pH of 5.9, which is consistent with the pH optima of the carbohydrases, verifying that it is well adapted for extracellular digestion and acts optimally under acidic conditions (Johnston & Yellowlees 1998). To fully understand the digestive physiology of slipper lobsters, it is clear that considerable work is required to determine the digestive enzymes present and their respective activities in a greater number of species. In comparison, the digestive physiology of spiny and nephropid lobsters has been studied in more detail, and may provide future directions for slipper lobster research (see, for example, Johnston 2003; Johnston et al. 2004).

6.4.3   Phyllosoma/Nisto Digestive System Structure

Most studies on slipper lobster phyllosoma are general morphological descriptions of the body and appendages during development. The structural and functional development of the alimentary tract has received very little attention, with only two dedicated studies on the development of the proventriculus of *Ibacus ciliatus* phyllosomas (Mikami & Takashima 1993; Mikami et al. 1994). In contrast, the foregut of palinurid lobster larvae has been studied in great detail, with the complete development of the foregut documented for all larval stages of *Panulirus argus* (Latreille, 1804) (Wolfe & Felgenhauer 1991), *Jasus edwardsii* (Hutton, 1875) (Johnston & Ritar 2001), and *Jasus verreauxi* (H. Milne Edwards, 1851) (Cox & Johnston 2004). Multiple or single larval stages of some other species have also been described (Nishida et al. 1990; Lemmens & Knott 1994; Mikami et al. 1994; Macmillan et al. 1997). There are significant similarities in alimentary tract structure between these palinurid and scyllarid phyllosomas. This section will concentrate on those descriptions documented for *I. ciliatus* (Mikami & Takashima 1993; Mikami et al. 1994).

The alimentary tract of the scyllarid phyllosoma is divided into a foregut, midgut, and hindgut, extending posteriorly from the mouthparts in the center of the cephalic shield through the abdomen to the anus. It comprises a relatively large cuticularized foregut, a short tube-like midgut (with digestive glands), and a long hindgut extending the length of the abdomen. Distinct digestive gland lobes lie on either side of the tract that increase in number and size during development.

6.4.3.1   *Foregut*

The esophagus is a short, narrow, tube with setae projecting into its anterior lumen. The foregut (or proventriculus) consists of a single chamber that is not divided into cardiac and pyloric chambers, lacks a cardiopyloric valve, and does not contain gastric mill teeth (Mikami & Takashima 1993; Mikami et al. 1994) (Figure 6.10). However, there are differences between the anterior and posterior foregut, with a series of lateral teeth brushes (setae) anteriorly and a series of lateral folds posteriorly (Mikami & Takashima 1993; Mikami et al. 1994). The posterior ends of the lateral setae are connected to the anterior edge of the lateral folds, and the latter divides the proventriculus into dorsal and ventral compartments. Well-developed spines are present in the dorsal compartment, which also has a dorsal groove. In the ventral compartment,

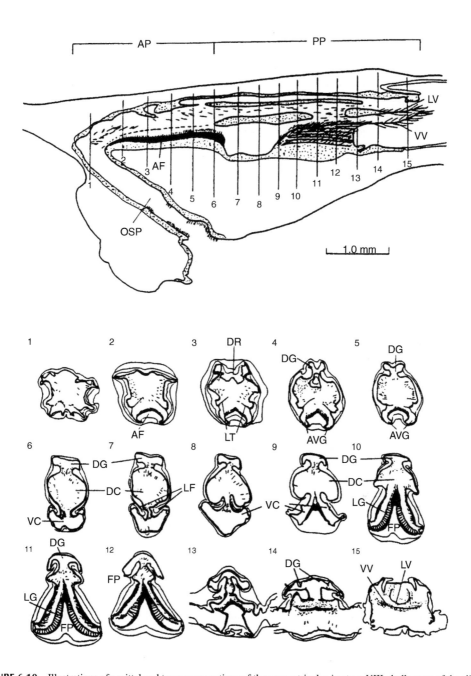

FIGURE 6.10 Illustrations of sagittal and transverse sections of the proventriculus in stage VIII phyllosoma of the slipper lobster, *Ibacus ciliatus*. Each number on the transverse section corresponds to the number of the sagittal section illustrated below. AF, anterior floor; AVG, anterior ventral grooves; AP, anterior portion; DC, dorsal compartment; DG, dorsal groove; DR, dorsal ridge; FP, filter press; LF, lateral fold; LG, longitudinal grove; LT, lateral teeth; LV, lateral valves; OSP, esophagus; PP, posterior portion; VC, ventral compartment; VV, ventral valve. From Mikami & Takashima 1993; used with permission.

the anterior floor is covered in setae and has two longitudinal grooves (Figure 6.10). Posteriorly, these longitudinal grooves form the filter press that is well formed in all but the first larval stage (Mikami & Takashima 1993). The digestive gland opens under the posterior edge of the longitudinal grooves. Four cuticular setose valves (one dorsal, two lateral, one ventral) extend posteriorly into the lumen of the midgut

(Figure 6.10). During development, the setae and brushes become larger and increase in number, and the grooves become larger and more distinct (Mikami & Takashima 1993; Mikami et al. 1994).

The proventriculus of the nisto is transitional between phyllosomas and benthic juveniles. It does not contain gastric mill teeth, but a cardiopyloric valve divides the proventriculus into an anterior chamber and large posterior chamber, as in adult lobsters. The anterior chamber is a simple sac-like structure with poorly developed lateral teeth, a cardiac floor and grooves, spines, and a filter press. The wall of the posterior chamber is thickened and the filter press is flattened on the ventral surface (Mikami & Takashima 1993).

6.4.3.2   Midgut

The digestive gland is the most obvious structure in the cephalothorax and is composed of a series of ceca, which contain four epithelial cell types (E, R, F, and B). Cytoplasmic features of these cells are generally similar in structure to other decapods. The midgut tube is lined by a single epithelial cell layer, similar in structure to the R-cells of the digestive gland (Mikami et al. 1994).

6.4.3.3   Hindgut

The hindgut is a simple tube that runs through the abdomen to the anus and is lined by a chitinous cuticle. It does not appear to have any secretory or absorptive functions (Mikami et al. 1994).

6.4.4   Phyllosoma/Nisto Digestive Function and Physiology

The phyllosoma proventriculus has no gastric mill, and it is clear that the larval proventriculus does not grind food, but functions mainly to sort and filter the food particles previously masticated by the mouthparts. Food particles enter the dorsal compartment and are squeezed through spines and dorsal ridges, continue over the lateral teeth, and then pass into the posterior portion of the ventral compartment through the anterior dorsal grooves (Mikami & Takashima 1993). Minute particles and fluids enter the longitudinal channels of the filter press and pass into the midgut gland. Large food particles that cannot pass between the lateral teeth pass directly into the lumen of the posterior portion of the dorsal chamber and then flow posteriorly into the midgut proper (Mikami & Takashima 1993). Prior to the development of the filter press, the lateral teeth possibly function to filter food particles.

Food particles are prevented from re-entering the posterior chamber by the cuticular setose valves. Digestive enzymes produced by the digestive gland do not appear to pass into the proventriculus; therefore, the majority of chemical degradation occurs in the gland itself (Mikami & Takashima 1994). Based on the proventricular structures of scyllarid phyllosomas, it is hypothesized that they are best equipped to utilize soft-bodied foods, and the greatest success in culture may be achieved if moist, soft-bodied foods are used, possibly including microencapsulated or micro-bound diets (Mikami et al. 1994; Johnston & Ritar 2001; Cox & Johnston 2003a, 2003b).

The development of digestive function by early life-history stages of lobsters is largely unstudied, with only a few published studies on the larval stages of *Homarus americanus* H. Milne Edwards, 1837, *H. gammarus* (Linnaeus, 1758), and *Procambarus clarkii* (Girard, 1852) (Biesiot & Capuzzo 1990; Kurmaly et al. 1990; Hammer et al. 2000). There have been no studies examining the digestive physiology of slipper lobster phyllosomas. However, a comprehensive analysis on the digestive physiology of spiny lobster phyllosoma, *Jasus edwardsii*, found that the digestive capacity of these phyllosomas was well-developed from posthatch (stage I), suggesting that they are capable of completely hydrolyzing ingested food items from the commencement of feeding (Johnston et al. 2004). Protease, trypsin, amylase, α-glucosidase, chitinase, and lipase were detected in all stages of development, suggesting that phyllosomas can readily digest dietary protein, lipid, and carbohydrate, as well as chitin, throughout larval development. Protease and lipase activities were considerably higher than amylase and α-glucosidase activities, indicating that dietary protein and lipid is more important than carbohydrate, and suggesting a carnivorous diet. Enzyme profiles suggest that spiny lobster phyllosomas are capable of digesting a wide range of zooplankton prey, but they make the best use of high protein and lipid dietary items (Johnston et al. 2004). Further research on at least some stages of scyllarid larvae will ascertain whether the digestive physiology of slipper lobster phyllosomas is similar to that of palinurid phyllosomas.

6.5   Conclusions

The feeding and digestive systems of slipper lobsters have unique structural and functional characteristics that are closely related to their habitat and dietary preferences. Ingestion mechanisms have been well described in two species, and both macrophagous and microphagous ingestion strategies are practiced, revealing that scyllarids are capable of efficiently ingesting a wide range of food sizes and types. However, bivalve mollusks are thought to be their preferred diet. Compared with other lobsters, scyllarid adult mouthparts are not heavily calcified, but are highly effective at masticating soft fleshy prey, such as the flesh from bivalve mollusks, using the shearing action of the asymmetrical mandibles. In turn, scyllarids have developed a system of wedging, using their pereiopod dactyls to efficiently open the shell valves on these bivalve mollusks. A large prominent membranous lobe (a modified metastome) appears to be unique to scyllarids and plays an important role in ingestion to lubricate food and allow quick and efficient swallowing of large items by retraction of its anterior lip.

The cardiac stomach of adult slipper lobsters is simplified compared with palinurid and nephropid lobsters, with reduced ossicles and a simplified ventral filtration system. It is essentially a soft, membranous sac that allows considerable food storage by extension of its walls, a beneficial feature for scavengers that may feed only intermittently. The gastric mill teeth are well developed with large incisor processes and chitinous plates. However, they do not have the accessory teeth common on the cardiopyloric valve of other carnivorous decapods, which reflects the primarily soft flesh diet of scyllarids, as additional mastication by the valve is not needed. The digestive glands are similar in structure to other decapods, and F-cells have been definitively identified as the only site of digestive enzyme production in scyllarids. High concentrations of proteolytic enzymes are produced, which is consistent with a carnivorous diet, as proteases are indicative of the capacity to hydrolyze protein.

Phyllosoma mouthparts are well-developed from stage I, with sharp incisor processes on the mandibles and well developed setation on the first maxilla and second maxillipeds. Although the ingestion process has not been described, it is clear that the main mastication in these larvae occurs through the action of their well developed incisor and molar mandibular processes. Nephropid lobster larvae lack the teeth of the gastric mill until stage III, when they gain heavily cuticularized medial and lateral teeth; in earlier stages, they have a series of pads and ridges that may be rudiments of the teeth (Factor 2005). In contrast, palinurid and scyllarid larvae lack a gastric mill throughout the larval stages and the gastric mill area does not grind food, but functions mainly to sort and filter food particles previously masticated by the mandibles. This is suggestive of a gelatinous diet that requires little additional mastication by internal teeth.

Compared to their palinurid and nephropid lobster counterparts, slipper lobster feeding and digestion remains relatively unstudied. In particular, the dietary preferences and structure and function of feeding and digestion in phyllosoma are poorly understood. These areas require further study if we are to fully understand the feeding and digestive requirements of slipper lobsters.

Acknowledgments

Thanks to Martin Lourey and two anonymous referees for critically reviewing this chapter and to Kari Lavalli and Ehud Spanier for assistance with editing. I thank Dr. Satoshi Mikami and Dr. Lau for use of their material and figures within this chapter.

References

Anger, K. 2001. *The Biology of Decapod Crustacean Larvae. Crustacean Issues, Vol. 14*: pp. 13–79; 109–146. The Netherlands: A.A. Balkemia.

Atkinson, J.M. & Boustead, N.C. 1982. The complete larval development of the scyllarid lobsters *Ibacus alticrenatus* Bate, 1888, in New Zealand waters. *Crustaceana* 42: 275–287.

Barnett, B.M., Hartwick, R.F., & Milward, N.E. 1984. Phyllosoma and nisto stage of the Morteon Bay Bug *Thenus orientalis* (Lund) (Crustacea: Decapoda: Scyllaridae), from shelf waters of the Great Barrier Reef. *Aust. J. Mar. Freshw. Res.* 35: 143–152.

Barnett, B.M., Hartwick, R.F., & Milward, N.E. 1986. Descriptions of the nisto stage of *Scyllarus demani* Holthuis, two unidentified *Scyllarus* species and the juvenile of *Scyllarus martensi* Pfeffer (Crustacea: Decapoda: Scyllaridae), reared in the laboratory and behavioural observations of the nistos of *S. demani, S. martensii* and *Thenus orientalis* (Lund). *Aust. J. Mar. Freshw. Res.* 37: 595–608.

Biesiot, P.M. & Capuzzo, J.M. 1990. Changes in the digestive enzyme activities during early development of the American lobster *Homarus americanus* Milne Edwards. *J. Exp. Mar. Biol. Ecol.* 136: 107–122.

Coutures, E. 2001. On the first phyllosoma stage of *Parribacus caledonicus* Holthuis, 1960, *Scyllarides squammosus* (H. Milne-Edwards, 1837) and *Arctides regalis* Holthuis, 1963 (Crustacea, Decapoda, Scyllaridae) from New Caledonia. *J. Plankton Res.* 23: 745–751.

Cox, S.L. & Bruce, M.P. 2002. Feeding behaviour and associated sensory mechanisms of stage I-III phyllosoma of *Jasus edwardsii* and *Jasus verreauxi*. *J. Mar. Biol. Assoc. U.K.* 83: 465–468.

Cox, S.L. & Johnston, D.J. 2003a. Feeding biology of spiny lobster larvae and implications for culture. *Rev. Fish. Sci.* 11: 89–106.

Cox, S.L. & Johnston, D.J. 2003b. Developmental changes in the structure and function of mouthparts of phyllosoma larvae of the packhorse lobster, *Jasus verreauxi* (Decapoda: Palinuridae). *J. Exp. Mar. Biol. Ecol.* 296: 35–47.

Cox, S.L. & Johnston, D.J. 2004. Developmental changes in foregut functioning of packhorse lobster, *Jasus (Sagmariasus) verreauxi* (Decapoda: Palinuridae) phyllosoma larvae. *Mar. Freshw. Res.* 55: 145–153.

Dall, W. & Moriarty, J.W. 1983. Functional aspects of nutrition and digestion. In: Mantal, L.H. (ed.), *The Biology of Crustacea, Vol. 5: Internal Anatomy and Physiological Regulation*: pp. 215–251. New York: Academic Press.

Dall, W., Hill, B.J., Rothlisberg, P.C., & Staples, D.J. 1990. The biology of the Penaeidae. *Adv. Mar. Biol.* 27: 1–461.

Derby, C.D. 1982. Structure and function of cuticular sensilla of the lobster *Homarus americanus*. *J. Crust. Biol.* 2: 1–21.

Factor, J.R. 1989. Development of the feeding apparatus in decapod crustaceans. In: Felgenhauer, B.E., Watling, L., & Thistle, A.B. (eds.), *Functional Morphology of Feeding and Grooming in Crustacea. Crustacean Issues, Vol. 6*: pp. 185–203. Rotterdam: A.A. Balkema.

Factor, J.R. 1995. The digestive system. In: Factor, J.R. (ed.), *Biology of the Lobster Homarus americanus*: pp. 395–440. New York: Academic Press.

Factor, J.R. 2005. Development and metamorphosis of the digestive system of larval lobsters, *Homarus americanus* (Decapoda: Nephropidae). *J. Morph.* 169: 225–242.

Felgenhauer, B.E. 1992. Internal anatomy of the Decapoda: an overview. In: Harrison, F.W. & Humes, A.G. (eds.), *Microscopic Anatomy of the Invertebrates, Vol. 10: Decapod Crustacea*: pp. 147–201. New York: Wiley-Liss.

Hammer, H.S., Bishop, C.D., & Watts, S.A. 2000. Activities of three digestive enzymes during development in the crayfish *Procambarus clarkii* (Decapoda). *J. Crust. Biol.* 20: 614–620.

Herrnkind, W., Halusky, J., & Kanciruk, P. 1976. A further note on phyllosoma larvae associated with medusae. *Bull. Mar. Sci.* 26: 110–112.

Icely, J.D. & Nott, J.A. 1992. Digestion and absorption: digestive system and associated organs. In: Harrison, F.W. & Humes, A.G. (eds.), *Microscopic Anatomy of the Invertebrates, Vol 10: Decapod Crustacea*: pp. 147–201. New York: Wiley-Liss.

Ito, M. & Lucas, J.S. 1990. The complete larval development of the scyllarid lobster *Scyllarus demani* Holthuis, 1946 (Decapoda, Scyllaridae) in the laboratory.*Crustaceana* 58: 144–167.

Johnston, D.J. 1994. Functional morphology of the membranous lobe within the preoral cavity of *Thenus orientalis* (Crustacea: Scyllaridae). *J. Mar. Biol. Assoc. U.K.* 74: 787–800.

Johnston, D.J. 2003. Ontogenetic changes in digestive enzymology of the spiny lobster, *Jasus edwardsii* Hutton (Decapoda, Palinuridae). *Mar. Biol.* 143: 1071–1082.

Johnston, D.J. & Alexander, C.G. 1999. Functional morphology of the mouthparts and alimentary tract of the slipper lobster *Thenus orientalis* (Decapoda: Scyllaridae). *Mar. Freshw. Res.* 50: 213–223.

Johnston, D.J. & Ritar, A. 2001. Mouthpart and foregut ontogeny in phyllosoma larvae of the lobster *Jasus edwardsii* (Decapoda: Palinuridae). *Mar. Freshw. Res.* 52: 1375–1386.

Johnston, D.J. & Yellowlees, D. 1998. Relationship between the dietary preferences and digestive enzyme complement of the slipper lobster *Thenus orientalis* (Decapoda: Scyllaridae). *J. Crust. Biol.* 18: 656–665.

Johnston, D.J., Hermans, J.M., & Yellowlees, D. 1995. Isolation and characterisation of a trypsin from the slipper lobster *Thenus orientalis* Lund. *Archiv. Biochem. Biophys.* 324: 35–40.

Johnston, D.J., Alexander, C.G., & Yellowlees, D. 1998. Epithelial cytology and function in the digestive gland of *Thenus orientalis* (Decapoda: Scyllaridae). *J. Crust. Biol.* 18: 271–278.

Johnston, D.J., Ritar, A., Thomas, C., & Jeffs, A. 2004. Digestive enzyme profiles of spiny lobster (*Jasus edwardsii*) phyllosoma larvae. *Mar. Ecol. Prog. Ser.* 275: 219–230.

Jones, C.M. 1988. *The Biology and Behaviour of Bay Lobsters, Thenus spp. (Decapoda: Scyllaridae), in Northern Queensland*, Australia. Ph.D. thesis: University of Queensland, Australia.

King, R.A. & Alexander, C.G. 1994. Fluid extraction and circulation in the proventriculus of the banana prawn *Penaeus merguiensis* De Man. *J. Crust. Biol.* 14: 497–507.

Kittaka, J. 1994. Culture of phyllosomas of spiny lobster and its application to studies of larval recruitment and aquaculture. *Crustaceana* 66: 258–270.

Kittaka, J. 1997. Culture of larval spiny lobsters: A review of work done in northern Japan. *Mar. Freshw. Res.* 48: 923–930.

Kurmaly, K., Jones, D.A., & Yule, A.B. 1990. Acceptability and digestion of diets fed to larval stages of *Homarus gammarus* and the role of dietary conditioning behaviour. *Mar. Biol.* 106: 181–190.

Lau, C.J. 1987. Feeding behaviour of the Hawaiian slipper lobster, *Scyllarides squamosus,* with a review of decapod crustacean feeding tactics on molluscan prey. *Bull. Mar. Sci.* 41: 378–391.

Lau, C.J. 1988. Dietary comparison of two slow-moving crustacean (Decapoda: Scyllaridae) predators by a modified index of relative importance. In: *Proceedings of the 6th International Coral Reef Symposium, Vol. 2*: pp. 95–100. Townsville, Australia, August 8–12, 1988.

Lavalli, K.L. & Factor, J.R. 1992. Functional morphology of the mouthparts of juvenile lobsters, *Homarus americanus* (Decapoda: Nephropidae), and comparison with the larval stages. *J. Crust. Biol.* 12: 467–510.

Lavalli, K.L. & Factor, J.R. 1995. The feeding appendages. In: Factor, J.R. (ed.), *Biology of the Lobster Homarus americanus*. pp. 349–393. New York: Academic Press.

Lemmens, J.W.T.J. & Knott, B. 1994. Morphological changes in external and internal feeding structures during the transition phyllosoma-puerulus-juvenile in the Western rock lobster (*Panulirus cygnus*, Decapoda: Palinuridae). *J. Morph.* 220: 271–280.

Macmillan, D.L., Sandow, S.L., Wikeley, D.M., & Frusher, S. 1997. Feeding activity and the morphology of the digestive tract in stage-I phyllosoma larvae of the rock lobster *Jasus edwardsii*. *Mar. Freshw. Res.* 48: 19–26.

Marinovic, B., Lemmens, J.W.T.J., & Knott, B. 1994. Larval development of *Ibacus peronii* Leach (Decapoda: Scyllaridae) under laboratory conditions. *J. Crust. Biol.* 14: 80–96.

Martins, H.R. 1985. Biological studies on the exploited stock of the Mediterranean locust lobster *Scyllarides latus* (Latreille, 1803) (Decapoda, Scyllaridae) in the Azores. *J. Crust. Biol.* 5: 294–305.

Mikami, S. & Greenwood, J.G. 1997. Complete development and comparative morphology of larval *Thenus orientalis* and *Thenus* sp. (Decapoda: Scyllaridae) reared in the laboratory. *J. Crust. Biol.* 17: 289–308.

Mikami, S. & Takashima, F. 1993. Development of the proventriculus in larvae of the slipper lobster, *Ibacus ciliatus* (Decapoda: Scyllaridae). *Aquaculture* 116: 199–217.

Mikami, S. & Takashima, F. 2000. Functional morphology of the digestive system. In: Phillips, B.F. & Kittaka, J. (eds.), *Spiny Lobsters: Fisheries and Culture*, 2nd Edition: pp. 601–610. Oxford, England: Blackwell Science.

Mikami, S., Greenwood, J.G., & Takashima, F. 1994. Functional morphology and cytology of the phyllosomal digestive system *Ibacus ciliatus* and *Panulirus japonicus* (Decapoda, Scyllaridae and Palinuridae). *Crustaceana* 67: 212–225.

Mitchell, J.R. 1971. Food Preferences, Feeding Mechanisms and Related Behavior in Phyllosoma Larvae of the California Spiny Lobster, Panulirus interruptus (Randall). Masters Thesis: San Diego State College, San Diego.

Nelson, M.M., Cox, S.L., & Ritz, D. 2002. Function of mouthparts in feeding behaviour of phyllosoma larvae of the packhorse lobster, *Jasus verreauxi* (Decapoda: Palinuridae). *J. Crust. Biol.* 22: 595–600.

Nishida, S., Quigley, B.D., Booth, J.D., Nemoto, T., & Kittaka, J. 1990. Comparative morphology of the mouthparts and foregut of the final-stage phyllosoma, puerulus, and postpuerulus of the rock lobster *Jasus edwardsii* (Decapoda: Palinuridae). *J. Crust. Biol.* 10: 293–305.

Phillips, B.F. & McWilliam, P.S. 1986. Phyllosoma and nisto stages of *Scyllarus matensenii* Pfeffer (Decapoda, Scyllaridae) from the Gulf of Carpentaria, Australia.*Crustaceana* 51: 133–153.

Powell, R.R. 1974. The functional morphology of the foreguts of the thalassinid crustaceans, *Callianassa californiensis* and *Upogebia pugettensis. Univ. Calif. Publ. Zool.* 102: 1–41.

Shyamasundari, K. & Hanumantha Rao, K. 1978. Studies on the Indian sand lobster *Thenus orientalis* (Lund): Mucopolysaccharides of the tegumental glands. *Folia Histochemica et Cytochemica* 16: 247–254.

Snodgrass, R.E. 1951. *Comparative Studies on the Head of Mandibulate Arthropods.* New York: Comstock Publishing Company Inc.

Spanier, E. 1987. Mollusca as food for the slipper lobster (*Scyllarides latus*) in the coastal waters of Israel. *Levantina* 68: 713–716.

Spanier, E., Tom, M., Pisanty, S., & Almog, G. 1988. Seasonality and shelter selection by the slipper lobster *Scyllarides latus* in the southeastern Mediterranean. *Mar. Ecol. Prog. Ser.* 42: 247–255.

Suthers, I.M. & Anderson, D.T. 1981. Functional morphology of mouthparts and gastric mill of *Ibacus peronii* (Leach) (Palinura: Scyllaridae). *Aust. J. Mar. Freshw. Res.* 32: 931–944.

Tong, L.J., Moss, G.A., Paewai, M.M., & Pickering, T.D. 1997. Effect of brine-shrimp numbers on growth and survival of early-stage phyllosomal larvae of the rock lobster *Jasus edwardsii. Mar. Freshw. Res.* 48: 935–940.

Webber, W.R. & Booth, J.D. 2001. Larval stages, developmental ecology and distribution of *Scyllarus* sp. Z (probably *Scyllarus aoteanus* Powell 1949) (Decapoda: Scyllaridae) *N. Z. J. Mar. Freshw. Res.* 35: 1025–1056.

Wolfe, S.H. & Felgenhauer, B.E. 1991. Mouthpart and foregut ontogeny in larval, postlarval, and juvenile spiny lobster, *Panulirus argus* Latreille (Decapoda, Palinuridae). *Zool. Scr.* 20: 57–75.

7

Behavior and Sensory Biology of Slipper Lobsters

Kari L. Lavalli, Ehud Spanier, and Frank Grasso

CONTENTS

Abstract

Scyllarid lobsters are a diverse group of animals (85 species) that possess varied early life-history strategies and live in dissimilar habitats and depths — as such, they possess diverse behaviors, and no "stereotypical" behavioral descriptions are possible. For species that specialize on bivalves, feeding behavior is probably the closest to being stereotypical, with the sharp-pointed dactyls of the pereiopods acting to pry open bivalves or remove the flesh from gastropods. Antennules, which are normally organs of distant chemoreception (olfaction) appear to play a contact chemoreception role (taste) in the species of this family. Setal types on mouthparts and pereiopods are less diverse than in the nephropids and palinurids, but are likely to represent bimodal chemo-mechanosensors and mechanosensors. Substrate preferences are varied, depending on final habitat of the adults. Some species (e.g., *Ibacus* and *Thenus*) are well adapted for digging into the substrate and show an advanced digging behavior compared to others that are adapted for sheltering in crevices or under living structures (corals, sponges) and do so either solitarily or communally. Only a few species apparently undertake migrations from adult grounds, while the majority appear to make nomadic movements only within their adult habitats. Solitary species and social species seem to have little agonism toward conspecifics, except over food items. Mating behavior has not been observed; however, here again, there is diversity among species in spermatophore placement (internal or external) and persistence. *Scyllarides, Ibacus*, and *Thenus* species appear capable of burst-and-coast swimming to effect what are presumed to be foraging movements within their habitats or to escape from predator threats. Neurophysiological control over such swimming has only been investigated in *Ibacus* species and demonstrates loss of giant interneurons that are used by other decapod forms for escape swimming. Little is known of the sensory inputs affecting scyllarid behaviors.

7.1   Introduction

Despite being recognized for thousands of years as a decapod lobster, slipper lobster behavior and sensory capabilities, like so many other aspects of their biology, have not been well studied. It is argued that this is due to their relative insignificance in commercial fisheries operations, even though they have been locally consumed in many cultures for hundreds of years (see Holthuis 1991; Spanier & Lavalli, Chapter 18). Only within the last two decades have researchers turned their eyes toward these lobsters and, of these, only limited studies have focused on their basic biology. The few studies on behavior that currently exist focus mainly on several species in three subfamilies: the Arctidinae — *Scyllarides aequinoctialis* (Lund, 1793), *Scyllarides astori* Holthuis, 1960, *Scyllarides latus* Latreille, 1802, and *Scyllarides nodifer* (Stimpson, 1866); the Ibacinae — usually *Ibacus peronii* Leach, 1815, but occasionally *Ibacus chacei* Brown & Holthuis, 1998 and *Ibacus ciliatus* (Von Siebold, 1824); and the Theninae — both *Thenus orientalis* (Lund, 1793) and *Thenus indicus* Leach, 1816.

In terms of sensory biology, numerous anatomical drawings exist in taxonomic descriptions of the more than 80 species (e.g. see the works of Holthuis: Holthuis 1985, 1991, 1993a, 1993b, 2002; Brown & Holthuis 1998); these drawings provide information regarding setal positions on pereiopods, but there is little information on setal types, their functions, or the information they provide for the guidance of behavior. Even in studies directly examining scyllarid feeding appendages (e.g., Suthers & Anderson 1981; Ito & Lucas 1990; Mikami et al. 1994; Johnston & Alexander 1999; Malcom 2003; Weisbaum & Lavalli 2004), only a few have provided a detailed analysis of setation — fewer still have provided information on the development of the setation pattern from larval to adult stages (Ito & Lucas 1990; Mikami et al. 1994). In the related and well-studied nephropid lobster, *Homarus americanus* H. Milne Edwards, 1837 setation developmental patterns, innervation, and responsiveness have been the focus of numerous studies on mouthparts, pereiopods, and antennules (Factor 1978; Derby 1982; Derby & Atema 1982a, 1982b; Derby et al. 1984; Moore et al. 1991a, 1991b; Lavalli & Factor 1992, 1995; Gomez & Atema 1996; Guenther & Atema 1998; Kozlowski et al. 2001), which have had important impacts in every aspect of nephropid lobster biology and its large fishery. A similar situation exists in the well-studied palinurid lobster, *Panulirus argus* (Latreille, 1804) where setal innervation and responsiveness have been a major focus (Derby & Ache 1984; Daniel & Derby 1988; Derby 1989; Steullet et al. 2000a, 2000b, 2001, 2002; Cate & Derby 2001; Derby & Steullet 2001; Derby et al. 2001; Garm et al. 2004). Currently, only several species of *Scyllarides* and *Ibacus ciliatus* have been examined specifically for purported sensory structures on appendages typically used for feeding (pereiopods) and orientation (antennules) (Mikami et al. 1994; Malcom 2003; Weisbaum & Lavalli 2004).

This lack of knowledge, not only of adult, but also of larval, postlarval, and juvenile behavior and sensory capabilities is problematic because these species are increasingly being exploited as by-product in many localities and as targeted fisheries in certain places in the world (see Chapters 11 to 18 for information on scyllarid fisheries). Behavior and sensory input affecting behavior contributes to the survival, growth, and ultimately, the reproductive success of individuals — thus, this information is vital to an understanding of why certain individuals, populations, and species are more successful than others in adapting to their environments and changes within those environments. Our lack of understanding of behavioral patterns of all life history stages, as well as the basic biology of how (or if) they migrate, when they mature, how they find mates and food, how and why they are distributed as larvae, postlarvae, juveniles, and adults, and the kinds of natural challenges the larvae, postlarvae, juveniles, and adults face, can lead to overexploitation and population depletion, as has already occurred in several regions (see, for example, DiNardo & Moffitt, Chapter 12; Hearn et al., Chapter 14; Radhakrishnan et al., Chapter 15; Haddy et al., Chapter 17; and Spanier & Lavalli, Chapter 18).

Because these animals are held readily in laboratory settings, there is potential for understanding many behaviors and sensory abilities, as has been done for nephropid lobsters, *Homarus americanus* and *H. gammarus* (Linnaeus, 1758), and for many of the palinurid lobster species. This chapter summarizes the limited behavioral and sensory research on the few species that have been studied and aims to stimulate in-depth research focused on these topics for more scyllarid species in both naturalistic laboratory settings (*sensu*, Atema 1986; Jones 1988; Cowan & Atema 1990; Barshaw & Spanier 1994a) and in field settings (*sensu*, Karnofsky et al. 1989a, 1989b; Wahle & Steneck 1991, 1992; Barshaw et al. 2003).

7.2   Larval (Naupliosomas and Phyllosomas) Behavior

The life history of scyllarids is similar to that of both nephropids and palinurids and can be divided into a series of developmental phases that occupy different ecological niches (Table 7.1; see Figure 1.2 in Lavalli & Spanier, Chapter 1). Compared to information accumulated for such phases in nephropid lobsters, especially *Homarus americanus*, and palinurid lobsters, especially *Panulirus argus*, we know very little about the corresponding phases in scyllarids. Much of the lack of knowledge is due to lack of success in sampling the different phases, particularly the larval, postlarval, and juvenile stages. This is clearly an area where more research is needed.

Observations of larvae are made from sampling of wild specimens (via plankton tows) and laboratory culture. In the laboratory, females have hatched their embryos during night or early morning hours (Sims 1966; Robertson 1968). Scyllarids typically begin their pelagic lives as phyllosomal larvae that are flattened, leaf-like, transparent, planktonic, zoeal forms with long appendages and long cephalic shields (Phillips et al. 1981). Unlike the postlarvae, juveniles, or adults, the second antennae of early phyllosomal instars are initially small compared to the first antennae (antennules) and grow to a length similar to the antennules; they are not broad and flat (Robertson 1969a). However, some species of scyllarids hatch as a naupliosoma (prelarva or prezoea), and remain in this form for a few minutes to several hours, depending on species, before molting into the first-stage phyllosoma (Booth et al. 2005; Sekiguchi et al., Chapter 4). The species that apparently hatch as this early form include: *Scyllarides aequinoctialis* (Robertson, 1968, 1969a), *S. herklotsii* (Herklots, 1851) (Crosnier 1972), *S. latus* (Martins 1985a), *Ibacus alticrenatus* (Bate, 1888) (Lesser 1974), *I. ciliatus* (Harada, 1958, although the observed individuals may have been *I. novemdentatus* Gibbes, 1850 instead, as per Holthuis, 1985). It is not clear if other scyllarids complete this stage within the egg or if the naupliosoma stage has simply not been observed given its short existence (Sekiguchi et al., Chapter 4). Some rearing studies indicate that certain species simply, however, do not hatch as naupliosomas (e.g., *I. peronii* — Stewart et al., 1997).

Dispersal of phyllosomas varies among species and depends largely on whether the parental stock is found within lagoons formed by coral island barrier reefs or in deeper waters (Baisre 1994; Johnson 1971a, 1971b; Yeung & McGowan 1991; Coutures 2000). Those hatched in coastal lagoons tend to remain there, while those hatched in deeper waters gradually move shoreward, such that final-stage phyllosomas of some scyllarids are found much closer to shore than is typical for palinurid phyllosomas (Booth et al. 2005; Sekiguchi et al., Chapter 4). *Scyllarides aequinoctialis, S. astori, S. herklotsii* (Herklots, 1851), *S. nodifer*, and *S. squammosus* (H. Milne Edward, 1837) all have oceanic distributions of their phyllosomas and are presumed to be dispersed in a manner similar to that for palinurids, since few mid- to late-stage larvae are found in inshore regions (Robertson 1969a; Johnson 1971b; Phillips et al. 1981; McWilliam & Phillips

TABLE 7.1

Summary of Life-History Characteristics of Scyllaridae and Associations with Environment and Habitat

	Naupliosoma	Phyllosoma	Nisto	Juvenile to Subadult	Adult
Location	Near hatching site	Oceanic, beyond the shelf break and over ocean basins (*Arctides guineensis, Scyllarides aequinoctialis, S. astori, S. herklotsii, S. nodifer, S. squammosus, Parribacus antarcticus, Antipodarctus aoteanus*) Oceanic and within and beyond the shelf break (*Ibacus peronii, Crenarctus bicuspidatus, Eduarctus modestus, Galearctus timidus, Scyllarus depressus*) Coastal, over the continental shelf, as well as within and beyond the shelf break (*Evibacus princeps, Ibacus ciliatus, Ibacus novemdentatus, Ibacus peronii, Biarctus sordidus, Chelarctus cultrifer, Crenarctus bicuspidatus, Eduarctus martensii, Galearctus kitanoviriosus, Petrarctus demani, P. rugosus, Scammarctus batei, Scyllarus americanus, S. chacei, S. depressus*) Coastal (*Thenus orientalis*, some *Scyllarus* species)	Pelagic to benthic; when pelagic, tends to be located over adult grounds (e.g., *Scyllarides astori*) or, alternatively in some species, swims towards such grounds from offshore (except *Ibacus chacei*)	Benthic	Benthic *Arctides* — tropical to subtropical localities in all oceans except the eastern Atlantic *Scyllarides* — western, eastern, & central Atlantic, Indo-Pacific, Mediterranean & Red Seas *Parribacus* — tropical to subtropical waters in Pacific, Atlantic, and Indian Oceans *Evibacus* — Pacific Ocean (west coast of Central America) *Ibacus* — Indo-West Pacific *Thenus* — Indo-West Pacific *Acantharctus* — one species each in Atlantic, Pacific, & Indian Oceans Remaining scyllarinid species — Indo-West Pacific
Depth (m)	Unknown	Pelagic, but depth varies from surface waters to deeper waters (some reports state that phyllosomas have been found to depths of 2000 m, with heavy concentrations at 500–600 m)	Pelagic to benthic (burying during day in sand but swimming in the water column at night: *Thenus orientalis, Eduarctus martensii*)	Usually the same as adults, at least for species that settle closer inshore; unknown for species that may settle offshore	*Arctides* — 5–146 *Scyllarides* — 0–380 *Parribacus* — 0–20 *Evibacus* — 2–90 *Ibacus* — 20–750 *Thenus* — 8–100 *Acantharctus* — 20–57 *Antarctus* — 122–440 *Antipodarctus* — <90 *Bathyarctus* — 26–800 *Biarctus* — 0–73 *Chelarctus* — 100–333 *Crenarctus* — 0–108 *Eduarctus* — 4–112 *Galearctus* — 30–390 or more *Gibbularctus* — 12–57 *Petrarctus* — 5–282 *Remiarctus* — 18–260 *Scammarctus* — 152–531 *Scyllarus* — 0–329

Habitat Selection	Unknown	Choice of depth may be the result of negative phototaxis and responses to depth of the mixed layer	Varies considerably from spatially complex reefs to sandy or muddy habitats Recruitment in to benthic habitat: *Thenus orientalis* and *T. indicus* — primarily in summer, but can occur throughout year	Usually the same as adults, but may differ in species that settle offshore and later return to adult grounds	*Arctides* — rocky habitats *Scyllarides* — rocky to shelly sand or mud *Parribacus* — rocky to sandy *Evibacus* — shelly sand to mud *Ibacus* — shelly sand to mud on continental shelves and slopes *Thenus* — shelly sand to mud *Acantharctus* — rubble, sand, shelly sand *Antarctus* — unknown *Antipodarctus* — rocky, and among sponges and corals *Bathyarctus* — rubble, shelly sand to mud *Biarctus* — reefs, shelly sand to mud *Chelarctus* — rock, mud *Crenarctus* — rock, coral reefs, shelly sand *Eduarctus* — rubble, sand, mud *Galearctus* — reefs, corals, mud *Gibbularctus* — shelly sand to mud *Scammarctus* — shelly sand to mud *Scyllarus* — highly variable, but includes habitats with algal vegetation
Food	Unknown	Most likely soft foods, such as larvae, small zooplankton, jellyfish	Nonfeeding at least in *Eduarctus martensii, Petarctus demani, Thenus orientalis*; if feeding, most likely it is on small, soft foods	Unknown; presumed to be similar to adults based on structure of mouthparts and gut	Small benthic invertebrates: mollusks, polychaetes, crustaceans. *Arctides, Scyllarides, Parribacus* specialize on bivalves (families Arcidae, Mytilidae, Isognomonidae, Pinnidae), limpets, and sea urchins (*Tripneutes*) while *Thenus* prefers scallops, but also consumes mollusks (including cephalopods and gastropods), fish and fish eggs, crustaceans (shrimp and barnacles), ostracods, polychaetes, and detritus
Predators	Unknown	Despite being transparent to blend in with water column, they are consumed by pelagic and coastal fish, such as pilot fish, albacore, rudderfish, sunfish (*Ibacus alticrenatus*); also found in stomachs of zooplankton feeding fish and barracuda (*Scyllarus* spp.)	Unknown; cryptically colored and capable of pigment changes throughout day (transparent while in water column; pigmented while on benthos)	Largely unknown; each species tends to have cryptic coloration for their preferred substrate. *Scyllarides*: dusky grouper, combers, and rainbow wrasse	Each species has cryptic coloration for their preferred substrate. *Scyllarides*: queen triggerfish: spotted gully shark; tiger shark: dusky, red, gag, and goliath groupers *Scyllarus*: scorpionfish, dusky flounder; snakefish; high hat drum; clearnose skata *Unidentified*: blackbar soldierfish, bigeye, yellowtail snapper, great barracuda, gray snapper; dog snapper; black margate

(Continued)

TABLE 7.1
(Continued)

	Naupliosoma	Phyllosoma	Nisto	Juvenile to Sub-adult	Adult
Movements	Typically nonswimming because legs are bound	Diel vertical migrations (surface waters at night; deeper waters during day) — seen in *Crenarctus bicuspidatus*, *Scyllarus* and *Scyllarides* species off the Florida Keys, USA, and *Chelarctus cultrifer*	Forward swimming via vigorous beating of pleopods; backward swimming via abdominal flexion; passive sinking	Some species may exhibit horizontal migratory movements to adult grounds (e.g., *Ibacus chacei*)	Nomadism: *Ibacus*, *Thenus* Migratory movements: *Ibacus chacei* Capable of walking long distances; capable of alternating walking with "burst-and-coast" swimming for relatively long durations (up to 40 min in *Thenus* spp.)
Growth	one molt to first phyllosoma stage	Number of molts is highly variable among subfamilies: oceanic species tend to have more molts (11–13), those found within and beyond the shelf break to oceanic areas have an intermediate number of molts (7–10), those found within and beyond the shelf break to coastal regions are highly variable regarding the number of molts (4–12)	one molt to juvenile stage	Largely unknown *Ibacus* — likely >5 molts before attaining physiological maturity (~2 years): molt increments = 20–35% premolt size *Thenus* — growth rate rapid in first year, slowing for *T. indicus* in second year after reaching 70 mm CL (size of 50% maturity is ~52 mm CL) *Scyllarides* — *S. nodifer* can attain 30 cm TL in 16–18 mo in laboratory; *S. latus* — molt increments = 24%	Variable. *Scyllarides* — *S. astori* molts every 18–24 months; growth increment = ~6% of premolt size; females attain sexual maturity between 21 and 23 cm TL (7–8 years); *S. latus* molt increments = 6%; *S. squammosus* attains sexual maturity at ~66–67 mm CL or 47.6 mm TW *Ibacus* — yearly molts; some individuals apparently molt once every two years; molt increments = 11–15% premolt size; females grow larger than males; lifespan ≥15 years *Thenus* — growth rate slows for *T. indicus* in second year after attaining sexual maturity and reaching 70 mm CL; in *T. orientalis*, growth rate remains high for several years until reaching 120 mm CL (50% sexual maturity is reached at ~58 mm CL in Australian waters and 107 mm TL in Indian waters; smallest mature female in Indian waters is 93.5 mm CL);

females grow larger, but more slowly than males; in Indian waters, size at 50% maturity is reached in ~1.5 years; lifespan ~4–5 years *Scyllarus* — *S. arctus* females are mature at 70 mm TL; *S. pygmaeus* females reach sexual maturity at ~23 mm TL; *S. rugosus* attains sexual maturity at 17–25.2 mm CL

| **Citations** | See Sekiguchi et al., Chapter 4 | Distribution: Gurney (1936), Prasad & Tampi (1960), Johnson (1968, 1971a, 1971b), Robertson (1968, 1969a, 1969b, 1971), Crosnier (1972), Ritz & Thomas (1973), Shojima (1973), Prasad et al. (1975), Ritz (1977), Phillips et al. (1981), Atkinson & Boustead (1982), McWilliam & Phillips (1983), Barnett et al. (1984), Olvera Limas & Ordonez Alcala (1988), Yeung & McGowan (1991), Rothlisberg et al. (1994), Coutures (2000), Inoue et al. (2001), Minami et al. (2001), Webber & Booth (2001), Sekiguchi & Inoue (2002), Higa & Shokita (2004)
Food: Shojima (1963), Sims & Brown (1968), Mikami et al. (1994), Mikami & Takashima (2000); see also Mikami & Kuballa, Chapter 5; Johnston, Chapter 6
Predators: Lyons (1970); Bailey & Habib (1982) | Distribution: Barnett et al. (1986), Jones (1988); Webber & Booth (2001)
Food: Barnett et al. (1986); Mikami & Greenwood (1997); Mikami & Takashima (1993, 2000); Johnston, Chapter 6; Jones, Chapter 16 | Distribution: Stewart & Kennelly (1998), Haddy et al. (2005), Sekiguchi et al., Chapter 4
Food: Johnston, Chapter 6
Predation: Martins (1985b), see Webber & Booth, Chapter 2
Growth: Rudloe (1983), Stewart & Kennelly (1998), Haddy et al. (2005) | Distribution: Holthuis (1985, 1991, 2002), Chan & Yu (1986, 1993), Brown & Holthuis (1998), Spanier & Lavalli (1998), Davie (2002), Webber & Booth, Chapter 2
Food: Cau et al. (1978), Suthers & Anderson (1981), Martins (1985b), Lau (1987, 1988), Spanier (1987), Jones (1988), Johnston & Yellowlees (1998), Martínez (2000); see also Radhakrishnan et al., Chapter 15 and Jones, Chapter 16
Predation: Lyons (1970), Martins (1985b), Smale & Goosen (1999)
Movements: Jones (1988), Stewart & Kennelly (1998), Haddy et al. (2005)
Growth: Kagwade & Kabli (1996a, 1996b), Relini et al. (1999), Stewart & Kennelly (2000), Kizhakudan et al. (2004), Subramanian (2004), DeMartini et al. (2005), Hearn (2006), Hearn et al., Chapter 14; Radhakrishnan et al., Chapter 15; Jones, Chapter 16; Bianchini & Ragonese, Chapter 9 |

TL = total length; CL = carapace length; TW = tail (abdomen) width.

1983; Yeung & McGowan 1991; Coutures 2000; see Table 1 in Booth et al. 2005). Likewise, larvae of *Arctides* and *Parribacus* species, as well as several genera of scyllarinids (*Antipodarctus aoteanus* (Powell, 1949), *Crenarctus bicuspidatus* (De Man, 1905), *Galearctus timidus* (Holthuis, 1960), and *Scyllarus depressus* (S.I. Smith, 1881)) have oceanic distributions that often place them in waters that are deeper than the depth at which their corresponding adult forms are found (Johnson 1971b; Robertson 1971; Phillips et al. 1981; Webber & Booth 2001). In contrast, species of *Evibacus, Ibacus, Biarctus, Chelarctus, Eduarctus, Petrarctus, Scammarctus, Thenus*, and most *Scyllarus* are not widely dispersed from their parental grounds. While swimming behavior of phyllosomas has not been assessed for most scyllarid species, it may be better developed in species that are dispersed farther offshore (e.g., *C. bicuspidatus*) than in species closer to shore (e.g., *Thenus* spp.), as determined by development and setation of the pereiopods and their associated exopodites in various instars of phyllosomas, and the pleopods in the nisto stage (Williamson 1969; Phillips et al. 1981; Minami et al. 2001; see Robertson 1971; Ito & Lucas 1990; Marinovic et al. 1994; Mikami & Greenwood 1997 for descriptions and diagrams of these appendages in various species and in different instars).

Coutures (2000) noted that, at least in *S. squammosus*, first-stage phyllosomas reach the surface rapidly post-hatching and are positively phototropic. Likewise, Stewart et al. (1997) noted that first-stage phyllosomas of *I. peronii* were active swimmers that were strongly attracted to light. Some larvae are known to undertake diel vertical migrations (e.g., *C. bicuspidatus* and *Chelarctus cultrifer* (Ortmann, 1897)), but data are limited as to the extent of these migrations and the species-specific preferences for various depths (Phillips et al. 1981; Minami et al. 2001; Booth et al. 2005; Sekiguchi et al., Chapter 4), as well as the efficacy of their swimming behavior. It is likely that smaller instars with less developed pereiopods vertically migrate less than later, larger instars with better developed pereiopods (Yeung & McGowan 1991). Those species or instars that do exploit diel vertical migrations may use passive transport by occupying vertical strata that move them in specific directions (Booth et al. 2005; Sekiguchi et al., Chapter 4). Thus, if the larvae are positively phototropic, they may be transported back toward parental grounds or to juvenile grounds from offshore areas by using surface drift to move inshore. Alternatively, if negatively phototropic, larvae may use specific subsurface countercurrents or gyres to avoid displacement by surface currents and to maintain their position over deeper waters (Johnson 1971b; Berry 1974; Sekiguchi 1986a, 1986b; Yeung & McGowan 1991; Lee et al. 1992, 1994; Fiedler & Spanier 1999; Inoue et al. 2000, 2001; Sekiguchi & Inoue 2002). However, Fiedler & Spanier (1999) found that phyllosoma larvae of *Scyllarus arctus* (Linneaus, 1758) in the Eastern Mediterranean were sampled considerable distances offshore and in long-term, persistent or recurrent eddies. Data from physical oceanography studies imply an increased probability that phyllosomas can be "trapped" in these eddies for relatively long periods of time, leading to offshore drift (but see McWilliam & Phillips (1983) who argue that macrozooplanktonic forms, including *Crenarctus* (formerly *Scyllarus*) *bicuspidatus*, are more abundant in surface waters than in subsurface eddies). Such offshore drift within a mostly closed basin, such as the Mediterranean, would have advantages — the larvae could possibly reach unexploited or mostly open coastline habitats where they would have little competition for food and shelter with other decapods. But, the larvae also risk being "lost at sea" and perishing if they do not reach a proper settlement habitat by the time they metamorphosed into the nisto stage; this would be particularly problematic for species in large, open ocean basins. Fiedler & Spanier (1999) suggested that longer planktonic lives and wide dispersal should be correlated with wider niche breadth of juveniles and adults, and might be more important in species that live in environments with limited resources, such as those found in the Mediterranean.

Some phyllosomas (various species of *Ibacus, Scyllarides, Parribacus*, and *Scyllarus*, as well as *Thenus orientalis, Petrarctus demani* (Holthuis, 1946), and *Eduarctus martensii* (Pfeffer, 1881)) even travel attached to the aboral surface of jellyfish medusae (Shojima 1963, 1973; Thomas 1963; Herrnkind et al. 1976; Barnett et al. 1986; Booth & Mathews 1994), which may affect their dispersal or allow them to remain relatively near shore where such medusae are common (Booth et al. 2005; Sekiguchi et al., Chapter 4). A clearer understanding of the phototropic responses of the different instars of the phyllosomas, their sensory abilities with regard to gravity reception and orientation, as well as more information on the actual depth ranges of their vertical distribution, would further understanding of the recruitment processes of these larval forms.

Phyllosomas appear to be raptorial predators, using their pereiopods to hold onto food items, which are then shreaded by the maxillipeds and masticated by molar processes of the mandibles (Mikami &

Takashima 1994). Scyllarid phyllosomas lack an exopodite on the third maxilliped — a diagnostic feature that clearly distinguishes these phyllosomas from all palinurid phyllosomas except *Jasus* spp. and which suggests differences in feeding strategies or abilities (Booth et al. 2005). Mostly fleshy foods are ingested corresponding to the masticatory abilities of the mouthparts and cardiac stomach; such food types should be readily available in the water column (Mikami et al. 1994; Booth et al. 2005; Sekiguchi et al., Chapter 4). Some *Scyllarus* larvae have been observed holding hydromedusae, but it is not known if these were subsequently ingested (Shojima 1963). Nematocysts have been found in the fecal matter of large, unidentified phyllosomas (Sims & Brown 1968), which suggests that at least some species may use the hydromedusae for food.

The developmental period for scyllarid larvae is far more variable than that for palinurids, and can last from a few weeks to at least nine months (Booth et al. 2005). Nauplisomas are small (1 to 2 mm total length, TL) and short-lived forms. While first-stage phyllosomas can be quite small (see Sekiguchi et al., Chapter 4 for size information on many species), the size of phyllosomas at final stage is highly variable among the species, ranging from 10 to 80 mm TL (Booth et al. 2005; Sekiguchi et al., Chapter 4). Final-stage phyllosomas typically have shorter, broader second antennae and well-developed, feathery exopodites on their pereiopods (Robertson 1971; Ritz & Thomas 1973). Table 7.1 summarizes the distribution, movement patterns, food preferences, potential predators, and growth for larval forms.

7.3   Postlarval (Nisto) Behavior

The final-stage phyllosoma metamorphoses into the nisto, or postlarval (megalopa) stage, which, like spiny lobster pueruli and clawed lobster postlarvae, recruits into the benthic environment. Nistos are neither completely planktonic nor completely benthic — they are caught in plankton tows demonstrating that they are pelagic at least part of the duration of this phase (Booth et al. 2005). Some scyllarid nistos are excellent swimmers (using their abdominal pleopods — Robertson, 1968), while other species are poor swimmers; some are also capable of executing tail flips (backward swimming) as a means of escape when disturbed (Lyons 1970; Higa & Saisho 1983; Barnett et al. 1984, 1986; Ito & Lucas 1990). Webber & Booth (2001) suggest that these swimming differences exist due to marked differences in the size of pleopods among different species. However, this suggestion has not been adequately tested.

Like the phyllosoma, the nisto is initially completely transparent, which makes it cryptic in the water column and, no doubt, helps it to avoid predation. In many species of scyllarids, the nisto appears to bury into soft substrates during the day and swim actively at night; some species even change coloration daily between these two habitats to remain cryptically colored in both (Barnett et al. 1986; Booth et al. 2005). Nistos look like adult scyllarid lobsters — the pereiopod exopodites are lost in the metamorphosis, the second antennae have become flattened, broad, and plate-like, and the abdomen varies in length to about twice the length of the carapace (e.g., *Scyllarus* spp.) or about half of the length of the carapace (e.g., *Ibacus peronii*) with functional pleopods (swimmerets) that bear natatory (swimming) setae (Williamson 1969; Robertson 1971; Ritz & Thomas 1973). Some researchers noted the similarity of nistos to adult *Ibacus* and *Parribacus* forms (rather than adult *Scyllarides* forms) and, as such, nistos were formerly referred to as a "pseudibacus" (Chace 1966; Crosnier 1972; Holthuis 1993b). The duration of the nisto phase lasts from seven to 24 days (see Table 1 in Booth et al., 2005). The nisto is typically nine to 13 mm in carapace length (CL) (Michel 1968; Lyons 1970; Crosnier 1972), but can reach a size of 20 mm CL in some species (Booth et al. 2005).

As with spiny lobster pueruli, the nisto appears to rely on energy reserves, rather than to actively feed (Sekiguchi et al., Chapter 4; Mikami & Kuballa, Chapter 5). However, its proventriculus has features that are transitional between the phyllosoma and the juvenile (Johnston, Chapter 6) and suggest that the ability to process and sort ingested food particles is more advanced than it is in phyllosomas. While the phyllosoma lacks a cardio-pyloric valve that divides the anterior cardiac chamber from the posterior chamber, the nisto has this feature. But, like the phyllosoma, the nisto lacks a gastric mill, suggesting that food, if consumed, is similar in softness to that of the phyllosoma, and is primarily masticated by the mouthparts prior to ingestion (Mikami & Takashima 1993; Johnston, Chapter 6).

Nistos are taken in plankton tows and occasionally are found on spiny lobster pueruli collectors. Better methods need to be developed to sample this phase in both the pelagic and benthic realms. Additional studies examining the substratum preferences and methods of substrate sampling could easily be conducted in laboratory settings to better understand how the nisto makes the transition to juvenile or adult habitat and to better elucidate the sensory modalities involved in this transition. Table 7.1 summarizes the little information available about nisto distribution, habitat selection, movement patterns, predators, and growth patterns.

7.4   Juvenile Behavior

In contrast to an ever-increasing body of knowledge of the juvenile life of clawed and spiny lobsters, almost nothing is known about the habits of juvenile slipper lobsters. The nisto stage in most species lasts approximately a half-month to a month. The size of the first juvenile, when known, is dependent on species, but varies from about 10 to 60 mm CL. In the few species where the molt increment of juveniles has been determined, juveniles appear to have greater growth increments than adults and more frequent molting (Stewart & Kennelly 2000; Haddy et al. 2005). In some species (e.g., *Ibacus* spp., *Thenus* spp., *Scyllarides nodifer*), growth is rapid and sexual maturity is attained at a relatively young age (two to three years) (Rudloe 1983; Stewart & Kennelly 2000; Courtney 2002; Haddy et al., Chapter 17); in other species (e.g., *S. astori*), growth is slower with sexual maturity occurring at an age >6 years (Hearn 2006).

Large juveniles (subadults) of some species are sampled in pots and trawl nets as by-catch (e.g., *Ibacus* spp. — Graham et al., 1993a, 1993b; Graham & Wood, 1997; *Thenus* spp. — Kizhakudan et al., 2004; Radhakrishnan et al., Chapter 15), but apparently have not been retained for laboratory studies. In other species, juveniles have not been sampled at all (e.g., *S. latus* — Spanier & Lavalli, 1998). A small preserved male *S. latus* of 34.3 mm CL was recorded in the Museum of Zoology of the University of Florence, "La Specola" — it had been collected by a scientific trawl in Italian waters possibly at a depth >400 m. This record may suggest that, at least in this species, the nisto settles on the substrate in deep water and the juveniles develop there. Similar suggestions have been made for other scyllarids. For example, *S. astori* displays a narrow size range in fishery samples, with almost a complete absence of juveniles (few individuals smaller than 20 cm TL). Hearn (2006; see also Hearn et al., Chapter 14) suggested that the juveniles occupy a different spatial niche from adults and are far more cryptic than adults. *Scyllarides nodifer* juveniles have not been sampled with adults in the month of April in the Gulf of Mexico, suggesting that they may have different spring movement patterns than adults (Hardwick & Cline 1990). Likewise, *Ibacus peronii* juveniles appear to live in a different habitat than adults and migrate shoreward from offshore waters to recruit into adult grounds (Stewart & Kennelly 1997). Interestingly, swimming abilities may differ between nistos and juveniles: in *S. nodifer*, *Scyllarus americanus* (S.I. Smith, 1869), *S. chacei* Holthuis, 1960, and *S. depressus*, the relative size of the pleopods is reduced in the juvenile compared to the nisto and remains reduced until the genital pores develop (Lyons 1970). Since the animals are experiencing growth in both their carapace and abdomen at this time, but the pleopods initially are about the same size as those found in the nisto, it is likely that, at least in these species, swimming abilities are reduced (Lyons 1970). However, as the animals approach sexual maturity, the pleopods increase in size, presumably for the purpose of oviposition in females.

Table 7.1 summarizes what is known about the distribution, food preferences, predators, and growth patterns of the juvenile forms. As can be seen in this table, most of what is known about this phase is speculative, based on assumptions made about adult distribution and biology. Thus, this phase of slipper lobster life is in great need of further study. However, it is clear that in order to obtain sufficient numbers of small individuals for such studies, specific sampling techniques must be developed that target the juveniles, which may prove difficult if many of the species have juvenile development in deep, oceanic waters.

7.5   Adult Behavior

Adults are captured more frequently than other life stages, either by diving, trawling, or pots (see Spanier & Lavalli, Chapter 18 for detailed information on fishing techniques) and, as a result, adults of several species have been brought into the laboratory or in controlled field settings for the purposes of studying growth (see Bianchini & Ragonese, Chapter 9), larval biology and aquaculture (see Mikami & Kuballa, Chapter 5; Haddy et al., Chapter 17), and behavior (see Martins 1985b; Lau 1987, 1988; Spanier 1987; Almog-Shtayer 1988; Jones 1988; Spanier et al. 1988, 1990, 1993; Spanier & Almog-Shtayer 1992; Barshaw & Spanier 1994a, 1994b; Barshaw et al. 2003; Jones, Chapter 16). One result of this access is that more is known about adult behavior than any other ontogenetic phase. Various species have also been captured, tagged, and released to determine wild-growth patterns and movements (see Bianchini & Ragonese, Chapter 9; DiNardo & Moffitt, Chapter 12; Hearn et al., Chapter 14; Jones, Chapter 16; and Haddy et al., Chapter 17 for descriptions of such studies). Even so, these studies have tended to focus only on a few species and, given the highly variable distribution and habitat preference among all species of scyllarids, care should be taken before generalizing among species within a subfamily or across subfamilies. This section identifies the species examined and focuses on the behaviors that have been examined in the laboratory, namely feeding behaviors, shelter preferences and substrate selection, mating behavior, intraspecific interactions, diel activity patterns, movement patterns, and antipredator behavior.

7.5.1   Feeding Behavior

Much of the work on external feeding mechanisms has focused on the *Scyllarides* spp. — *Scyllarides aequinoctialis, S. haanii* (De Haan, 1841), *S. nodifer, S. latus, S. squammosus, S. tridacnophaga* Holthuis, 1967 (Holthuis 1968; Lau 1987, 1988; Malcom 2003; Malcom & Lavalli, personal observations), and the two *Thenus* spp. — *Thenus indicus* and *T. orientalis* (Jones, 1988). One study has used *Parribacus antarcticus* (Lund, 1793) and *Arctides regalis* Holthuis, 1963 (Lau 1988). It is important to characterize the feeding behavior for specific genera within a subfamily. The descriptions that follow are based on these studies and should not be overgeneralized for two established reasons. First, the various species within the four subfamilies differ in food preferences. Some species appear to be invertebrate generalists (*Parribacus antarcticus*), while others appear to be molluskan specialists (*Scyllarides* spp.). Secondly, fundamental differences exist in the structure of the mouthparts of genera among different subfamilies: the Arctidinae (*Arctides* and *Scyllarides* species) and Ibacinae (*Evibacus, Ibacus*, and *Parribacus* species) possess multi-articulated flagella on their maxillipeds, whereas the Scyllarinae (*Acantharctus, Antarctus, Antipodarctus, Bathyarctus, Biarctus, Chelarctus, Crenarctus, Educartus, Galearctus, Gibbularctus, Petrarctus, Remiarctus, Scammarctus*, and *Scyllarus* species) and Theninae (*Thenus* species) lack such a flagellum on their first and third maxillipeds and bear only a single-segmented flagellum on their second maxilliped (Webber & Booth, Chapter 2). Ibacinae species have a mandibular palp of only two segments, while Arctidinae have a three-segmented mandibular palp (Webber & Booth, Chapter 2). There is also a fundamental difference that separates the Theninae from the other subfamilies — the fifth pereiopod of the female is achelate in *Thenus* spp., but is chelate in genera within the other subfamilies (Webber & Booth, Chapter 2; Jones, Chapter 16). Such differences can affect the use of appendages during the feeding sequence.

Scyllarides spp. appear to have become specialized for feeding on bivalves (usually scallops, clams, mussels, or oysters). Bivalves have existed since the pre-Cambrian era and have colonized much of the world's marine and aquatic environments. Given their specialization for feeding on bivalves, it seems almost certain that the radiation of *Scyllarides* spp. into the central, eastern, and Western Atlantic Ocean, as well as into the Indo-Pacific (east Asia-Australasia region, east Pacific, and western Indian Ocean) (Webber & Booth, Chapter 2) followed the beds of bivalves around the world. Where clawed lobsters crush bivalve shells with their claws and spiny lobsters use their mandibles to crack and chip away at bivalve shells to access the meat, *Scyllarides* spp. have evolved an elegant feeding mechanism that involves using the physics of the bivalve shell to their advantage, while, at the same time, overcoming the disadvantage of the extremely effective adductor muscles that keep molluskan valves closed. In essence, they "shuck" bivalves (Lau 1988; Spanier 1987), using tactile and olfactory senses, as well as a guided mechanical advantage, to avoid the cost that a "brute" force mechanism would require. In contrast, clawed lobsters use repetitive loading via their crusher claw to cause fracture lines in the rigid shell of bivalves (Moody & Steneck 1993). This works because there is little organic matrix that would blunt the cracks created by repetitive loading. Spiny lobsters lack claws but have, in their place, strong mandibles, which they use to bite cracks into the valve edges of bivalves (Randall 1964; Lavalli, personal observations). Once a sufficiently large hole is bitten into the edge, the lobster can dig molluskan flesh out of the shells with its more-rounded pereiopod dactyl tips (Smale 1978). In the case of both claw-loading and biting of bivalves, the process to crack the shell takes some time; probably, this time exceeds that needed by slipper lobsters to *shuck* shells. *Scyllarides* spp., like spiny lobsters, lack claws, and apparently also lack strong mandibles; thus, they resort to using their pereiopods — specifically the sharp, pointed, and chitin-reinforced dactyl tips — to open bivalves (Spanier 1987; Lau 1988).

During the feeding sequence, *Scyllarides* spp. typically approach a bivalve with their antennules down near the substrate. Upon encountering the mollusk, the lobster picks up the shell with the first two to three pairs of pereiopods and repetitively probes the outer valves with its antennules, as though "smelling" and assessing the shell for its possible value (Lau 1987; Malcom 2003). This behavior contrasts with that of clawed and spiny lobsters that do not use their antennules for such probing activities, but instead use them to distantly "chemo-orient" to the food source (Devine & Atema 1982; Zimmer-Faust & Spanier 1987; Moore & Atema 1991; Moore et al. 1991a, 1991b; Beglane et al. 1997; Nevitt et al. 2000; Derby et al. 2001). In clawed and spiny lobsters, the setose pereiopods are used for initial assessment of food items (Derby & Atema 1982b). However, after the initial assessment of the shell by the antennules, lobsters (*S. aequinoctialis* and *S. nodifer*) grasp the bivalve and orient its shell rim properly for access by the pereiopods. With the bivalve thus firmly held with either the first, third, and fourth pairs, or the second, third, and fourth pairs of pereiopods, the lobster then uses the dactyl tips of either the second or first pair of pereiopods to repetitively probe the edges of the valves (Malcom 2003). By such repetitive probing, they eventually wedge the dactyl tips into the shell edge and then insert the tips further and further into the shell — a process known as "wedging" (Lau 1987). Once one pair of pereiopod dactyls is inserted, another pair — usually those of the second and the third pereiopods — is used to cut the mantle tissue along the pallial line (line of attachment to the valve). Then the lobster uses a "scissoring" motion of the first two pairs of pereiopods to increase the opening angle and to provide access to the adductor muscles (Figure 7.1A and Figure 7.1B; Malcom 2003). The second or third pereiopod cuts the adductor muscles, so that the valves open freely. With the valve rims separated, the meat is repetitively scraped from the surface of the valves and passed directly to the third maxillipeds (Figure 7.1C and Figure 7.1D; Lau 1987; Malcom 2003). These appendages are used to stretch the flesh and pass the strands back to the subsequent five pairs of mouthparts for ingestion. Until the molluskan flesh is actually passed back to the third maxillipeds, the antennules make repeated downward motions to probe inside the valves, to touch the flesh, and to touch the shell as the pereiopods scrap the flesh from it (Malcom 2003).

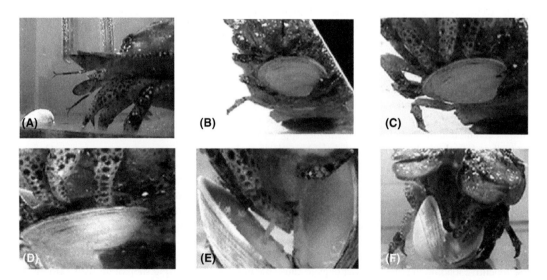

FIGURE 7.1   The feeding process of scyllarid lobsters. (A) investigative behavior, using the antennules; (B) probing of the shell by the first two sets of pereiopods; (C) insertion of the pereiopods into the shell; (D) wedging and subsequent cutting of the adductor muscles; (E) opening of the shell and removal of the bivalve flesh; (F) scraping of the empty shell. (Video stills by C. Malcom; used with permission.)

In *S. squammosus* this scheme is modified (Lau 1987): the lobster uses the dactyls of the first two pairs of pereiopods to pry open shell lips; the dactyls of the third pair (or sometimes the second pair) are used to sever the adductor muscle, and the fourth and fifth pair of pereiopods brace the bivalve against the lobster and the lobster against the substrate. If the shell is cemented onto a substrate (e.g., oysters), then the lobster inserts dactyls into the shell while rocking back and forth, presumably to exhaust the adductor muscle with the repetitive pulls and pushes. For shells attached to substrates via byssal threads, the lobster simply pulls the shells off and then wedges. If the initial attempt at wedging fails, *S. squammosus* may attempt to chip the edge of the shell with its mandibles. Occasionally, lobsters will sample the shell with their antennules and then simply hold the shell with their dactyls poised two to three cm above the opening and will wait until the bivalve reopens, whereupon they will plunge the dactyls into the opening, wedging the valves fully open and cutting the adductor muscles. Typically these different schemes range in duration from 10 to >40 min (Lau 1987).

The opening of giant clams (*Tridacna* spp.) by *S. tridacnophaga* was observed by C. Lewinsohn and reported in Holthuis (1968). The giant clam uses byssal threads to attach itself to a substrate. The lobster manipulates the clam to expose its dorsal surface from whence the byssal thread attachments protrude and then it plunges its dactyls into this exposed and vulnerable region, which causes the clam to gape. At this point, the lobster turns the shell over and inserts its pereiopods to further wedge the valves open.

Additional observations by Lau (1987) on *S. haanii* show that this species carries out a scheme similar to *S. squammosus* and has similarly shaped, blunt dactyls. In contrast, *A. regalis* and *P. antarcticus* have more tapered, sharper dactyls that appeared, at least to Lau, to be less suited to wedging behaviors. His observations indicate that these species can perform a type of wedging behavior to open bivalves, but the time required to do so is longer than that for *Scyllarides* species. In addition, *P. antarcticus* appears to consume a wider variety of prey species than *S. squammosus* and, while mollusks were still the most important prey group, they tended to be better represented by gastropods and chitons than by bivalves (Lau 1987). Soft, fleshy, sea anemones represented the second largest proportion of the diet of *P. antarcticus*, with annelids, sipunculids, echinoderms, and small arthropods also being present (Lau 1987).

Thenus spp. employ a tactic more similar to clawed lobsters to locate food — they nocturnally search an area of soft sediment by repetitive probing of their first two pairs of pereiopods, while continuously moving their antennules up and down and flicking the lateral filaments (Jones 1988). Food odors elicit chemo-orientation behaviors and directed movements to the source. Contact with the food source results in grasping and manipulation by the pereiopods (Jones 1988). Presumably they open bivalves in a similar manner to *Scyllarides*, *Arctides*, and *Parribacus* species, but the exact sequence of leg and mouthpart movements has not been described (see Jones, Chapter 16 for additional information). In the laboratory, *Thenus* spp. readily take scallops (*Amustium* and *Chlamys* spp.), goatfish (*Upeneus* spp.), shrimp (*Metapenaeopsis* and *Trachypenaeus* spp.), and lizardfish (*Synodus*, *Saurida*, and *Tracinocephalus* spp.), but avoid urchins and cuttlefish (Jones 1988; see Table 7 in Jones, Chapter 16).

While the preferred food of *Scyllarides* spp. appears to be molluskan bivalves, these lobsters are also known to take sea urchins, crustaceans, sponges, gastropods, barnacles, sea squirts, algae (*Ulva* spp.), and fish flesh from gut content analyses. Different species may well prefer different food items, and setting (laboratory vs. nature) may exert an effect. For example, wild *S. latus* appear to consume bivalves (*Venus verrucosa* Linnaeus, 1758, *Glycymeris pilosa* (Linnaeus, 1767), *Pinctada radiata* (Leach, 1814), *Brachidontes semistriatus* (Krauss, 1848), *Spondylus spinsosus* Schreibers, 1793, and limpets (Martins 1985b; Spanier 1987; Spanier, personal observations)). Laboratory tests with *S. latus* demonstrate that they prefer soft flesh to crabs and bivalves to soft flesh or snails, while no preference was shown between choices of oysters, clams, and limpets (Almog-Shtayer 1988). In the laboratory, *S. squammosus* prefers *Ostrea* spp. and *Isognomon* spp., but will take limpets (Lau 1988), and gut-content analyses of wild-caught *S. squammosus* indicate that gastropods are also taken. Wild *S. astori* prefer white sea urchins, *Tripneustes depressus* (A. Agassiz, 1863) (Martínez 2000), but, nonetheless, display a varied diet of molluskan prey (mostly mussels, pen shells, oysters, arks and cockles) as seen by stomach content analyses (Martínez, 2000).

Parribacus antarcticus gut contents include several species of chitons and gastropods, red algae, sediment (also seen in *Thenus* spp.), polychaetes, sabellids, nemertines, sipunculids, sea cucumbers, sea stars, ostracods, copepods, amphipods, stomatopods, small anomurans, carideans, and brachyurans (Lau 1988). *Thenus* spp. also seem to have a broad diet with gut contents of wild-caught animals showing a distinct preference for mollusks, followed by sediments, fishes, crustaceans, and polychaetes, with occasional incidences of sipunculids, sponges, and formaniferans (Kabli 1989). Apparently, wild *T. orientalis* will take cuttlefish (*Sepia* spp.) and squid (*Lologio* spp.), as well as arrow worms, fish eggs, and barnacles, but these items may be taken more in scavenging activities, rather than as directly, sought-after, and captured items (Kabli 1989; see Radhakrishnan et al., Chapter 15 for more information). Kabli's (1989) data conflict with laboratory findings of Jones (1988), in which *Thenus* spp. always avoided cephalopods.

Differences in preferred prey and methods of consumption may allow for niche separation in the cases where multiple species within a subfamily or multiple genera among the subfamilies overlap in distribution. They may also allow for niche separation in cases where slipper lobster species are sympatric with palinurids and nephropids. It is clear from what already is known that certain species may be generalists that maximize intake of a variety of invertebrates (and some vertebrates), while others may be highly specialized to feed on a few molluskan species or echinoderm taxa (e.g. *S. astori* and urchins). A better understanding of diet and feeding mechanics in the Ibacinae, Scyllarinae, and Theninae would improve our ability to manage entire ecosystems, especially in cases where multiple taxa are targeted for fisheries exploitation (e.g., in the Galápagos where spiny, slipper, and urchin fisheries exist) and where there is an important trophic relationship among the targeted species.

7.5.2 Sheltering Behavior and Substrate Preferences

From species descriptions and various sampling programs/cruises, the general habitat and depth preferences of most adult scyllards is known (see Table 7.1 for a listing of such preferences and appropriate references). However, few studies exist that examine the sheltering behavior and preferred physical properties of scyllarid shelters. Only Jones (1988), working on *Thenus* spp., and Spanier & Almog-Shtayer (1992) and Barshaw & Spanier (1994a), working on *Scyllarides latus*, have conducted laboratory assessments of sheltering behavior and preferred configurations of shelters. Only Spanier et al. (1988, 1990) and Spanier & Almog-Shtayer (1992) have assessed sheltering behavior in wild settings, although Sharp et al. (Chapter 11) describe habitat preferences and sheltering behavior of *S. aequinoctialis*, *S. nodifer*, and *Parribacus antarctus* and Jones (1993) examined abundance and population structure of *Thenus* spp. over different habitats and depths to determine if there was a species-specific distribution pattern. This section summarizes these studies; however, generalizations to other genera based on these studies are unwise given the broad range of habitats exploited by scyllarid species.

Microhabitat preferences have been determined for *S. latus* by providing lobsters with a variety of shelter designs in artificial reefs made of used car tires (for a review of such studies, see Spanier & Lavalli 1998). *Scyllarides latus* significantly preferred horizontally oriented dens to vertically oriented dens where light levels were higher. They also preferred shelters with small, multiple openings, like those between tires, over those with larger entrances (in the tires themselves) (Spanier et al. 1990; Spanier & Almog-Shtayer 1992). When additional "back doors" were experimentally blocked, lobsters stopped using the single-opening dens. These preferences are translated in the use of naturally constructed dens: all natural dens had two openings and most had a top covering and side walls, as well as some kind of a back structure (Spanier & Almog-Shtayer 1992). During daylight hours, light in these natural shelters was 10 to 20 times less than that in the open-reef habitat. In laboratory choice tests, using opaque or transparent plexiglass pipes for shelters, lobsters significantly preferred opaque to transparent shelters of the same shape and size. They also preferred medium-sized shelter diameters (20 to 30 cm) that were open on both ends (Spanier & Almog-Shtayer 1992). A similar preference was shown in the field in an artificial tire reef where lobsters preferentially chose to live in medium-sized diameter shelters formed between adjacent tires, rather than in the large, central hole of the tires themselves (Spanier et al. 1988; Spanier & Almog-Shtayer 1992).

When held in captivity, even in large, naturalistic aquaria with biogenic rocks and sand, *S. latus* ceased to show substrate preferences within two months, after initially spending significantly more time on rocky substrates (Barshaw & Spanier 1994a). In contrast, Chessa et al. (1996) reported that in laboratory experiments, lobsters preferred rough artificial substrates (plastic carpet) over smooth ones only after having experience with each for some time. The rougher substrates allowed the lobsters to cling with their nail-like dactyls, which protects their vulnerable ventral surface from predators (see Section 7.5.6 for additional details). Other than these two reports — one using natural structures and one using artificial structures, substrate preferences have not been well studied in other *Scyllarides* spp. and can only be inferred from areas where they are caught.

The adults of most species of *Scyllarides* are found on hard substrates, with some species also inhabiting soft-bottom habitats or shifting between soft bottom and rocky habitats (see Table 7.1 and Webber & Booth, Chapter 2). In dive surveys, *S. aequinoctialis* tended to be more common in high-relief, coral habitats with ready-made shelters, while *S. nodifer* was found on both patch reefs with high-relief and unconsolidated sediments (Sharp et al., Chapter 11). It is likely that *Scyllarides* species sampled both on hard and soft substrates result when lobsters that usually shelter in hard substrates are collected in soft substrates during their short- and long-term movements (see Section 7.5.7 for descriptions of these movements) — such a suggestion was made by Ogren (1977) to explain the distribution of *S. nodifer*. Nevertheless, a few species have been reported only on soft substrates (e.g., *S. elisabethae* (Ortmann, 1894)). Holthuis (1991) states that *S. elisabethae* seems to dig into the mud; he also mentions that *S. aequinoctialis* buries in the sand, although others report that this species is a reef dweller that shelters within coral-rock caves and under coral heads (Moe 1991). Likewise, Hardwick & Cline (1990) have reported that because they were caked in mud, *S. nodifer* in the northern Gulf of Mexico may bury in sediments during the daylight. It is assumed that such digging into soft substrates is an antipredator adaptation, similar to that seen in other lobster genera living long term or temporarily on soft substrates (e.g., *Homarus americanus* in offshore habitats — Cooper & Uzmann, 1977, 1980).

When in shelters, *S. aequinoctialis* and *S. nodifer* rarely occupy the den floor, but instead hang from the ceilings and walls. This is also common in *S. latus* that are occupying caves (Spanier, personal observations). Likewise, *Parribacus antarcticus*, which prefers similar high-relief habitats to those preferred by *S. aequinoctialis*, also occupies the den wall, rather than the floor. *Scyllarides aequinoctialis*, in particular, appears to have a high degree of den fidelity and the same individual can be found day after day occupying the same den (Sharp et al., Chapter 11).

Thenus spp. live on soft-bottom substrates into which they bury during the day (Jones 1988; see Jones, Chapter 16 for a complete description of their burial patterns and substrate use). However, there is a differential spatial distribution of the two currently recognized valid species (that have yet to be fully described in the literature; see Webber & Booth, Chapter 2 and Jones, Chapter 16), *T. indicus* and *T. orientalis*. In the laboratory, *T. indicus* prefers more fine-grained sediments (mud/silt) and *T. orientalis* prefers coarser particle sizes (sand) (Jones 1988); these preferences are carried over into wild distributions of the two species which spatially separate not only by depth, but also by the coarseness of the substrate particles (Jones 1993). Because the preferred substrate is soft and ready-made shelters are not available, as they would be in rocky or coral habitats, *Thenus* spp. bury themselves into the substrate, usually following nocturnal activity periods. Burial behavior involves backward movements and repetitive extensions of the abdomen, followed by rapid abdominal flexions — these movements cover the dorsal surface of the carapace and antennae with ~1 cm of sediment, leaving only the antennules and eyes exposed (Jones 1988; see Jones, Chapter 16). The entire sequence takes ~2 min (Jones 1988). Antennular flicking occurs while buried, albeit at a low frequency, and a gill current is effected by scaphognathite beating that draws water beneath the eyes and out at the base of the antennules (Jones 1988). Partial burial is also observed, but this generally occurs during short resting sessions throughout the nocturnal activity period. Contrary to expectations from the name "shovel-nosed" lobster, the broad, flat, antennae are not used for burial purposes (Jones 1988), but in ovigerous females of *Scyllarus arctus*, these antennae of *Thenus* spp. are apparently used for burial (see Pessani & Mura, Chapter 13).

Ibacus spp. also are found on soft-bottom substrates and are presumed to bury into those sediments, much in the same manner as *Thenus* spp. (see Haddy et al., Chapter 17 for more information). A recent study (Faulkes 2006) of digging in *I. peronii* provided a description of the sequence of behaviors involved. Normally pereiopods three to five are used in a typical alternating tripod gait (Wilson 1966; Johnston & Yellowlees 1998). Digging is initiated when all pereiopods are inserted into the substrate and the abdomen is fully flexed and pressed down such that the tailfan contacts the substrate. The abdomen is then extended which causes the substrate (usually sand) to be pushed backward and results in the abdomen submerging. The abdomen then flexes and reextends several times, and the pereiopods may be repositioned, by lifting out the anterior to posterior legs in a metachronal wave and reinserting them in a more posterior position than before. The sequence ends with a few tail flips that cover the exposed portions of the carapace completely with sand. *Ibacus peronii* is a slow digger, taking >4 min and an average of 17 cycles of abdominal extension and flexion to completely submerge into the sand. As with *Thenus* spp., the broad, flat antennae are not used in this behavior.

7.5.3   Mating Behavior

Because scyllarids copulate, spawn, and brood readily in the laboratory, some aspects of their reproductive biology are known; however, less is known of the actual mating behavior or the rituals involved during the mating process. It is likely that the permanence of the spermatophore differs among the genera; it is also likely that fertilization is internal in some genera and external in others.

In *Scyllarides latus*, males produce white, gelatinous spermatophores, which they carry around on the base of their fourth and fifth pereiopods (Almog-Shtayer 1988; Spanier, personal observations). These are transferred externally to females. It is not clear whether females retain the spermatophores externally and fertilize their eggs externally, or whether they somehow manipulate the spermatophore and store it internally. In some *Scyllarides* species, females have been observed carrying spermatophores externally 6 to 10 days or less prior to egg extrusion (*S. latus* — Martins 1985b; Almog-Shtayer, 1988), while in others, the lack of observable spermatophores prior to egg extrusion has led to a belief that the spermatophore is stored internally and fertilization is internal (*S. nodifer* — Lyons 1970; and *S. squammosus* — DeMartini et al. 2005). Females of most *Scyllarides* species can spawn multiple broods in a season due to short brooding periods (but see Hearn et al., Chapter 14 for a contrasting view, in which *S. astori* broods once annually), and these broods are usually carried during warm (spring and summer) months. Only in *S. latus* have both eggs and spermatophores been observed simultaneously (Almog-Shtayer 1988); however, in this species, multiple broods have not been observed, with only one peak spawning occurring in spring and summer months of April to July (Spanier, personal observations).

Ibacus species deposit persistent, gelatinous spermatophores near the genital openings and it is thought that egg extrusion and fertilization (external) occurs shortly after the spermatophore is deposited on the female (Stewart & Kennelly 1997; Haddy et al. 2005). *Ibacus chacei* and *I. peronii* appear capable of spawning at least twice per year; however, there is no evidence that *I. alticrenatus* and *I. brucei* Holthuis, 1977 have multiple broods, as females in these species possess inactive ovaries at various times during a year (Haddy et al. 2005). Broods are usually spawned and carried in the colder months (autumn to spring) with hatching in warmer months (spring through summer).

Parribacus antarcticus has been observed bearing spermatophores that were hard and black, somewhat similar in form to those typically seen on most spiny lobsters species (reported in Lyons 1970). Apparently they are capable of carrying both eggs and a spermatophore simultaneously, and can fertilize multiple broods with the same spermatophore (Lyons 1970; Sharp et al., Chapter 11). They can also repetitively mate and have been observed with a fresh spermatophore atop a spent one (Sharp et al., Chapter 11). It is not clear what other species within the genus do with regard to spermatophores.

Thenus species do not deposit a persistent spermatophore, so egg extrusion typically occurs within hours of mating. Even after 2000 h of remote video observations, no precopulatory courtship behavior has ever been observed (Jones 1988). The manner of spermatophore transfer is unknown, although Kneipp (1974) noted that spermatophores were deposited on both sides of the female's sternum below the genital openings after a rapid (few seconds) embracement of the ventral surfaces of the animals. Fertilization is presumed to be external and to occur shortly after mating (Jones 1988; Kizhakudan et al. 2004; see Radhakrishnan et al., Chapter 15 and Jones, Chapter 16 for more information), and in *T. orientalis*, the spermatophoric mass appears to be lost 12 h after mating (Kizhakudan et al. 2004). Spawning in *T. orientalis* apparently occurs between September and April (Kabli & Kagwade 1996), with the highest frequency of ovigerous females occurring November to January (spring and summer). A second spawning peak can occur in March, so it is likely that this species may be capable of spawning multiple broods each year (Kabli & Kagwade 1996); however, in the Red Sea, only a single spawning period was noted (Branford 1982).

Scyllarus species mating habits are not well known. In the laboratory, *S. depressus* was seen to hatch larvae and three days later to have extruded another brood (Robertson 1971). No external spermatophore was noted on the sternum of the female and fertilization was presumed to be internal. Lyons (1970) sampled numerous *S. depressus* females in the Gulf of Mexico, but also never saw the presence of a spermatophoric mass. However, in *S. rugosus*, spermatophoric masses were seen on intermolt females after nocturnal mating and ovipositioning occurred several hours later (Kizhakudan et al. 2004).

7.5.4 Intra- and Interspecific Interactions

In those species for which diving censuses have been conducted, it is clear that slipper lobsters can range from being solitary to highly gregarious. From observations on species that are gregarious (e.g., *S. latus*, Barshaw & Spanier, 1994a), there is intraspecific competition over food items and, generally, the largest females are dominant over other lobsters (Barshaw & Spanier 1994a). The least aggressive of such encounters is the "approach/retreat" sequence common to all lobsters (Atema & Cobb 1980) — one lobster walks toward the other, which responds by walking away or otherwise avoiding the approaching lobster. A more aggressive encounter involves use of the flattened second antennae and is called a "flip." Here the lobster jerks up its flattened second antennae under the opponent's carapace, attempting to dislodge it. In the most intense aggressive behavior, the attacking lobster grabs the anterior portion of the opponent and holds on with the dactyls of the pereiopods. The opponent usually tail flips, as does the attacking lobster, which causes the opponent to end up on its back. Often after this "face grab" maneuver, both lobsters are holding onto the same food item, ventral side to ventral side, and they continue in this fashion until one lobster finally relinquishes its hold on the food item (Barshaw & Spanier 1994a).

Only one observation of cannibalism was observed in *S. latus* where an individual was feeding on a conspecific postmolt, and it was not clear if the cannibalizing individual killed the conspecific during or after its molt, or if it was simply opportunistically feeding on an individual that died during ecydsis (Spanier, personal observation).

A tendency for gregarious sheltering among *S. latus* was observed in a survey of natural dens where 95% of lobsters cohabited with one or more conspecifics (Spanier & Almog-Shtayer 1992). This gregarious behavior also was demonstrated by very large lobsters (>100 mm CL) in the field, which differs from gregarious spiny lobster species where the tendency for cohabitation is greater in smaller and medium-sized animals (Spanier & Zimmer-Faust 1988). Fishermen have reported aggregations of 50 to 60 lobsters in the same shelter or shelter-providing structure (reported in Spanier & Almog-Shtayer, 1992).

Similar clustering behavior was observed among *S. latus* individuals in naturalistic habitats and with artificial shelters in laboratory tanks. In the absence of a predator, freshly caught, laboratory-held lobsters significantly preferred an opaque artificial shelter compared to a transparent shelter of the same shape and size, and formed clusters within this opaque shelter. When they were supplied with no shelter but with shade, the lobsters concentrated under the shade (Spanier & Almog-Shtayer 1992). When neither shelter nor shade was supplied, they showed distinct gregarious behavior similar to the defensive "rosette" observed in migrating Caribbean spiny lobsters under attack by a triggerfish (Herrnkind 1980; Kanciruk 1980; Herrnkind et al. 2001). However, field predation studies indicate this gregarious behavior does not confer any advantage on exposed individuals within the group who are under attack by fish predators. They suffer an equal amount of predation as do solitary animals exposed to the same fish predators and gain only a small advantage of time, as predatory attack patterns are less focused when lobsters are grouped (Lavalli & Spanier 2001).

Reports of gregarious behavior also exist for *S. nodifer* (Moe 1991), but these may be question-able. Rudloe (1983) reported that individuals established their own residences in the laboratory and that social interactions were limited to displacement of other individuals on mussel clumps. In diver surveys, Sharp et al. (Chapter 11) report that *S. nodifer* was observed sheltering alone 100% of the time, and *S. aequinoctialis* was observed sheltering alone 84 out of 86 instances, and with a second conspecific only two out of 86 times. In the same survey, *Parribacus antarcticus* was observed sheltering twice — once with a spiny lobster and once with a conspecific. In similar surveys of *S. astori* in the Galápagos archipelago, divers found that these lobsters were primarily solitary (Hearn et al., Chapter 14). In a study of the population biology of congeners in the Hawaiian archipelago, Morin & MacDonald (1984) reported that *S. haanii* was a solitary species, whereas *S. squammosus* tended to occur in groups. It seems that, at least in the genus *Scyllarides*, there is great variability in the sociality of the species, and this suggests that other genera may be equally variable in their sociality.

While *Thenus* spp. appear to be solitary, they are capable of being held together in tanks, but display little to no interest in other individuals. Agonistic encounters only appear to arise over food items (as is the case for *S. latus*, described earlier), and involve maneuvering for better mechanical advantage in possession of the food item (via pereiopod movements and tail flipping) (Jones 1988; see Jones, Chapter 16 for more details). Mikami (1995) noted that the phyllosomal stages of *Thenus* spp. can be cannibalistic, but only if insufficient food is present in the rearing chambers.

Obviously, more work is needed to understand the nature of associations between conspecifics when they occur, as well as to determine their value to the individual lobster, and how widespread they are through the various genera of scyllarids. It is possible that in those species that display communal sheltering or sheltering with at least one conspecific, the "guide effect," seen in juvenile *Panulirus argus* (Childress & Herrnkind 1996, 2001), may play a role. In the *guide effect*, individuals are attracted to shelters that harbor conspecifics, as chemo-orientation to the odor of the conspecific reduces the time necessary to find an appropriate shelter. Collective denning was highly correlated with conspecific density and scarcity of local shelters, rather than with lobster size, molt condition, shelter type, or predator density. Childress & Herrnkind (1996, 2001) suggested that this *guide effect* benefited the shelter-seeking individual by reducing the time of exposure on the substrate with its associated predator risk. Once sheltered with conspecifics (particularly those with weapons, as is the case with spiny lobsters and their spinose, long antennae), the risk of predation is reduced upon the individual via collective prey vigilance and defense, making gregarious behavior a beneficial trait (e.g., *P. argus*: Zimmer-Faust & Spanier 1987; Spanier & Zimmer-Faust 1988; Butler et al. 1997; Herrnkind et al. 2001; *Jasus edwardsii* (Hutton, 1875): Butler et al. 1999). It is likely that collective denning may provide a similar benefit to scyllarids, at least in those species that live in rocky habitats and for which large shelters are available.

7.5.5   Diel-Activity Patterns

Adult specimens of all scyllarids appear to be well camouflaged due to their flattened morphology and various coloration patterns that blend into their preferred substrates (see Webber & Booth, Chapter 2 for an expanded discussion of distribution and morphological features). However, for those species that live in shallow, brightly illuminated waters, this camouflage provides only limited concealment against diurnal predators. Therefore, like most other lobsters, scyllarids, by and large, appear to be nocturnally active. However, it is not clear whether deeper-water scyllarids display activity patterns different from shallow-water species during daytime hours, given the reduction in light at depth. Again, very few studies examining diel activity patterns exist for scyllarids — only *Scyllarides* and *Thenus* spp. have been examined in laboratory settings (Jones 1988; Spanier & Almog-Shtayer 1992; Barshaw & Spanier 1994a).

Scyllarides spp. forage during the night and shelter during the day on the ceilings of caves, in crevices in vertical rocky walls (e.g., Barr, 1968; Martínez et al., 2002 for *S. astori*; Spanier & Lavalli, 1998 for *S. latus*), and in other natural dens, as well in artificial reefs in the field (including ship wrecks), or in man-made shelters in the laboratory (*S. nodifer*: Rudloe, 1983; *S. latus*: Spanier et al., 1988, 1990; Spanier & Almog-Shtayer, 1992; Spanier, 1994). However, in laboratory holding tanks, where predators are not encountered for a long time, they tend to shift to diurnal activity (Spanier et al. 1988, 1990; Spanier & Almog-Shtayer 1992), foraging for bivalves and even carrying bivalves back to their shelters to consume at a later time (Spanier et al. 1988). In addition, when visibility is low during the day, as is common during and following a storm, lobsters can be detected in the entrances of their shelters and exposed on substrates (Spanier & Almog-Shtayer 1992; Spanier, personal observations). This latter behavior suggests, as has been found for *Thenus* spp. (Jones 1988, see below), that increases in *S. latus* activity may be linked to crepuscular periods of the diel cycle.

Based on trawl surveys, it appears that the shallow-water species, *Ibacus chacei* and *I. peronii*, also are largely inactive during the day — catches per 60 min tow during the day averaged $\sim 1.8 \pm 2.0$ (normalized) and 4.9 ± 4.4 individuals, respectively, vs. 37.8 ± 28.4 (normalized) and 9.3 ± 7.8 at night, respectively (Graham et al. 1993a, 1993b, 1995, 1996). However, catch rates for *I. alticrenatus* and *I. brucei*, both deep water species, are high during the day (see Haddy et al., Chapter 17 for more information). This result suggests that diel patterns may vary depending on the preferred depth of a species.

Jones (1988) has provided the most detailed analysis of diel activity patterns to date for any scyllarid species (see Jones, Chapter 16 for details). *Thenus indicus* and *T. orientalis* are shallow-water species that vary their activities around the diurnal phase, the nocturnal phase, and the crepuscular phases. During daylight hours, they remain buried with only eyes and antennules exposed and the antennules maintain a low-frequency flicking rate. During twilight (the first crepuscular phase), lobsters unbury and begin to forage for food via nomadic movements on the substrate punctuated by brief bursts of swimming activity. This activity drops off after about 4 h into the nocturnal phase, with lobsters remaining still on the substrate or partially buried. The major activity here is grooming. At dawn (the second crepuscular phase), activity again increases and consists primarily of foraging behavior. Lobsters then rebury completely into the sediment and remain inactive during the day (Jones 1988).

7.5.6 Predators and Antipredator Behavior

By and large, predators of slipper lobsters are currently known only from gut-content analyses of sampled fish or invertebrate species, but little is known of how various genera escape predation and how the predators subdue the prey. The general assumption is that the blunt tubercular (or smooth) surface of a flattened carapace, along with the cryptic coloration patterns of each species provides for a highly cryptic lifestyle that provides excellent concealment of individuals against their preferred habitat. Barshaw & Spanier (1994b) noted that lobsters tethered in the open placed themselves alongside rocks, where their cryptic coloration made them visually difficult to detect; alternatively, they tucked under whatever macroalgae was in the vicinity in an attempt to hide. In addition, an activity pattern that directly avoids diurnal predators is assumed to reduce potential predator risk. Finally, in the event that an individual is detected and attacked, the escape response — rapid, abdominal flexions, resulting in swimming movements — aid the lobster in escaping to an area where it can reconceal itself. However, none of these assumptions has been well studied and, given that some scyllarids may undergo migrations over terrains that may or may not be as cryptic, it is clear that antipredator behaviors need to be more carefully examined.

Scyllarides latus represents the one species in which the response to predator attack has been well studied (Spanier et al. 1988, 1991, 1993; Barshaw & Spanier 1994b; Lavalli & Spanier 2001; Barshaw et al. 2003). This response consists of three strategies, two of which are typically executed in sequence: (1) the "fortress strategy" in which the animal grasps the bottom and attempts to outlast its attacker's motivation to penetrate its hard shell (described in Barshaw et al., 2003); (2) the "swimming escape" response (described in Barshaw & Spanier, 1994a, 1994b, and Barshaw et al., 2003); and (3) remaining sheltered in dens (Spanier et al. 1988; Spanier & Almog-Shtayer 1992). Lacking claws (like *Homarus* spp.) or long, spinose antennae (like spiny lobsters; see Zimmer-Faust & Spanier 1987; Spanier & Zimmer-Faust 1988; Lozano-Alvarez & Spanier 1997; Herrnkind et al. 2001) with which to fend off swimming predators, *S. latus* has developed a shell that is at least twice as thick and more durable to mechanical insult than clawed or spiny lobsters that live in the same general region (Barshaw et al. 2003; Tarsitano et al. 2006). They use their short, strong legs to grasp the substrate and resist being dislodged (Barshaw & Spanier 1994a, 1994b). This clinging force can reach maximum magnitudes of 3 to 15 kg (∼8 to 29 times the body weight of the lobster) and linearly correlates with lobster size (Spanier & Lavalli 1998). When this "fortress defense" fails, they are exceptionally deft swimmers capable of evasive maneuvers like barrel rolls (presumably using their flat, broad antennae like reciprocal aileron stabilizers on an airplane wing) en route to a shelter (Spanier et al. 1991). Also they may suddenly change the direction of their swimming, presumably to confuse the chasing predator, a tactic known as "protean" behavior. This is an energetically costly response to a threat and is generally used as a last resort. Barshaw et al. (2003) argue that *S. latus* has matched the energy invested by clawed lobsters in claws and spiny lobsters in antennae by increasing only moderately the thickness of their shells and bettering their swimming-escape behavior. If this strategy is shared among all genera within the family, it would appear to be highly successful, as slipper lobsters are the most diverse group of lobsters with more than 85 species distributed worldwide (Booth et al. 2005; Webber & Booth Chapter 2).

As demonstrated with *S. latus*, slipper lobsters also display a variety of shelter-related behaviors that provide a third highly effective survival strategy (Barshaw & Spanier 1994a). By combining nocturnal foraging with diurnal sheltering, as well as carrying food to their shelters for later consumption rather than remaining exposed while feeding, slipper lobsters may fully minimize their exposure to diurnal predators. Open bivalves have been found in shelters occupied by lobsters, as well as in a 40 m diameter surrounding the occupied shelters. When such empty shells are removed by divers, new ones appear in a manner of days (Spanier & Almog-Shtayer 1992). Horizontally oriented shelters supply shade and reduce visual detection by diurnal predators. Small shelter openings also supply shade but, in addition, increase physical protection against large diurnal predators, especially fish with high body profiles, such as the gray triggerfish. Clinging may enable the lobsters to survive an attack inside a den and even in open areas. Multiple shelter openings enable escape through a "back door" if a predator is successful in penetrating the den. They can then escape by using their fast tail-flip swimming capability (Spanier & Almog-Shtayer 1992). The tendency for cohabitation with conspecifics may be adaptive because of collective "prey vigilance" and defense or concealment among cohorts ("selfish herd" response or the "dilution effect" *sensu* Hamilton, 1971). If all else fails, their thick carapace, designed to effectively blunt cracks, may serve them in times of exposure to attacking predators.

The function of sheltering as a predator-avoidance mechanism against diurnal fish was tested in a series of field-tethering experiments on *S. latus* by Barshaw & Spanier (1994b). Tagged lobsters were tethered with monofilament line inside and outside an artificial reef. All predation events occurred only during the day. Predation by the gray triggerfish, *Balistes carolinensis* Gmelin, 1789, a high-body-profile, large, diurnal fish, was significantly less on lobsters tethered in the reefs compared to those tethered in open areas. The lobsters tried to cling to the substrate, relying on their armature and lack of movement to protect them. Because the lobster could not tail flip to a shelter, the fish was eventually able to turn the lobster over and consume it by biting through its thinner, vulnerable, ventral surface.

There have been some observations (in natural habitats) of specimens of *S. latus* sharing dens with the Mediterranean moray eel, *Muraena helena* Linnaeus, 1758, with no apparent predator–prey interactions between the fish and the lobsters (Spanier & Almog-Shtayer 1992; Martins, personal communication). Sharing shelters with a moray eel may have mutual benefits: the octopus, *Octopus vulgaris* Cuvier, 1797, is prey for the moray eel and may be a predator of the lobster (at least it is in laboratory settings; Spanier, personal observations). Thus, by cohabiting, the lobster may be protected by a moray which preys on octopus, and the moray eel may take advantage of any octopus attracted to the shared den by the presence of its prey — the lobster (Spanier & Almog-Shtayer 1992). However, this association needs to be further studied to elucidate any such mutualistic interaction.

Besides the grey triggerfish, dusky groupers (*Epinephelus guaza* Linnaeus, 1758) have been reported as predators of adult and juvenile *S. latus* and combers (*Serranus* spp.) and rainbow wrasse (*Coris julis* Linnaeus, 1758) apparently prey on juvenile *S. latus* (Martins 1985b). The spotted gully shark, *Triakis megalopterus* (Smith, 1839), has been reported to feed on *S. elisabethae* in South Africa (Smale & Goosen 1999), and tiger sharks, *Galeocerdo cuvieri* (Peron & LeSueur, 1822), red grouper, *Epinephelus morio* (Valenciennes, 1828), and gag grouper, *Mycteroperca microlepis* (Goode & Bean, 1879), have been reported as predators of *S. nodifer* (Lyons, 1970). Queen triggerfish, *Balistes vetula* Linnaeus, 1758, and Goliath groupers or jewfish, *Epinephelus itajara* (Lichtenstein, 1822), are the main predators of *S. aequinoctialis* (Lyons, 1970). Scorpionfish, *Scorpaena brasiliensis* Cuvier, 1829, dusky flounder, *Syacium papillosum* (Linnaeus, 1758), high hat drum, *Equetus acuminatus* (Bloch & Schneider, 1801), and clearnose skate, *Raja eglanteria* Bosc, 1802, feed on various *Scyllarus* species (*S. americanus*, *S. chacei*, *S. depressus*) (Lyons, 1970). Unidentified scyllarids have also been found in the stomachs of various zooplankton feeding fish (blackbar soldierfish, *Myripristis jacobus* Cuvier, 1829; bigeye, *Priacanthus arenatus* Cuvier, 1829; and yellowtail snapper, *Ocyurus chrysurus* (Bloch, 1791)); great barracuda, *Sphyraena barracuda* (Walbaum, 1792); gray snapper, *Lutjanus griseus* (Linnaeus, 1758); dog snapper, *L. jocu* (Bloch & Schneider, 1801); and black margate, *Anisotremus surinamensis* (Bloch, 1791) (Lyons 1970).

7.5.7   Movement Patterns

Slipper lobsters demonstrate two modes of movement: (1) slow, benthic walking movements that may be nomadic within a small home range or migratory from inshore, shallow waters to offshore, deep waters; and (2) swimming movements that can be used for rapid escape or, as some have suggested, for vertical movements (swimming) in the water column.

7.5.7.1   Daily and Seasonal Horizontal Patterns

Tagging studies of only a few species (*Scyllarides astori*, *S. latus*, and *Thenus* spp.) have been used not only to determine growth and maturity indices, but also to determine movement patterns. These studies demonstrate that some species remain within local grounds year-round, while other species demonstrate two annual patterns of movement: local nomadic movements made while inshore and migratory offshore movements. Such studies are summarized below.

In a study of *S. latus*, 314 lobsters caught at an artificial tire-reef complex were tagged between the carapace and the abdomen, using numbered spaghetti tags (Spanier et al. 1988). Lobsters were also marked by puncturing small holes in the telson and were released at their site of capture. In a later phase of this study, they were also tagged between the third and fourth abdominal segments. Thirty-two percent were recaptured at least once, 9% were recaptured twice, and 2.6% were recaptured thrice. All but 3% of the recaptured tagged lobsters retained their spaghetti tags. These remaining 3% were identified by the holes punctured in their telsons or the scars left by these holes after molting. However, a later laboratory study of tag retention in these lobsters (Spanier & Barshaw 1993) found that only 40% of the animals retained tags positioned between the carapace and abdomen after molting, indicating that many individuals may not have been properly reidentified if captured. During the inshore lobster season (February to June in the southeastern Mediterranean), lobsters left their shelters at night to make short-term movements to forage and bring back food (mostly bivalves). More than 71% of tagged lobsters were recaptured in the artificial reef site repeatedly during the season. Time between repeat captures was one to 17 weeks (mean 29 days) and this rate probably represents short-term movements for foraging or local nomadism (Spanier et al. 1988).

In contrast, only 7.2% of the tagged lobsters were recaptured in the same man-made shelter site after more than half a year. Time between these captures ranged between 10 and 37 months (mean 338 days) and may represent long-term movements or migration. Returning lobsters have to orient and locate the small artificial reef site in the widespread continental shelf. Due to limited cooperation with local fishermen, only 11 tagged lobsters were reported outside the artificial shelter site. Six were caught by divers about 300 to 800 m off the reef site. The rest were caught by fishermen in the late part of the summer two to three months after tagging, 20 to 35 km north of the site and at depths >50 m (Spanier et al. 1988).

For several years, a seasonal survey of all lobsters was conducted in the earlier-mentioned artificial tire-reef complex (Spanier et al. 1988, 1990). Lobsters appeared in the reefs in early winter (December to early February), with their numbers peaking in the spring (March to May). From June onward, their numbers decreased in the shallow part of the continental shelf and they disappeared from shallow water in August/September until the beginning of the following winter. This seasonality correlates with water temperature. Lobsters appear in the shallow part of the continental shelf (15 to 30 m depth) when water temperatures are the lowest for the southeastern Mediterranean region (15 to 16°C), and their numbers decrease when water temperatures rise to 26 to 27°C (Spanier et al. 1988). A similar trend is seen in the yield of the commercial fisheries off the Mediterranean coast of Israel.

Several traps (see Figure 18.1 for an example of the type) set offshore, off-season, and at a depth of 48 m caught lobsters in October. Water temperature at these trap sites was 23.6°C, while in the much shallower artificial reef site at 18.5 m where no lobsters were detected, it was 27.7°C. Also lobsters were caught during the fall at depths greater than 50 m by a rough bottom trawl (Spanier et al. 1988). This limited information suggests that slipper lobsters off the coast of Israel seasonally move to deeper and more northerly waters (i.e., colder water). By migrating, the colder-water lobsters may avoid the high, and perhaps unfavorable, summer and autumn temperatures in the shallow waters of the Levant basin of the Mediterranean. Today these temperatures may rise as high as 31°C, which may cause molting difficulties or abnormalities (also reported by Spanò et al., 2003). Some lobsters kept in the laboratory with ambient water supply in the fall died while molting or only incompletely molted. This happened after they were exposed for over two months to water temperatures of 26°C and higher (Spanier, personal observations). Thus, one possible function of the seasonal shallow-to-deep migration may be to meet physiological and behavioral requirements for molting. Molting in deeper habitats may also be a predator-avoidance strategy during this vulnerable period. Berry (1971) pointed out that while most spiny lobsters are associated with hard substrates that supply shelters, some deep water species appear to exhibit behavioral adaptations for soft substrates, perhaps because of fewer predators in greater depths. Slipper lobsters could switch to soft bottoms at greater depths for the same purpose.

Of 115 *S. latus* tagged by Bianchini et al. (2001) in 1995 to 1997, 29 individuals were later recovered up to 70 weeks after tagging. One tagged female was caught by a trammel net, after being 1575 days at large, at ~5 km from the place of release. Contrary to the findings of Spanier et al. (1988), the recurrence of tagged specimens during every period of the year, at least in Sicily, reduces the possibility of widespread seasonal horizontal migrations, although some specimens might displace vertically. However, surface temperatures in Sicilian waters range from 14 to 23°C and differ considerably from those in Israeli waters (15 to 31°C); thus, movement pattern differences may vary in a temperature-dependent fashion.

In a mark-recapture program in the Galápagos Islands, Hearn et al. (see Chapter 14 for more details) reported that of a total of 1926 *S. astori* tagged and released back into the wild, 116 (6 %) were recaptured and reported by the local fishermen. No information was reported on distances between release and capture locations; thus, it is unknown whether *S. astori* migrates to deeper water to molt, or simply behaves in a more cryptic fashion during this vulnerable period (as reported for *S. latus* by Spanier et al., 1988).

Via repetitive diving surveys of the same dens, it appears that *S. aequinoctialis* displays den fidelity (Sharp et al., Chapter 11), but it is not clear if they remain in a single locale over long periods of time or if they migrate in a manner similar to *S. latus*. It is also not clear how they are able to orient back to the same shelter after foraging movements.

Jones (1988) and Courtney et al. (2001) conducted a rather ambitious tag and release program involving > 13,000 lobsters with recapture rates ranging from 7 to 10%. From these studies, the following picture of *Thenus* spp. movements emerges: lobsters are capable of moving 24 km from point of release, but do not do so in any particular direction, as would be expected if these movements represented migration. Their movement pattern is clearly nomadic, which, given their preference for spatially simple, sedimentary substrates, is logical, as there is no shelter to repetitively locate. It also appears that these lobsters do not migrate, as *Thenus orientalis* sexes are not segregated at any time during the year, indicating that females do not leave the fishing grounds for either shallower or deeper waters (Kagwade & Kabli 1996a).

Likewise, in a tagging program involving nearly 3900 *Ibacus peronii*, with recapture rates of 14.3%, the movement pattern also appeared to be nondirectional nomadism, with lobsters being recaptured close to their point of release. One cannot generalize, however, about *Ibacus* spp. because 94 (13.1%) of 557 tagged *I. chacei* demonstrated a northward migration from released locations and covered approximately 0.15 to 0.71 km/day for a total of traveling time of 310 km in 655 days (Stewart & Kennelly 1998; see Haddy et al., Chapter 17 for more information).

For those lobsters that engage in a migration involving a return to a specific location (e.g., *S. latus*), individuals seem to be capable of shoreward homing movements to arrival back at to their preferred location (e.g., artificial reef site for *S. latus* in Israeli waters). The advantage of this homing ability is obvious. Natural rocky outcrops that supply shelters with the physical parameters preferred by lobsters constitute a very small portion of the shallow continental shelf along many coasts. Thus, it may be advantageous for lobsters to "recall" these preferred natural sites or dens (or artificial sites/dens that would be found in artificial reefs and ship wrecks; see Figure 18.3 in Chapter 18 for an example) and to return to them after short- as well as long-term movements. It seems that they just walk or walk and swim relatively long distances to return. The mechanism by which they orient and locate such preferred habitats is completely unknown. They may use some geomagnetic cues or magnetic maps, as has been reported for spiny lobsters (Lohman 1984, 1985; Boles & Lohman 2003). Lohman (1984, 1985) found magnetic remenance in the Caribbean spiny lobster, *Panulirus argus*, along with the ability to detect geomagnetic fields, while Boles & Lohman (2003) demonstrated that spiny lobsters could orient homeward and navigate without any cues from their outbound, displacement trips. Some species of slipper lobsters may have similar abilities.

7.5.7.2 *Swimming Behavior (Vertical Movements)*

In mechanical terms, tail-flip swimming in crustaceans constitutes locomotion in which a single "append-age" — the abdomen — produces thrust by a combination of a rowing action and a final "squeeze" force when the abdomen presses against the cephalothorax (Neil & Ansell 1995). Although the tail-flip response is known in adults and juveniles of all three major taxonomic group of lobsters (e.g., Ritz & Thomas 1973; Jones 1988; Newland & Neil 1990; Newland et al. 1992; Jacklyn & Ritz 1986; Jackson & Macmillan 2000; Jeffs & Holland 2000), as well as in other crustaceans (e.g., crayfish, shrimp, squat lobsters, stomatopods), it is best developed in slipper lobsters. Tail flipping is first developed in the nisto phase, where it can vary among species in strength (Robertson 1968; Lyons 1970; Barnett et al. 1986; Higa & Saisho 1983).

The hydrodynamics of swimming in slipper lobsters has been well studied in *Scyllarides latus* (Spanier et al. 1991; Spanier & Weihs 1992, 2004), *Ibacus peronii* and *I. alticrenatus* (Jacklyn & Ritz 1986; Faulkes 2004), and *Thenus orientalis* (Jacklyn & Ritz 1986; Jones 1988). In only one of these species (*Ibacus peronii*) has the neural circuitry controlling the tail flip been studied (Faulkes 2004). Faulkes (2004) argues that the external morphology of the more flattened species of scyllarids, combined with their proclivity for holding their abdomen in a flexed position, is likely to have resulted in the loss of giant interneurons that are involved in the control circuitry of escape behavior. While the neuroanatomy in *Ibacus peronii* supports this idea to the extent that these neurons appear absent and a similar argument has been made for hermit crabs and galathean crabs (squat lobsters) (Faulkes & Paul 1997, 1998) this idea remains to be more fully examined at the behavioral and neuroanatomical levels in the *Ibacus* spp., as well as in the allied genera of *Parribacus*, *Scyllarides*, and *Thenus*.

Scyllarides latus uses a "burst-and-coast" type of swimming (see Weihs 1974) in response to a predator (e.g., triggerfish) or harassment from divers. Large-amplitude movements of the tail propel the lobster quickly backward, with periods of acceleration reaching top velocities of three body lengths per sec; these movements are followed by periods of powerless gliding, decelerating to velocities of less than one body length per sec (Figure 7.2). The force per tail beat ranges between 1.25 to more than 3.00 N and correlates with body length because additional force is needed to move the greater mass of larger animals, rather than to increase speed and acceleration. The intermittent fast-escape swimming is only of short duration and does not appear to be used for foraging or long-range movements; instead it is an emergency response, whereby the animal invests considerable energy resources to reduce its exposure time in the open area until it can reach safety.

Spanier & Weihs (1992) suggested that the flattened second antennae of *S. latus* (mistakenly called "shovels" or "flippers") along with their movable joints, play an important hydrodynamic role in controlling the swimming movement. Essentially, they serve as stabilizers and rudders in "take off," acceleration, gliding, turning, and landing. Significant lift is created during backward tail flips, and articulation of the flattened second antennae (*a.k.a.* "rudders") alters the distributions of this lift so that pitching and rolling movements are possible. Essentially the flattened antennae are articulated hydrofoils that move in different horizontal and vertical planes; they can form continuous or separated surfaces via an overlapping or spreading out of the component segments (Spanier 2004; Spanier & Weihs 2004). When spreading, the flattened antennae can increase surface area by 56.2 to 79.5%. Exact movements and resultant forces

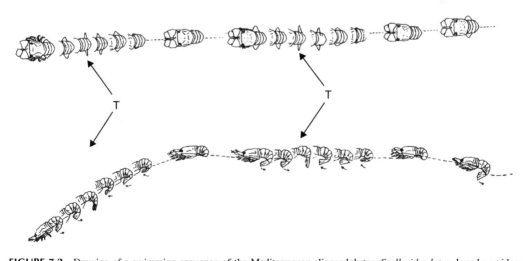

FIGURE 7.2 Drawing of a swimming sequence of the Mediterranean slipper lobster, *Scyllarides latus,* based on video recordings; top view: above, side view: below (reprinted from Spanier, E., Weihs, D., & Almog-Shtayer, G. 1991. *J. Exp. Mar. Biol. Ecol.* 145: 15–31, with permission from Elsevier).

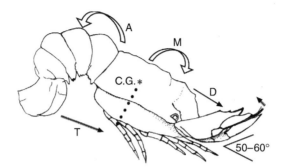

FIGURE 7.3   Forces and counterforces produced during take off in swimming behavior of *Scyllarides latus*. The abdominal flexion results in a propulsive force (T) that produces a head-up (A) movement due to a misalignment in the animal's center of gravity (C.G.). To compensate, the lobster deflects the 2nd antennae 50 to 60° upward from the body axis. This produces a drag force (D) that results in a counterbalancing head-down movement (M). (Drawing based on videorecordings by R. Pollak; used with permission.)

are described via videorecording analysis of take off, swimming, and turning. In take off, propulsive abdominal flexion produces a head-up moment due to a misalignment in the lobster's center of gravity. To compensate for this force and its misalignment, the lobster deflects its flattened antennae 50 to 60° upward from the body axis (known as "rotation" in aircraft); this, in turn, produces a drag force that results in a counterbalancing head-down movement (Figure 7.3).

This alteration in lift via the second antennae is also seen in *Ibacus* spp. and *Thenus* spp. (Jacklyn & Ritz 1986). In contrast, spiny lobsters found in similar habitats and ranges (e.g., *Jasus edwardsii* and *J. novaehollandiae* Holthuis, 1963), produce a negligible amount of lift during each tail flip and do not possess antennae that were shaped or positioned properly to control any created lift (Jacklyn & Ritz 1986). As a result, spiny lobster tail flips are not efficient for continuous swimming or maneuvering (Ritz & Jacklyn 1985; Jacklyn & Ritz 1986).

Neil & Ansell (1995) compared data of swimming performance for a number of decapod, mysid, and euphausid crustaceans. Maximum velocity of body movement achieved during the tail flip was similar across the adults in each group, and ranged from 10 to 300 mm body lengths, although this represented a 30-fold difference in the velocities expressed as body lengths per second. *Scyllarides latus* was ranked as the fastest of those tested, with a maximum velocity of close to 1 m/sec compared to 0.6 m/sec in the clawed lobster *Nephrops norvegicus* (Linnaeus, 1758) (Newland et al. 1988). In a more recent analysis, Spanier & Weihs (2004) identified the contribution of the tail as the propulsor, the legs as landing gear, and the second antennae as control surfaces. They also examined secondary hydrodynamic effects of carapace curvature and the longitudinal ridge associated with vortex production and control. A possible function has been postulated for this ventrolateral curvature (keel) of the lobster carapace. It may be similar to the scale armature of rigid-body boxfishes. In a detailed study of these fish, Bartol et al. (2002) found that the ventral keels produce vortices that serve to stabilize motion, resulting in a smooth swimming trajectory. This could be the case for slipper lobsters as well, and might explain why certain species of scyllarids have a more pronounced keel than others. Furthermore, while in the coast portion of swimming, deflection of the second antennae would increase drag; thus, the lobster returns its second antennae (and antennules) to a position that is in line with the body, and shifts its legs forward in an effort to minimize drag. With the advent of another abdominal flexion (to create acceleration), the lobster must repeat the deflection of the second antennae, but does so at a smaller deflecting angle of 10 to 30°. Turning is achieved by differentially tilting and spreading the second antennae (Figure 7.4).

Ibacus peronii and *I. alticrenatus* have been examined both for the hydrodynamics of swimming as well as for the neural circuitry controlling swimming. In their description of the tail-flip response, Jacklyn & Ritz (1986) argued that because the abdomen was fully extended before being flexed, and because some flexions occurred with only a partial extended abdomen, these flexions must be mediated by the lateral giant (LG) interneurons. Typically the decapod startle response involves a circuit in which medial giant (MG) interneurons and LG interneurons trigger a single, short-latency, powerful abdominal flexion; fast

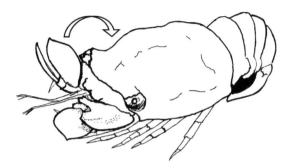

FIGURE 7.4 An example of differential tilting and spreading of the 2nd antennae by *Scyllarides latus* to produce turning movements during swimming. (Drawing based on videorecordings by R. Pollak; used with permission.)

flexor motor giant neurons (MoG) receive synaptic input from the LG and MG interneurons and control the exact flexor response of the abdominal musculature (Wine & Krasne 1972; Mittenthal & Wine 1978). The MG interneurons fire when stimuli come from an anterior direction and their firing results in a direct, backward movement. In contrast, the LG interneurons fire when the stimulus is at the posterior end of the animal and cause the abdomen to lift off the surface, dragging the animal into the water column (Wine & Krasne 1972; Cooke & Macmillan 1985; Newland & Neil 1990). However, swimming by abdominal extensions and flexions can be mediated by nongiant neurons (Reichert et al. 1981; Reichert & Wine 1983; Faulkes 2004).

Ibacus spp. do not respond to sudden tactile stimuli with tail flips — they simply flatten their eyestalks into their eye cups; however, attempts to turn them over result in swimming via repeated tail flips (Faulkes 2004). Histological analysis of the abdominal ventral nerve cord showed that *I. peronii* lacked both MG and LG interneurons, as well as the MoG neurons. This result is surprising because these neurons modulate the fast-escape response in crayfishes, spiny lobsters, and clawed lobsters and have been the subject of considerable research (Wine & Krasne 1982). In the place of these highly conserved control circuits, *I. peronii* has fast flexor neurons and flexor inhibitor motor neurons that appear to mediate the tail-flip response. *Ibacus alticrenatus*, although not as well studied as *I. peronii*, also apparently lacks the tail-flip escape circuitry (Faulkes 2004). Faulkes (2004) argues that the adaptive value of the giant-mediated escape tail-flip response should be inversely proportional to the amount of time the abdomen is flexed and, because it is flexed normally in *Ibacus* spp., such a tail flip response would only produce minimal movement and thus be rather ineffective as an escape mechanism. While the paucity of data currently do not permit a thorough evaluation of this provocative idea, it is another example of the degree to which the specializations of slipper lobsters offer valuable avenues for basic research.

Jones (1988) distinguished two types of swimming activity in *Thenus* species: locomotion or free swimming and escape swimming, based on the presence/absence of an overt stimulus and the speed of the swimming response (29 cm/sec for the first type; 1 m/sec for the latter). In free swimming, lift is generated by the body shape (aerofoil) and by the downward thrust of the abdomen, while drag is reduced by all pereiopods being extended anteriorly (Jacklyn & Ritz 1986). Height is controlled by the second antennae and each flexion helps maintain the animal above the sediment (Jacklyn & Ritz 1986; Jones 1988). In contrast, escape swimming was always effected after a direct stimulus or threat was applied and consisted of an abdominal flexion that was proportional to the magnitude of the stimulus (see Jones, Chapter 16 for more information). Faulkes (2004) argues that these two separate responses need not be the result of separate and distinct neural circuitry, but could simply represent two extremes of the nongiant flexor motor neuron effected tail flips.

7.6 Sensory Biology of Scyllarids

The sensory world of the slipper lobster unfortunately has received little attention when compared to that world for nephropid and palinurid lobsters, as well as other decapods such as crayfish and certain species of crabs. Extrapolating from morphological homologies with these other genera, it is likely that chemical and mechanical stimuli will be of comparable central importance to the natural history of slipper lobsters. However, the one likely exception is that vision may make a greater contribution to the umveldt of many species within this family that live in shallow waters. Like other lobsters, slipper lobsters have many appendages (cephalic, thoracic, abdominal) that appear to bear many different types of setae which may or may not have a sensory function. Unlike other lobsters, many species of slipper lobsters have been described as having a fine "pubescence" covering their cephalothorax that may provide those species with additional sensory information. Few studies have actually examined the sensory function of appendages or the setae borne upon such appendages; therefore, this section will describe what is known about the different sensory modalities that are likely to be used by slipper lobsters. Also, where possible, this section will describe potential appendages and setae that, by likely morphological homology, can reasonably be expected to serve as primary receptors for sensory modalities typical in other lobsters. It is our hope that by laying out potential functions for appendages and the setae borne upon them, future research will be directed toward a systematic study of neurophysiological responses of those appendages and setae to a variety of sensory stimuli. With this purpose in mind, Table 7.2 provides a summary of appendages found in slipper lobsters for all life-history stages described, setae borne upon them when identified by various researchers, purported sensory functions, and associated references.

7.6.1 Chemoreception

Like many other animals, lobsters use chemoreception to orient to and locate food and conspecifics. Extensive research in the nephropid and palinurid families of lobsters, has distinguished "distance" chemoreception (smell) and "contact" chemoreception (or taste) on morphological, neurophysiological and behavioral grounds (Atema 1977, 1980; Schmidt & Ache 1992, Voigt & Atema 1994, Atema & Voigt 1995). In these lobsters, chemo- and mechanoreceptors on the antennules guide the animal to the vicinity of odor sources; those on the distal ends of the walking legs are used in local probing searches of the substrate, and those of the mouthparts evaluate the "palatability" of the objects lifted by the legs. Descriptions of feeding behavior in various species of *Scyllarides* (Lau 1987; Malcom 2003) and *Thenus* (Jones 1988), and differences in morphology of the feeding appendages make the utility of these distinctions questionable for slipper lobsters.

In nephropids and palinurids, the antennules are thought to be essential for distance chemoreception and help the animal orient to a food, conspecific, or substrate source (Reeder & Ache 1980; Devine & Atema 1982; Moore et al. 1991b; Boudreau et al. 1993; Basil & Atema 1994; Beglane et al. 1997; Koehl et al. 2001; Goldman & Patek 2002; Koehl 2006). A key feature of antennule morphology subserving this function is the ability to reach beyond the fluid-boundary layer surrounding the animal's own body into the potentially odor-bearing free stream of the ambient current. The long antennules of clawed and spiny lobsters are well suited to this function, but the short antennules of slipper lobsters are not (Grasso & Basil 2002). *Scyllarides aequinoctialis*, *S. notifer*, and *S. latus* performed poorly compared to *H. americanus* and *P. argus* when challenged to locate food sources under flow conditions identical to those under which the latter easily tracked free-stream borne food odors over distances of two meters (Grasso, unpublished observations). In slipper lobsters, antennules obviously play a role in distance chemoreception as part of the detection process, but based on morphological considerations and behavioral observations, the sensing strategy is clearly different.

Once the (clawed or spiny) lobster reaches the proximity of an odor source, the pereiopods take over the decision-making and act as "near-field" and contact chemoreceptors. This function is mediated by numerous chemo- and mechanosensory setae covering their broad, blunt dactyls. Such setae are largely absent from the dactyls of slipper lobsters (discussed later) and, in their place, are epicuticular caps tapering to sharp points — well adapted to the mechanical task of shucking bivalve shells, but not for chemosensing. In spiny and clawed lobsters, the antennules have little to no participation in contact chemoreception which the pereiopods are morphologically specialized to serve.

When searching for buried food items, it is unclear how slipper lobsters detect such items, as they lack setal tufts on the tips of their pereiopod dactyl segments. Nonetheless, they are capable of digging bivalves out from 3.5 cm of sediment (Almog-Shtayer 1988), and increase flicking rates of the antennules

TABLE 7.2

Possible Behavioral and Sensory Functions of Scyllarid Appendages

Appendage	Possible Function	References
Antennules (= 1st antennae)		
Lateral flagellum	• Olfaction (tracking odor plumes) – aesthetascs present in two rows per segment	• Odor-plume tracking not observed, but orientation to food source observed by Jones (1988) and increased antennule flicking observed by Jones (1988) & Lau (1987) – scyllarids: Weisbaum & Lavalli (2004)
	• Chemical food recognition (observed to touch surface and interior of bivalve shells) or debris recognition – have toothbrush setae (similar to hooded setae in palinurids) – have asymmetric setae (similar to those seen in palinurids) that initiate grooming in palinurids in response to specific chemicals	• Lau (1987, 1988); Jones (1988); Malcom (2003) – scyllarids: Weisbaum & Lavalli (2004) – palinurids: Cate & Derby (2002), Schmidt et al. (2003); Schmidt & Derby (2005)
	• Mechanoreception (for current information) – simple setae present as "guard hairs" and form a water channel – have asymmetric setae — in palinurids, these are innervated by both mechano- and chemosensory neurons	• No neurophysiological studies done on scyllarids – scyllarids: Weisbaum & Lavalli (2004) – palinurids: Schmidt et al. (2003); Schmidt & Derby (2005) – nephropids: Guenther & Atema (1998)
Medial flagellum	Unknown; chemoreceptive in nephropids	Tierney et al. (1988)
Basal segment	Unknown; location of statocyst in nephropids	Cohen (1955, 1960)
Antennae (= 2nd antennae)		
Flagella	• Rudders and stabilizers (articulated hydrofoils) used while swimming – Flagellar segments have great range of movement	• Spanier & Weihs (1992, 2004); Spanier (2004)
	• Mechanoreception (for water flow information) – Setae present on most edges; unknown types	• See various figures in Holthuis (1985, 1991, 2002) for setal arrangements
Bases	Unknown; location of nephropore in nephropids	Bushmann & Atema (1994)
Mandibles		
Endopod	Mastication of food; grasping, shearing, and ripping of food; in some species can be used to chip edges of bivalve shells • Larvae: divided into three portions: Upper portion = blunt, multipronged, canine-like process Middle portion = sharp incisor teeth along inner, medial edge Lower portion = flattened, tuberculate process	Lau (1987) • Mikami et al. (1994)
	• Adults: asymmetric; left mandible larger with calcified molar process that extends over right mandible; both left and right mandibles have a serrated, tooth-like, incisor process ventromedially	• Suthers & Anderson (1981); Johnston & Alexander (1999)

(Continued)

TABLE 7.2

(Continued)

Appendage	Possible Function	References
Palp	Unknown; presumably aids in directing food into mouth	
	• Adult: noncalcified; bears densely arranged pappose setae along its upper margins (setal barriers?)	• Suthers & Anderson (1981); Johnston & Alexander (1999)
First maxillae		
Endopod [Biramous (distal endite + proximal endite)]	Unknown; presumably involved in food handling and grooming based on setal types present	
	• Larvae: bears "spines" and setae, the number of which varies with larval instar and species; some species are described as having "masticatory" spines (cuspidate setae?)	• Robertson (1968); Johnson (1971b); Barnett et al. (1986); Ito & Lucas (1990); Mikami & Greenwood (1997); Webber & Booth (2001)
	• Adult: distal endite large and covered with simple, stout setae; proximal endite small with two multidenticulate, cuneate setae projecting terminally	• Johnston & Alexander (1999)
Exopod	No information	
Second maxillae		
Endopod	Unknown; presumably used to handle food; has potential chemo-mechanosensory setae (simple) as well as setae thought to seal or act as barrier (pappose)	
	• Larvae: paddle-shaped lobe with several plumose setae that gradually expands distally, loses the plumose setae, and becomes flatter with anterior lobes	• Barnett et al. (1984); Phillips & McWilliam (1986); Ito & Lucas (1990); Mikami & Greenwood (1997); Webber & Booth (2001)
	• Adults: distal and proximal endites fused; inner margin of distal endite covered with pappose setae; inner margin of proximal endite covered with simple setae	• Johnston & Alexander (1999)
Exopod (Scaphognathite)	Gill bailer; drives gill current	
	• Larvae: develops slowly as a posterior lobe	• Robertson (1968); Ito & Lucas (1990); Webber & Booth (2001)
	• Adults: fringed with plumose setae; aboral surface covered with simple setae; oral surface covered with tapered setae	• Johnston & Alexander (1999)
First maxillipeds		
Endopod	Unknown; presumably used to handle food; has potential chemo-mechanosensory setae (simple), mechanosensory setae (plumose); and setae thought to act as a barrier (pappose)	
	• Larvae: not present or simply a rudimentary bud in early instars of some species	• Robertson (1968); Johnson (1970, 1971b); Ito & Lucas (1990)
	• Adults: distal and proximal endites reduced and fused to exopod; lined with simple, pappose, and plumose setae	• Suthers & Anderson (1981); Johnston & Alexander (1999)

Appendage	Description	References
Exopod	Unknown; fusion with endopod probably limits use as a current generating structure	
	• Adults: terminal flagellum in *I. peronii*, but not in *T. orientalis*; lined with simple, pappose, and plumose setae[a]	• Suthers & Anderson (1981); Johnston & Alexander (1999)
Second maxillipeds		
Endopod (5 segments)	Unknown; presumably used to handle food; has setae that are likely chemo-mechanosensory	
	• Larvae: spines and setae on distal segments only (dactyl, propus, carpus)	• Barnett et al. (1984); Webber & Booth (2001)
	• Adults: fused basis and coxa; reduced ischium; carpus and propus have short, simple, stout setae on upper margins; dactyl bears simple, stout setae (hooklike) along edges	• Suthers & Anderson (1981); Johnston & Alexander (1999)
Exopod	Unknown; lacks a flagellum, so cannot generate or redirect currents[a]	• Suthers & Anderson (1981); Johnston & Alexander (1999)
Third maxillipeds		
Endopod (5 segments)	Presumably used to recognize food, to tear food, to groom antennules; has setae for these purposes and crista dentate for maceration of food	
	• Larvae: setae distributed on all segments, but most numerous on distal-most segments (dactyl, propus)	• Barnett et al. (1984); Webber & Booth (2001)
	• Adults: ischium has only small, blunt, crista dentate; merus and carpus have numerous setae covering the oral surface; inner margin of merus bears multidenticulate setae; dactyl has simple, stout setae covering entire oral and portions of aboral surface	• Suthers & Anderson (1981); Johnston & Alexander (1999)
Exopod	Unknown; may be involved in current generation in some species	
	• Larvae: lacking in early larvae; bud present in later larval stages	• Webber & Booth (2001)
	• Adults: reduced and lacking flagellum in *T. orientalis*; elongate with flagellum in *I. peronii*[a]	• Suthers & Anderson (1981); Johnston & Alexander (1999)
Pereiopods		
First pereiopods	Used to grasp food in larvae and to "wedge" open bivalves in adults; have setal types that are likely chemo- and mechanosensory; in larval instars, exopods are used as swimming organs and then lost in metamorphic molt to nisto (recognizable as vestiges in some species)	Robertson (1971); Lau (1987); Ito & Lucas (1990)
	• Larvae: well developed in early instars with exopods bearing paired natatory setae that increase in number in successive instars; curving subexopodal spines present in latter instars — these become fine, short setae; dactyl is claw-like	• Robertson (1968, 1971); Johnson (1971b); Phillips & McWilliam (1986); Ito & Lucas (1990)
	• Adults: large and thick; right and left are symmetrical; dactyl bears cuspidate and simple setae in rows on both oral and aboral surfaces; propus, carpus, and ischium bear cuspidate setae in rows (longer on aboral surface; shorter on oral surface); some species have conate setae on merus, others have teasel-like or paintbrush setae	• Malcom (2003)

(Continued)

TABLE 7.2

(Continued)

Appendage	Possible Function	References
Second pereiopods	Used to grasp food in larvae and to *wedge* open bivalves in adults; have setal types that are likely chemo- and mechanosensory; in larval instars, exopods are used as swimming organs and then lost in metamorphic molt to nisto (recognizable as vestiges in some species)	Robertson (1971); Lau (1987); Ito & Lucas (1990)
	• Larvae: well developed in early instars with exopods bearing paired natatory setae that increase in number in successive instars; curving subexopodal spines present in latter instars — these become short, fine setae; dactyl is claw-like	• Robertson (1968, 1971); Johnson (1971b); Phillips & McWilliam (1986); Ito & Lucas (1990)
	• Adults: longest and slimmest leg; right and left are symmetrical; dactyl bears cuspidate and simple setae in rows on both oral and aboral surfaces; propus, carpus, and ischium bear cuspidate setae in rows (longer on aboral surface; shorter on oral surface); some species have conate setae on merus, others have teasel-like or paintbrush setae and cuspidate setae	• Malcom (2003)
Third pereiopods	Used in walking gait; probing of substrate, and digging sequence in adults; in larval instars, probably used to grasp food; exopods are used as swimming organs and then lost in metamorphic molt to nisto (recognizable as vestiges in some species)	Robertson (1971); Jones (1988); Ito & Lucas (1990); Faulkes (2004)
	• Larvae: in first instars of some species, less well developed than legs one and two with only an exopodal bud; curving subexopodal spines present in latter instars; dactyl is claw-like	• Robertson (1968, 1971); Johnson (1971b); Phillips & McWilliam (1986); Ito & Lucas (1990)
	• Adults: shorter in length but wider; two tufts of cuspidate and simple setal on dactyl, with rows of cuspidate and simple setae present on oral and aboral surface; propus, carpus, merus, and ischium bear rows of cuspidate setae that are longer on the aboral surface and shorter on the oral surface; conate setae on portions of merus in some species, cuspidate or paintbrush setae in other species	• Malcom (2003)
Fourth pereiopods	Used in walking gait; probing of substrate, and digging sequence; in larval instars, probably used to grasp food; exopods are used as swimming organs and then lost in metamorphic molt to nisto (recognizable as vestiges in some species)	Robertson (1971); Jones (1988); Ito & Lucas (1990); Faulkes (2004)
	• Larvae: rudimentary or absent in earliest instars of some species; develops exopod with natatory setae; dactyl is claw-like	• Robertson (1968); Johnson (1971b); Phillips & McWilliam (1986); Ito & Lucas (1990)
	• Adults: smaller than 3rd leg; two tufts of cuspidate and simple setal on dactyl, with rows of cuspidate and simple setae present on oral and aboral surface; propus, carpus, merus, and ischium bear rows of cuspidate setae that are longer on the aboral surface and shorter on the oral surface; conate setae on portions of merus in some species, cuspidate or paintbrush setae in other species	• Malcom (2003)

Fifth pereiopods	Grooming of abdomen (and eggs in females); chemoreception(?); walking; probing the substrate; exopods are used as swimming organs and then lost in metamorphic molt to nisto (recognizable as vestiges in some species)	Robertson (1971); Jones (1988); Ito & Lucas (1990); Faulkes (2004)
	• Larvae: rudimentary in earliest instars of some species with no trace of an exopod; in later instars, leg may be present but is initially nonsetose and then develops with exopod and natatory setae	• Robertson (1968); Johnson (1971b); Ritz & Thomas (1973); Berry (1974); Ito & Lucas (1990)
	• Adults: sexual dimorphism present[b]	• Malcom (2003)
	– Male: leg is achelate; two tufts of cuspidate and simple setal on dactyl, with rows of cuspidate and simple setae present on oral and aboral surface; propus, carpus, merus, and ischium bear rows of cuspidate setae that are longer on the aboral surface and shorter on the oral surface	
	– Female: leg is chelate or subchelate and larger than that in males; distal tip of dactyl and of propus articulate and bear a structure ("brush pad") set into a semicircular groove and made of long cuspidate and simple setae in some species and of only long cuspidate setae in others; this structure is flanked by simple setae at the proximal end rows of cuspidate and simple setae present on oral and aboral surface; propus, carpus, merus, and ischium bear rows of cuspidate setae that are longer on the aboral surface and shorter on the oral surface	
Pleopods	Swimming current generation in forward swimming (nistos); bear plumose setae along fringes	Robertson (1971)
Uropods	Unknown; used to cover ventral abdomen in some species	Faulkes (2004)
Telson	Unknown; used to cover ventral abdomen in some species	Faulkes (2004)

Information on appendages is based on a variety of scyllarid species: (1) antennules = *Scyllarides* species (*S. aequinoctialis, S. latus, & S. nodifer*); (2) second antennae = *Scyllarides latus*; (3) mouthparts, larvae = a variety of species, see references for specifics; (4) mouthparts, adults = *Ibacus peronii & Thenus orientalis*; (5) pereiopods = *Scyllarides* species (*S. aequinoctialis, S. latus, & S. nodifer*). Refer to Johnston, Chapter 6 for detailed descriptions and figures of mouthparts and Jones, Chapter 16 for detailed descriptions of feeding behavior and use of appendages in *Thenus* spp.

[a] Exopods or flagella of exopods of the maxillipeds are missing in the Theninae and Scyllarinae, but present in the Arctidinae and Ibacinae.

[b] Not clear if sexual dimorphism exists for all species, because some species descriptions are based on males alone (e.g., Holthuis 1993; Tavares 1997).

while doing so, or when food items are deposited in their holding tanks (Jones 1988; Lavalli, personal observations). Jones (1988) reports that *Thenus* spp. probe the substrate with their first two pairs of pereiopods, while at the same time raising and lowering the antennules above the substrate surface as described by Lau (1987). Given that epicuticular caps cover the distal portion of the dactyls, it is unclear how the pereiopods detect buried food, and more observations are needed to elucidate food detection mechanisms of scyllarids.

If feeding on bivalves, leg motions are suspended several times during the manipulation and initial probing phases while the lobster brings its distance chemoreceptor organs (antennules) beneath the carapace and into contact with the bivalve (Malcom 2003). This is surprising for two reasons. First, the chemosensory maxillipeds are better positioned to reach the shell. Second, as mentioned above, the tactile and proprioceptive sense from the legs should be sufficient to provide feedback information during opening. But, because slipper lobsters are unique in lacking distally placed dactyl setal tufts, they may not be able to obtain sufficient sensory feedback without using the antennules.

The above makes clear that chemo-orientation in slipper lobsters deviates from the model that emerged from extensive studies of food-searching behavior in clawed and spiny lobsters. Compared against this model, three interlocking hypotheses are suggested: (1) periopod chemosensory ability has been replaced by epicuticular caps suitable for shucking bivalves; (2) the antennules have, at least partially, filled this sensory void and perform the functional role of the dactyls in clawed and spiny lobsters; (3) the distance chemoreceptive function of the antennules has been reduced to accommodate the added role of contact chemoreception. Despite these drastic modifications, the antennules continue to be used to examine the odor source directly, in contact chemoreception, thereby seemingly blurring the distinction between "smell" and "taste." Given that the chemosensory areas of the brains of slipper lobsters do not differ dramatically in architecture or volume from those of spiny or clawed lobsters (Sandeman et al. 1993; Sandeman 1999), studies of chemoreception focused on these hypotheses are likely to put the smell and taste distinction into context and reveal unexpected insights into the organization of the crustacean brain.

The annuli of the lateral flagella of the antennules are covered with a variety of different types of setae (Figure 7.5H and Figure 7.5I); many of these have already been identified as chemoreceptors and mechanoreceptors in clawed and spiny lobsters. The positioning of the different types of setae is highly organized and appears not to differ significantly among the several *Scyllarides* species examined by Weisbaum & Lavalli (2004): aesthetascs, asymmetric, modified simple, and hemiplumose setae (Figure 7.5H) are only found in the tuft region between the distal and proximal ends of the lateral flagellum. Simple setae are found on all regions of all annuli of the lateral flagellum, and toothbrush setae (Figure 7.5I) are concentrated on the annuli of the base region or on the proximal annuli of the tuft region (Weisbaum & Lavalli 2004). This positioning of the setae is ideal as they occur where stimuli carried by currents, possibly generated by the exopodites of the mouthparts, can reach them and provide an assessment of the shell and flesh within or where bottom currents can provide information about distant sources. It is likely that aesthetascs on the slipper lobster antennule serve the same chemosensory function that has been demonstrated for nephropid and palinurid lobsters (Spencer 1986; Derby & Atema 1988; Michel et al. 1991; Ache & Zhainazarov 1995; Cate et al. 1999; Steullet et al. 2000a), that simple setae are bimodal (chemo- and mechanosensory) as they are in palinurids (Cate & Derby 2001), that asymmetric setae are mechanosensory as they are in blue crabs (Gleeson 1982) or bimodal (chemo- and mechanosensory) as they are in palinurids (Schmidt et al. 2003; Schmidt & Derby 2005), and that toothbrush setae are bimodal chemo-mechanosensory homologues of hooded sensilla in palinurids (Cate & Derby 2002). The asymmetric setae apparently are necessary to elicit antennular grooming in palinurids, but, thus far, no studies on antennular grooming in scyllarids have been conducted.

The sensory-motor mechanisms that slipper lobsters use in shucking are unknown at this time. Visual cues can be excluded because the shucking process takes place beneath the animal and is outside the field of view of its dorsally placed eyes, except, possibly, in *Thenus* spp. in which the eyes are at the lateral-most edges of the carapace. However, Jones (1988) tested the response of *Thenus* spp. to the introduction of their favorite food item and found that visual cues alone did not elicit searching responses — only cues with a chemical signal did. Touch, proprioceptive, and chemosensory modalities are the most likely candidates and there is evidence to suggest that each is involved. The touch and proprioceptive senses that could control this task are located in the pereiopods. For manipulation, the positions of each pereiopod

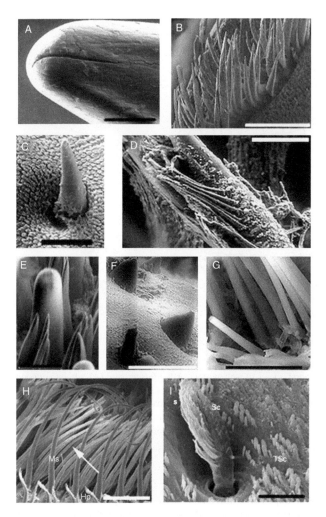

FIGURE 7.5 Setal types found on scyllarid pereiopods and antennules. (A) dactyl tip of left 2nd pereiopod of *Scyllarides aequinoctialis*; (B) simple setae set in groove on oral surface of left 5th pereiopod propus of *S. nodifer*; (C) cuspidate seta of right oral surface of 3rd pereiopod propus of *S. aequinoctialis*; (D) teasel-like seta with setules from left 1st pereiopod, aboral surface of *S. aequinoctialis*; (E) cuspidate and simple setae of female's right 5th pereiopod dactyl, oral surface of *S. aequinoctialis*; (F) conate setae on merus shield of 3rd right pereiopod, oral surface of *S. nodifer*; (G) miniature simple-type setae covering the cuticular surface of the aboral surface of propus of *S. aequinoctialis*; (H) aesthetasc (arrow), modified simple setae (Ms), and hemiplumose setae (Hp) found on the ventral surface of the antennular flagellum of *Scyllarides* species; (I) toothbrush setae found on the dorsal surface of the antennular flagellum of *Scyllarides* species showing scale (Sc), setules (S), and textured scaling on flagellar surface (TSc). Scale bars: A = 200 μm; B = 200 μm; C = 200 μm; D = 20 μm; E = 50 μm; F = 250 μm; G = 350 μm (box = 20 μm); H = 273 μm; I = 20 μm. (Modified from Malcom, C. 2003. Description of the setae on the pereiopods of the Mediterranean Slipper Lobster, *Scyllarides latus*, the Ridged Slipper Lobster, *S. nodifer*, and the Spanish Slipper, *S. aequinoctialis*. M.Sc. thesis: Texas State University at San Marcos, San Marcos, Texas. (A–F); Weisbaum & Lavalli 2004 (G,H); used with permission.)

segment, as well as each pereiopod in its entirety, informed by proprioception, are likely to signal the size, orientation, and location of the bivalve shell through signaling of joint angles. Points of contact for application of manipulative forces may be sensed through tactile sensors on the dactyl tips or through the tension in muscle organs. The pereiopods of *Scyllarides* species do bear numerous tufts of setae (Malcom 2003), but most are placed on the more proximal segments of the propus, carpus, merus, and ischium, rather than the distal dactyl (see Figure 7.5A, Figure 7.6A to Figure 7.6D). Those that are present on the proximal edge of the dactyl segment are usually damaged, most likely due to the abrasive action they suffer

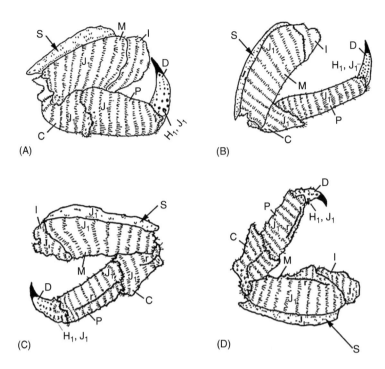

FIGURE 7.6 Outer, or aboral, views of walking legs of *Scyllarides* spp. (A) left 1st pereiopod; (B) left 2nd pereiopod; (C) right 3rd pereiopod; (D) right 4th pereiopod. D — dactyl; P — propus; C — carpus; M — merus; I — ischium; S — meral shield; J_1 — cuspidate setae; H_1 — simple setae, • — stripped dactyl setal pit. (Modified from Malcom, C. 2003. Description of the setae on the pereiopods of the Mediterranean Slipper Lobster, *Scyllarides latus*, the Ridged Slipper Lobster, *S. nodifer*, and the Spanish Slipper, *S. aequinoctialis*. M.Sc. thesis: Texas State University at San Marcos, San Marcos, Texas; used with permission.)

while probing and shucking. However, the setae may pick up information from the surrounding water as the flesh of the bivalve is exposed to that water after valve opening. Setal types present consist mostly of simple, cuspidate, teasel-like, conate, and miniature simple (Figure 7.5B to Figure 7.5G) (Malcom 2003). Simple setae have been demonstrated to have both chemo- and mechanoreceptive functions in clawed (Derby 1982) and spiny lobsters (Cate & Derby 2001); cuspidate setae are implicated in grasping food (Farmer 1974). However, a more recent study on the mouthparts of spiny lobsters (mandibular palp, medial rim of the basis in the first maxilla and first maxilliped, propus and dactyl of the second maxilliped, and dactyl of the third maxilliped), found that simple setae apparently respond to displacement of the setal shaft and cuspidate setae are highly sensitive mechanoreceptors (Garm et al. 2004). It is not clear if such setae on the pereiopods would serve similar functions, but it is likely given that three pairs of mouthparts (the maxillipeds) are thoracic-derived appendages and, essentially, highly modified legs. Malcom (2003) observed some trends while documenting the setal morphology of *S. latus*, *S. nodifer* and *S. aequinoctialis*. Pereiopods grow progressively smaller toward the posterior end of the lobster's body, with the fifth male pereiopods being the smallest. Male and females are dimorphic with regard to the fifth pereiopods, with the female having a subchelate structure on which sits a pad of stiff, short setae (a "brush pads"; Figure 7.7). This structure no doubt is involved in the grooming of eggs. The propus, carpus, merus, and ischium become more ridged along their lateral and medial edges, starting with the third pereiopods, which is where the pereiopods start becoming smaller in length and width. The oral surfaces tend to have shorter cuspidate setae, often much shorter than those found on the aboral surfaces. Lateral edges of the aboral surface tend to have longer cuspidate setae; these become progressively shorter as one moves medially. At the distal end of each segment (propus, carpus, merus, and ischium only), long cuspidate setae are present. Miniature setae are widespread on the pereiopods and result in a sculpturing look on the shell surface.

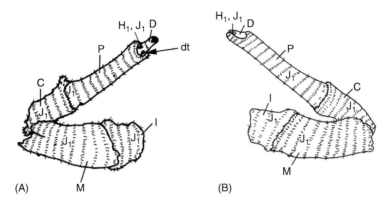

FIGURE 7.7 Inner, or oral, views of the left 5th pair of walking legs of *Scyllarides* spp. (A) male; (B) female, showing false chelae formed by dactyl and propus segments. D — dactyl; P — propus; C — carpus; M — merus; I — ischium; S — meral shield; J_1 — cuspidate setae; H_1 — simple setae, dt — dactyl tuft. (Modified from Malcom, C. 2003. Description of the setae on the pereiopods of the Mediterranean Slipper Lobster, *Scyllarides latus*, the Ridged Slipper Lobster, *S. nodifer*, and the Spanish Slipper, *S. aequinoctialis*. M.Sc. thesis: Texas State University at San Marcos, San Marcos, Texas; used with permission.)

From numerous descriptions and illustrations of phyllosomal stages and nistos, it is clear that all stages bear setae on both pereiopods and antennules (see Sims 1966; Johnson 1970, 1971b, 1975; Crosnier 1972; Ritz & Thomas 1973; Berry 1974; Phillips & McWilliam 1986, 1989; Robertson 1968, 1971; Martins 1985a; Ito & Lucas 1990; Sekiguchi 1990); however, only a few studies have actually described those setae. Typically, phyllosomal pereiopods are described as bearing natatory setae on the exopods (Robertson 1971). Martins (1985a) states that these setae bear fine setules in *S. latus* phyllosomas and these are illustrated by Harada (1958, his Figure 1) for *Ibacus ciliatus* and by Johnson (1968, his Figure 1 and Figure 2) for *Evibacus princeps* S.I. Smith, 1869 — the setules would increase the surface area of the setae, presumably making them more like paddles than rods, and thereby aid in swimming. Pereiopods are also described as bearing spines, most likely for the purpose of repelling a predator attack, and hook-like setae on their dactyls that probably grasp food. As for the antennules, it appears that the lateral and medial filaments increase in size by increasing the number of segments present, and also increase in the amount of setation borne upon them (Figures 45, 53, 59, 63, 66, 70 for *Parribacus antarcticus* in Johnson 1971b; Figures 77, 80, 86 for *S. squammosus* in Johnson 1971b; Figure 4 for *S. astori* and Figure 10 for *E. princeps* in Johnson 1975; Figure 2 to Figure 9 for *Eduarctus* (formerly *Scyllarus*) *martensii* in Phillips & McWilliam 1986; Figures 1d, 2d, 3d, 4d, 5d, 6d, 7d, 8d, 9a for *Petarctus* (formerly *Scyllarus*) *demani* in Ito & Lucas 1990). Robertson (1968; Figure 16) states that the number of aesthetasc tufts increases with each successive phyllosomal stage in *S. americanus* and provides figures illustrating these changes, while Robertson (1971; Figure 14 to Figure 23) shows a nice sequence for the development of the antennules in the phyllosoma of *S. depressus*.

As with nephropids and palinurids, the six pairs of mouthparts bear numerous setae on their various segments. Adult mouthpart structure has only been described in *Ibacus peronii* (Suthers & Anderson 1981) and *Thenus orientalis* (Johnston, 1994; Johnston & Alexander 1999), while phyllosomal mouthpart structure has been investigated in *I. ciliatus* (Mikami et al. 1994) and *Petarctus* (formerly *Scyllarus*) *demani* (Ito & Lucas 1990). Scanning electron microscopy has not been used for a complete analysis of setal types and distribution; thus, only descriptions from specimens viewed under light microscopy are available and many of these are incomplete.

Adult mouthparts tend to have a simple setation pattern with fusion of endites — this makes them considerably less complex than the mouthparts of nephropid or palinurid lobsters. A detailed description of the mouthparts is provided in Johnston, Chapter 6 and will not be repeated here. Instead, only the setal types, where known, will be described. Mandibular palps are lined along their upper margins with pappose setae (Suthers & Anderson 1981; Johnston & Alexander 1999). The first maxilla bears simple setae on the outer (aboral) surface of the distal endite and bears two multidenticulate, cuneate setae on the proximal endite. The proximal endite of the second maxilla is covered with simple setae and the inner margin of the distal endite is covered with pappose setae (Johnston & Alexander 1999). The margin of the scaphognathite is fringed with plumose setae (Johnston & Alexander 1999), as it is for nephropid lobsters (Lavalli & Factor 1992, 1995), which presumably aid in the production of the current by increasing the surface area of the beating scaphognathite. The endite and exopod of the first maxilliped are fused and both are lined with simple, pappose, and plumose setae. A terminal flagellum of the exopod is present in some species, but not in others (see Webber & Booth, Chapter 2; Johnston, Chapter 6 for more information). In the second maxilliped, short simple setae are found along the upper margins of the propus and carpus, while longer, hook-like setae are found along the edges of the dactyl (Suthers & Anderson 1981; Johnston & Alexander 1999). As in the case of the first maxillipeds, the third maxilliped exopod is lacking or severely reduced in some species, but present in others; additionally the teeth of the ischium are small, blunt, and reduced compared to those found in nephropid or palinurid lobsters. Multidenticulate setae are borne upon the merus and simple setae cover the entire inner (oral) surface and part of the outer (aboral) surface of the dactyl (Suthers & Anderson 1981; Johnston & Alexander 1999).

In the phyllosomal stages, the mouthparts generally increase in the number of setae (often called spines) present on various surfaces. Little information on the specific types of setae are available, although Robertson (1968) and Ito & Lucas (1990) state that the lobe constituting the second maxilla bears plumose setae in the early stages on the proximal but not basal segment, but loses these in later stages as the endites fully form and separate from the paddle-like scaphognathite.

Evidence from nephropid and palinurid lobsters suggests that simple setae are most likely to be bimodal chemo- and mechanoreceptors (Derby 1982; Cate & Derby 2001), although those present on mouthparts appear to be highly sensitive to bend (Garm et al. 2004) and may allow for fine-tuning of manipulation of the food bolus as ingestion proceeds. Cuspidate setae seem to have multiple functions — from their robust appearance, they appear to be well suited for stabbing, tearing, and shredding of food particles and may serve a mechanosensory function (see Garm 2004; Garm et al. 2004 for a discussion) to allow feedback to the lobster concerning the food-handling process. Plumose setae are not implicated as a sensory structure, but are thought to be advantageous on appendage segments that create currents (see Lavalli & Factor 1992, 1995; Garm 2004 for a discussion of current generation). Alternatively, they may act as gaskets, sealing the space between the edge of the scaphognathite and the branchial chamber (Farmer 1974; Factor 1978) or as filters to prevent entry of particulate matter into the branchial chamber (Farmer 1974). The function of pappose setae and multidenticulate setae has yet to be determined for any lobster species; however, Farmer (1974) suggested that pappose setae might be used for grooming purposes, while Garm (2004) suggests that they act as setal barriers. Setae described as stout spines or spinous processes are likely to be used for the purpose of gripping particles or pieces of food (Farmer 1974) or may have a mechanosensory purpose in the same manner that cuspidate setae do (Garm et al. 2004). The other setal types encountered on the pereiopods, mouthparts, and antennules may be implicated in grooming, touch or hydrodynamic reception, or fine-food handling.

7.6.2 Mechanoreception

The carapace and abdomen of slipper lobsters, like those of other decapod crustaceans, are covered with stiff, presumably mechanosensory setae (Figure 7.8), and in some species, these setae are so numerous and long that the species are said to be covered in a fine *pubescence*. In *Scyllarides* species, this basic pubescent pattern shows two distinct elaborations. First, the hairs are arranged in groups of 7 to 12 on small islands of hard, protruding cuticle (about 500 islands are present on the carapace, depending on the species). Second, this pattern continues out onto the surfaces of the flattened antennae to the extent that the antennae, when extended, appear to form a continuous surface with the carapace. These structures likely serve a hydrodynamic sensing role that provides important sensory feedback while swimming, or they could be used to detect motion above the lobster while it is walking or still upon the substrate.

The setae present on the second antennae are likely mechanoreceptors used for the purpose of detecting water currents or vortices as the animal is engaged in swimming activity. Since these appendages act as rudders that alter the pitch and roll of the lobster's body while engaged in tail flipping or coast-and-burst swimming (see Section 7.5.7.2), the numerous setae are likely to represent an important feedback

FIGURE 7.8 Setal tufts found covering the surface of the carapace of *Scyllarides latus*. (A) setal tufts of tubercles; (B) arrangement of setae (likely cuspidate or simple setae) within carapace tufts; (C) single tuft, showing distribution of setae around tubercle. (Photo by Lavalli, Spanier & Grasso.)

system to the lobster. It is possible that setae on the legs and the carapace may also provide important sensory feedback during swimming. However, significantly more work is needed to elucidate the feedback mechanisms that control the swimming response in these lobsters.

While not investigated in slipper lobsters, statocysts are present in other families (Cohen 1955) and are used by individuals within those families to maintain balance. Slipper lobsters are known to initiate tail flips in response to being flipped over (or even in response to the attempt to flip them over) (Barshaw & Spanier 1994a; Faulkes 2004), so it is likely that their statocysts are well developed. It is also likely that the statocyst is heavily relied upon during swimming, so that the second antennae can adjust the pitch of the animal.

7.6.3 Vision

The eyes of slipper lobsters are stalked, which should provide a broader field of vision. For *Thenus* spp., this field of view is extended as the eyes are located on the lateral edges of the carapace. In all other species, eyes are more intermediately placed and fit into cups or orbits, which presumably protect the eyes from mechanical damage. Currently, no descriptions of the compound eyes, their ommatidia, or the visual pigments exist. No studies have systematically examined the importance of visual signals in various slipper lobster species. It is likely that vision plays some role in swimming, as blindfolded animals are not as easily stimulated to initiate swimming when harassed. In addition, while swimming, the eyes apparently are directed in such a way to provide backward vision (Spanier, personal observations). Finally, when *Thenus* and *Ibacus* spp. bury, they do not cover up their eyes, but leave them exposed, along with the antennules (Jones 1988; Faulkes 2006), a behavior that strongly suggests that vision may be important. Clearly, this sensory modality needs to be better investigated.

7.6.4   Control Mechanisms and Sensory-Motor Integration

Studies of the advanced sensory-motor integration and precise control in shucking by slipper lobsters promise insights into the functional organization of the bäuplan of the crustacean central nervous system. Although no specific investigations of the central mechanisms that control feeding behavior in slipper lobsters have been conducted, comparisons of what is known from well-studied clawed and spiny lobsters, and freshwater crayfishes suggest what we will learn. For all three of the latter taxa, the location of the eyes means that vision can play no roll in the positioning and manipulation of the bivalve during feeding (although some role of vision is possible for clawed lobsters during crushing of the shell). However, at least in *Thenus* spp., vision may play a role as the eyes are situated on the lateral margins of the carapace (see Jones, Chapter 16 for more information). The precise positioning and manipulation of the shell by slipper lobsters requires them to make finer sensory discriminations and to have more precise motor control of dactyl tip force and positioning and tighter feedback between these sensory and motor functions than in spiny or clawed lobsters consuming bivalves. Considerable evidence suggests that neutral control programs for the legs are situated in the thoracic ganglia (Ayers & Davis 1977, 1978; Ayers & Crisman 1993; Clarac et al. 1987; Cruse & Saavedra 1996; Jamon & Clarac 1995, 1997; Domenici et al. 1998, 1999). Elucidation of the neural mechanisms controlling the dexterous feeding behavior in slipper lobsters will lead to a deeper understanding of sensory-motor control in crustaceans and arthropods in general.

    Observations of shucking behavior and comparative neuranatomy raise further questions about the neural control of the behavior that can profitably, and uniquely, be explored with slipper lobsters. For example, slipper lobsters are observed to suspend leg activity as the antennules are brought into contact with the bivalve during initial assessment of the shell (Malcom 2003). Antennular chemo- and mechanoreceptors synapse in the cerebral ganglion in decapod crustaceans (Sandeman 1990; Sandeman et al. 1992). Does the cerebral ganglion inhibit the motor programs controlling leg movement? How, if at all, does the information collected by the antennules and processed in the cerebral ganglion affect the thoracic ganglia? These are fundamental questions of nervous organization and function that apply through the homologous arthropod bäuplan to lobsters (Fraser 1982; Sandeman et al. 1993).

7.7   Conclusions

Like so many other aspects of their basic biology, little is known of the behavior of the different scyllarid species. Where studies exist, the focus has been on species within three subfamilies: *Scyllarides*, *Ibacus*, and *Thenus*, most likely because these are relatively large lobsters that are by-product in other fisheries or that are specifically targeted. While many *Parribacus* species are targeted by the aquarist trade, their commercial importance is significantly less and subsequently little is known of their behavior. The picture that emerges from the few studied species suggests that, with the possible exceptions of feeding behavior on bivalves and swimming behavior, there is no "stereotypical" pattern of behavior that describes the "general" scyllarid — this is because there is no *general* scyllarid. Bivalve feeding behavior and burst-and-coast swimming behavior has only been investigated in a few species; thus, it is not clear if the behavioral patterns seen thus far are common to all scyllarids. The 85 species that have been described differ widely in their depth and substrate preferences, their early larval life-history strategy, their dietary preferences, and their reproductive strategies.

    Behavioral studies, in and of themselves, should not be difficult to undertake on these species. More and more scyllarids are being successfully cultured in the laboratory from the naupliosoma or first-stage phyllosoma. These successes should allow for detailed studies of depth preferences, phototactic responses, responses to thermoclines, and swimming behavior in flow. If phyllosomas are reared to the nisto stage, studies using techniques that have been employed for nephropids and palinurids should allow the determination of how nistos sample the benthos, what causes them to choose specific substrates, and how they avoid predators. Most adults and juveniles are readily held in the laboratory, and lessons learned from nephropid and palinurid lobsters demonstrate that many "natural" behaviors can be examined if the study animals are provided with sufficient space within a naturalistic setting. Videotaped analysis removes the need for tedious, in-person observations, and allow for a more detailed analysis of both the behavior studied and the order in which behavioral sequences are carried out.

While behavioral research on these lobsters is limited, studies on the sensory structures and their functions are almost nonexistent. This is remarkable because of the unique behaviors observed in this family. The dramatic change in the dactyls of the pereiopods with the concomitant loss of setae at the distal end of the dactyl suggests that the mode of finding food buried within the benthos differs greatly in scyllarids from that seen in nephropids and palinurids. Furthermore, the observations by a number of workers that the antennules are used to apparently directly sample the food item (i.e., contact chemoreception) during its manipulation, suggests that the sensory structures of the antennule may differ in their responsiveness to various odor stimulants. Additional observations of distance orientation to food-odor plumes, something that is well studied and easily performed by nephropids and palinurids, suggest that fundamental differences exist in the central nervous control of feeding behavior in scyllarids.

The flattened second antennae, which are very important in the swimming response of several species thus far studied, must have dynamic hydroreceptors that provide feedback to the lobster during swimming, so that it can effect turning maneuvers and change its body's pitch angle. The numerous surface setae on the carapace may also provide a feedback role here. Likewise, the statocyst of slipper lobsters is probably extremely well developed to provide rapid feedback to the lobster while it engages in coast-and-burst swimming movements. In addition, while vision has been discounted compared to chemoreception and mechanoreception in other species, the swimming abilities of these lobsters and the fact that species that bury do not cover over their eyes, suggests that it may be a more important and a better developed sense in scyllarids.

As stated by Faulkes (2004), the loss of the giant interneuron that effect the escape tail-flip response of scyllarids begs numerous questions concerning the evolution of decapod nervous systems and the response of those nervous systems to deleted or altered components. It also provides a comparative approach because some scyllarids spend much of their time with their abdomen mostly tucked under their body, while others leave the abdomen extended. The nervous system of the former species could be compared to other decapods which have similar postures and have lost the giant interneurons, while the latter species could be compared to others that have not lost the interneurons.

Behavior is, ultimately, the result of the interplay of genes, physiology, and the environment. Sensory input affects and finely tunes behavioral output. Together, the sensory biology and behavior of individuals affects the success of the populations comprising the species and the ability of those populations to adapt to their environments. A failure to understand these aspects of the animal's biology can have profound effects upon the ability of the species to survive and thrive under exploitation. As slipper lobsters are becoming more and more the target of fisheries, it is critical that we begin to focus our attention upon them, not only because they are exploited, but also because they are, in and of themselves, interesting animals that can provide evolutionary insights into other decapod species and arthropods as a whole.

References

Ache, B.W. & Zhainazarov, A.B. 1995. Dual second-messenger pathways in olfactory transduction. *Curr. Opin. Neurobiol.* 5: 461–466.

Almog-Shtayer, G. 1988. Behavioural-ecological aspects of Mediterranean slipper lobsters in the past and of the slipper lobster *Scyllarides latus* in the present. M.A. thesis: University of Haifa, Israel. 165 pp.

Atema, J. 1977. Functional separation of smell and taste in fish and Crustacea. In: Le Magnen, J. & MacLeod, P. (eds.), *Olfaction and Taste*, Vol. 6: pp. 165–174. London: Information Retrieval.

Atema, J. 1980. Smelling and tasting underwater. *Oceanus* 23: 4–18.

Atema, J. 1986. Review of sexual selection and chemical communication in the lobster, *Homarus americanus*. *Can. J. Fish. Aquat. Sci.* 43: 2283–2290.

Atema, J. & Cobb, J.S. 1980. Social behavior. In: Cobb, J.S. & Phillips, B.F. (eds.), *The Biology and Management of Lobsters*, Vol. 2: pp. 409–450. New York: Academic Press.

Atema, J. & Voigt, R. 1995. Behavior and sensory biology. In: Factor, J.R. (ed.), *The Biology of the American Lobster Homarus americanus*: pp. 313–344. New York: Academic Press.

Atkinson, J.M. & Boustead, N.C. 1982. The complete larval development of the scyllarid lobster *Ibacus alticrenatus* Bate, 1888 in New Zealand waters. *Crustaceana* 42: 275–287.

Ayers, J.L. & Crisman, J. 1993. Lobster walking as a model for omnidirectional robotic ambulation architecture. In: Beer, R., Ritzmann, R., & McKenna, T. (eds.), *Biological Neural Networks in Invertebrate Neurothology and Robotics*: pp. 287–316. New York: Academic Press.

Ayers, J.L. & Davis, W.J. 1977. Neuronal control of locomotion in the lobster *Homarus americanus*. II. Types of walking leg reflexes. *J. Comp. Physiol.* 115: 29–46.

Ayers, J.L. & Davis, W.J. 1978. Neuronal control of locomotion in the lobster, *Homarus americanus*. III. Dynamic organization of walking leg reflexes. *J. Comp. Physiol.* 123: 289–298.

Bailey, K.N. & Habib, G. 1982. Food of incidental fish species taken in the purse-seine skipjack fishery, 1976–81. *Fish. Res. Div. Occas. Publ. Data Ser.* 6: 1–24.

Baisre, J.A. 1994. Phyllosoma larvae and the phylogeny of Palinuroidea (Crustacea: Decapoda): A review. *Aust. J. Mar. Freshw. Res.* 45; 925–944.

Barnett, B.M., Hartwick, R.F. & Milward, N.E. 1984. Phyllosoma and nisto stage of the Morteon Bay Bug *Thenus orientalis* (Lund) (Crustacea: Decapoda: Scyllaridae), from shelf waters of the Great Barrier Reef. *Aust. J. Mar. Freshw. Res.* 35: 143–152.

Barnett, B.M., Hartwick, R.F., & Milward, N.E. 1986. Descriptions of the nisto stage of *Scyllarus demani* Holthuis, two unidentified *Scyllarus* species, and the juvenile of *Scyllarus martensii* Pfeffer (Crustacea: Decapoda: Scyllaridae), reared in the laboratory; and behavioural observations of the nistos of *S. demani*, *S. martensii*, and *Thenus orientalis*. *Aust. J. Mar. Freshw. Res.* 37: 595–608.

Barr, L. 1968. Some aspects of the life history, ecology and behaviour of the lobsters of the Galápagos islands. *Stanford Oceanogr. Exped.* 17: 254–262.

Barshaw, D.E. & Spanier, E. 1994a. The undiscovered lobster – A first look at the social behaviour of the Mediterranean slipper lobster, *Scyllarides latus* Decapoda, Scyllaridae. *Crustaceana* 67: 187–197.

Barshaw, D.E. & Spanier, E. 1994b. Anti-predator behaviours of the Mediterranean slipper lobster *Scyllarides latus*. *Bull. Mar. Sci.* 55: 375–382.

Barshaw, D.E., Lavalli, K.L., & Spanier, E. 2003. Is offence the best defence: The response of three morphological types of lobsters to predation. *Mar. Ecol. Prog. Ser.* 256: 171–182.

Bartol, I.K., Gordon, M.S., Ghrib, M., Hove, J.R., Webb, P.W., & Weihs, D. 2002. Flow patterns around the carapaces of rigid-bodies, multi-propulsor boxfishes (Teleostei Ostraciidae). *Integr. Comp. Biol.* 42: 971–980.

Basil, J. & Atema, J. 1994. Lobster orientation in turbulent odor plumes: Simultaneous measurement of tracking behavior and temporal odor patterns. *Biol. Bull.* 187: 272–273.

Beglane, P.F., Grasso, F.W., Basil, J.A., & Atema, J. 1997. Far field chemo-orientation in the American Lobster, *Homarus americanus*: Effects of unilateral ablation and lesion of the lateral antennule. *Biol. Bull.* 193: 214–215.

Berry, P.F. 1971. The spiny lobsters (Palinuridae) of the east coast of Southern Africa: Distribution and ecological notes. *Ocean Res. Inst. Invest.* R27: 1–23.

Berry, P.F. 1974. Palinurid and scyllarid lobster larvae of the Natal Coast, South Africa. *Investig. Rep. Oceanogr. Res. Inst.* 34: 1–44.

Bianchini, M., Bono, G., & Ragonese, S. 2001. Long-term recaptures and growth of slipper lobsters, *Scyllarides latus*, in the Strait of Sicily (Mediterranean Sea). *Crustaceana* 74: 673–680.

Boles, L.C. & Lohman, K.J. 2003. True navigation and magnetic maps in spiny lobsters. *Nature* 421: 60–63.

Booth, J.D. & Matthews, R. 1994. Phyllosomas riding jellyfish. *Lobster Newslett.* 7(1): 12.

Booth, J.D., Webber, W.R., Sekiguchi, H., & Coutures, E. 2005. Diverse larval recruitment strategies within the Scyllaridae. *N. Z. J. Mar. Freshw. Res.* 39: 581–592.

Boudreau, B. Bourget, E., & Simard, Y. 1993. Behavioural responses of competent lobster postlarvae to odor plumes. *Mar. Biol.* 117: 63–69.

Branford, J.R. 1982. Notes on scyllarid lobster *Thenus orientalis* (Lund, 1793) off Tokar Delta (Red Sea). *Crustaceana* 38: 21–224.

Brown, D.E. & Holthuis, L.B. 1998. The Australian species of the genus *Ibacus* (Crustacea: Decapoda: Scyllaridae), with the description of a new species and addition of new records. *Zool. Med. Leiden* 72: 113–141.

Butler, M.J., IV, Herrnkind, W.F., & Hunt, J.H. 1997. Factors affecting the recruitment of juvenile Caribbean spiny lobsters dwelling in macroalgae. *Bull. Mar. Sci.* 61: 3–19.

Butler, M.J., IV, MacDiarmid, A.B., & Booth, J.D. 1999. Ontogenetic changes in social aggregation and its adaptive value for spiny lobsters in New Zealand. *Mar. Ecol. Prog. Ser.* 188: 179–191.

Bushmann, P. & Atema, J. 1994. Nephropore glands in the lobster, *Homarus americanus*: A source of urine-carried chemical signals? *Chem. Senses* 19: 448.

Cate, H.S. & Derby, C.D. 2001. Morphology and distribution of setae on the antennules of the Caribbean spiny lobster *Panulirus argus* reveal new types of bimodal chemo-mechanosensilla. *Cell Tissue Res.* 304: 439–454.

Cate, H.S. & Derby, C.D. 2002. Hooded sensilla homologues: Structural variations of a widely distributed bimodal chemomechanosensillum. *J. Comp. Neurol.* 444: 345–357.

Cate, H.S., Gleeson, R.A. & Derby, C.D. 1999. Activity-dependent labeling of the olfactory organ of blue crabs suggests that pheromone-sensitive and food odor-sensitive receptor neurons are packaged together in aesthetasc sensilla. *Chem. Senses* 24: 559.

Cau, A., Deiana, A.M. & Mura, M. 1978. Sulla frequenza e bionomia di *Scyllarus pygmaeus* (Bate) in acque neritiche sarde. *Natura Milano* 69: 118–124.

Chace, F.A. 1966. Decapod crustaceans from St. Helena Island, South Atlantic. *Proc. U.S. Nat. Mus.* 118: 623–661.

Chan T.-Y. & Yu, H.-P. 1986. A report on the *Scyllarus* lobsters (Crustacea: Decapoda: Scyllaridae) from Taiwan. *J. Taiwan Mus.* 39: 147–174.

Chan T.-Y. & Yu, H.-P. 1993. *The Illustrated Lobsters of Taiwan*. Taipei: SMC Publishing Inc. 251 pp.

Chessa, L.A., Pais, A., & Serra, S. 1996. Behavioural observations on slipper lobster *Scyllarides latus* (Latreille, 1803) (Decapoda, Scyllaridae) reared in laboratory. In: *Proceedings of the 6th Colloquium Crustacea Decapoda Mediterranea*, Florence, September 12–15, 1996, pp. 25–26.

Childress, M.J. & Herrnkind, W.F. 1996. The ontogeny of social behavior among juvenile Caribbean spiny lobsters. *Anim. Behav.* 51: 675–687.

Childress, M.J. & Herrnkind, W.F. 2001. The influence of conspecifics on the ontogenetic habitat shift of juvenile Caribbean spiny lobsters. *Mar. Freshw. Res.* 52: 1077–1084.

Clarac, F., Libersat, F. Pfluger, H.G. & Rathmayer, W. 1987. Motor pattern analysis in the shore crab (*Carcinus maenus*) walking freely in water and on land. *J. Exp. Biol.* 133: 395–414.

Cohen, M.J. 1955. The function of receptors in the statocyst of the lobster, *Homarus americanus*. *J. Physiol. (London)* 130: 9–34.

Cohen, M.J. 1960. The response pattern of single receptors in the crustacean statocyst. *Proc. R. Soc. London* B152: 30–49.

Cooke, I.R.C. & Macmillan, D.L. 1985. Further studies of crayfish escape behaviour. I. The role of the appendages and the stereotyped nature of non-giant escape swimming. *J. Exp. Biol.* 118: 351–365.

Cooper, R.A. & Uzmann, J.R. 1977. Ecology of juvenile and adult clawed lobsters, *Homarus americanus*, *Hommarus gammarus*, and *Nephrops norvegicus*. *Div. Fish. Oceanogr. Circ.* (Aust. C.S.I.R.O.) 7: 187–208.

Cooper, R.A. & Uzmann, J.R. 1980. Ecology of juvenile and adult *Homarus*. In: Cobb, J.S. & Phillips, B.F. (eds.), *The Biology and Management of Lobsters*, Vol. 2: pp. 97–142. New York: Academic Press.

Courtney, A.J. 2002. *The Status of Queensland's Moreton Bay Bug (Thenus spp.) and Balmain Bug (Ibacus spp.) Stocks*. Brisbane, AU: Queensland Government: Department of Primary Industries.

Courtney, A.J., Cosgrove, M.G., & Die, D.J. 2001. Population dynamics of Scyllarid lobsters of the genus *Thenus* spp. on the Queensland (Australia) east coast: I. Assessing the effects of tagging. *Fish. Res.* 53: 251–261.

Coutures, E. 2000. Distribution of phyllosoma larvae of Scyllaridae and Palinuridae (Decapoda: Palinuridae) in the south-western lagoon of New Caledonia. *Mar. Freshw. Res.* 51: 363–369.

Cowan, D.F. & Atema, J. 1990. Moult staggering and serial monogamy in American lobsters, *Homarus americanus*. *Anim. Behav.* 39: 1199–1206.

Crosnier, A. 1972. Naupliosoma, phyllosomes et pseudibacus de *Scyllarides herklotsi* (Herklots) (Crustacea, Decapoda, Scyllaridae) recoltes par l'ombango dans le sud du Golfe de Guinee. *Cahiers O.R.S.T.O.M. Oceanogr.* 10: 139–149.

Cruse, H. & Saavedra, M.G.S. 1996. Curve walking in crayfish. *J. Exp. Biol.* 199: 1477–1482.

Daniel P.C. & Derby C.D. 1988. Behavioral olfactory discrimination of mixtures in the spiny lobster (*Panulirus argus*) based on a habituation paradigm. *Chem. Senses* 13: 385–395.

Davie, P.J.F. 2002. *Crustacea: Malacostraca, Ilocarida, Hoplocarida, Eucarida (Part 1)*. Zoological Catalogue of Australia 19.3A. Canberra, AU: CSIRO Publishing / Australian Biological Resources Study (ABRS).

DeMartini, E.E., McCracken, M.L., Moffitt, R.B., & Wetherall, J.A. 2005. Relative pleopod length as an indicator of size at sexual maturity in slipper (*Scyllarides squammosus*) and spiny Hawaiian (*Panulirus marginatus*) lobsters. *Fish. Bull.* 103: 23–33.

Derby, C.D. 1982. Structure and function of cuticular sensilla of the lobster *Homarus americanus*. *J. Crust. Biol.* 2: 1–21.

Derby, C.D. 1989. Physiology of sensory neurons in morphologically identified cuticular sensilla of crustaceans. In: Felgenhauer, B.E., Watling, L., & Thistle, A.B. (eds.), *Functional Morphology of Feeding and Grooming in Crustacea: Crustacean Issues*, Vol. 6: pp. 27–47. Rotterdam, The Netherlands: Balkema.

Derby, C.D. & Ache B.W. 1984. Electrophysiological identification of stimulatory and interactive components of a complex odorant. *Chem. Senses* 9: 201–218.

Derby, C.D. & Atema, J. 1982a. Chemosensitivity of walking legs of the lobster, *Homarus americanus*: Response spectrum and thresholds. *J. Exp. Biol.* 98: 303–315.

Derby, C.D. & Atema, J. 1982b. Function of chemo- and mechanoreceptors in lobster (*Homarus americanus*) feeding behavior. *J. Exp. Biol.* 98: 317–327.

Derby, C.D. & Atema, J. 1988. Chemoreceptor cells in aquatic invertebrates: Peripheral mechanisms of chemical signal processing in decapod crustaceans. In: Atema, J., Popper, A.N., Fay, R.R., & Tavolga, W.N. (eds.), *Sensory Biology of Aquatic Animals*: pp. 365–385. New York: Springer-Verlag.

Derby, C.D. & Steullet, P. 2001. Why do animals have so many receptors? The role of multiple chemosensors in animal perception. *Biol. Bull.* 200: 211–215.

Derby, C.D., Reilly, P.M., & Atema, J. 1984. Chemosensitivity of the lobster *Homarus americanus* to secondary plant compounds: unused receptor capabilities. *J. Chem. Ecol.* 10: 879–892.

Derby, C.D., Steullet, P., Horner, A.J., & Cate, H.S. 2001. The sensory basis of feeding behaviour in the Caribbean spiny lobster, *Panulirus argus*. *Mar. Freshw. Res.* 52: 1339–1350.

Devine, D. & Atema, J. 1982. Function of chemoreceptor organs in spatial orientation of the lobster *Homarus americanus*: differences and overlap. *Biol. Bull.* 163: 144–153.

Domenici, P., Jamon, M., & Clarac, F. 1998. Curve walking in freely moving crayfish (*Procambarus clarkii*). *J. Exp. Biol.* 201: 1315–1329.

Domenici, P., Schmitz, J., & Jamon, M. 1999. The relationship between leg stripping pattern and yaw torque oscillations in curve walking of two crayfish species. *J. Exp. Biol.* 202: 3069–3080.

Factor, J.R. 1978. Morphology of the mouthparts of larval lobsters, *Homarus americanus* (Decapoda: Nephropidae), with special emphasis on their setae. *Biol. Bull.* 154: 383–408.

Farmer, A.S. 1974. The functional morphology of the mouthparts and pereiopods of *Nephrops novegicus* (L.) (Decapoda: Nephropidae). *J. Nat. Hist.* 8: 121–142.

Faulkes, Z. 2004. Loss of escape responses and giant neurons in the tailflipping circuits of slipper lobsters, *Ibacus* spp. (Decapoda, Palinura, Scyllaridae). *Arthrop. Struct. Develop.* 33: 113–123.

Faulkes, Z. 2006. Digging mechanisms and substrate preferences of shovel nosed lobsters *Ibacus peronii* (Decapoda: Scyllaridae). *J. Crust. Biol.* 26: 69–72.

Faulkes, Z. & Paul, D.H. 1997. Coordination between the legs and tail during digging and swimming in sand crabs. *J. Comp. Physiol.* 180A: 161–169.

Faulkes, Z. & Paul, D.H. 1998. Digging in sand crabs: Coordination of joints in individual legs. *J. Exp. Biol.* 201: 2139–2149.

Fiedler, U. & Spanier, E. 1999. Occurrence of *Scyllarus arctus* (Crustacea, Decapoda, Scyllaridae) in the eastern Mediterranean — Preliminary results. *Ann. Istrian Med. Stud.* 17: 153–158.

Fraser, P.J. 1982. Views on the nervous control of complex behavior. In: Sandeman, D.C. & Atwood, H.L. (eds.), *Neural Integration and Behavior*: pp. 293–319. New York: Academic Press.

Garm, A. 2004. Mechanical functions of setae from the mouth apparatus of seven species of decapod crustaceans. *J. Morphol.* 260: 85–100.

Garm, A., Derby, C.D., & Høeg, J.T. 2004. Mechanosensory neurons with bend- and osmo-sensitivity in mouthpart setae from the spiny lobster *Panulirus argus*. *Biol. Bull.* 207: 195–208.

Gleeson, R.A. 1982. Morphological and behavioral identification of the sensory structures mediating pheromone reception in the blue crab, *Callinectes sapidus*. *Biol. Bull.* 163: 162–171.

Goldman, J.A. & Patek, S.N. 2002. Two sniffing strategies in palinurid lobsters. *J. Exp. Biol.* 205: 3891–3902.

Gomez, G. & Atema, J. 1996. Temporal resolution in olfaction: stimulus integration time of lobster chemoreceptor cells. *J. Exp. Biol.* 199: 1771–1779.

Graham, K.J. & Wood, B.R. 1997. *Kapala Cruise Report No. 116.* Sydney: NSW Fisheries. pp. 86.

Graham, K.J., Liggins, G.W., Wildforster, J., & Kennelly, S.J. 1993a. *Kapala Cruise Report No. 110.* Sydney: NSW Fisheries. pp. 69.

Graham, K.J., Liggins, G.W., Wildforster, J., & Kennelly, S.J. 1993b. *Kapala Cruise Report No. 112.* Sydney: NSW Fisheries. pp. 74.

Graham, K.J., Liggins, G.W., & Wildforster, J. & Wood, B. 1995. *Kapala Cruise Report No. 114.* Sydney: NSW Fisheries. pp. 63.

Graham, K.J., Liggins, G.W., & Wildforster, J. 1996. *Kapala Cruise Report No. 115.* Sydney: NSW Fisheries. pp. 63.

Grasso, F.W. & Basil, J.A. 2002. How lobsters, crayfishes, and crabs locate sources of odor: Current perspectives and future directions. *Curr. Opin. Neurobiol.* 12: 721–727.

Guenther, C. & Atema, J. 1998. Distribution of setae on the *Homarus americanus* lateral antennular flagellum. (Abstr.) *Biol. Bull.* 195: 182–183.

Gurney, R. 1936. Larvae of decapod Crustacea: Part III Phyllosoma. *Discovery Rep.* 12: 379–440.

Haddy, J.A., Courtney, A.J., & Roy, D.P. 2005. Aspects of the reproductive biology and growth of Balmain bugs (*Ibacus* spp.) (Scyllaridae). *J. Crust. Biol.* 25: 263–273.

Hamilton, W.D. 1971. Geometry for the selfish herd. *J. Theor. Biol.* 31: 295–311.

Harada, E. 1958. Notes on the naupliosoma and newly-hatched phyllosoma of *Ibacus ciliatus* (von Siebold). *Publ. Seto Mar. Biol. Lab.* 7: 173–179.

Hardwick, C.W., Jr. & Cline, C.B. 1990. Reproductive status, sex ratios and morphometrics of the slipper lobster *Scyllarides nodifer* (Stimpson) (Decapoda: Scyllaridae) in the northeastern Gulf of Mexico. *N.E. Gulf Sci.* 11: 131–136.

Hearn, A. 2006. Life history of the slipper lobster *Scyllarides astori* Holthuis 1960, in the Galápagos Islands, Ecuador. *J. Exp. Mar. Biol. Ecol.* 328(1): 87–97.

Herrnkind, W.F. 1980. Spiny lobsters: Patterns of movement. In: Cobb, J.S. & Phillips, B.F. (eds.), *The Biology and Management of Lobsters*, Vol. 1: pp. 349–401. New York: Academic Press.

Herrnkind, W.F., Halusky, J., & Kanciruk, P. 1976. A further note on phyllosoma larvae associated with medusae. *Bull. Mar. Sci.* 26: 110–112.

Herrnkind, W.F., Childress, M.J., & Lavalli, K.L. 2001. Defense coordination and other benefits among exposed spiny lobsters: inferences from mass migratory and mesocosm studies of group size and behavior. *Mar. Freshw. Res.* 52: 1133–1143.

Higa, T. & Saisho, T. 1983. Metamorphosis and growth of the late-stage phyllosoma of *Scyllarus kitanoviriosus* Harada (Decapoda, Scyllaridae). *Mem. Kagoshima Univ. Res. Ctr. S. Pac.* 3: 86–98.

Higa, T. & Shokita, S. 2004. Late-stage phyllosoma larvae and metamorphosis of a scyllarid lobster, *Chelarctus cultrifer* (Crustacea: Decapoda: Scyllaridae), from the northwestern Pacific. *Spec. Diversity* 9: 221–249.

Holthuis, L.B. 1968. The Palinuridae and Scyllaridae of the Red Sea. The second Israel South Red Sea expedition, 1965, report no. 7. *Zool. Med. Leiden* 42: 281–301.

Holthuis, L.B. 1985. A revision of the family Scyllaridae (Crustacea: Decapoda: Macrura). I. Subfamily Ibacinae. *Zool. Verhand.* 218: 1–130.

Holthuis, L.B. 1991. *Marine Lobsters of the World. An Annotated and Illustrated Catalogue of the Species of Interest to Fisheries Known to Date.* FAO Species Catalogue No. 125, Vol. 13: pp. 1–292. Rome: Food and Agriculture Organization of the United Nations.

Holthuis, L.B. 1993a. *Scyllarus rapanus*, a new species of locust lobster from the South Pacific (Crustacea, Decapoda, Scyllaridae). *Bull. Mus. Nat. d'Hist. Natur. Paris* 4e ser. 15, section A, no. 1–4: 179–186.

Holthuis, L.B. 1993b. *Scyllarides obtusus* spec. nov., the scyllarid lobster of Saint Helena, Central South Atlantic (Crustacea: Decapoda Reptantia: Scyllaridae). *Zool. Med.* 67: 505–515.

Holthuis, LB. 2002. The Indo-Pacific scyllarinid lobsters (Crustacea, Decapoda, Scyllaridae). *Zoosystema* 24: 499–683.

Inoue, N., Sekiguchi, H., & Nagasawa, T. 2000. Distribution and identification of phyllosoma larvae in the Tsushima Current region. *Bull. Japan. Soc. Fish. Oceanogr.* 64: 129–137.

Inoue, N., Sekiguchi, H., & Shinn-Pyng, Y. 2001. Spatial distributions of phyllosoma larvae (Crustacea: Decapoda: Palinuridae and Scyllaridae) in Taiwanese waters. *J. Oceanogr.* 57: 535–548.

Ito, M. & Lucas, J.S. 1990. The complete larval development of the scyllarid lobster *Scyllarus demani* Holthuis, 1946 (Decapoda, Scyllaridae) in the laboratory. *Crustaceana* 58: 144–167.

Jacklyn. P.M. & Ritz, D.A. 1986. Hydrodynamics of swimming in scyllarid lobsters. *J. Exp. Mar. Biol. Ecol.* 101: 85–89.

Jackson, D.J. & Macmillan, D.L. 2000. Tailflick escape behavior in larval and juvenile lobsters (*Homarus americanus*) and crayfish (*Cherax destructor*). *Biol. Bull.* 198: 307–318.

Jamon, M. & Clarac, F. 1995. Locomotor patterns in freely moving crayfish (*Procambarus clarkii*). *J. Exp. Biol.* 198: 683–700.

Jamon, M. & Clarac, F. 1997. Variability of leg kinematics in free-walking crayfish, *Procambarus clarkii*, and related inter-jointed coordination. *J. Exp. Biol.* 200: 1201–1213.

Jeffs, A.G. & Holland, R.C. 2000. Swimming behaviour of the puerulus of the spiny lobster, *Jasus edwardsii*. *Crustaceana* 73: 847–856.

Johnson, M.W. 1968. The phyllosoma larvae of scyllarid lobsters in the Gulf of California and off Central America with special reference to *Evibacus princeps* (Palinuridea). *Crustaceana Suppl.* 2: 98–116.

Johnson, M.W. 1970. On the phyllosoma larvae of the genus *Scyllarides* Gill (Decapoda, Scyllaridae). *Crustaceana* 18: 13–20.

Johnson, M.W. 1971a. The palinurid and scyllarid lobster larvae of the tropical eastern Pacific and their distribution as related to the prevailing hydrography. *Bull. Scripps Inst. Oceanogr., Univ. Calif.* 19: 1–36.

Johnson, M.W. 1971b. The phyllosoma larvae of slipper lobsters from the Hawaiian Islands and adjacent areas (Decapoda, Scyllaridae). *Crustaceana* 20: 77–103.

Johnson, M.W. 1975. The postlarvae of *Scyllarides astori* and *Evibacus princeps* of the eastern tropical Pacific (Decapoda, Scyllaridae). *Crustaceana* 28: 139–144.

Johnston, D.J. 1994. Functional morphology of the membranous lobe within the preoral cavity of *Thenus orientalis* (Crustacea: Scyllaridae). *J. Mar. Biol. Assoc. U.K.* 74: 787–800.

Johnston, D.J. & Alexander, C.G. 1999. Functional morphology of the mouthparts and alimentary tract of the slipper lobster *Thenus orientalis* (Decapoda: Scyllaridae). *Mar. Freshw. Res.* 50: 213–223.

Johnston, D.J. & Yellowlees, D. 1998. Relationship between the dietary preferences and digestive enzyme complement of the slipper lobster *Thenus orientalis* (Decapoda: Scyllaridae). *J. Crust. Biol.* 18: 656–665.

Jones, C. 1988. *The Biology and Behaviour of Bay Lobsters, Thenus spp. (Decapoda: Scyllaridae) in Northern Queensland*, Australia. Ph.D. dissertation: University of Queensland, Brisbane, Australia.

Jones, C.M. 1993. Population structure of two species of *Thenus* (Decapoda: Scyllaridae), in northeastern Australia. *Mar. Ecol. Prog. Ser.* 97: 143–155.

Kabli, L.M. 1989. *Biology of Sand Lobster, Thenus orientalis* (Lund). Ph.D. thesis: University of Bombay, Bombay, India. 378 pp.

Kabli, L.M. & Kagwade, P.V. 1996. Morphometry and conversion factors in the sand lobster *Thenus orientalis* (Lund) from Bombay waters. *Indian J. Fish.* 43: 249–254.

Kagwade, P.V. & Kabli, L.M. 1996a. Age and growth of the sand lobster *Thenus orientalis* (Lund) from Bombay waters. *Indian J. Fish.* 43: 241–247.

Kagwade, P.V. & Kabli, L.M. 1996b. Reproductive biology of the sand lobster *Thenus orientalis* (Lund) from Bombay waters; *Indian J. Fish.* 43: 13–25.

Kanciruk, P. 1980. Ecology of juvenile and adult Palinuridae (spiny lobsters). In: Cobb, J.S. & Phillips, B.F. (eds.), *The Biology and Management of Lobsters*, Vol. 2: pp. 59–96. New York: Academic Press.

Karnofsky, E.B., Atema, J., & Elgin, R.H. 1989a. Field observations of social behavior, shelter use, and foraging in the lobster, *Homarus americanus. Biol. Bull.* 176: 239–246.

Karnofsky, E.B., Atema, J., & Elgin, R.H. 1989b. Natural dynamics of population structure and habitat use of the lobster, *Homarus americanus*, in a shallow cove. *Biol. Bull.* 176: 247–256.

Kizhakudan, J.K., Thirumilu, P., Rajapackiam, S. & Manibal, C. 2004. Captive breeding and seed production of scyllarid lobsters — Opening new vistas in crustacean aquaculture. *Mar. Fish. Infor. Serv. T & E Ser.* 181: 1–4.

Kneipp, I.J. 1974. A Preliminary Study of Reproduction and Development in *Thenus orientalis* (Crustacea: Decapoda: Scyllaridae). B.Sc. honours thesis: James Cook University, Townsville, Australia.

Koehl, M.A.R. 2006. The fluid mechanics of arthropod sniffing in turbulent odor plumes. *Chem. Senses* 31: 93–105.

Koehl, M.A.R., Koseff, J.R., Crimaldi, J.P., McCay, M.G., Cooper, T., Wiley, M.B., & Moore, P.A. 2001. Lobster sniffing: antennule design and hydrodynamic filtering of information in an odor plume. *Science* 294: 1948–1951.

Kozlowski, C. Yopak, K., Voigt, R., & Atema, J. 2001. An initial study on the effects of signal intermittency on the odor plume tracking behavior of the American lobster, *Homarus americanus*. *Biol. Bull.* 201: 274–276.

Lau, C.J. 1987. Feeding behavior of the Hawaiian slipper lobster, *Scyllarides squammosus*, with review of decapod crustacean feeding tactics on molluscan prey. *Bull. Mar. Sci.* 41: 378–391.

Lau, C.J. 1988. Dietary comparison of two slow-moving crustacean (Decapoda: Scyllaridae) predators by a modified index of relative importance. In: *Proceedings of the 6th International Coral Reef Symposium*, Vol. 2: pp. 95–100. Townsville, Australia, August 8–12, 1988.

Lavalli, K.L. & Factor, J.R. 1992. Functional morphology of the mouthparts of juvenile lobsters, *Homarus americanus* (Decapoda: Nephropidae), and comparison with larval stages. *J. Crust. Biol.* 12: 467–510.

Lavalli, K.L. & Factor, J.R. 1995. The feeding apparatus. In: Factor, J.R. (ed.), *The Biology of the Lobster Homarus americanus*: pp. 467–510. New York: Academic Press.

Lavalli, K.L. & Spanier, E. 2001. Does gregariousness function as an antipredator mechanism in the Mediterranean slipper lobster, *Scyllarides latus*? *Mar. Freshw. Res.* 52: 1133–1143.

Lee, T.N., Rooth, C., Williams, E., McGowan, M., Szmant, A.F., & Clarke, M.E. 1992. Influence of Florida Current, gyres and wind-driven circulation on transport of larvae and recruitment in the Florida Keys coral reefs. *Cont. Shelf Res.* 12: 971–1002.

Lee, T.N., Clarke, M.E., Williams, E., Szmant, A.F., & Berger, T. 1994. Evolution of the Tortugas Gyre and its influence on recruitment in the Florida Keys. *Bull. Mar. Sci.* 54: 621–646.

Lesser, J.H.R. 1974. Identification of early larvae of New Zealand spiny and shovel-nosed lobsters (Decapoda, Palinuridae and Scyllaridae). *Crustaceana* 27: 259–277.

Lohman, K.S. 1984. Magnetic remnants in the Western Atlantic spiny lobster, *Panulirus argus*. *J. Exp. Biol.* 113: 29–41.

Lohman, K.S. 1985. Geomagnetic field detection by the Western Atlantic spiny lobster, *Panulirus argus*. *Mar. Behav. Physiol.* 12: 1–17.

Lozano-Alvarez, E. & Spanier, E. 1997. Behavior and growth of captive sub-adults spiny lobsters, *Panulirus argus*, under the risk of predation. *Mar. Freshw. Res.* 48: 707–713.

Lyons, W.G. 1970. Scyllarid lobsters (Crustacea, Decapoda). *FL Mar. Res. Lab., Mem. Hourglass Cruises* 1: 1–74.

Malcom, C. 2003. Description of the Setae on the Pereiopods of the Mediterranean Slipper Lobster, *Scyllarides latus*, the Ridged Slipper Lobster, *S. nodifer*, and the Spanish Slipper Lobster, *S. aequinoctialis*. M.Sc. thesis: Texas State University at San Marcos, San Marcos, TX.

Marinovic, B., Lemmens, J.W.T.J., & Knott, B. 1994. Larval development of *Ibacus peronii* Leach (Decapoda: Scyllaridae) under laboratory conditions. *J. Crust. Biol.* 14: 80–96.

Martínez, C.E. 2000. *Ecología trófica de Panulirus gracilis, P. penicillatus y Scyllarides astori (Decapoda, Palinura) en sitios de pesca de langosta en las islas Galápagos*. Tesis de Licenciatura: Universidad del Azuay, Ecuador. 102 pp.

Martínez, C.E., Toral, V., & Edgar, G. 2002. Langostino. In: Danulat, E. & Edgar, G.J. (eds.), *Reserva Marina de Galápagos, Línea Base de la Biodiversidad*: pp. 216–232. Santa Cruz, Galápagos, Ecuador: Fundación Charles Darwin y Servicio Parque Nacional Galápagos.

Martins, H.R. 1985a. Some observations on the naupliosoma and phyllosoma larvae of the Mediterranean locust lobster, *Scyllarides latus* (Latreille, 1803), from the Azores. *ICES C.M.K.* 52 Shellfish Committee, 13 pp.

Martins, H.R. 1985b. Biological studies of the exploited stock of the Mediterranean locust lobster *Scyllarides latus* (Latreille, 1803) (Decapoda: Scyllaridae) in the Azores. *J. Crust. Biol.* 5: 294–305.

McWilliam, P.S. & Phillips, B.F. 1983. Phyllosoma larvae and other crustacean macrozooplankton associated with Eddy J, a warm-core eddy off south-eastern Australia. *Aust. J. Mar. Freshw. Res.* 34: 653–663.

Michel, A. 1968. Les larves phyllosomes et la post-larve de *Scyllarides squammosus* (H. Milne Edwards) — Scyllaridae (Crustaces Decapodes). *Cah. O.R.S.T.O.M. Ser. Oceanogr.* 6: 47–53.

Michel, W.C., McClintock, T.S., & Ache, B.W. 1991. Inhibition of lobster olfactory receptor cells by an odor activated potassium conductance. *J. Neurophysiol.* 64: 446–453.

Mikami, S. 1995. *Larviculture of Thenus (Decapoda, Scyllaridae), the Moreton Bay Bugs*. Ph.D. thesis: University of Queensland, Brisbane, Australia.

Mikami, S. & Greenwood, J.G. 1997. Complete development and comparative morphology of larval *Thenus orientalis* and *Thenus* sp. (Decapoda: Scyllaridae) reared in the laboratory. *J. Crust. Biol.* 17: 289–308.

Mikami, S. & Takashima, F. 1993. Development of the proventriculus in larvae of the slipper lobster, *Ibacus ciliatus* (Decapoda: Scyllaridae). *Aquaculture* 116: 199–217.

Mikami, S. & Takashima, F. 1994. Functional morphology and cytology of the phyllosomal digestive system of *Ibacus ciliatus* and *Panulirus japonicus* (Decapoda, Scyllaridae and Palinuridae). *Crustaceana* 67: 212–225.

Mikami, S. & Takashima, F. 2000. Functional morphology of the digestive system. In: Phillips, B.F. & Kittaka, J. (eds.), *Spiny Lobsters: Fisheries and Culture*, 2nd Edition: pp. 601–610. Oxford, England: Blackwell Science.

Mikami, S., Greenwood, J.G., & Takashima, F. 1994. Functional morphology and cytology of the phyllosomal digestive system *Ibacus ciliatus* and *Panulirus japonicus* (Decapoda, Scyllaridae and Palinuridae). *Crustaceana* 67: 212–225.

Minami, H., Inoue, N., & Sekiguchi, H. 2001. Vertical distributions of phyllosoma larvae of palinurid and scyllarid lobsters in the western North Pacific. *J. Oceanogr.* 57: 743–748.

Mittenthal, J.E. & Wine, J.J. 1978. Segmental homology and variation in flexor motorneurons of the crayfish abdomen. *J. Comp. Neurol.* 177: 311–334.

Moe, M.A., Jr. 1991. *Lobsters: Florida, Bahamas, the Caribbean*. Plantation, FL: Green Turtle Publications.

Moody, K.E. & Steneck, R.S. 1993. Mechanisms of predation amoung large decapod crustaceans of the Gulf of Maine Coast: Functional vs. phylogenetic patterns. *J. Exp. Biol. Ecol.* 168: 111–124.

Moore, P.A. & Atema, J. 1991. Spatial information in the three-dimensional fine structure of an aquatic odor plume. *Biol. Bull.* 181: 408–418.

Moore, P.A., Atema, J., & Gerhardt, G.A. 1991a. Fluid dynamics and microscale chemical movement in the chemosensory appendages of the lobster, *Homarus americanus. Chem. Senses* 16: 663–674.

Moore, P.A., Scholz, N., & Atema, J. 1991b. Chemical orientation of lobsters, *Homarus americanus*, in turbulent odor plumes. *J. Chem. Ecol.* 17: 1293–1307.

Morin, T.D. & MacDonald, C.D. 1984. Occurrence of the slipper lobster *Scyllarides haanii* in the Hawaiian archipelago. *Proc. Biol. Soc. Wash.* 97: 404–407.

Neil, D.M. & Ansel, A.D. 1995. The orientation of tail-flip escape swimming in decapod and mysid crustaceans. *J. Mar. Biol. Assoc. U.K.* 75: 55–70.

Nevitt G., Pentcheff, N.D., Lohmann, K.J., & Zimmer, R.K. 2000. Den selection by the spiny lobster *Panulirus argus*: Testing attraction to conspecific odors in the field. *Mar. Ecol. Prog. Ser.* 203: 225–231.

Newland, P.L. & Neil, D.M. 1990. The tail flip of the Norway lobster, *Nephrops norvegicus*. I. Giant fibre activation in relation to swimming trajectories. *J. Comp. Physiol.* 166A: 517–536.

Newland, P.L., Chapman, C.J., & Neil, D.M. 1988. Swimming performance and endurance of the Norway lobster, *Nephrops norvegicus. Mar. Biol.* 98: 345–350.

Newland, P.L., Neil, D.M., & Chapman, C.J. 1992. Escape swimming of the Norway lobster. *J. Crust. Biol.* 12: 342–353.

Ogren, L.H. 1977. Concealment behaviour of the Spanish lobster, *Scyllarides nodifer* (Stimpson), with observations on its diel activity. *N.E. Gulf Sci.* 1: 115–116.

Olvera Limas, R.Ma. & Ordonez Alcala, L. 1988. Distribution, relative abundance and larval development of the lobsters *Panulirus argus* and *Scyllarus americanus* in the Exclusive Economic Zone of the Gulf of Mexico and the Caribbean Sea. *Ciencia Pesquera* 6: 7–31.

Phillips, B.F. & McWilliam, P.S. 1986. Phyllosoma and nisto stages of *Scyllarus martensii* Pfeffer (Decapoda: Scyllaridae) from the Gulf of Carpentaria, Australia. *Crustaceana* 51: 133–154.

Phillips, B.F. & McWilliam, P.S. 1989. Phyllosoma larvae and the ocean currents off the Hawaiian Islands. *Pac. Sci.* 43: 352–361.

Phillips, B.F., Brown, P.A., Rimmer, D.W., & Braine, S.J. 1981. Description, distribution and abundance of late larval stages of the Scyllaridae (slipper lobsters) in the south-eastern Indian Ocean. *Aust. J. Mar. Freshw. Res.* 32: 417–437.

Prasad, R.R. & Tampi, P.R.S. 1960. Phyllosomas of scyllarid lobsters from the Arabian Sea. *J. Mar. Biol. Assoc. India* 2: 241–249.

Prasad, R.R., Tampi, P.R.S., & George, M.J. 1975. Phyllosoma larvae from the Indian Ocean collected by Dana Expedition 1928–1930. *J. Mar. Biol. Assoc. India* 17: 56–107.

Randall, J.E. 1964. Contributions to the biology of the queen conch, *Strombus gigas. Bull. Mar. Sci.* 14: 246–295.

Reeder, P.B. & Ache, B.W. 1980. Chemotaxis in the Florida spiny lobster, *Panulirus argus. Anim. Behav.* 28: 831–839.

Reichert, H. & Wine, J.J. 1983. Coordination of lateral giant and non-giant systems in crayfish escape behavior. *J. Comp. Physiol.* A153: 3–15.

Reichert, H., Wine, J.J., & Hagiwara, G. 1981. Crayfish escape behavior: Neurobehavioral analysis of phasic extension reveals dual systems for motor control. *J. Comp. Physiol.* 142A: 281–294.

Relini, G., Bertrand, J., & Zamboni, A. 1999. Synthesis of the knowledge on bottom fishery resources in central Mediterranean (Italy and Corsica). *Biol. Mar. Medit.* 6 suppl.: 593–600.

Ritz, D.A. 1977. The larval stages of *Scyllarus demani* Holthuis, with notes on the larvae of *S. sordidus* (Stimpson) and *S. timidus* Holthuis (Decapoda, Palinuridea). *Crustaceana* 32: 229–240.

Ritz, D.A. & Jacklyn, P.M. 1985. Believe it or not — bugs fly through water. *Aust. Fish.* 44: 35–37.

Ritz, D.A. & Thomas, L.R. 1973. The larval and postlarval stages of *Ibacus peronii* Leach (Decapoda, Reptantia, Scyllaridae). *Crustaceana* 24: 5–16.

Robertson, P.B. 1968. *The Larval Development of Some Western Atlantic Lobsters of the Family Scyllaridae.* Ph.D. dissertation: University of Miami, Coral Gables, Florida.

Robertson, P.B. 1969a. The early larval development of the scyllarid lobster *Scyllarides aequinoctialis* (Lund) in the laboratory, with a revision of the larval characters of the genus. *Deep-Sea Res.* 16: 557–586.

Robertson, P.B. 1969b. Biological investigations of the deep sea. No. 48. Phyllosoma larvae of a scyllarid lobster, *Arctides guineensis*, from the western Atlantic. *Mar. Biol.* 4: 143–151.

Robertson, P.B. 1971. The larvae and postlarva of the scyllarid lobster *Scyllarus depressus* (Smith). *Bull. Mar. Sci.* 21: 841–865.

Rothlisberg, P.C., Jackson, C.J., Phillips, B.F., & McWilliam, P.S. 1994. Distribution and abundance of scyllarid and palinurid lobster larvae in the Gulf of Carpentaria, Australia. *Aust. J. Mar. Freshw. Res.* 45: 337–349.

Rudloe, A. 1983. Preliminary studies of the mariculture potential of the slipper lobster, *Scyllarides nodifer. Aquaculture* 34: 165–169.

Sandeman, D. 1990. Structural and functional levels In: Wiese, K., Krenz, W.-D., Tautz, J., Reichert, H. & Mulloney, B. (eds.), *Frontiers in Crustacean Neurobiology: The Organization of Decapod Crustacean Brains*: pp. 223–239. Basel: Birkhauser Verlag.

Sandeman, D. 1999. Homology and convergence in vertebrate nervous systems. *Die Natur Wissenschaften* 86: 378–387.

Sandeman, D., Sandeman, R., Schmidt, M., & Derby, C. 1992. Morphology of the brain of crayfish, crabs, and spiny lobsters: A common nomenclature for homologous structures. *Biol. Bull.* 183: 304–326.

Sandeman, D.C., Scholtz, G., & Sandeman, R.E. 1993. Brain evolution in decapod Crustacea. *J. Exp. Zool.* 265: 112–133.

Schmidt, M. & Ache, B.W. 1992. Antennular projections to the midbrain of the spiny lobster. II. Sensory innervation of the olfactory lobe. *J. Comp. Neurol.* 318: 291–303.

Schmidt, M. & Derby, C.D. 2005. Non-olfactory chemoreceptors in asymmetric setae activate antennular grooming behavior in the Caribbean spiny lobster *Panulirus argus. J. Exp. Biol.* 208–233–248.

Schmidt, M., Gruenert, U., & Derby, C.D. 2003. Non-olfactory sensilla mediate chemically induced antennular grooming in the spiny lobster, *Panulirus argus*. Program No. 595.5. 2003 Abstract Viewer/Itinerary Planner. Washington, DC: Society for Neuroscience. Online: http://sfn.scholarone.com.itin2003/index.html.

Sekiguchi, H. 1986a. Identification of late-stage phyllosoma larvae of the scyllarid and palinurid lobsters in the Japanese waters. *Bull. Japan. Soc. Sci. Fish.* 52: 1289–1294.

Sekiguchi, H. 1986b. Spatial distribution and abundance of phyllosoma larvae in the Kumano- and Enshu-nada seas north of the Kuroshio Current. *Bull. Japan. Soc. Fish. Oceanogr.* 50: 289–297.

Sekiguchi, H. 1990. Four species of phyllosoma larvae from the Mariana waters. *Bull. Japan. Soc. Fish. Oceanogr.* 54: 242–248.

Sekiguchi, H. & Inoue, N. 2002. Recent advances in larval recruitment processes of scyllarid and palinurid lobsters in Japanese waters. *J. Oceanogr.* 58: 747–757.

Shojima, Y. 1963. Scyllarid phyllosomas' habit of accompanying the jelly-fish. *Bull. Japan. Soc. Sci. Fish.* 29: 349–353.

Shojima, Y. 1973. The phyllosoma larvae of Palinura in the East China Sea and adjacent waters 1. *Ibacus novemdentatus*. *Bull. Seikai Reg. Fish. Res. Lab.* 43: 105–115.

Sims, H.W., Jr. 1966. Notes on the newly hatched phyllosoma of the sand lobster *Scyllarus americanus* (Smith). *Crustaceana* 11: 288–290.

Sims, H.W., Jr. & Brown, C.L. 1968. A giant scyllarid phyllosoma larva taken north of Bermuda (Palinuridea). *Crustaceana Suppl.* 2: 80–82.

Smale, M.J. 1978. Migration, growth and feeding of the Natal rock lobster *Panulirus homarus* (Linnaeus). *Invest. Rep. Oceanogr. Res. Inst. Durban* 47: 1–56.

Smale, M.J. & Goosen, A.J.J. 1999. Reproduction and feeding of spotted gully shark, *Triakis megalopterus*, off the Eastern Cape, South Africa. *Fish. Bull.* 97: 987–998.

Spanier, E. 1987. Mollusca as food for the slipper lobster *Scyllarides latus* in the coastal waters of Israel. *Levantina* 68: 713–716.

Spanier, E. 1994. What are the characteristics of a good artificial reef for lobsters? *Crustaceana* 67:173–186.

Spanier, E. 2004. The Mediterranean slipper lobster, *Scyllarides latus*, as a "flying machine" — Hydrodynamic aspects of locomotion — Preliminary results. *RIMS Newsltr.* 30: 29–30.

Spanier, E. & Almog-Shtayer, G. 1992. Shelter preferences in the Mediterranean slipper lobster: Effects of physical properties. *J. Exp. Mar. Biol. Ecol.* 164: 103–116.

Spanier, E. & Barshaw, D.E. 1993. Tag retention in the Mediterranean slipper lobster. *Israel J. Zool.* 39: 29–33.

Spanier, E. & Lavalli, K.L. 1998. Natural history of *Scyllarides latus* (Crustacea Decapoda): A review of the contemporary biological knowledge of the Mediterranean slipper lobster. *J. Nat. Hist.* 32: 1769–1786.

Spanier, E. & Weihs, D. 1992. Why do shovel-nosed (slipper) lobsters have shovels? *The Lobster Newsletter* 5(1): 8–9.

Spanier, E. & Weihs, D. 2004. Hydrodynamic aspects of locomotion in the Mediterranean slipper lobster, *Scyllarides latus*. In: *Proceedings of the 7th International Conference & Workshop on Lobster Biology and Management*. Tasmania, Australia, 8–13 February, 2004, p. 61.

Spanier, E. & Zimmer-Faust, R.K. 1988. Some physical properties of shelter that influence den preference in spiny lobsters. *J. Exp. Mar. Biol. Ecol.* 122: 137–149.

Spanier, E., Tom, M., Pisanty, S., & Almog, G. 1988. Seasonality and shelter selection by the slipper lobster *Scyllarides latus* in the southeastern Mediterranean. *Mar. Ecol. Prog. Ser.* 42: 247–255.

Spanier, E., Tom, M., Pisanty, S., & Almog-Shtayer, G. 1990. Artificial reefs in the low productive marine environment of the Southeastern Mediterranean. *P.S.Z.N.I. Mar. Ecol.* 11: 61–75.

Spanier, E., Weihs, D., & Almog-Shtayer, G. 1991. Swimming of the Mediterranean slipper lobster. *J. Exp. Mar. Biol. Ecol.* 145: 15–31.

Spanier, E., Almog-Shtayer, G., & Fiedler, U. 1993. The Mediterranean slipper lobster *Scyllarides latus*: The known and the unknown. *BIOS* 1: 49–58.

Spanò, S., Ragonese, S., & Bianchini, M. 2003. An anomalous specimen of *Scyllarides latus* (Decapoda, Scyllaridae). *Crustaceana* 76: 885–889.

Spencer, M. 1986. The innervation and chemical sensitivity of single aesthetasc hairs. *J. Comp. Physiol.* A158: 59–68.

Steullet P., Cate, H.S., & Derby, C.D. 2000a. A spatiotemporal wave of turnover and functional maturation of olfactory receptor neurons in spiny lobster *Panulirus argus*. *J. Neurosci.* 20: 3282–3294.

Steullet, P., Cate, H.S., Michel, W.C. & Derby, C.D. 2000b. Functional units of a compound nose: Aesthetasc sensilla house similar populations of olfactory receptor neurons on the crustacean antennule. *J. Comp. Neurol.* 418: 270–280.

Steullet, P., Dudar, O., Flavus, T., Zhou, M., & Derby, C.D. 2001. Selective ablation of antennular sensilla on the Caribbean spiny lobster *Panulirus argus* suggests that dual antennular chemosensory pathways mediate odorant activation of searching and localization of food. *J. Exp. Biol.* 204: 4259–4269.

Steullet, P., Kruetzfeldt, D.R., Hamidani, G., Flavus, T., & Derby, C.D. 2002. Dual antennular chemo-sensory pathways mediate odor-associative learning and odor discrimination in the Caribbean spiny lobster *Panulirus argus*. *J. Exp. Biol.* 205: 851–867.

Stewart, J. & Kennelly, S.J. 1997. Fecundity and egg-size of the Balmain bug *Ibacus peronii* (Leach, 1815) (Decapoda, Scyllaridae) off the east coast of Australia. *Crustaceana* 70: 191–197.

Stewart, J. & Kennelly, S.J. 1998. Contrasting movements of two exploited scyllarid lobsters of the genus *Ibacus* off the east coast of Australia. *Fish. Res.* 36: 127–132.

Stewart, J. & Kennelly, S.J. 2000. Growth of the scyllarid lobsters *Ibacus peronii* and *I. chacei*. *Mar. Biol.* 136: 921–930.

Stewart, J., Kennelly, S.J., & Hoegh-Guldberg, O. 1997. Size at sexual maturity and the reproductive biology of two species of scyllarid lobster from New South Wales and Victoria, Australia. *Crustaceana* 70: 344–367.

Subramanian, V.T. 2004. Fishery of sand lobster *Thenus orientalis* (Lund) along Chennai coast. *Indian J. Fish.* 51: 111–115.

Suthers, I.M. & Anderson, D.T. 1981. Functional morphology of mouthparts and gastric mill of *Ibacus peronii* (Leach) (Palinura: Scyllaridae). *Aust. J. Mar. Freshw. Res.* 32: 931–44.

Tarsitano, S.F., Lavalli, K.L., Horne, F. & Spanier, E. 2006. The constructional properties of the exoskeleton of homarid, scyllarid, and palinurid lobsters. *Hydrobiologia* 557: 9–20.

Tavares, M. 1997. *Scyllarus ramosae*, new species from the Brazilian continental slope, with notes on congeners occurring in the area (Decapoda: Scyllaridae). *J. Crust. Biol.* 17: 716–724.

Thomas, L.R. 1963. Phyllosoma larvae associated with medusae. *Nature* 198: 208.

Tierney, A.J., Voigt, R., & Atema, J. 1988. Response properties of chemoreceptors from the medial antennule of the lobster *Homarus americanus*. *Biol. Bull.* 174: 364–372.

Voigt, R. & Atema, J. 1994. Comparison of tuning properties of five chemoreceptor organs of the American lobster: Spectral filters. In: Kurihara, K., Suzuki, N., & Ogawa, H. (eds.), *Olfaction and Taste XI*: p. 787. Tokyo: Springer Verlag.

Wahle, R.A. & Steneck, R.S. 1991. Recruitment habitats and nursery grounds of the American lobster *Homarus americanus*: A demographic bottleneck? *Mar. Ecol. Prog. Ser.* 69: 231–243.

Wahle, R.A. & Steneck, R.S. 1992. Habitat restrictions in early benthic life. Experiments on habitat selection and *in situ* predation with the American lobster. *J. Exp. Mar. Biol. Ecol.* 157: 91–114.

Webber, W.R. & Booth, J.D. 2001. Larval stages, developmental ecology, and distribution of *Scyllarus* sp. Z (probably *Scyllarus aoteanus* Powell, 1949) (Decapoda: Scyllaridae). *N.Z. J. Mar. Freshw. Res.* 35: 1025–1056.

Weihs, D. 1974. Energetic advantages of burst swimming of fish. *J. Theor. Biol.* 48: 215–229.

Weisbaum, D. & Lavalli, K.L. 2004. Morphology and distribution of antennular setae of scyllarid lobsters (*Scyllarides aequinoctialis*, *S. latus*, and *S. nodifer*) with comments on their possible function. *Invert. Biol.* 123: 324–342.

Williamson, D.I. 1969. Names of larvae in the Decapoda and Euphausiacea. *Crustaceana* 16: 210–213.

Wilson, D.M. 1966. Insect walking. *Ann. Rev. Entomol.* 11: 103–122.

Wine, J.J. & Krasne, F.B. 1972. The organization of escape behaviour in the crayfish. *J. Exp. Biol.* 56: 1–18.

Wine, J.J. & Krasne, F.B. 1982. The cellular organization of crayfish escape behavior. In: Sandeman, D.C. & Atwood, H.L. (eds.), *Neural Integration and Behavior*: pp. 242–289. New York: Academic Press.

Yeung, C. & McGowan, M.F. 1991. Differences in inshore-offshore and vertical distribution of phyllosoma larvae of *Panulirus*, *Scyllarus* and *Scyllarides* in the Florida Keys in May–June, 1989. *Bull. Mar. Sci.* 49: 699–714.

Zimmer-Faust, R.K. & Spanier, E. 1987. Gregariousness and sociality in spiny lobsters: Implications for den habitations. *J. Exp. Mar. Biol. Ecol.* 105: 57–71.

8

The Mineralization and Biomechanics of the Exoskeleton

Francis R. Horne and Samuel F. Tarsitano

CONTENTS

Abstract

Site-directed mineralization is necessary to build complicated shapes seen in the skeletons of arthropods and other animals. Crustaceans may offer an insight into the process of mineralization, because they use a single tissue to mineralize the skeleton, and must do so periodically in order to grow. Along with chitin, proteins are also deposited with the fibrous portion of the exoskeleton. We found, using polyacrylamide gel electrophoresis (PAGE) and immunohistological techniques that carbonic anhydrase is deposited in the exoskeletons of slipper lobsters, particularly in the endocuticle. Carbonic anhydrase is capable of raising the bicarbonate concentrations in specific layers and areas of the exoskeleton, thus raising carbonate concentrations in the newly forming exoskeleton during molting. The biopolymers may act as a nucleating surface for calcium carbonate crystallization and, with a supply of calcium from seawater, gastroliths, and the carbonate via the action of carbonic anhydrase, crystallization can proceed until the enzyme becomes entombed in the crystalline phase. For this reason, skeletal layering becomes an inevitable consequence of such enzyme-mediated supply of the anion. Membranes or nonmineralized portions of the exoskeleton do not stain for carbonic anhydrase. Because slipper lobsters have the thickest shells and a complicated internal pit system, they have the strongest shells among lobsters. Punch test data showed that slipper lobsters resisted shell fracture better than clawed and spiny lobsters.

8.1   Introduction

A major difference between arthropods and other invertebrates is that arthropods must shed their exoskeleton (in the process of ecdysis) to grow. Instead of gradually building on the existing shell by accretionary growth, a process that occurs in mollusks, at each molt crustaceans manufacture and harden a new shell in a matter of days (Greenaway 1985). Therefore, in what is probably not an overstatement, Lowenstam & Weiner (1989) noted that "crustaceans must be the champions of mineral mobilization and deposition."

The arthropod cuticle is a composite material that is made up of calcium carbonate and a chitinous-protein matrix that often consists of more than 50% organic matrix (Simkiss & Wilbur 1989). Thus, decapod crustaceans initially harden their exoskeletons by sclerotinization of microfibers of chitin and proteins, and then further harden the cuticle with deposits of calcite (Vinogradov 1953; Travis 1963; Lowenstam & Weiner 1989). More pliant cuticles (arthrodal or intersegmental membranes) tend to have more chitin, less protein, less mineralization, and less sclerotinization than do more stiff cuticles like carapaces (Anderson 1998).

Although lobster cuticles have been extensively studied for many years, little is known about slipper lobster cuticle physiology and biochemistry. Since arthropod cuticles all seem to be constructed in a similar manner (Travis 1963; Travis & Friberg 1963; Neville 1975), slipper lobsters most likely form and mineralize their shell in the same manner as other marine crustaceans. Calcification of the crustacean cuticle varies greatly from thick, hard shells in some crabs and lobsters to thin, extremely pliant shells in shrimps (Richards 1951; Hackman 1971). Strong cuticles function in defense (Barshaw et al. 2003) and as attachment sites for powerful muscles (Luquet & Marin 2004). Bivalves also mineralize a shell and do so using the same enzyme system. However, heavily calcified molluskan shells differ from the lobster cuticle in that the organic matrix consists primarily of protein rather than chitin and usually accounts for only a small percent of the shell mass (Currey 1976; Belcher & Gooch 2000). In both types of exoskeletons, however, calcium carbonate deposits in association with highly organized, matrix fibrils (Travis 1955; Hegdahl et al. 1977a, 1977b, 1977c; Simkiss & Wilbur 1989; Nousiaine et al. 1998; Knoll 2003) and carbonic anhydrase is found in the shells of both of these taxa.

Among crustaceans, the degree of cuticle sclerotinization, thickness, and mineralization shows considerable diversity both intraspecifically and interspecifically (Greenaway 1985; Hunter et al. 1978; Kumari et al. 1995). Elasticity, flexibility, and rigidity of the cuticle are determined, in part, by the types of proteins present and the extent of sclerotinization and mineralization. For example, the ventral arthrodal membranes are flexible and poorly calcified to allow movement, whereas the majority of the cuticle is rigid and calcified (Kumari et al. 1995; Luquet & Marin 2004). In the case of the lobster, the arthrodal membrane is more similar to insect cuticles than to other portions of its own stiff cuticle (Anderson 1998).

Three layers of the cuticle — the epicuticle, exocuticle, and endocuticle — are at least partially mineralized with calcite, while the fourth, the inner membrane, is not calcified. Epicuticles are thin and not mineralized to the extent that the endocuticle and exocuticle are (Hegdahl et al. 1977a, 1977b, 1977c). Here, mineralization occurs within and around the microfibrils of protein-chitin and also in the pore canals. Calcite forming along the chitin-protein framework suggests that mineralization is controlled by these microfibrils, but not by the pore canals (Travis 1963; Yano 1980; Roer & Dillman 1984). In this chapter, we will discuss the role of the protein fraction of the exoskeleton in the mineralization process. We focus on the mechanism of supply of the anion since calcium is abundant in seawater and, thus, the controlling factor appears to be the carbonate ion.

8.2 Mineral Composition of Exoskeletons

Organisms have produced skeletons of calcium carbonate and calcium phosphate for over 500 million years (Lowenstam & Margulis 1980; Lowenstam 1981). Due to the natural abundance of calcium in marine waters and its high insolubility as a carbonate salt, calcium carbonate probably has been favored over strontium and magnesium salts for shell manufacture. Thus, a diversity of organisms use comparable mechanisms to mineralize their skeletons (Crenshaw 1990; Knoll 2003). Although magnesium may play some role in crystal formation (Raz et al. 2003), the lower natural abundance of strontium and the greater solubility of magnesium salts have likely limited their use as major constituents of shells. On a molar basis in the Earth's crust, calcium is slightly more abundant than magnesium ($1.04\times$), but 267 times more abundant than strontium. In contrast, where solubility plays a major role, seawater has about five times as much magnesium as calcium and 67 times as much calcium as strontium (Simkiss & Wilbur 1989). Even though magnesium and strontium availability is enhanced in water, salts of these alkaline earth cations are not frequently used in biomineralization.

The two common crystal forms of calcium carbonate used in shell mineralization differ by calcite: (1) having six vs. nine coordination sites for aragonite; (2) being less water soluble; (3) being more dense; (4) being harder; (5) containing less strontium; and (6) containing more magnesium than aragonite. Because of crystal morphology, more magnesium and less strontium is incorporated into calcite than aragonite. Slipper lobster (*Scyllarides latus* (Latreille, 1802)) cuticle calcite contains about 1.0% magnesium and 0.02% strontium (Table 8.1). Along with crystal morphology, temperature and pressure affect incorporation of strontium and magnesium into the calcite and aragonite. Even so, neither element plays an important role in exoskeletons, as only small quantities of strontium and magnesium are incorporated into decapod cuticle calcite (Table 8.1).

The ready availability of both calcium and carbonate ions in the alkaline oceans and the high water insolubility of their salt, calcium carbonate, make calcite and aragonite the prevalent minerals for shell biomineralization (Lowenstam & Weiner 1989). Although calcium deposition as carbonate may be limited in cold waters, deep waters, many freshwaters, and by low pH, none of these environmental restrictions on mineralization would affect slipper lobsters which live in relatively alkaline, temperate, subtropical,

TABLE 8.1

Major Cations in the Carapace of Intermolt Crustaceans

Species	Ca^{++}	Mg^{++}	Sr^{++}	Na^+	K^+	Source
Scyllarides latus Latreille, 1802[a] Dorsal Cuticle	16.7–16.78	1.4–1.00	0.26–0.19	0.35	0.04	This chapter
Scyllarides latus Ventral Cuticle	1.1	0.61	<0.01	—	—	This chapter
Cancer pagurus Linnaeus, 1758	29.4	2.0	0.35	—	—	Gibbs & Bryan (1972)
Nephrops norvegicus (Linnaeus, 1758)	18	1.8	—	—	—	Amills (1980)
Palaemontes pugio Holthuis, 1949	10.3	—	0.44	—	—	Brannon & Rao (1979)
Callinectes sapidus M.J. Rathbun, 1896	29.1	1.7	0.41	1.3	0.3	Cameron (1985)
Penaeus aztecus Ives, 1891[b]	13.6	0.5	0.16	—	—	This chapter

[a] Matrix = 34–36% dry wt.
[b] Matrix = 64.0% dry wt.

and tropical seas (Holthuis 1991, 2002; see Webber & Booth, Chapter 2 and Lavalli et al., Chapter 7 for more information about habitat selection and depth preferences of slipper lobsters). For those species living at depths greater than 300 m (a number of ~85 species within 8 of the 14 subfamilies), calcium deposition may occur at a slower rate than in species that live between 0 to 300 m, rendering the newly molted individual vulnerable for a longer period of time.

8.3   Molt

To grow, decapods must shed their exoskeleton and then regenerate and harden a new one, all within a few days (Greenaway 1974, 1985; Hegdahl et al. 1977a, 1977b, 1977c; Cameron 1989; Lowenstam & Weiner 1989). The new cuticle consists of an organic matrix that forms first during premolt and then is calcified later. Premolt involves the manufacture of a new nonmineralized cuticle before the old exoskeleton is shed, muscular atrophy, and reabsorption of areas of the old exoskeleton, particularly around joints (Drach 1939, 1944; Travis 1955, 1960; Hepper 1965). It is induced and regulated by the release of the ecdysteroid molting hormones once the neurosecretory cells that secrete several products involved in the molt process cease secretion of molt-inhibiting hormone (Skinner 1985; Waddy et al. 1995; see Mikami & Kubala, Chapter 5 for more information on hormonal control of ecdysis). While many environmental factors (temperature, photoperiod, season, etc.) can delay or increase the frequency of molting, poor nutrition usually retards molt (Cobb 1995) or causes difficulties during molt, such that tissue connections remain between the old and new cuticle, with the result that the animal cannot escape the old shell, and dies due to asphyxiation — this is commonly referred to as "molt death syndrome" (Conklin et al. 1980; Bowser & Rosemark 1981; Eagles et al. 1986; Wickins et al. 1995; Fiore & Tlusty 2005).

8.3.1   Behavior During and Following Ecdysis

As the old cuticle softens via decalcification, individuals are potentially more susceptible to predators and more helpless than in the hard-shell condition (Cobb 1995); however, it appears that slipper lobsters are capable of all normal movement patterns until moments before ecdysis and rapidly recover from molting (Jones 1988). To avoid predation during molting, most species of lobsters seek shelter, reduce activity patterns, and do not eat (Cobb 1995, Waddy et al. 1995). Few observations have been made on scyllarid molting — Jones (1988) provides the most detailed information for a sand-dwelling species that remains exposed during ecdysis (e.g., *Thenus orientalis* (Lund, 1793); see Chapter 16, Section 16.7.6). This species ceases feeding two days premolt when the characteristic ecdysial lines on the branchiostegites appear (Travis 1954; Jones 1988). Shortly before ecdysis, the lobsters begin "exercise movements" wherein they flex and extend various appendages (limb, abdomen, eyestalks), presumably to help separate the old exoskeleton from the new one underneath (originally described by Travis, 1954; see also Jones, 1988, and Jones, Chapter 16). The actual process of shedding the old exoskeleton is fast — in *Thenus orientalis* it occurs from between 60 sec to 3 min (Jones 1988). Hardening of the exoskeleton occurs within 3 h and lobster activity during this period of time increases and includes swimming and walking movements (Jones 1988) — similar hardening of the exoskeleton and actively is also seen in *Scyllarides latus* (Lavalli & Spanier, personal communication). Such movements may provide slipper lobsters with the opportunity to escape potential predators, and represent a departure from the highly weakened, prolonged soft condition of nephropid lobsters (Cobb 1995; Waddy et al. 1995).

8.3.2   Decalcification PreMolt and Recalcification Postmolt

During premolt, the epithelial cell layer immediately beneath the old cuticle/shell detaches and releases organic acids and enzymes that soften, thin, and dissolve the hard cuticle (Drach & Tchernigovtzeff 1967). Underneath the old cuticle, a soft new cuticle begins to form and is usually partially calcified in a matter of days postmolt. We know that in clawed lobsters (*Homarus* species) a new endocuticle is deposited with one to two days. In the premolting process, some of the mineral, mostly calcium, is dissolved and reabsorbed from the old exoskeleton, and is stored in gastroliths before it is lost in the exuvia (Travis 1955; Aiken 1980). Along with the uptake of calcium from the environment and from the calcium recovered from the gastroliths, a new mineralized cuticle is quickly rebuilt (Greenaway 1985). Calcium recaptured from the consumption of the exuvia is probably more important to freshwater crustaceans than to marine ones because of the limited availability of environmental calcium and, accordingly, the size of the gastroliths is far greater in freshwater forms as a result (Travis 1955; Aiken 1980; Greenaway 1985). However, it may also be important in colder and deeper water forms, where calcium carbonate deposition may occur at a slower rate.

Calcium is regulated via ion exchanges in the gills, antennal glands, and the midgut (Huner et al. 1978; Greenaway 1985; Ahearn & Zhuang 1996). Hemolymph calcium levels are held constant (Greenaway 1985; Zanotto & Wheatly 1993; Wheatly 1999); while total calcium climbs dramatically in premolt, it is primarily the bound form that increases (Wheatly & Hart 1995). This condition is most likely due to calcium uptake, turnover from the old exoskeleton, and reduced calcium excretion. Uptake of calcium from the environment via the gills, epipodites, and branchiostegites (Robertson 1960; Flik et al. 1994; Flik & Haond 2000) or from calcium-storage reservoirs, such as gastroliths and areas of the digestive system including the hepatopancreas (Cameron & Wood 1985; Greenaway 1985; Roer 1980), or from reabsorption of urinary calcium by the antennal gland (Wheatly 1999) is transferred ultimately from the hemolymph via the outer layer of hypodermis to the cuticle (Luquet & Marin 2004). Marine crabs generally store and utilize far less of their old cuticle calcium than do freshwater species like the crayfish, *Procambarus clarkii* (Girard, 1852) (<10% vs. 75%) (Chaisemartin 1964; Lowenstam & Weiner 1989), and it is likely that marine lobsters also store less of the old cuticular calcium. So prior to ecdysis, storage of calcium from the old exoskeleton as gastroliths or as crystals in the midgut is not as important in marine decapods as it is in freshwater species where calcium may be limiting (10 mM in marine environments as opposed to <3 mM in freshwater environments) (Boycott 1936; Greenaway 1985). While rebuilding and quickly mineralizing a new cuticle, marine species such as slipper lobsters, would likely sequester most of their calcium from seawater. Some species of lobster recover additional calcium via consumption of the shed exoskeleton (e.g., *Homarus americanus* H. Milne Edwards, 1837); however, some species of slipper lobsters do not appear to consume their exoskeletons (Jones 1988; Spanier & Lavalli 1998). The cuticle calcium content in the blue crab, *Callinectes sapidus* M.J. Rathbun, 1896, increases from 2–3% to 26% within 5 to 6 h following ecdysis (Coblentz et al. 1998), and given that slipper lobsters — at least those that have been observed — harden within 3 h postecdysis, their calcium content may increase at an even quicker rate. Bicarbonate would be supplied both via the environment and via metabolic carbon dioxide converted to bicarbonate by carbonic anhydrase.

Before cuticle calcification, water is rapidly absorbed osmotically, tissues swell, and the old cuticle splits (Mykles 1980). Fractures occur dorsally (along ecdysial lines) that allow the lobster to back out of the old cuticle. Once out of the cuticle, more water is osmotically moved into the tissues to inflate them by as much as 50% of premolt volume (Aiken 1973). Stretching of the new cuticle occurs in a matter of hours. Water in the tissues is slowly replaced by new tissue growth which occurs inside the new enlarged cuticle.

The epicuticle and exocuticle are formed preecdysis, whereas the endocuticle is formed postecdysis (Cameron 1985; Simkiss & Wilbur 1989). These layers are formed by the hypodermis. Calcification occurs postecdysis. In the endocuticle, but not in the exocuticle, a protein called crustocalcin seems to concentrate calcium with its acidic-rich region and also induces crystallization (Endo et al. 2004). While specifics are not known for lobsters per se, hemocytes also play a role in the hardening of the exoskeleton of other decapods, and are extremely abundant in the hemolymph immediately postmolt (Vacca & Fingerman 1983; Tsing et al. 1989; Hose et al. 1992). They move to the epidermis and secrete cytoplasmic constituents that help harden the exoskeleton (Vacca & Fingerman 1983) and they also release tanning substances into the hemolymph that bind to large proteins which are transported to the cuticle to attach to glucosides and cross-link the protein matrix. After these roles are achieved, hemocyte numbers fall dramatically in both the hemolymph and the epidermal tissue (Martin & Hose 1995).

8.4 Sources of Calcium and Bicarbonate

Because of the abundance of calcium in seawater, marine crustaceans, including slipper lobsters, usually do not need to store as much calcium as do freshwater species to satisfy the demand of new cuticle mineralization (Greenaway 1985; Wheatly 1999). Calcium-active-transport mechanisms and enzymes display higher affinities for calcium in freshwater crayfish than in marine lobsters (Greenaway 1974, 1983). Food, other than the exuvia, is usually not an important source of calcium during biomineralization, because the cuticle is soft, including the lining of the esophagus and cardiac stomach (Lowenstam & Weiner 1989); however, mandibles and maxillipeds are usually hardened first, and, at least in clawed lobsters, the teeth of the gastric mill remain hard (Waddy et al. 1995). Calcium may be food derived in the time leading up to premolt via the uptake of calcium from the intestine under the influence of ecdysone. The consumption of the calcium-rich exuvia is variable in marine species, unlike terrestrial and freshwater species (Greenaway 1985). Thus far, neither *Thenus orientalis* nor *Scyllarides latus* have been observed consuming their exuvia (Jones 1988; Spanier & Lavalli 1998).

The only available and significant source of carbonate is from bicarbonate. Although slow, the conversion of bicarbonate to carbonate in the extracellular cuticle is much less of a problem in the ocean than in freshwater because of the higher pH in oceans (pH $\sim$ 8.3). Using $NaH^{14}CO_3$ and $^{45}CaCl$, Roer & Dillaman (1984) concluded that *Carcinus* tissue and cuticle pools of bicarbonate are much larger and come to equilibrium much faster than those of calcium. Unlike calcium, which is supplied via the hypodermis, likely sources of bicarbonate and ultimately carbonate were assumed to be both metabolic and environmental. Uniform distribution of ^{14}C-bicarbonate throughout the cuticle indicated that bicarbonate comes from both internal/hypodermal and external/environmental sources. The enzyme carbonic anhydrase undoubtedly plays a major role in converting carbon dioxide to bicarbonate and thus indirectly provides carbonate to the site of mineralization (Henry & Kormanik 1985). Carbonic anhydrase has been detected in crustacean hypodermal and cuticle tissue (Chockalingham 1971; Hegdahl 1977a, 1977b, 1981; Lowenstam & Weiner 1989; Horne et al. 2002). Supporting this view, histochemically more enzyme was detected in postmolt crabs and along interprismatic septae (Giraud 1981; Giraud-Guille 1984).

8.5   Mineralization of Cuticles

The generally accepted biomineralization theory of calcareous cuticles and shells is that the organic matrix secreted extracellularly by epithelial cells directs mineralization via the protein fraction (Simkiss & Wilbur 1989; Lowenstam & Weiner 1989). The chitin and protein matrix act to attract and bind calcium and thus serve as nucleation sites. The enzymes present in the matrix act to raise the bicarbonate levels via the enzymatic formation of carbonate. Experiments *in vitro* by Falini et al. (1996) and Belcher et al. (1996) using supersaturated solutions of calcium carbonate support this view. Recently, Endo et al. (2004) isolated a protein, crustocalcin, which was secreted in the first 8 h postmolt that induced calcium carbonate crystallization in the endocuticle. Inactivation of inhibitory proteins as an early step in initiating calcification (Wilbur & Jodrey 1955; Pierce et al. 2001; Roer et al. 2001) appeared unnecessary in calcification of the endocuticle. Crustocalcin was not reported from the exocuticle, so mineralization here may be initiated by other mechanisms (Endo et al. 2004).

8.5.1   Proteins of the Matrix

Proteins of the matrix are judged to serve as sites for (1) ion concentration, (2) crystal nucleation and growth, and (3) control of crystal morphology. Experiments *in vitro* using shell insoluble matrix extracts and high ionic (calcium bicarbonate) concentrations show that deposition of calcareous crystals, such as aragonite or calcite, is species specific and probably biopolymer directed (Addadi & Weiner 1985; Addadi et al. 1987; Lowenstam & Weiner 1989; Belcher et al. 1996; Falini et al. 1996; Endo et al. 2004). Biopolymers may lower the activation energies of nucleation by lowering the interfacial energy and by increasing supersaturation (Belcher & Gooch 2000). However, the ion concentrations necessary to induce mineralization were no different from those of the controls (Mount 1999). The enzyme carbonic anhydrase also has been shown to play a role in cuticle mineralization via bicarbonate formation (Simkiss & Wilbur 1989; Weiner & Dove 2003). Giraud (1981) has shown that by blocking carbonic anhydrase activity with acetazolimide, cuticle growth of calcite is reduced by a factor of two. Other studies also support the likelihood that carbonic anhydrase plays an important role in mineralization since its inactivation hampers mineralization (Costlow 1959). Likely carbonic anhydrases have been identified in molluskan and crustacean shells (Miyamoto et al. 1996; Horne et al. 2002). It seems apparent that biomineralization is genetically directed through the type and arrangement of the secreted macromolecules (Weiner & Dove 2003). Local manipulation of pH and concentrations of carbon dioxide, calcium, and bicarbonate ions via metabolic processes would also be important in the mineralization process. The many sizes, shapes, elemental compositions, and crystal characteristics of shell minerals emphasize the differences between inorganic and biological mineralization, and further emphasizes biological control (Weiner & Dove 2003).

8.5.2   Role and Source of Calcium

Because calcium performs a wide variety of critically important biochemical and physiological intra-cellular functions, including neural, muscular, and signaling functions as transmitters, cells contain low concentrations of calcium ions (0.1 μM or 0.004 ppm). With little calcium available in the tissues, ions are pumped from the environment via gills into the hemolymph and then out again at the site of mineral deposition (Roer 1980). While intracellular calcium concentrations are kept low, calcium carbonate is supersaturated at matrix nucleation sites. Calcium for mineralization is ultimately supplied by the aquatic environment via food (and stored in the digestive tissues) and by absorption via the gills at considerable metabolic expense.

The tissue enzyme, carbonic anhydrase, provides a constant source of carbonate ion via bicarbonate for calcite or aragonite formation. The turnover number and the catalytic efficiency for vertebrate carbonic anhydrase of approximately $k_{cat}(sec^{-1}) = 10^7$ and 10^8, respectively, are high (Williams & Frolik 1991). Simply put, carbonic anhydrase is one of the faster reacting enzymes known (Garrett & Grisham 1999) and would supply bicarbonate much faster than the noncatalyzed reaction. Availability of carbonate ions for calcification is greatly dependent on pH (see equations provided) as the equilibrium dissociation constant (K) greatly favors bicarbonate over both carbon dioxide and carbonate. Carbonate becomes more available as conditions become more alkaline. Carbonic anhydrase assures that equilibrium is attained instantly and that bicarbonate concentrations are maximum under existing conditions. More bicarbonate assures formation of more carbonate by the noncatalyzed reaction.

$$At\ pH = 7.4: \quad 1\ \mu M \longleftrightarrow 6500\ \mu M$$

$$CO_2 + H_2O \longleftarrow CA \longrightarrow H^+ + HCO_3^-$$

$$At\ pH = 7.0: \quad 1800\ \mu M \longleftrightarrow 1\ \mu M$$

$$HCO_3^- \longleftrightarrow H^+ + CO_3^-$$

$$At\ pH = 8.0: \quad 180\ \mu M \longleftrightarrow 1\ \mu M$$

$$HCO_3^- \longleftrightarrow H^+ + CO_3^-$$

Epithelial tissue (hypodermal or mantle) is usually involved in pumping calcium ions into the matrix region where negatively charged proteins, high in acidic amino acids, concentrate the positively charged calcium ions (Simkiss & Wilbur 1989; Belcher et al. 1996; Falini et al. 1996; Endo et al. 2004). With bicarbonate ions also being secreted via epithelial tissues, carbonate is indirectly made available so calcium carbonate can form. Keeping the calcium and carbonate ion product elevated would be essential for crystal growth. Calcite or aragonite forms with the traces of carbonate provided via bicarbonate dehydration. Bicarbonate levels are maintained by the highly favored hydration over dehydration of carbon dioxide (6500 to 1, pH 7.4). The enzyme, carbonic anhydrase, quickly reestablishes the equilibrium as bicarbonate ion is converted to carbonate. Tissue carbonic anhydrase facilitates the formation of bicarbonate because the hydration of carbon dioxide to bicarbonate is slow via the noncatalyzed reaction. Miyamoto et al. (1996) reported a form of carbonic anhydrase as an extracellular matrix component and proposed its role in calcification of oyster pearls. A similar enzyme is also proposed in decapod crustaceans (Horne et al. 2002).

In alkaline environments, like the oceans (pH $= 8.3$), carbonate ions probably are not limiting, but as pH approaches 7.0, availability of carbonate ions would be greatly reduced (Williams 1984). Chemical equilibrium between bicarbonate and carbonate shifts away from carbonate as solutions become more acidic. Although pH is a serious problem in some freshwaters (Boycott 1936), the pH would not be a concern for slipper lobsters or other marine organisms. The problem of forming inorganic salts or minerals of calcium carbonate in cuticles is that the salts must form specifically on the extracellular matrix and nowhere else. Both calcium ions and the carbonate ions, therefore, must be in high enough concentration at the same time at the extracellular matrix site for precipitation to occur. Tissue fluid (pH 7.0) concentrations of carbonate ions are extremely low; in contrast, tissue calcium ion levels are much higher — in the several millimolar ranges. If the carbonate concentrations are only slightly elevated, a nonsite-specific, general precipitation of these highly insoluble calcium salts will likely occur. Because of the general precipitation dilemma, high concentrations of anions such as carbonate cannot be released into the tissue fluids along with calcium. Exceeding the solubility product of calcium carbonate could cause salt precipitation that would be problematic in tissue fluids. More likely, bicarbonate probably forms carbonate ion at the nucleation site. Along these lines, a role for granulocytic hemocytes has been implicated in oyster shell crystallization (Mount et al. 2004) and this type of hemocyte predominates just prior to ecdysis in many decapods (Martin & Hose 1995).

How, then, do the calcium salts precipitate in association with the matrix protein of the slipper lobster cuticle and nowhere else? Our hypothesis emphasizes the availability of carbonate ions that might be the limiting factor in calcite formation (Williams 1980). The slipper lobster hypodermal cells secrete carbonic anhydrase along with other ordinarily secreted matrix biopolymers. Calcium ions and bicarbonate ions would be secreted postecdysis. Being in the matrix, along with other proteins, the enzyme might assure that maximum quantities of bicarbonate ions are available for carbonate formation only at the matrix site and nowhere else. Extracellular carbonic anhydrase could also assist in removal of protons formed along with carbonate from bicarbonate. The protons would be removed as water molecules.

Calcium ion concentrations in the matrix are probably elevated due to their being pumped into the new uncalcified matrix via the hypodermis (Shafer et al. 1995). Bicarbonate ions formed by hypodermal tissue carbonic anhydrase could follow actively transported calcium passively to the negative sites on the matrix proteins (Williams 1984; Simkiss & Wilbur 1989; Falini et al. 1996; Belcher et al. 1996; Endo et al. 2004). Elevation of both calcium and carbonate at the matrix sites probably allows for the solubility product to be attained. Mineralization would then occur in and around the matrix biopolymers. Rapid mineralization following crustacean molt (Coblentz et al. 1998), as is also seen in bird eggs (3 g of calcium per day), illustrates both the importance of active pumping of calcium to the mineralization site and having a ready supply of carbonate ions.

8.5.3   Cuticular Data for *Scyllarides* Species

The slipper lobster (*Scyllarides latus*) cuticle tissue sections stain positive with antibodies to bovine RBC carbonic anhydrase II (Chemicon®) in and around the laminar lattice in both the exocuticle and the endocuticle, but not in the epicuticle (Figure 8.1A and Figure 8.1B). A lattice pattern is apparent

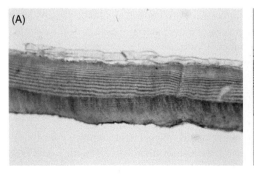
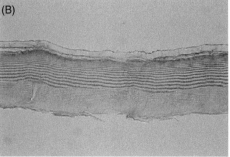

FIGURE 8.1   Slipper lobster (*Scyllarides latus*) intermolt cuticle. (A) immunohistologically stained for carbonic anhydrase. (B) Control tissue. Magnification = 40×.

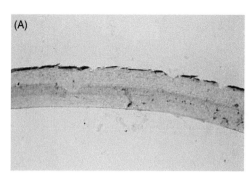
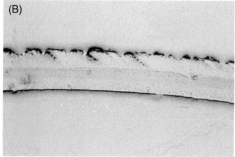

FIGURE 8.2 Brown shrimp (*Penaeus aztecus*) intermolt cuticle. (A) Immunohistologically stained for carbonic anhydrase. (B) Control tissue. Magnification = 40×.

TABLE 8.2

The Ratio of Decalcified Matrix to Calcium in Decapod Exoskeletons

	Carapace	Ventral Ridges	Ventral (No Ridges) Intersegmental Membrane
Slipper Lobster, *Scyllarides latus* (Latreille, 1802)			
Specimen 1	1.023/0.339 = 3.01	242/81 = 2.98	55/1.9 = 28.8
Specimen 2	0.381/0.178 = 2.14	154/44 = 3.51	35/0.587 = 59.6
Specimen 3	104/83.2 = 1.25	109/35 = 3.11	82/0.951 = 86.2
Means	2.13	3.20	58.2
Brown shrimp, *Penaeus aztecus* Ives, 1891			
Specimen 1	132/26 = 5.0		
Specimen 2	121/28 = 4.3		
Specimen 3	104/22 = 4.7		
Mean	4.7		

with staining in both the endocuticle and exocuticle. The exocuticle was more spacious and with larger microfibers than the endocuticle. Positive staining proteins appeared to be associated with the matrix fibrils in both layers, but not in the epicuticle (Figure 8.1A and Figure 8.1B). The poorly calcified flexible sternum of the ventral cuticle (1.07% Ca and 0.61% Mg) showed much less staining and no distinctly stained striations. Controls for both cuticles stained lightly. For comparison, the cuticle of the brown shrimp (*Penaeus aztecus*, Ives, 1891), which is poorly calcified in comparison to that of slipper lobster, was sectioned and found to have much weaker staining than that of the slipper lobster (Figure 8.2A and Figure 8.2B). The endocuticle is more highly mineralized than the exocuticle and is laid down after the animal molts. The exocuticle or most of this portion of the exoskeleton is laid down before the animal molts. Thus, late-molt shells show the highest concentrations of carbonic anhydrase (Henry & Kormanik 1985; Coblenz et al. 1998) and this may indicate that carbonic anhydrase is deposited in the endocuticle as it is being laid down. Carbonic anhydrase pervades the entire endocuticle and is not found in spotty areas — that would be the case if the presence of carbonic anhydrase was due to the presence of individual cells.

The decalcified matrix/Ca^{++} ratios in: (1) the slipper lobster dorsal carapace, (2) the slipper lobster ventral cuticle, and (3) ventral ridges were 2.1, 58.2, 3.2 respectively. Similar matrix /Ca^{++} values for the American lobster, *Homarus americanus*, were 1.1, 97, 1.49 (Horne, unpublished data 2005), while the brown shrimp dorsal carapace had a ratio of 4.7 (Table 8.2). Immunohistochemical staining for carbonic anhydrase (CA) appears to be associated with the degree of mineralization of the cuticle. That is, immunohistochemically stained cuticle sections from more flexible and less calcified portions of the exoskeleton give only a weak response or none to CA. These data illustrate that the degree of calcification in the cuticle is positively related to immunohistochemical staining for CA. Perhaps more CA is secreted

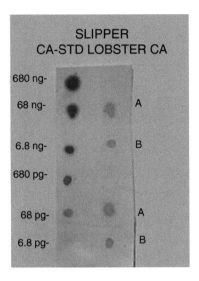

FIGURE 8.3 Dot blot comparing known concentrations of carbonic anhydrase (CA) with dilutions of extracts of slipper lobster (*Scyllarides latus*) cuticles. (A) lobster extract, (B) 0.1 lobster extract.

along with other biopolymers in the more calcified cuticles and it may, therefore, play some role in degree of mineralization.

Strontium and magnesium are of little importance in cuticle calcite formation. Strontium with high matrix/Sr ratios seems even less important than magnesium in the ventral cuticle. Magnesium is more readily incorporated into calcite than is strontium. Lower matrix/Sr ratios occur in the more calcified dorsal carapace (Table 8.1).

8.6   Protein Isolation

The affinity chromatography of isolated slipper lobster (*Scyllarides latus*) cuticular proteins (15 ng) run on sodium dodecyl sulphate/polyacrylamide gel electrophoresis (SDS/PAGE) gels showed one distinctly Coomasie Blue-stained protein with an estimated molecular weight of 56 kDa. A much more weakly staining protein of 58 kDa was also detected (Figure 8.3). Using a dot blot, serial dilutions of known concentrations of bovine RBC CA (Sigma CA II) were immunochemically compared with two dilutions (1.0 and 0.5 ng) of the total slipper lobster cuticular proteins to estimate quantities of enzyme present. Comparative estimates of the concentration of slipper lobster cuticle CA were estimated to be about 68 pg-CA/μg-total protein in our samples. Because this enzyme concentration is perhaps below detection limits of Coomasie Blue stain on SDS/PAGE gels, CA might not be detected on the gels. Thus the 56 kDa protein is probably not the enzyme. The 58 kDa may be a dimer of CA.

8.7   Biomechanics and Ecological Considerations

In looking at three major families of lobsters (Nephropidae, Palinuridae, and Scyllaridae) we see that defensive weapons when present are correlated with the type of shell construction used by the lobster. For example, clawed lobsters have the thinnest shells, but the most apparent and display-oriented defensive weapons against fishes and other predators (Barshaw et al. 2003). Essentially the lobsters are putting all of their energy into the size of the claws as a means to ward off predators by making threat gestures and attacks, as well as making themselves appear larger to predators. This may be due to the selectional pressure imposed by the particular predators that clawed lobsters face, as well as competing selective pressures from intraspecific competition for mates and shelters. If predators are mainly engulfers or swallowers,

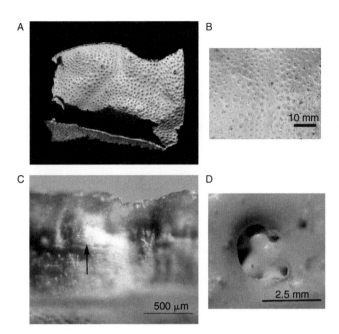

FIGURE 8.4 (A) Tuberculate surface of a slipper lobster carapace; (B) inner surface of carapace showing the pits formed by tubercles; (C) histological section of the exoskeleton, showing numerous laminate layers; (D) single tuberculate pit, showing inner strut structure that confers added strength to carapace. (Modified from Tarsitano, S.F., Lavalli, K.L., Horne, F., & Spanier, E. 2006. *Hydrobiologia* 557: 9–20; used with permission.)

appearing larger would be a key benefit in the clawed lobster's defensive strategy. Spiny lobsters also have defensive weapons: long, spinous antennae that they whip at their attackers. These weapons, though effective, do not have the mass and visually threatening appearance of claws. Moreover, the predators of spiny lobsters are more likely to be biters and rammers, so there seems to be a trade off between reliance on defensive weapons and improved armor (shell thickness), and spiny lobsters do have significantly thicker shells than clawed lobsters (Barshaw et al. 2003; Tarsitano et al. 2006). In addition, the armor design barely noticeable in clawed lobsters is enhanced by forming pits on the inner surface of the spiny lobster's exoskeleton (see Figure 8.4 for an example of such pits in slipper lobster shells). The pits have thickened rims that are associated with crack-blunting ability (Tarsitano et al. 2006). Essentially, the open pits increase the crack-tip diameter of a running crack to effectively blunting the crack's progress (Cook & Gordon 1964). As one would expect, the pitting on the inside of the shell corresponds to bumps or spines on the outer exoskeleton surface.

Both clawed and spiny lobsters actively defend themselves using the claws or antennae. Slipper lobsters employ another strategy toward their defense that includes blending into the environment and gripping the substrate tightly while trying to outlast the attack (i.e., the "fortress" strategy; see Barshaw et al., 2003 and Lavalli et al., Chapter 7). Slipper lobsters have no visibly defensive weapons although the dactyls of their pereiopods are sharp. Their flattened, bumpy, and thick shells, in addition to their cryptic coloration and ability to remain motionless, allow them to blend into the bottom landscape and avoid predation. The shells of slipper lobsters are significantly thicker than shells of nephropids or palinurids (Barshaw et al. 2003). In addition, their pit system on the inner surface of the shell is correspondingly deeper with supporting struts not found in other lobsters (Figure 8.4). Punch-test data showed that this shell construction could withstand greater stress than the exoskeletons of other lobster families (Barshaw et al. 2003). Thus, the strategy of slipper lobsters is to abandon defensive weapons, thicken and sculpture the shell to survive attacks until they can find shelter, rely upon blending in with the seafloor architecture in order to avoid predators, or use their escape-swimming response to gain access to shelter (see Lavalli et al., Chapter 7 for more details). This strategy, combined with the rapid hardening of their exoskeleton post-molt (more rapid that either nephropid or palinurid lobsters), suggests that predation may have played a significant role in both the mineralization pathway and the biomechanical design of the shells of these lobsters.

8.8   Conclusions

Data demonstrate that the enzyme CA is deposited in the extracellular portion of the cuticle along with other proteins. These results support our hypothesis that CA is secreted extracellularly to provide and maintain a ready supply of bicarbonate ions (Horne et al. 2002). Another likely and important role for CA would be removal of the protons formed during the dehydration reaction (Benesch 1984). To further demonstrate the role of CA, relative amounts of this enzyme need to be examined among the Crustacea to determine whether the amounts of CA observed in shell varies in poorly mineralized to heavily mineralized species. Gene expression for CA during the premolt is unknown, but would offer another avenue of evidence that CA was indeed crucial for the mineralization process. It was interesting to note that the exocuticle and membrane separating the hypodermis from the endocuticle had no appreciable amounts of CA as determined from antibody staining for this enzyme. Thus, CA may be necessary for mineralization as outlined by Horne et al. (2002). Moreover, by laying down the controlling enzyme that maintains the supply of the anions for mineralization, a mechanism for the shape of the shell and zones of flexibility can be devised. By depositing CA in the chitinous matrix, the shape of the exoskeleton can be guaranteed according to the shape of the matrix fibers. Modifications of the fibrous portion of the matrix associated with the enzyme can lead to shape changes in the exoskeleton over generations. The physiological role of CA, although hypothetical, must be an important one and eventually will be resolved. Hopefully, given the difference among scyllarid, nephropid, and palinurid lobsters, the study of physiological differences among these groups should lead to a better understanding of the mineralization process in decapods in general. The slipper lobster can thus serve as a new model for understanding decapod growth and molting.

References

Addadi, L. & Weiner, S. 1985. Interactions between acidic proteins and crystals: Stereochemical requirements for biomineralization. *Proc. Natl. Acad. Sci. USA* 82: 4110–4114.

Addadi, L., Moradian, J., Shay, E., Maroudas, N.G., & Weiner, S. 1987. A chemical model for the cooperation of sulfates and carboxylates in calcite crystal nucleation: Relevance to biomineralization. *Proc. Natl. Acad. Sci. USA* 84: 2732–2736.

Ahearn, G.A. & Zhuang, Z. 1996. Cellular mechanisms of calcium transport in crustaceans. *Phys. Zool.* 69: 383–402.

Aiken, D.E. 1973. Proecdysis, setal development and molt prediction in the American lobster (*Homarus americanus*). *J. Fish. Res. Board Can.* 30: 1337–1344.

Aiken, D.E. 1980. Molting and Growth. In: Cobb, J.S. & Philips, B.F. (eds), *The Biology and Management of Lobsters*, Vol. 1, pp. 91–163. New York: Academic Press.

Amills, F.S. 1980. Contribucoin al Concimiento d Bela Biologia j de *Nephrops norvegicus* (l. Estudio del Ciclo j. de Intermuda). D.Sc. thesis: University of Barcelona, Barcelona, Spain.

Anderson, S.O. 1998. Characterization of proteins from arthrodial membranes of the lobster, *Homarus americanus*. *Comp. Biochem. Physiol.* A121: 375–383.

Barshaw, D.E., Lavalli, K.L., & Spanier, E. 2003. Is offence the best defence: The response of three morphological types of lobsters to predation. *Mar. Ecol. Prog. Ser.* 256: 171–182.

Belcher, A.M. & Gooch E.E. 2000. Protein components and inorganic structure in shell nacre. In: Baeuerlein, E. (ed.), *Biomineralization: From Biology to Biotechnology and Medical Applications*: pp. 221–250. New York: Wiley-VCH.

Belcher, A.M., Wu, X.H., Christensen, R.J., Hansma, P.K., Stucky, G.D., & Morse, D.E. 1996. Control of crystal phase switching and orientation by soluble mollusc-shell proteins. *Nature, London* 381: 56–58.

Benesch, R. 1984. Carbonic anhydrase and calcification. In: Tashian, R.E. & Hewell-Emmett, D. (eds.), *Biology and Chemistry of the Carbonic Anhydrases,* Vol. 429: pp. 457–458. New York: Annals of the New York Academy Science.

Bowser, P.R. & Rosemark, R. 1981. Mortalities of cultured lobsters, *Homarus,* associated with a molt death syndrome. *Aquaculture* 23: 11–18.

Boycott, A.E. 1936. The habitats of freshwater Mollusca in Britain. *J. Anim. Ecol.* 5: 116–186.

Brannon, A.C. & Rao, R. 1979. Barium, strontium and calcium levels in the exoskeleton, hepatopancreas and abdominal muscle of the grass shrimp, *Palaemonetes pugio*: Relation to molting and exposure to barite. *Comp. Physiol.* 63A: 261–274.

Cameron, J.N. 1985. Postmoult calcification in the blue crab, *Callinectes sapidus*: Relationships between apparent net H^+ excretion, calcium and bicarbonate. *J. Exp. Biol.* 119: 275–285.

Cameron, J.N. 1989. Post-molt calcification in the blue crab *Callinectes sapidus*: Timing and mechanisms. *J. Exp. Biol.* 142: 285–304.

Cameron, J.N. & Wood, C.M. 1985. Apparent H^+ excretion and CO_2 dynamics accompanying carapace mineralization in the blue crab, *Callinectes sapidus* following molting. *J. Exp. Biol.* 114: 181–196.

Chaisemartin, C. 1964. Importance des gastrolithes dans l'econmomie du calcium chez *Astacus pallipes* Lereboullet. Bilan calcique de l'exuviation. *Vie Milieu* 15: 457–474.

Chockalingham, S. 1971. Studies on the enzymes associated with calcification of the cuticle of the hermit crab, *Clibanarius olivaceous. Mar. Biol.* 10: 169–182.

Cobb, J.S. 1995. Interface of ecology, behavior, and fisheries. In: Factor, J. (ed), *Biology of the Lobster Homarus americanus*: pp. 139–151. New York: Academic Press.

Coblentz, F.E., Shafer, T.H., & Roer, R.D. 1998. Cuticular proteins from the blue crab *in vitro* calcium carbonate mineralization. *Comp. Biochem. Physiol.* B121: 349–360.

Conklin, D.E., D'Abramo, L.R., Bordner, C.E., & Baum, N.A. 1980. A successful purified diet for the culture of juvenile lobsters: The effect of lecithin. *Aquaculture* 21: 243–249.

Cook, J. & Gordon, J.E. 1964. A mechanism for the control of crack propagation in all-brittle systems. *Proc. R. Soc. Lond.* A282: 508–520.

Costlow, J.D. 1959. Effect of carbonic anhydrase inhibitors on shell development and growth in *Balanus improvisus* Darwin. *Physiol. Zool.* 32: 177–174.

Crenshaw, M.A. 1990. Biomineralization mechanisms. In: Carter, J.G. (ed.), *Skeletal Biomineralization: Patterns, Processes and Evolutionary Trends*, Vol. 1: pp. 1–10. New York: Van Nostrand Reinhold.

Currey, J.D. 1976. Further studies on the mechanical properties of mollusc shell material. *J. Zool.* 180: 445–453.

Drach, P. 1939. Mue et cycle d'intermue chez les Crustacé Décapodes. *Ann. Inst. Océanogr. (Paris)* [N.S.] 18: 109–391.

Drach, P. 1944. Etude préliminarie sur le cycle d'intermue et son conditionnement hormonal chez *Leander serratus* (Pennant). *Bull. Biol. Fr. Belg.* 78: 40–62.

Drach, P. & Tchernigovtzeff, C. 1967. Sur le méthode de détermination des stades d'intermue et son application générale aux Crustacés. *Vie Milieu (Ser. A. Biol. Mar.)* 18: 595–610.

Eagles, M.D., Aiken, D.E., & Waddy, S.L. 1986. Influence of light and food on larval American lobsters, *Homarus americanus. Can. J. Fish. Aquat. Sci.* 43: 2303–2310.

Endo, H., Takagi, Y., Ozaki, N. & Watanabe, T. 2004. A crustacean Ca^{2+}-binding protein with a glutamate rich sequence promotes $CaCO_3$ crystallization. *Biochem. J.* 384: 150–167.

Falini, G., Albeck, S., Weiner, S., & Addadi, L. 1996. Control of aragonite or calcite polymorphism by mollusk shell macromolecules. *Science* 271: 67–69.

Fiore, D.R. & Tlusty, M.F. 2005. Use of commercial *Artemia* replacement diets in culturing larval American lobsters (*Homarus americanus*). *Aquaculture* 243: 291–303.

Flik, G. & Haond, C. 2000. Na^+ and Ca^+ pumps in the gills, epipodites and branchiostegites of the European lobster *Homarus gammarus*: Effects of dilute sea water. *J. Exp. Biol.* 203: 213–220.

Flik, G., Verbost, P.M., Atsma, W., & Lucu, C. 1994. Calcium transport in gill plasma membranes of the crab *Carcinus maenas*: Evidence for carriers driven by ATP and a Na^+ gradient. *J. Exp. Biol.* 195: 109–122.

Garrett. R.H. & Grisham, C.M. 1999. *Biochemistry.* 440 pp. New York: Saunders College Publishing.

Gibbs, P.E. & Bryan, G.W. 1972. A study of strontium, magnesium, and calcium in the environment and exoskeleton of decapod crustaceans, with special reference to *Uca burgersi* on Barbuda, West Indies. *J. Exp. Mar. Biol. Ecol.* 9: 97–110.

Giraud, M.M. 1981. Carbonic anhydrase activity in the integument of the crab, *Carcinus maenas* during the intermolt cycle. *Comp. Biochem. Physiol.* A69: 381–387.

Giraud-Guille, M.M. 1984. Fine structure of the chitin-protein system in the crab cuticle. *Tissue Cell.* 16: 75–92.

Greenaway, P. 1974. Calcium balance at the postmolt stage of the freshwater crayfish, *Austropotamobius pallipes* (Lereboullet). *J. Exp. Biol.* 61: 35–45.

Greenaway, P. 1983. Uptake of calcium at the postmoult stage of the marine crabs *Callinectes sapidus* and *Carcinus maenas*. *Comp. Biochem. Physiol.* A75: 181–184.

Greenaway, P. 1985. Calcium balance and molting in the Crustacea. *Biol. Rev.* 60: 425–454.

Hackman, R.H. 1971. The integument of Arthropoda. In: Florkin, M. & Scheer, B.T. (eds.), *Chemical Zoology, The Arthropods*, Part B, Vol. VI: pp. 1–60. New York: Academic Press.

Hegdahl, T., Silness, J., & Gustavsen, F. 1977a. The structure and mineralization of the carapace of the crab (*Carcinus pagurus* L.). I. The endocuticle. *Zool. Scr.* 6: 89–99.

Hegdahl, T., Gustavsen, F., & Silness, J. 1977b. The structure and mineralization of the carapace of the crab (*Carcinus pagurus* L.). II. The exocuticle. *Zool. Scr.* 6: 101–105.

Hegdahl, T., Gustavsen, F., & Silness, J. 1977c. The structure and mineralization of the carapace of the crab (*Carcinus pagurus* L.). III. The epicuticle. *Zool. Scr.* 6: 215–220.

Henry, R.P. & Kormanik, G.A. 1985. Carbonic anhydrase activity and calcium deposition during the molt cycle of the blue crab, *Callinectes sapidus*. *J. Crust. Biol.* 5: 234–241.

Hepper, B.T. 1965. Pre-moult changes in the structure of the integument of the lobster, *Homarus vulgaris*. *Rapp. P.-V. Reun., Cons. Int. Explor. Mer* 156: 1–14.

Holthuis, L.B. 1991. *Marine Lobsters of the World. An Annotated and Illustrated Catalogue of the Species of Interest to Fisheries Known to Date*. FAO Species Catalogue No. 125, Vol 13: pp. 1–292. Rome: Food and Agriculture Organization of the United Nations.

Holthuis, L.B. 2002. The Indo-Pacific scyllarinid lobsters (Crustacea, Decapoda, Scyllaridae). *Zoosystema* 24: 499–683.

Horne, F., Tarsitano, S.F., & Lavalli, K.L. 2002. Carbonic anhydrase in mineralization of crayfish cuticle. *Crustaceana* 75: 1067–1081.

Hose, J.E., Martin, G.G., Tiu, S., & McKrell, N. 1992. Patterns of hemocyte production and release throughout the molt cycle in the penaeid shrimp *Sicyonia ingentis*. *Biol. Bull.* 183: 185–189.

Huner J.V., Kowalczuk, J.G., & Avault, J.W. 1978. Postmolt calcification in subadult red swamp crayfish, *Procambarus clarkii* (Girard) (Decapoda, Cambaridae). *Crustaceana* 34: 275–280.

Jones, C.M. 1988. *The Biology and Behaviour of Bay Lobsters, Thenus spp. (Decapoda: Scyllaridae) in Northern Queensland*, Australia. Ph.D. dissertation: Brisbane, Australia: University of Queensland. 190 pp.

Knoll, A.H. 2003. Biomineralization and evolutionary history. *Rev. Mineral. Geochem.* 54: 329–356.

Kumari, S.S., Willis, H.J., & Skinner, D.M. 1995. Proteins of crustacean exoskeleton: IV, Partial amino acid sequence of exoskeleton proteins from the Bermuda land crab, *Gecarsinus lateralis* and comparison to certain insect proteins. *J. Exp. Zool.* 273: 389–400.

Lowenstam, H.A. 1981. Minerals formed by organisms. *Science* 21: 1126–1131.

Lowenstam, H.A. & Margulis, L. 1980. Calcium regulation and the appearance calcareous skeletons in the fossil record. In: Omori, M. & Watabe, N. (eds.), *The Mechanisms in Biomineralization in Animals and Plants*: pp. 289–300. Tokyo: Tokai University Press.

Lowenstam, H.A. & Weiner, S. 1989. *On Biomineralization*. New York: Oxford University Press.

Luquet, G. & Marin, F. 2004. Biomineralization in crustaceans: Storage strategies. *C. R. Palevol.* 3: 515–534.

Martin, G.G. & Hose, J.E. 1995. Circulation, the blood, and disease. In: Factor, J.R. (ed.), *The Biology of the Lobster Homarus americanus*: pp. 465–495. New York: Academic Press.

Miyamoto, H., Miyashita, T., Okushima, M., Nakano, S., Morita, T., & Matsushiro, A. 1996. A carbonic anhydrase from the nacreous layer in oyster pearls. *Proc. Natl. Acad. Sci. USA* 95: 9657–9660.

Mount, A.S. 1999. *Hemocyte-Mediated Shell Mineralization in the Eastern Oyster*. Ph.D. thesis, Clemson University, Clemson, South Carolina.

Mount, A.S., Wheeler, A.P., Paradkar, R.P., & Snider, D. 2004. Hemocyte-mediated shell mineralization in the eastern oyster. *Science* 304: 297–300.

Mykles, D.K. 1980. The mechanism of fluid absorption at ecdysis in the American lobster, *Homarus americanus*. *J. Exp. Biol.* 84: 89–101.

Neville, A.C. 1975. *Biology of the Arthropod Cuticle*. Berlin: Springer Verlag, 448 pp.

Nousiaine, M., Rafin, K., Skou, L., Roepstorff, P., & Anderson., S.O. 1998. Characterization of exoskeletal proteins from the American lobster, *Homarus americanus*. *Comp. Biochem. Physiol.* 119B: 189–199.

Pierce, D.C., Butler, K.D., & Roer, R.D. 2001. Effects of exogenous N-acetylhexosaminidase on the structure and mineralization of post-ecdysial exoskeleton of the blue crab, *Callinectes sapidus*. *Comp. Biochem. Physiol.* B128: 691–700.

Raz, P.C., Hamilton, F.H., Wilt, F.H., Weiner, S., & Addadi, L. 2003. The transient phase of amorphous calcium carbonate in sea urchin larval spicules: The involvement of proteins and magnesium ions in its formation and stabilization. *Adv. Funct. Mater.* 13: 480–486.

Richards, A.G. 1951. *The Integument of Arthropods*: pp. 1–411. Minneapolis: University of Minnesota Press.

Robertson, J.D. 1960. Ionic regulation in the crab *Carcinus maenus* (L.) in relation to the molting cycle. *Comp. Biochem. Physiol.* 1: 183–212.

Roer, R. 1980. Mechanisms of reabsorption and deposition of calcium in the carapace of the crab, *Carcinus maenas*. *J. Exp. Biol.* 88: 205–218.

Roer, R. & Dillaman, R. 1984. The structure and calcification of the crustacean cuticle. *Am. Zool.* 24: 893–909.

Roer, R.D., Holbrook, K.E., & Shafer, T.H. 2001. Glycosidase activity in the post-ecdysial cuticle of the blue crab, *Callinectes sapidus*. *Comp. Biochem. Physiol.* B128: 683–690.

Shafer, T.H., Roer, R.D., & Midgette-Luther, C. 1995. Postecdysial cuticle alternation in the blue crab, *Callinectes sapidus*: synchronous changes in glycoproteins and mineral nucleation. *J. Exp. Zool.* 271: 171–182.

Simkiss, K. & Wilbur, K.M. 1989. *Biomineralization, Cell Biology and Mineral Deposition*. San Diego, CA: Academic Press.

Skinner, D.M. 1985. Moulting and regeneration. In: Bliss, D.E., (ed.), *The Biology of Crustacea: Integument, Pigments, and Hormonal Processes*, Vol. 9: pp. 44–128. New York: Academic Press.

Spanier, E. & Lavalli, K.L. 1998. Natural history of *Scyllarides latus* (Crustacea: Decapoda): A review of the contemporary biological knowledge of the Mediterranean slipper lobster. *J. Nat. Hist.* 32: 1769–1786.

Tarsitano, S.F., Lavalli, K.L., Horne, F., & Spanier, E. 2006. The constructional properties of homarid, palinurid and scyllarid lobsters. *Hydrobiologia* 557: 9–20.

Travis, D.F. 1954. The molting cycle of the spiny lobster *Panulirus argus* (Latreille). I. Molting and growth in laboratory maintained animals. *Biol. Bull.* 107: 433–450.

Travis, D.F. 1955. The molting cycle of the spiny lobster, *Panuluris argus* Latreille. II. Pre-ecdysial histological and histochemical changes in the hepatopancreas and integumental tissue. *Biol. Bull.* 108: 88–112.

Travis, D.F. 1960. Matrix and mineral deposition in skeletal structures of the decapod Crustacea. In: Sognnaes, R.F. (ed.), *Calcification in Biological Systems*: pp. 57–116. Washington, D.C.: Am. Assoc. Adv. Sci. Publ. No. 64.

Travis, D.F. 1963. Structural features of mineralization from tissue to macromolecular levels of organization in the decopod Crustacea. *Ann. N. Y. Acad. Sci.* 109: 177–245.

Travis, D.F. & Friberg, U.A. 1963. The deposition of skeletal structures in Crustacea. VI. Microradiographic studies of exoskeleton of the crayfish, *Orconectes virilis* Hagen. *J. Ultrastr. Res.* 9: 285–301.

Tsing, A., Arcier, J.-M., & Brehelin, M. 1989. Hemocytes of penaeid and palaemonid shrimps: Morphology, cytochemistry, and hemograms. *J. Invert. Pathol.* 53: 64–77.

Vacca, L.L. & Fingerman, M. 1983. The roles of hemocytes in tanning during the molting cycle: A histochemical study of the fiddler crab, *Uca pugilator*. *Biol. Bull.* 165: 758–777.

Vinogradov, A.P. 1953. The elementary chemical composition of marine organisms. In: *Memoir 2: Memoirs of the Sears Foundation for Marine Research*, Vol. 2: pp. 1–647. New Haven, CT: Yale University.

Waddy, S.L., Aiken, D.E., & De Kleijn, D.P.V. 1995. Control of growth and reproduction. In: Factor, J. (ed.), *Biology of the Lobster Homarus americanus*: pp. 217–266. New York: Academic Press.

Weiner, S. & Dove, P.M. 2003. An overview of biomineralization processes and the problem of the vital effect. In: Dove, P.M., DeYoreo, J.J., & Weiner, S. (eds.), *Biomineralization. Reviews in Mineralogy and Geochemistry*, Vol. 54: 1–29. Washington, D.C.: Mineralogical Society of America.

Wheatly, M.G. 1999. Calcium homeostasis in Crustacea: The evolving role of branchial, renal, digestive and hypodermal epithelia. *J. Exp. Zool.* 283: 620–640.

Wheatly, M.G. & Hart, M.K. 1995. Hemolymph ecdyses and electrolytes during the molting cycle of crayfish: A comparison of natural molts with those induced by eyestalk removal or multiple limb autonomy. *Physiol. Zool.* 68: 583–607.

Wickins, J.F., Beard, T.W., & Child, A.R. 1995. Maximizing lobster, *Homarus gammarus* (L.), egg and larval viability. *Aquat. Res.* 26: 379–392.

Wilbur, K.M. & Jodrey, L. 1955. Studies on shell formation. V. The inhibition of shell formation by carbonic anhydrase inhibitors. *Biol. Bull.* 108: 82–112.

Williams. R.J.P. 1980. A general introduction to the special properties of the calcium ion and their deployment in biology. In: Siegal F.L., Carafoli, E., Kretsinger, R.H., MacLennan, D.H., & Wasserman, D.H. (eds.), *Calcium-Binding Proteins: Structure and Function*: pp. 3–10. New York: Elsevier-North-Holland.

Williams, R.J.P. 1984. An introduction to biominerals and the role of organic molecules in their formation. *Phil. Trans. R. Soc. Lond., Ser. B* 304: 411–424.

Williams, R.J.P. & Frolik, C.A. 1991. Physiological and pharmacological regulation of biological calcification. *Int. Rev. Cytol.* 126: 195–292.

Yano, I. 1980. Calcification of crab exoskeleton. In: Masae, O. & Norimitsu, W. (eds.), *The Mechanisms in Biomineralization in Animals and Plants*: pp. 187–196. Tokyo, Japan: Tokai University Press.

Zanotto, F.P. & Wheatly, M.G. 1993. The effect of water pH on postmolt fluxes of calcium and associated electrolytes in the freshwater crayfish (*Procambarus clarkia*). In: Romaire, R. (ed.), *Freshwater Crayfish*: pp. 437–450. Baton Rouge: Louisiana State University.

9

Growth of Slipper Lobsters of the Genus Scyllarides

Marco L. Bianchini and Sergio Ragonese

CONTENTS

Abstract

Growth is a diachronic process that influences many other processes; thus, the knowledge of growth parameters is a prerequisite for a better understanding of lobster biology and for applying sensible fisheries regulations for commercially important stocks. Lobsters can only increase in length through molting, and aging is difficult as the frequency of molting varies throughout an individual's lifetime. There are no known specific differences in molting physiology or in growth modalities of scyllarid lobsters compared with those of other lobsters. The growth phenomenon depends on external variables, including the influence of nutrition, season, temperature, light, and photoperiod, as well as other environmental or stressful parameters that may be found in artificial settings; in cases where conditions are suboptimal, reductive molts can occur. Literature specifically dealing with adult crustacean growth is limited, and information on scyllarids is even scarcer. Some data exist for the slipper lobsters of the genus *Scyllarides*, mainly of the Mediterranean species, which may constitute a case study for the whole family. These data are discussed and limitations of data obtained both in the laboratory and in the wild are reviewed. The inadequacy of the usual models, in particular the von Bertalanffy growth function (vBGF), to actually describe decapod growth is also examined. The parameters of the seasonalized vBGF obtained in the laboratory for *Scyllarides latus* (Latreille, 1802) are $CL_\infty = 127.2$ mm, $k = 0.20$, $C = 1.0$, $t_s = 0.83$, which are in agreement with data from long-term recaptures in the wild, and not far from the values for the Galápagos species *S. astori* Holthuis, 1960. The vBGF suggests that *Scyllarides* species are long lived, with a yearly molt in the adult phase, and incremental increases in carapace length per molt of around 6%. Other information is presented on morphology, morphometry, seasonality, and length-frequency distributions from commercial catches in Italian waters.

9.1   Introduction

Growth is a diachronic process, whose realization is size; it depends on physiological events and influences, directly or indirectly, many other processes, among which maturity (Bianchini et al. 1998a; De Martini & Kleiber 1998), fecundity (Stewart et al. 1997; De Martini & Williams 2001), survival (Carrasco & Barros 1996), and mortality (Ragonese et al. 1994) are most important. The knowledge of the growth parameters is therefore a "prerequisite" for biological modeling (Pollock 1995a), understanding population dynamics (Courtney et al. 2001), providing stock assessment (Punt 2003), and developing resource management (Stewart et al. 1997; Cruywagen 1997; Brandão et al. 2004).

Covered by a hard, rigid exoskeleton, lobsters can only increase in length through molting — the shedding of their shell — which is replaced, generally, by a larger one at the end of ecdysis (see Mikami & Kuballa, Chapter 5 and Horne & Tarsitano, Chapter 8 for a discussion of the process). Larval lobsters must pass through multiple stages to allow for appendage growth and increases in sensory structures, and then must metamorphose to change body shape in preparation for a change in habitat (see Sekiguchi et al., Chapter 4 for a discussion of these changes). Postlarval, adolescent, and adult lobsters require multiple ecdyses to accommodate the modifications that are the consequence of the production of new body mass, that is, of growth. Up to a certain limit, increases of body mass can proceed inside the exoskeleton; however, when body growth is restricted by the shell, it becomes necessary to replace the old shell with a new, larger one. This replacement involves every external surface, from the massive cephalothorax to the smallest sensory hair and from the tip of the antennae to the lining of the gills and of the digestive tract (Aiken 1980). Since no hard structure is left from molt to molt, it is very difficult, if not impossible, to assign an actual absolute age to decapods collected from the wild; the accumulation with age of lipofuscin in postmitotic tissues (e.g., the brain) seems to offer a new route for aging crustaceans (Sheehy et al. 1998).

There are no known specific differences in the molting physiology of scyllarid lobsters and that of clawed or spiny lobsters. All share a rigid exoskeleton, in contrast to some other decapod crustaceans (e.g., the shrimps' exoskeleton is flexible enough to allow a degree of length expansion without molting; see Choe 1971) and all lack a "terminal" ecdysis (Hartnoll 1982, 2001). Rapid ecdysis events are followed by intermolt periods that may last many months. As such, the molt cycle may be divided into phases (Drach 1939), which in the simplest system are: stage A (postmolt 1), with the animal still soft shelled; stage B (postmolt 2), when the exoskeleton begins to harden; stage C (intermolt, strictly speaking), during which new body mass is produced; stage D (premolt), in preparation of the impending ecdysis; and stage E (molt), the ecdysis itself (Dall 1977). The timing of ecdysis is controlled by an endocrine system involving at least two hormones with antagonistic effects (Aiken 1980; Waddy et al. 1995; for a recent, essential overview of the hormonal control of molting, see Hartnoll 2001). The whole process depends also on external variables (Aiken & Waddy 1988; Waddy et al. 1995), such as quality and quantity of feed; "stressing" conditions, such as lost appendages (Juanes & Smith 1995), injuries (e.g., tagging and marking, Cooper, 1970; Spanier & Barshaw, 1993), crowding (Cobb & Tamm 1974; Aiken & Waddy 1976; Van Olst et al. 1980); isolation (Chittleborough 1975); season (Aiken & Waddy 1976); water temperature (Hughes et al. 1972); light and photoperiod (Aiken & Waddy 1976); oxygen concentrations; or other environmental parameters (see Mikami & Kuballa, Chapter 5 for more information). In artificial, nonoptimal conditions, it is not uncommon to have reductive molts, when the animal decreases in weight and length (Bianchini et al. 1997). Individual variability also plays an important role, as Sastry & French (1977) demonstrated in singly reared American clawed lobsters, whose growth increment range varied by a factor of five. Thus, physiological status (maturation, reproduction, etc.) of the animal, as well as its phenotypic plasticity (Pollock 1995b), plays a role in molting (Kulmiye & Mavuti 2004). The incremental increase in length and weight after ecdysis relates to the size of the lobster and to the time elapsed between molts; as a broad rule, the larger the animal, the less the percent increase (Mauchline 1977) and the longer the intermolt period (Cooper & Uzmann 1977).

Oversimplifying the actual growth process, a lobster slightly varies in weight during the intermolt; at ecdysis it loses the exuvial mass, but regains some weight by hydration of the soft tissues. The swelling of the body modifies its linear measurements, and the calcification of the shell produces another weight increase. These changes in size and weight can be used to model the growth process.

This chapter reviews some growth models commonly used to describe the growth of decapod crustaceans, and the limitations of growth data obtained both in the laboratory and in the wild. The remaining part of the chapter focuses on the available data dealing with the growth of slipper lobsters; it puts a greater emphasis on the genus *Scyllarides*, mainly the Mediterranean species, *Scyllarides latus* (Latreille, 1802), caught in the Strait of Sicily. The approach followed with the Mediterranean slipper lobster could be used as a broad basis for examining growth in other scyllarid species.

9.2 Growth Models for Lobsters

Studying growth in aquaria introduces errors due to the artificial conditions in which animals are held. In fact, any nonnatural environment produces stresses which may affect the rate of growth: even the nature and size of the container can affect the growth of lobsters (Waddy et al. 1995). On the other hand, field experiments on lobster growth are difficult to perform, and introduce different types of errors (Gonzáles-Cano & Rocha 1995). These errors include tag-related problems (Spanier & Barshaw 1993; Spanier & Lavalli 1998; Groeneveld 2004), not knowing the time lapsed from the previous molt, and having to estimate the number of molt cycles completed since the preceding capture (Hartnoll 2001). Moreover, even within a population of animals, different subpopulations may exist, each presenting its own vital parameters; this internal variability increases the uncertainty of the growth estimates even at small-scale geographical levels.

9.2.1 Common Growth Models

Despite the considerable body of knowledge existing on the biology of lobsters, only a small fraction of papers addresses growth issues. Literature dealing with adult growth is even more limited, as many of the studies focus attention on larvae and early juveniles (see reviews in Aiken 1980; Waddy et al. 1995; Booth & Kittaka 2000). Moreover, some of the studies now have only historical value (e.g., Hadley 1906; Marshall 1945; Templeman 1948; Burkenroad 1951; Wilder 1953; Backus 1960; Fielder 1964). Present research interest has shifted from frequency of molting to the intrinsic mechanisms controlling ecdysis and biomineralization (e.g., van Olst et al. 1980; Skinner 1985; Laufer et al. 1987; Fingerman 1992; Hartnoll 2001; see Mikami & Kuballa, Chapter 5 and Horne & Tarsitano, Chapter 8). The majority of recent publications on lobster growth are based on data from clawed lobsters, in particular *Homarus americanus* H. Milne Edwards, 1837, with the remainder on data from palinurid lobsters. Aside from methodologies, almost nothing from these papers can be specifically applied to the slipper lobsters; in fact, Fielder (1964), McKoy (1985), and Booth & Kittaka (2000) and many others stressed that growth rates can vary enormously from species to species, not to say among genera or families.

To describe the growth process, several models have been developed, which could be and have been applied to crustacean growth. The most common growth model, the von Bertalanffy (1934) growth function (vBGF), is given in its simplest formulation as:

$$\mathrm{CL}_t = \mathrm{CL}_\infty^*(1 - e^{-k(t-t_0)}),$$

or modified to consider seasonal variations of growth (Gayanillo et al. 1994) as:

$$\mathrm{CL}_t = \mathrm{CL}_\infty^*(1 - e^{-k*(t-t_0)-(C*k/2\pi)*\sin[2\pi*(t-t_s)]}),$$

where CL_t is the expected mean carapace length at time t, CL_∞ is the mean asymptotic length, k represents the time to reach CL_∞, t_0 is the hypothetical age when the length should be zero, C is the seasonal oscillation amplitude, and t_s is the start of the oscillation with respect to t_0. Growth increment data can be fitted to estimate the parameters of a "seasonalized" vBGF using the Appeldoorn's (1987) approach, a derivation of the Gulland and Holt method (Gulland & Holt 1959). Sometimes, bootstrap procedures are used to obtain an indirect measure of variability (Ragonese & Bianchini 1992; Verdoit et al. 1999). For decapod crustaceans, the carapace length (CL) is deemed to be a more consistent measure than the total length (TL) (Farmer 1986).

Decapod crustacean adult growth data seems to conform to the von Bertalanffy equation (Courtney et al. 2001), but the model is sensitive to biased length data, as is usually seen in commercial catches, and is not robust when the actual age of individuals is not known (Fogarty 1995). Therefore, it is seldom used, and always with some "disclaimers" when applied to macrurid decapods. Conversely, the latter, seasonalized formula may be useful for length-frequency distributions (LFD) that are collected independently from actual measurements of growth (Ragonese et al. 1994), but it does not apply well to animals, such as lobsters, that grow in "stanzas," that is, by discrete varying increments during uneven periods. The LFD must be "deconvoluted" in its components (Battacharya 1967), which are then considered as year-classes in the following modal progress analysis. Many programs, including Mix (MacDonald & Green 1985) and Multifan (Fournier et al. 1991), perform such a separation, on the assumption of normality of the year-class distribution. Recently, Caddy (2003) suggested a discrete gnomonic growth function, which is based on the assumption that the successive molt intervals follow a geometric rhythm; he also proposed to switch the usual dependent and independent variables, considering size as the error-free variable (x-axis) and time as the dependent one (y-axis).

In 1948, Hiatt published a direct graphic method (the Hyatt growth diagram) for brachyurans by plotting preexuvial vs. postexuvial measurements (length and weight) that yielded approximated straight-line relationships, which reduces to the form:

$$L_{n+1} = a + b*L_n \quad \text{and} \quad W_{n+1} = a + b*W_n,$$

where L_n and L_{n+1} are pre- and postmolt lengths, respectively; W_n and W_{n+1} are pre- and postmolt weights, respectively; and a and b are constants that describe the rates at which the size or weight varies at successive molts. Because changes in the relationship between premolt and postmolt size generally occur at the transitions from larval to juvenile forms and from immature to sexually mature forms (detected initially by Kurata [1962] for clawed lobsters), Hiatt (1948) suggested that joined linear models should be used, with inflection points at settlement and at sexual maturity.

Mauchline (1976) used the Hiatt growth diagrams produced by Kurata (1962) to demonstrate that the inflection points described by Hiatt (1948) represented a hyperbola, such that size increase at ecdysis is generally greater in younger individuals than in older (reviewed in Aiken, 1980). He implemented a method that used a semilogarithmic graph to show that the relationship between premolt and postmolt size changed at various points throughout the lobster's life cycle. The Hiatt growth diagrams produced by Kurata (1962) from *H. americanus* were later revised by Mauchline (1976) to take into account this nonlinearity; however, Mauchline believed that the inflections detected at maturity were artifacts of the shape of the hyperbola. Mauchline (1976) also devised linear regressions to express the log percent length increase against the body length or molt number and the log intermolt period against the body length or molt number. These regression lines were used to generate two constants: a molt slope factor (the percent increase or decrease in size during successive molts) and an intermolt-period slope factor (the increase in intermolt period for successive molts).

9.2.2   Growth Studies on Clawed Lobsters

Cooper & Uzmann (1977) used data from recaptures of several hundred tagged *H. americanus* lobsters released in the wild to estimate the yearly probability of molting and yearly average size increments. Using LFDs collected and estimations of age at these sizes, they also computed von Bertalanffy growth equations for male and female lobsters that are summarized in Aiken (1980), and determined maximum attainable sizes that came within 10 mm of the largest recorded female lobster (Wolff 1978). Ennis (1972), fitting a Hiatt diagram, measured growth-per-molt of *H. americanus* in the wild, and found that the growth of internally tagged animals was representative of the growth of untagged and wild specimens. McLeese (1972), Hedgecock et al. (1976), and Hedgecock & Nelson (1978) examined the main factors affecting growth of captive *H. americanus*, that is, food, light, and temperature; moreover, McLeese (1972) found that growth rates in the laboratory were positively correlated with the space available to the lobsters. Nelson et al. (1980) found density-dependent growth in this species and, most recently, Cowan et al. (2001) examined patterns in abundance and growth of juvenile American clawed lobsters, which appeared to be strongly influenced by season (growth occurred almost exclusively from late spring to late autumn).

Simpson (1961) applied these techniques to *Homarus gammarus* (Linnaeus, 1758) and suggested that while both male and females of this species have smaller linear increases per molt than *H. americanus*, they grow to larger sizes. His data were suspect, however, because historical records indicated that *H. americanus* always attained larger sizes than *H. gammarus* (Wolff, 1978). In later studies on *H. gammarus*, Hepper (1967) found linear relationships between the body length and the logarithm of the intermolt period, while Hewett (1974) determined that yearly growth ranged from 7 to 20% for adult lobsters. More recently, Bannister et al. (1994) obtained direct measurements of growth of European lobsters in the wild by microwire tagging and releasing hatchery-reared early juveniles, finding that the legal fishing size (85 mm CL) was reached in four to five years, but with substantial individual growth variation. Finally, Jørstad et al. (2001) studied artificial conditions that affected lobster juvenile growth, and found that substrates mimicking natural bottoms promoted better growth in laboratory aquaria.

9.2.3   Growth Studies on Spiny Lobsters

For spiny lobsters, Mohamed & George (1968) obtained growth data from tagged *Panulirus homarus* (Linneaus, 1758). Chittleborough (1976) derived a von Bertalanffy curve from *Panulirus cygnus* George, 1962 reared in aquaria, while Campillo et al. (1979) estimated the growth of *Palinurus elephas* (Fabricius, 1787) in captivity. Phillips et al. (1992) compared growth of *P. cygnus, Panulirus argus* (Latreille, 1804), and *Panulirus ornatus* (Fabricius, 1798), finding that, besides a large individual variability, the temperate species *P. cygnus* grew much more slowly than the other two tropical species. Dennis et al. (1997) linked growth and shelter preferences in *P. ornatus*; Jones et al. (2001) studied growth of aquaculture-reared *P. ornatus* specimens from small juveniles to half-pounders; and Lozano-Alvarez & Spanier (1997) examined growth of captive subadult *P. argus* under predation risk.

A few studies examined growth of *Jasus lalandii* (H. Milne Edwards, 1837): Newman & Pollock (1974) related growth and pabulum; Cockcroft & Goosen (1995) studied reductive molts in the wild; Goosen & Cockcroft (1995) measured annual growth increments in this species; and Hazell et al. (2001, 2002) examined factors affecting juvenile growth in aquaria, as well as in the wild (i.e., season, temperature, and diet). The latter authors found that while lower temperatures increased the intermolt period, the mean increment per molt (within the normal temperature range) did not increase, provided enough food was available. Cruywagen (1997) used a general linear model (GLM) to study an apparent decline of *J. lalandii* growth rates in South Africa; more recently, a stochastic approach (GLMM, general linear mixed model) to take into account temporal and spatial random effects on growth has been applied by Brandão et al. (2004) on the same stock. In recent years, more knowledge on growth has been accumulated on other species: Pollock & Roscoe (1977) studied *Jasus tristani* Holthuis, 1963; McKoy & Esterman (1981),

McKoy (1985), and Hooker et al. (1997) studied *Jasus edwardsii* (Hutton, 1875). Those last authors, in an area with annual temperatures ranging from 23.3°C to 13.4°C, estimated an average daily growth rate of 0.07, 0.036, and 0.031 mm for 0+, 1+, and 2+ year old animals, respectively, such that lobsters reached 200 g in three years. McKoy (1985), however, argued that growth rates are the result of such factors as environment, locality, habitat, temperature, and captivity.

9.2.4   Growth Studies on Slipper Lobsters

The available bibliography on growth of adult scyllarids is extremely scarce. Rahman & Subramoniam (1989) described the shedding phases in the Asian slipper lobster, *aka* the sand lobster, *Thenus orientalis* (Lund, 1793), and stated that color changes preceded the actual ecdysis and that the species exhibited temporal "peaks" of molting, a sort of growth synchronism. Kagwade & Kabli (1996) estimated the von Bertalanffy growth parameters for *T. orientalis*, showing that its growth pattern was retrogressive and geometric throughout life. Growth in the genus *Ibacus* also has been studied, both in captivity and in the wild. Stewart & Kennelly (1997, 2000) found that while males and females of *Ibacus peronii* Leach, 1815 differ in growth rates, no sex-related differences exist in *Ibacus chacei* Brown & Holthuis, 1998. Stewart (2003) used recaptures (up to almost 10 years) to estimate long-term growth in *I. peronii*, which confirmed a prior von Bertalanffy function determined from short-term recaptures. Groenevelt & Goosen (1996) reported some morphometric data of *Scyllarides elisabethae* (Ortmann, 1894). Rudloe (1983) examined the aquaculture potential of *Scyllarides nodifer* (Stimpson, 1866), hypothesizing that growth from a postlarva to a 300 g animal should require 9 to 10 molts and approximately 18 months. Hardwick & Cline (1990) examined the morphometrics of *S. nodifer* in the northeastern Gulf of Mexico, finding sex-related differences. Most recently, Hearn (2006) studied growth of recaptured tagged *Scyllarides astori* Holthuis, 1960 in the Galápagos Islands, Ecuador (for details see Hearn et al., Chapter 14). For *Scyllarides latus*, Martins (1985) reported measures of landed animals in the Azores Islands, basic information about molting, reproduction, and larval development, and described their morphometrics (TL vs. CL) and weight–length relationships. A similar study, as well as some tagging and recapture experiments of lobsters in artificial reefs and natural habitats, was done with the same species in Israel (Almog-Shtayer 1988; Spanier et al. 1988, 1990, 1993), again without providing details on growth. Spanier & Lavalli (1998) reviewed the available knowledge on *S. latus*, and described observations on the molting process. Bianchini et al. (2001), using medium- and long-term recaptures, combined with laboratory data, derived the parameters of the vBGF, and compared observed and expected measurements for *S. latus*.

The remainder of this chapter is mainly based on the Mediterranean slipper lobster, *S. latus*, considered as a case study in growth of scyllarid lobsters. In fact, it is the only species for which a relative amount of knowledge has accumulated, and other species within the family Scyllaridae may differ in their natural history and growth patterns.

9.3   Molting Under Laboratory Conditions

Slipper lobsters collected from the wild may molt while kept in confined conditions; therefore, it is possible to check them before and after ecdysis to obtain direct measures of their growth, as well as other useful data. These results are, nevertheless, influenced by many factors which may alter the "quality" of the data. The adult Mediterranean slipper lobster, *Scyllarides latus*, has not been the subject of specific studies on molting physiology or of the effect of external parameters on the process of ecdysis (except tagging, e.g., Spanier & Barshaw 1993). While some insight could be derived from studies on other macrurids, the distinctiveness of the slipper lobster natural history does not lend itself to direct comparison. Furthermore, the Mediterranean slipper lobster is seldom caught in the Strait of Sicily below a weight of 100 g (Bianchini et al. 1997; Hearn 2006; and all the available LFDs). At this size, the molting frequency should be already low (Tremblay & Eagles 1997), since the duration of the intermolt is in general positively correlated with size (Hartnoll 2001), and it is rare to record multiple molting events with animals kept for a relatively short periods of time in the laboratory (Bianchini et al. 1998b).

The first problem encountered is completely out of the experimenter's control. If the captured animals are in molt stage C (intermolt) or stage D (premolt), it is difficult, if not impossible, to estimate the time of their last ecdysis, and the amplitude of the growth stanza. Moreover, the longer the time in captivity, the higher the chances are that artificial conditions have influenced the molt. Temperature, different light conditions, unsuitable feed, lack of suitable shelter, crowding or isolation, water quality, etc. may affect ecdysis (see Mikami & Kuballa, Chapter 5 for more information on the effect of these factors on larval scyllarids). Even capture and handling can be stressful, hampering the natural molting process, and invasive tagging methods may increase the risk of disruption of the molt cycle (Spanier & Lavalli 1998; Courtney et al. 2001).

9.3.1   Effect of Temperature and Light

As a rule, growth rates rise with temperature, but mortality also increases at the highest viable temperatures (Hartnoll 1982). This growth-rate increase is due to larger molt increments or shorter intermolt periods (or both, often as antagonistic processes), the former effect being, in general, greater (Nair & Anger 1979).

Spanier et al. (1988) and Spanier & Lavalli (1998) pointed out that in the wild, Mediterranean slipper lobsters could migrate offshore to avoid the extreme temperatures found in shallow waters. Thus, they position themselves in a preferred range of offshore temperatures during the sensitive period of molting. Incidentally, these authors further suggested another advantage for molting in such deeper habitats: lack of illumination and fewer potential predators. This is not the case in controlled situations, however, where the water temperature is influence by the weather directly or as a result of the water in open-systems being pumped from surface waters, with temperatures too low in winter and too high in summer. Conversely, if one attempts to control the temperature in laboratory settings, the physiological activity of the animals may be artificially increased, which may affect the number of molts per unit of time and the incremental change of dimensions compared to the values obtained in the wild.

Slipper lobsters of the species *S. latus* are often caught in caves and crevices (Spanier & Lavalli 1998), with limited or no illumination. As is the case with most other lobsters, they are typically nocturnally active (Bianchini et al. 1997). Thus, the intensity of light to which they are exposed in aquaria and open-air tanks, even if diffuse, is many times that of natural conditions. Moreover, the photoperiod in captivity may follow unnatural rhythms, due to artificial lighting.

9.3.2   Influence of Water Quality and Environment

In artificial conditions, the water quality is different from natural conditions, with an increased concentration of catabolites in closed-water-systems (Bianchini 1982), or the presence of pollutants in open-circuit tanks, or varying amounts of dissolved oxygen. Some slipper lobster species are typically gregarious, as is the case with *S. latus* (Spanier & Almog-Shtayer 1992); thus, the high density typically present in the tanks should not be too stressful, but such may not be true for other species, such as the more solitary *Scyllarides astori* (Hearn, 2006) or the solitary sand lobster, *Thenus orientalis* (see Jones, Chapter 16). Furthermore, the suitability of available shelter and substrate for burial purposes may be important in reducing stress levels. Therefore, dissolved oxygen levels, catabolite levels, crowding or isolation, and shelter may play a role in the well-being and molting behavior of animals held in captivity.

9.3.3   Influence of Nutrition

It is obvious that a food supply below the optimum level produces a reduction in growth rates (Hartnoll 1982), through extended intermolt periods and smaller molt increments (Chittleborough 1975). The food given to lobsters in captivity differs in quality and quantity from that available in the wild. Even if fed *ad libitum*, animals do not have access to the same variety of food present in their natural habitat, some of which may have nutritional elements that are necessarily for the physiological processes associated with molting. The qualitative and quantitative nutritional requirements of the slipper lobsters are not known in sufficient detail to avoid the risk of providing unbalanced diets. That is ever more true with formulated diets, where even the feed presentation plays a role in the lobster nutrition (Tolomei et al. 2003). In any event, it seems that in decapod crustaceans, contrary to fish (Zoccarato et al. 1993), multiple daily feeding (to lower antagonism and the stress of competition) does not stimulate growth (Thomas et al. 2003).

9.3.4   Stress and Reductive Molts

In aquaria, more than a few ecdyses are reductive (Bianchini et al. 1997): apparently healthy animals, fed *ad libitum* on natural foods, kept in conditions that do not seem stressful, "shrink" to smaller sizes. It seems likely that if the animal loses tissues for whatever reason, the exoskeleton might shrink at molt because the body can swell only so much with water. But why molt in such conditions, without waiting longer to allow recovery? This phenomenon raises several questions. Could molting be an overwhelming physiological necessity, like the need to rid the body of catabolites? Is there a yet unknown external triggering factor, which induces molting even if the animal body is not physically ready, or is there an underlying internal clock that stimulates the ecdysis process independent of body conditions? In the case of the Mediterranean slipper lobster, molting happens almost in every season (Bianchini et al. 1997); therefore, photoperiod and temperature, notwithstanding their recognized importance in the hormonal control of molting (Hartnoll 2001), are not the triggering factors. Still, some synchronizing mechanism seems to operate, because more occurrences of molting are recorded in the warmer months (Bianchini et al. 1997).

Aside from eyestalk ablation, specifically intended to induce a precocious molt, and intentional removal of limbs (e.g., for genetic studies), mutilations can occur as incidents during capture and captivity, as well as during inter- and intraspecific aggressive interactions in the wild. Such incidences may shorten the intermolt/premolt period, depending on the molt stage the animal has reached, and reduce the extent of growth (given the energy shift from growth to regeneration).

In order to keep track of each individual separately, animals have normally been tagged with invasive methods, such as clipping, "finning," inserting streamer tags, introducing internal microwires and passive induced transponder (PIT) tags, and placing bands around the antennal bases, etc. (for an overview of some classical methods, cfr. Farmer 1981; Wickins et al. 1986; Krouse & Nutting 1990; Prentice et al. 1990; Bianchini et al. 1992). These tagging techniques may affect molting (Courtney et al. 2001; Dubula et al. 2005); even the use of paint may be harmful, since chemicals could be subtoxic.

9.3.5   Actual Growth in Aquaria

While the above questions have not been addressed in many studies, some limited knowlegde is available from a study conducted in the late 1990s (Bianchini et al. 1997, 1998b). During the program "Feasibility and assessment of stock enhancement by restocking for the Mediterranean slipper lobster, *Scyllarides latus*," breeding and restocking experiments were carried out on a small, pilot-study scale. Close to 300 animals (individual body weight ranged between 100 to 1800 g, with the majority weighing 400 to 800 g each) were captured in Sicilian waters, held for at least one month in tanks, and fed with blue mussels and fish discards. Specimens were sexed, measured (carapace length, CL, and width, CW; antennal (IV segment) length, AL, and width, AW; to the accuracy of 0.1 mm), weighed (body weight, BW; to the accuracy of 1 g), and tagged with both external and internal tags (t-tag and PIT-tag, respectively). Some tagged specimens were subsequently released to the wild. Morphological changes associated with molting were recorded in retained lobsters that molted in aquaria, as well as those released and recaptured in the wild (Bianchini et al. 2001).

Fifty-nine ecdyses were recorded; six reoccurred in the same animals (i.e., these particular animals molted more than once). Most of the molting occurred in late autumn and early spring, but occasional molts occurred year round (except in summer months). Reductive molts were observed in some cases, especially in animals with multiple molts. With only one recorded exception, shedding of the exoskeleton always happened during the night.

Molting data were analyzed using the Appeldoorn method (after Gayanillo et al. 1994), although this method is not fully appropriate on animals with incremental growth. The parameters of von Bertalanffy asymptotic growth function were estimated, taking into consideration a seasonal oscillation; thus,

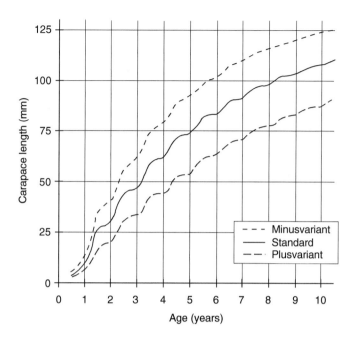

FIGURE 9.1 Von Bertalanffy growth functions for *Scyllarides latus* obtained in captivity ($N = 59$), with overestimated (CL_∞ +s.e.; k +s.e.) and underestimated (CL_∞ −s.e.; k −s.e.) parameters ("true" $CL_\infty = 127.2$ mm, $k = 0.20$).

the following values and associated coefficients of variation (CV, as percentage) were obtained:

$$CL_\infty = 127.2 \text{ mm (CV} = 6\%),$$

$$k = 0.20 \text{ (CV} = 29\%),$$

$$C = 1.0 \text{ (CV} = 52\%)$$

$$t_s = 0.83 \text{ (CV} = 11\%).$$

Figure 9.1 presents the vBGF for *S. latus*, and its minus- and plus-variant realizations. The actual sizes of the largest carapaces (a maximum of 160 mm in the largest lobster) are higher than the calculated average CL_∞ (127.2 mm CL). Nevertheless, it seems reasonable to assume, also from the vBGF, that medium to large slipper lobsters are 10 to 15 years old.

It appears that small damages to the exoskeleton are quickly repaired (Hopkins 1993), as seen in mutilations (by forced autotomy under traction) carried out for electrophoretic studies (Spanò et al. 2003). However, full regeneration is seldom completed with the first molt following the injury (Skinner 1985), and may produce monstrosities (Spanò et al. 2003).

9.4 Growth Increments

The growth of the slipper lobsters can be described with simple regression models, which take into account the increments in dimensions following the ecdysis. In *Scyllarides latus* from Sicilian waters, the premolt vs. postmolt regressions are similar in both sexes (Figure 9.2 and Figure 9.3, respectively for CL and total weight); the $\log_e$-$\log_e$ parameters for the combined sexes are $a' = 1.01$ and $b' = 0.79$ for the CL, and $a' = 1.03$ and $b' = 0.84$ for the total weight (Table 9.1).

A recently molted Mediterranean slipper lobster, *S. latus*, weighs, on average, 12% more than its premolt weight, mainly as the result of water absorption by tissues (care should be taken, since the premolt weight may have been measured weeks before the ecdysis); however, the coefficient of variation

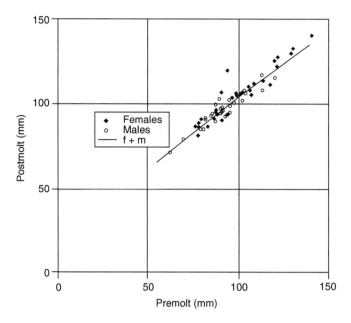

FIGURE 9.2   Premolt vs. postmolt length interpolation in *Scyllarides latus*, by combined sexes.

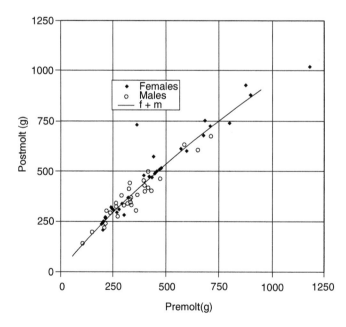

FIGURE 9.3   Premolt vs. postmolt weight interpolation in *Scyllarides latus*, by combined sexes. (From Bianchini et al., 1998b; used with permission.)

is almost 100%, as a result of increments ranging from −13 to 47%. A correlation with size is evident, since mean increments range from 24% for the smallest individuals to 6% for animals over 500 g of weight. In the first few days after molting, weight increases an additional 4 to 5%. In general, the exuvia represents 26% of the premolt body weight, with an almost linear relationship with the total premolt weight (Figure 9.4). Thus, when considering the shed exoskeleton, the average animal encompasses an actual weight increase/recovery of almost 40%. Reductive ecdyses, probably a consequence of substandard feed and of captivity stresses, reached almost −15% in weight and −5% in carapace length losses. Such data

TABLE 9.1

Premolt vs. Postmolt Regressions in *Scyllarides latus*

	Regression	Pearson's Coefficient (r^2)	Standard Error of the Estimate (S.E.E.)
Sex combined	$\log_e(CL_{post}) = 1.01 + 0.79* \log_e(CL_{pre})$	0.927	0.031
	$\log_e(BW_{post}) = 1.03 + 0.84* \log_e(BW_{pre})$	0.953	0.079
Females	$\log_e(CL_{post}) = 1.04 + 0.79* \log_e(CL_{pre})$	0.922	0.027
	$\log_e(BW_{post}) = 1.12 + 0.83* \log_e(BW_{pre})$	0.961	0.079
Males	$\log_e(CL_{post}) = 1.096 + 0.77* \log_e(CL_{pre})$	0.927	0.035
	$\log_e(BW_{post}) = 1.13 + 0.82* \log_e(BW_{pre})$	0.944	0.072

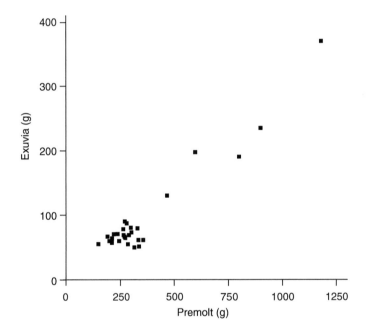

FIGURE 9.4 Scatter plot of the exuvia vs. premolt weight in *Scyllarides latus*.

for *S. latus* are similar to those for *Ibacus* spp. where growth increments are higher in juveniles than mature adults (20–35% vs. 11–15%, respectively) (Haddy et al. 2005).

Ancillary information could show differences in the relative growth of body parts. From the length–weight regressions (Table 9.2), the growth of the adult Mediterranean slipper lobster seems to be almost isometric in both sexes (males: $b = 2.97$; females: $b = 3.01$). Additionally, animals obtained from the edges of the species' geographical distribution, Azores and Israel, seem to show a similar isometry (Bianchini et al. 1997), but with a lower intercept (parameter a). Analyses of correlations among various measurements also suggest that growth does not alter the relative proportions of the body. That notwithstanding, a canonical analysis on the shape of the carapace and of the antenna (Figure 9.5) allows one to discriminate the actual form among different sizes, suggesting that growth is effectively allometric (Table 9.3). Incidentally, the same approach shows a sexual dimorphism in the Mediterranean slipper lobster.

TABLE 9.2

Length-Weight Regressions in *Scyllarides latus*, by Sex

	n	CL_{min} (in mm)	CL_{max} (in mm)	W_{min} (in g)	W_{max} (in g)	a	b	r^2	Standard error
Females	118	69.0	160.5	136	1800	−7.8707 (±0.2489)	3.0126 (±0.05304)	0.965	0.09003
Males	118	62.6	136.5	103	965	−7.7126 (±0.2460)	2.9718 (±0.05314)	0.964	0.06987

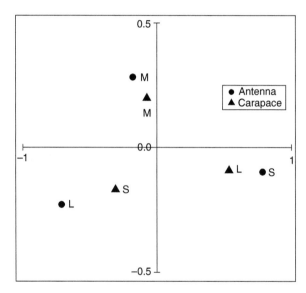

FIGURE 9.5 Results of the canonical discriminant analysis, showing the shape separations of carapace and antenna for small (S, up to 400 g), medium (M, between 400 and 600 g), and large (L, over 600 g) *Scyllarides latus*.

9.5 Molting in the Wild

Even when considering slipper lobsters tagged in the laboratory and released in the wild, growth results are influenced by many factors that may alter the quality of the data. In fact, if the released animals are in intermolt or in premolt, it is almost impossible to know the date of their last ecdysis, and thus the temporal amplitude of the initial growth stanza cannot be established. Moreover, time spent in captivity, handling, and tagging might have produced unknown effects on the natural molting process; the tag itself might alter or otherwise affect, directly or indirectly, the natural molting process (Dubula et al. 2005).

Over long periods of time, it is probable that multiple ecdyses occur, the number of which can only be a guesstimate at recapture. Such guesswork has obvious consequences on the per-molt estimates in any of the growth models typically used in analysis. The location of release, which is seldom the same as the site of capture, could be less-than-optimal for molting (absence of essential food, lack of suitable shelter, strong currents, etc.), thus hampering the process. Additionally, using many different release areas (as in restocking trials) may increase the variability of the response (Skewes et al. 1997).

As in the laboratory, environmental parameters, such as temperature, salinity, and illumination affect molting in nature too; there is anecdotal evidence that even depth, and thus pressure, influences the molt cycle. Some of these parameters may differ among seasons, such that molting may peak in certain seasons. Spanier & Lavalli (1998) state that *Scyllarides latus* molts in Israeli waters during winter months, while

TABLE 9.3

Canonical Discriminant Analysis of Shape in *Scyllarides latus*, by Size or Sex

Body Feature			Size	Mahalanobis'#	P	Significance
Size (3)	Carapace (4 points)	I° autovalue = 80%	Small–Large	0.71	0.15	n.s.
	Antenna (5 points)	I° autovalue = 88%	Small–Large	2.21	0.005	**
			Small–Medium	1.06	0.04	*
		Canonical Variables				
Sex (2)	Carapace (4 points)	M = −0.32; F = 0.39		0.51	0.04	*
	Antenna (5 points)	M = 0.60; F = −0.57		1.36	0.002	**

P, probability of obtaining a larger Mahalanobis' number, with associated level of statistical significance, i.e., * = significant and ** = highly significant.

TABLE 9.4

Pre- and Postmolt Measurements for Seven *Scyllarides latus* Specimens Which Molted in the Wild

Specimen #	Years at Large	Sex	CL_{pre} (mm)	CL_{post} (mm)	BW_{pre} (g)	BW_{post} (g)	$\Delta CL_{post/pre}$ (%)	$\Delta BW_{post/pre}$ (%)
1	0.90	F	96.9	104.0	397.0	480.0	7.3	20.9
2	1.04	F	105.5	110.0	440.0	573.0	4.3	30.2
3	1.07	F	106.0	108.0	482.0	521.0	1.9	8.1
4	1.33	M	95.9	103.3	330.0	439.0	7.7	33.0
5	1.34	M	95.6	102.2	329.0	413.0	6.9	25.5
6	1.34	F	86.4	91.3	240.0	322.0	5.7	34.2
7	4.34	F	94.0	119.9	365.0	730.0	27.5	100.0

CL = carapace length; BW = body weight.

Source: Modified from Bianchini, M.L., Bono, G., & Ragonese, S. 2001. *Crustaceana* 74: 673–680; with permission.

Stewart & Kennelly (2000) found ecdysis seasonality (spring to summer) in the Australian lobster, *Ibacus peronii*.

When the animal is free-ranging, many important events may remain undetected by investigators, such as reproduction, predator–prey interactions, seasonal "hibernation," migrations, and so on. The energy expenditures associated with such activities could compete with growth demands and also impact molting. Finally, slipper lobsters do move considerable distances for foraging and migration, both vertically and horizontally, making their recapture a low-probability event. Yet, emigration of tagged lobsters, as well as loss of tags or lack of reporting, does not undermine the results of growth studies in the field.

In the tag-and-release study in Sicilian waters (Bianchini et al. 1997, 1998b), 30 of a total of 115 slipper lobsters captured, tagged, and released, were later recaptured, seven of which seemed to have gone through at least one incremental molt (Table 9.4). The first six specimens were recaptured after a period of approximately one year at large. It seems that they had molted only once, gaining, on the average, a length increment of just around 6%, which may be reasonable for animals of medium size.

The seventh tagged lobster, however, was recaptured after more than four years from release, showing an increment increase of 27.5% of the carapace length and a doubling of its weight; its light-colored and clean carapace indicated that it had molted recently (Bianchini et al. 2001). Table 9.5 reports in more detail the observed and expected measurements of this individual. The observed length increments are in full agreement with the growth expected after four yearly ecdyses (119.9 mm vs. 120.7 mm CL, calculated from the premolt/postmolt regression). The same seems to hold also for the observed vs. expected weight increase (730 g vs. 714 g, calculated from the length/weight regression). If, on the other hand, the premolt weight is used as regressor, the actual weight gain is higher than expected (730 g vs. 565 g). This may suggest greater increments per molt or more molts than estimated from laboratory studies, that is, a possible better growth performance in the wild.

TABLE 9.5

Measurements and Growth Increments of a *Scyllarides latus* Female Recaptured After 1575 Days at Large

	Carapace Length (CL; mm)	Carapace Width (CW; mm)	Antenna Length (AL; mm)	Antenna Width (AW; mm)	Body Weight (BW; g)
Release	94.0	76.6	61.4	39.5	365
Recapture	119.9	98.1	72.5	47.0	730
Absolute and % Δ	+25.9 (+27.5%)	+21.5 (+28.1%)	+11.1 (+18.1%)	7.5 (+19.0%)	+365 (+100%)
Expected values	120.7 [a]	95.8 [b]	n.e.	n.e.	714 [c]

The expected values are computed using the following parameters:
[a] $\log_e(CL_{post}) = 1.04 + 0.79*\log_e(CL_{pre})$ and considering 4 molts;
[b] $\log_e(CW_{post}) = -0.085 + 0.969*\log_e(CL_{post})$;
[c] $\log_e(BW_{post}) = -7.871 + 3.013*\log_e(CL_{post})$;
n.e.: not estimated

Source: Modified from Bianchini, M.L., Bono, G., & Ragonese, S. 2001. *Crustaceana* 74: 673–680; with permission.

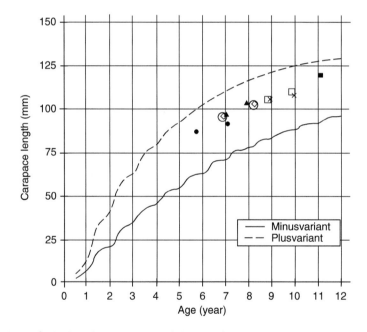

FIGURE 9.6   Initial and final values of the carapace length (in mm) of 7 recaptured *Scyllarides latus* (records in Table 9.4). Equal symbols represent the same animal; starting position on the time scale is calculated from the hypothetical vBGF of $CL_\infty = 127.2$ mm and $k = 0.20$ estimated from laboratory growth. Lines represent overestimated (CL_∞ +s.e.; k +s.e.) and underestimated (CL_∞ −s.e.; k −s.e.) vBGF. (From Bianchini, M.L., Bono, G., & Ragonese, S. 2001. *Crustaceana* 74: 673–680; used with permission.)

The actual size increases of the recaptured Sicilian slipper lobsters are presented (Figure 9.6) as superimposed points on the vBGF "existential space" already defined (minus- and plus-variant realizations based on the "true" von Bertalanffy curve of $CL_\infty = 127.2$ mm and $k = 0.20$). It is evident that the postmolt measurements fall fairly well near the previously predicted values.

Hearn (2006) had 20 medium-term recaptures of tagged *Scyllarides astori* (from seven to 37 months at large) and, having previously determined L_∞ with a Powell-Wetherall plot, used a Gulland-Holt plot "forced" through $CL_\infty = 175.3$ mm (males) and $CL_\infty = 163.8$ mm (females) to obtain a little lower, but similar k of 0.153 and 0.162 for the Galápagos slipper lobster.

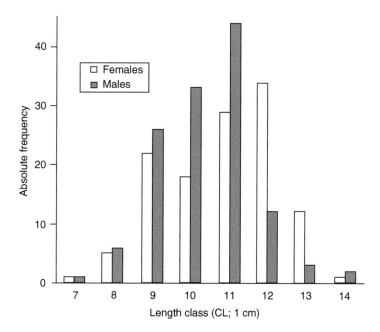

FIGURE 9.7 Length-frequency distribution of *Scyllarides latus* recovered from Sicilian waters, by sex.

9.6 Length-Frequency Distributions (LFDs) in the Commercial Catch

In general, the Mediterranean slipper lobster in Sicilian waters is not common enough to support an *ad hoc* commercial fishery, and it is impossible, therefore, to obtain sufficient data for growth studies. Moreover, the size structure of the Sicilian slipper lobster stock is not known, because the examined animals are not a random sample of the population-at-large; the same problem exists for the Galápagos fishery data presented by Hearn et al. (see Chapter 14). Thus, methods using LFDs cannot be applied. With such data, only a description of the collection of the slipper lobsters examined during the study is possible, without even trying to discern any diachronic processes, like growth and age.

The present shape of the distribution (unimodal for males, quasi-bimodal for females) does not allow one to deconvolute it in its significant modes (Figure 9.7). Males and females are equally present in the catch, in a 0.49 sex-ratio. The average carapace length is 106.8 mm, with a maximum of 160.5 mm. The weight (at capture) of the Sicilian slipper lobsters averages 513.5 g, with a difference of 32% in favor of females (582 g vs. 442 g). Only 2.5% of the animals, all females, exceeded 1 kg (the biggest male weighed 965 g), while the two largest animals caught weighed 1800 g and 1514 g, well below the reported historical maxima of 2.5 to 3 kg. Small specimens, below 200 g, are rare too.

9.7 Final Remarks

Increasingly complex and sophisticated ways of analyzing crustacean growth processes have been implemented over the past years, but our basic understanding has not developed; in fact, only a small proportion of the available literature addresses decapod growth beyond its mere description (Hartnoll 2001). One of the major problems for growth studies resides in accurately aging lobsters: tagging offers only a partial solution, because the absolute age of captured lobsters remains unknown. Deconvolution of LFDs is seldom of use; new methods still need validation, and are expensive in time and money. Another unresolved problem rests with the inadequacy of the mathematical models that are commonly used to describe crustacean growth — a problem largely due to our poor understanding of the growth process.

Regarding the growth of slipper lobsters, there is a lack not only of scientific knowledge, but even of raw data: this situation is true for information arising both from field studies and from experiments in laboratory conditions. Still, adult slipper lobsters are sturdy animals that seem suitable for withstanding handling, tagging, and many other stresses which are often related with growth experiments. Unluckily, some field studies carried on slipper lobsters, while recording length and weight measurements, did not focus on growth issues.

Apart from their insufficiency, available data are restricted to a narrow section of the scyllarid family (primarily *Scyllarides astori*, *Scyllarides latus*, and *Ibacus* spp.), with results that cannot be extrapolated to the whole family. The growth of many of the Scyllaridae taxa has never been studied, not even in the few commercial species. That situation should be of concern because, in general, slipper lobsters are long-lived, slow-growing, natural resources and their exploitation should require careful fishery management based on sound biological data. Thus, this remains an important research area for further development in the coming years.

References

Aiken, D.E. 1980. Molting and growth. In: Cobb, J.S. & Phillips, B.F. (eds.), *The Biology and Management of Lobsters*, Vol. 1: pp. 91–163. New York: Academic Press.

Aiken, D.E. & Waddy, S.L. 1976. Controlling growth and reproduction in the American lobster. *Proc. Annu. Meet. World Maric. Soc.* 7: 415–430.

Aiken, D.E. & Waddy, S.L. 1988. Strategies for maximizing growth of communally reared juvenile American lobsters. *World Aquat. Rept.* 19: 61–63.

Almog-Shtayer, G. 1988. Behavioural-Ecological Aspects of Mediterranean Slipper Lobsters in the Past and of the Slipper Lobster *Scyllarides latus* in the Present. M.A. thesis: University of Haifa, Israel. 165 pp.

Appeldoorn, R. 1987. Modifications of a seasonally oscillating growth function for use with mark-recapture data. *J. Cons. CIEM* 43: 194–198.

Backus, J. 1960. Observations on the growth rate of the spiny lobster. *Calif. Fish Game* 46: 177–181.

Bannister, R.C.A., Addison, J.T., & Lovewell, S.R.J. 1994. Growth, movement recapture rate and survival of hatchery-reared lobsters *Homarus gammarus* (Linnaeus, 1758) released into the wild on the English East Coast. *Crustaceana* 67: 156–172.

Battacharya, C.C. 1967. A simple method of resolution of a distribution into Gaussian component. *Biometrics* 23: 115–135.

Bianchini, M. 1982. Una metodica semplice ed economica per il filtraggio ed il riciclo di acqua di mare per esperimenti di laboratorio. *Quad. Ist. Idrobiol. Acquacolt. Brunelli* 2: 40–48.

Bianchini, M.L., Casolino, G., Ferrone, A., Grieco, R., Palmegiano, G.B., Scovacricchi, T., & Villani, P. 1992. Esperimenti di marcaggio su Peneidi. *Oebalia* 17: 211–213.

Bianchini, M.L., Ragonese, S., Greco, S., Chessa, L., & Biagi, F. 1997. Valutazione della fattibilità e potenzialità del ripopolamento attivo per la magnosa, *Scyllarides latus* (Crostacei Decapodi). *Final Rep. MiRAAF (Pesca Marittima)*: 1–145.

Bianchini, M.L., Di Stefano, S., & Ragonese, S. 1998a. Size and age at onset of sexual maturity of female Norway lobster *Nephrops norvegicus* L. (Crustacea: Nephropidae) in the Strait of Sicily (Central Mediterranean Sea). *Sci. Mar.* 62: 151–159.

Bianchini, M.L., Greco, S., & Ragonese, S. 1998b. Il progetto "Valutazione della fattibilità e potenzialità del ripopolamento attivo per la magnosa, *Scyllarides latus* (Crostacei Decapodi)": sintesi e risultati. *Biol. Mar. Medit.* 5: 1277–1283.

Bianchini, M.L., Bono, G., & Ragonese, S. 2001. Long-term recaptures and growth of slipper lobsters, *Scyllarides latus*, in the Strait of Sicily (Mediterranean Sea). *Crustaceana* 74: 673–680.

Booth, J. & Kittaka, J. 2000. Spiny lobster growout. In: Phillips, B.F. & Kittaka, J. (eds.) *Spiny Lobsters: Fisheries and Culture*: pp. 556–585. Oxford: Blackwell Scientific.

Brandão, A., Butterworth, D.S., Johnston, S.J., & Glazer, J.P. 2004. Using a GLMM to estimate the somatic growth rate trend for male South African west coast rock lobster, *Jasus lalandii*. *Fish. Res.* 70: 335–345.

Burkenroad, M.D. 1951. Measurement of the natural growth rates of decapod crustaceans. *Proc. Gulf Carib. Fish. Inst. Univ. Miami* 3: 25–26.

Caddy, J.F. 2003. Scaling elapsed time: an alternative approach to modeling crustacean moulting schedules? *Fish. Res.* 63: 73–84.

Campillo, A., Amadei, J., & de Reynal, L. 1979. Croissance en captivité de la langouste rouge *Palinurus elephas* Fabr. *Rapp. Proc.-Verb. Réun. CIESM* 25/26: 237–238.

Carrasco, J.F. & Barros, A.R. 1996. Cultivo de juveniles de bogavante europeo *Homarus gammarus* (Linnaeus, 1758): dados sobre su crecimiento y supervivencia. *Bol. Inst. Espanol Oceanogr.* 12: 91–97.

Chittleborough, R.G. 1975. Environmental factors affecting growth and survival of juvenile western rock lobsters *Panulirus longipes* (Milne-Edwards). *Aust. J. Mar. Freshw. Res.* 26: 177–196.

Chittleborough, R.G. 1976. Growth of juvenile *Panulirus longipes cygnus* George on coastal reefs compared with those reared under optimal environmental conditions. *Aust. J. Mar. Freshw. Res.* 27: 279–295.

Choe, S. 1971. Body increases during molt and molting cycle of the oriental brown shrimp *Penaeus japonicus*. *Mar. Biol.* 8: 31–37.

Cobb, J.S. & Tamm, G.R. 1974. Social conditions increase intermolt period in juvenile lobsters, *Homarus americanus*. *J. Fish. Res. Board Can.* 32: 1941–1943.

Cockcroft, A.C. & Goosen, P.C. 1995. Shrinkage at moulting in the rock lobster *Jasus lalandii* and associated changes in reproductive parameters. *S. Afr. J. Mar. Sci.* 16: 195–203.

Cooper, R.A. 1970. Retention of marks and their effects on growth, behavior and migrations of the American lobster, *Homarus americanus*. *Trans. Am. Fish. Soc.* 95: 239–247.

Cooper, R.A. & Uzmann, J.R. 1977. Ecology of juvenile and aduly clawed lobsters, *Homarus americanus*, *Homarus gammarus* and *Nephrops norvegicus*. *Circ.–CSIRO, Div. Fish. Oceannogr. (Aust.)* No. 7: 187–208.

Courtney, A.J., Cosgrove, M.G., & Die, D.J. 2001. Population dynamics of scyllarid lobsters of the genus *Thenus* spp. on the Queensland (Australia) east coast — I: Assessing the effects of tagging. *Fish. Res.* 53: 251–261.

Cowan, D.F., Solow, A.R., & Beet, A. 2001. Patterns in abundance and growth of juvenile lobster, *Homarus americanus*. *Mar. Freshw. Res.* 52: 1095–1102.

Cruywagen, G.C. 1997. The use of generalized linear modeling to determine inter-annual and inter-area variation of growth rates: The Cape rock lobster as example. *Fish. Res.* 29: 119–131.

Dall, W. 1977. Review of the physiology of growth and moulting in rock lobsters. *Circ.–CSIRO, Div. Fish. Oceannogr. (Aust.)* No. 7: 75–81.

De Martini, E.E. & Kleiber, P. 1998. Estimated body size at sexual maturity of slipper lobster *Scyllarides squamosus* at Maro Reef and Necker Island (Northwestern Hawaiian Islands), 1986–97. Honolulu Lab., Southwest Fisheries Science Center, National Marine Fisheries Service, NOAA. *Southwest Fish. Sci. Cent. Admin. Rep. H-98–02*: 14 pp.

De Martini, E.E. & Williams, H.A. 2001. Fecundity and egg size of *Scyllarides squammosus* (Decapoda: Scyllaridae) at Maro Reef, Northwestern Hawaiian Islands. *J. Crust. Biol.* 21: 891–896.

Dennis, D.M., Skewes, T.D., & Pitcher, C.R. 1997. Habitat use and growth of juvenile ornate rock lobsters, *Panulirus ornatus* (Fabricius, 1798), in Torres Strait, Australia. *Mar. Freshw. Res.* 48: 663–670.

Drach, P. 1939. Mue et cycle d'intermue chez les Crustacés Décapodes. *Ann. Inst. Oceanogr. (Paris) [N.S.]* 19: 103–391.

Dubula, O., Groeneveld, J.C., Santos, J., van Zyl, D.L., Brouwer, S.L., van den Heever, N., & McCue, S.A. 2005. Effects of tag-related injuries and timing of tagging on growth of rock lobster, *Jasus lalandii*. *Fish. Res.* 74: 1–10.

Ennis, G.P. 1972. Growth per moult of tagged lobsters (*Homarus americanus*) in Bonivista Bay, Newfoundland. *J. Fish. Res. Board Can.* 29: 143–148.

Farmer, A.S.D. 1981. A review of crustacean marking methods with particular reference to penaeid shrimp. *Kuwait Bull. Mar. Sci.* 2: 167–183.

Farmer, A.S.D. 1986. Morphometric relationships of commercially important species of penaeid shrimp from the Arabian Gulf. *Kuwait Bull. Mar. Sci.* 7: 1–21.

Fielder, D.R. 1964. The spiny lobster, *Jasus lalandii* (H. Milne-Edwards) in South Australia. I: Growth of captive animals. *Aust. J. Mar. Freshw. Res.* 15: 77–92.

Fingerman, M. 1992. Glands and secretion. In: Frederick, W. & Humes, A.G. (eds.), *Microscopic Anatomy of Invertebrates, Vol. 10, Decapod Crustacea*: pp. 345–394. New York: Wiley-Liss.

Fogarty, M.J. 1995. Populations, fisheries, and management. In: Factor, J.R. (ed.), *The Biology of the Lobster Homarus americanus*: pp. 111–137. New York: Academic Press.

Fournier, D.A., Sibert, J.R., & Terceiro, M. 1991. Analysis of length frequency samples with relative abundance data for the Gulf of Maine northern shrimp (*Pandalus borealis*) by the MULTIFAN method. *Can. J. Fish. Aquat. Sci.* 48: 591–598.

Gayanillo F.C., Jr., Sparre P., & Pauly D. 1994. The FAO-ICLARM stock assessment tools (FiSAT) user's guide. *FAO Computer Inf. Ser. (Fisheries)* 6: 1–186.

Gonzáles-Cano, J. & Rocha, C.A.S. 1995. Problems in the estimation of growth parameters for the spiny lobster *Panulirus argus* in the Caribbean and Northeast Brazil. *Proc. II World Fish. Congr. (Athens)* 5: 145–157.

Goosen, P.C. & Cockcroft, A.C. 1995. Mean annual growth increments for male west coast rock lobster *Jasus lalandii*, 1969–1993. *S. Afr. J. Mar. Sci.* 16: 377–386.

Groeneveld, J. 2004. Effects of tagging on the somatic growth rate of rock lobster *Jasus lalandii*. Abstracts of the *VII International Conference on Lobster Biology and Management*, Hobart, Tasmania, 9–13 February 2004, pp. 70.

Groeneveld, J.C. & Goosen, P.C. 1996. Morphometric relationships of palinurid lobsters *Palinurus delagoae* and *P. gilchristi* and a scyllarid lobster *Scyllarides elisabethae* caught in traps off the south and east coast of South Africa. *South Afr. J. Mar. Sci.* 17: 329–334.

Gulland, J.A. & Holt, S.J. 1959. Estimation of growth parameters for data at unequal time intervals. *Journal du Conseil. Cons. Int. pour l' Explor. De la Mer.* 25: 47–49.

Haddy, J.A., Courtney, A.J., & Roy, D.P. 2005. Aspects of the reproductive biology and growth of Balmain bugs (*Ibacus* spp.) (Scyllaridae). *J. Crust. Biol.* 25: 263–273.

Hadley, P.B. 1906. Regarding the growth rate of the American lobster. *Biol. Bull. (Woods Hole, Mass.)* 10: 233–241.

Hardwick, C.W. & Cline G.B. 1990. Reproductive status, sex ratios and morphometrics of the slipper lobster *Scyllarides nodifer* (Stimpson) (Decapoda: Scyllaridae) in the northeastern Gulf of Mexico. *Northeast Gulf Sci.* 11: 131–136.

Hartnoll, R.G. 1982. Growth. In Bliss, D.E. & Abele, L.G. (eds.), *The Biology of Crustacea — 2: Embryology, Morphology and Genetics*: pp. 111–196. New York: Academic Press.

Hartnoll, R.G. 2001. Growth in Crustacea: twenty years on. *Hydrobiol.* 449: 111–122.

Hazell, R.W.A. Cockcroft, A.C., Mayfield S., & Noftke M. 2001. Factors influencing the growth rate of juvenile rock lobsters, *Jasus lalandii*. *Mar. Freshw. Res.* 52: 1367–1373.

Hazell, R.W.A., Schoeman, D.S., & Noftke, M. 2002. Do fluctuations in the somatic growth rate of rock lobster encompass all size classes? A re-assessment of juvenile growth of *Jasus lalandii*. *Fish. Bull. U.S.* 100: 510–518.

Hearn, A. 2006. Life history of the slipper lobster *Scyllarides astori* Holthuis, 1960, in the Galápagos Islands, Ecuador. *J. Exp. Mar. Biol. Ecol.* 328: 87–97.

Hedgecock, D. & Nelson, K. 1978. Components of growth rate variation among laboratory cultured lobsters (*Homarus*). *Proc. World Maric. Soc.* 9: 125–137.

Hedgecock, D., Nelson, K. & Shleser, R.A. 1976. Growth differences among families of the lobster *Homarus americanus*. *Proc. Annu. Meet. — World Maric. Soc.* 7: 347–361.

Hepper B.T. 1967. On the growth at moulting of lobsters (*Homarus vulgaris*) in Cornwall and Yorkshire. *J. Mar. Biol. Assoc. U.K.* 47: 629–643.

Hewett, C.J. 1974. Growth and moulting in the common lobster (*Homarus vulgaris* Milne-Edwards). *J. Mar. Assoc. U.K.* 54: 379–391.

Hiatt, R.W. 1948. The biology of the lined shore crab, *Pachygrapsus crassipes* Randall. *Pac. Sci.* 2: 135–213.

Hooker, S.H., Jeffs, A.J., Creese, R.G., & Sivaguru, K. 1997. Growth of captive *Jasus edwardsii* (Hutton) (Crustacea: Palinuridae) in north-eastern New Zealand. *Mar. Freshw. Res.* 48: 903–909.

Hopkins, P.M. 1993. Regeneration of walking legs in the fiddler crab *Uca pugilator*. *Am. Zool.* 33: 348–356.

Hughes, J.T., Sullivan J.J., & Shleser R. 1972. Enhancement of lobster growth. *Science* 177: 1110–1111.

Jones, C.M, Linton, L., Horton, D., & Bowman, W. 2001. Effect of density on growth and survival of ornate rock lobster, *Panulirus ornatus* (Fabricius, 1798), in a flow-through raceway system. *Mar. Freshw. Res.* 52: 1425–1429.

Jørstad K.E.,Agnalt, A.-L., Kristiansen, T.S., & Nostvol, E. 2001. High survival and growth of European lobster juveniles (*Homarus gammarus*) reared communally on a natural-bottom substrate. *Mar. Freshw. Res.* 52: 1431–1438.

Juanes, F. & Smith, L.D. 1995. The ecological consequences of limb damage and loss in decapod crustaceans: a review and prospectus. *J. Exp. Mar. Biol. Ecol.* 193: 197–223.

Kagwade, P.V. & Kabli, L.M. 1996. Age and growth of the sand lobster *Thenus orientalis* (Lund) from Bombay waters. *Indian J. Fish.* 43: 241–247.

Krouse, J.S. & Nutting, G.E. 1990. Effectiveness of the Australian western rock lobster tag for marking juvenile American lobster along the Maine coast. *Am. Fish. Soc. Symp.* 7: 94–100.

Kulmiye, A.J. & Mavuti, K.M. 2004. Growth and moulting of captive spiny lobster, *Panulirus homarus homarus* in Kenya, Western Indian Ocean. Abstracts of the *VII International Conference on Lobster Biology and Management,* Hobart, Tasmania, February 9–13, 2004, pp. 71.

Kurata, H. 1962. Studies on age and growth of Crustacea. *Bull. Hokkaido Reg. Fish. Res. Lab.* 24: 1–115.

Laufer, H., Landau, M., Homola, E., & Borst, D.W. 1987. Methyl farnesoate: its site of synthesis and regulation of secretion in a juvenile crustacean. *Insect Biochem.* 17: 1129–1131.

Lozano-Alvarez, E. & Spanier, E. 1997. Behavior and growth of captive sub-adults spiny lobsters, *Panulirus argus,* under the risk of predation. *Mar. Freshw. Res.* 48: 707–713.

MacDonald, P.D.M. & Green, P.E.J. 1985. *User's Guide to Program MIX: An Interactive Program for Fitting Mixtures of Distribution.* Hamilton, Ontario: Ichthus Data Systems.

Marshall, N. 1945. The molting without growth of spiny lobsters *Panulirus argus* kept in a live car. *Trans. Am. Fish. Soc.* 75: 267–269.

Martins, H.R. 1985. Biological studies of the exploited stock of the Mediterranean locust lobster *Scyllarides latus* (Latreille, 1802) (Decapoda: Scyllaridae) in the Azores. *J. Crust. Biol.* 5: 294–305.

Mauchline, J. 1976. The Hiatt growth diagram for Crustacea. *Mar. Biol.* 35: 79–84.

Mauchline, J. 1977. Growth of shrimps, crabs and lobsters: An assessment. *J. Cons. Int. Explor. Mer.* 37: 162–169.

McKoy, J.L. 1985. Growth of tagged rock lobsters (*Jasus edwardsii*) near Stewart Island, New Zealand. *N.Z. J. Mar. Freshw. Res.* 19: 457–466.

McKoy, J.L. & Esterman, D.B. 1981. Growth of rock lobsters (*Jasus edwardsii*) in the Gisborne region, New Zealand. *N.Z. J. Mar. Freshw. Res.* 15: 123–136.

McLeese, D.W. 1972. Effects of several factors on the growth of the American lobster (*Homarus americanus*) in captivity. *J. Fish. Res. Board Can.* 29: 1725–1730.

Mohamed, K.H. & George, M.J. 1968. Results of tagging experiments on the Indian spiny lobster *Panulirus homarus* (Linnaeus): Movement and growth. *Indian J. Fish.* 15: 15–21.

Nair, K.K.C. & Anger, K. 1979. Life cycle of *Corophium insidiosum* (Crustacea, Amphipoda) in laboratory culture. *Helg. Wiss. Meeresunters.* 32: 279–294.

Nelson K., Hedgecock, D., Borgeson, W., Johnson, E., Daggett, R., & Aronstein, D. 1980. Density-dependent growth inhibition in lobsters, *Homarus* (Decapoda, Nephropidae). *Biol. Bull. (Woods Hole)* 159: 162–176.

Newman, G.G. & Pollock D.E. 1974. Growth of the rock lobster *Jasus lalandii* and its relationship to benthos. *Mar. Biol.* 24: 339–346.

Phillips, B.F., Palmer, M.J., Cruz, R., & Trendall, J.T. 1992. Estimating growth of the spiny lobsters *Panulirus cygnus*, *P. argus* and *P. ornatus*. *Aust. J. Mar. Freshw. Res.* 43: 1177–1188.

Pollock, D.E. 1995a. Changes in maturation ages and sizes in crustacean and fish populations. *S. Afr. J. Mar. Sci.* 15: 99–103.

Pollock, D.E. 1995b. Notes on phenotypic and genotypic variability in lobsters. *Crustaceana* 68: 193–202.

Pollock, D.E. & Roscoe, M.J. 1977. The growth at moulting of crawfish *Jasus tristani* at Tristan da Cunha, South Atlantic. *J. Cons. Int. Explor. Mer.* 37: 144–146.

Prentice, E.F., Flagg, T.A., McCutcheon, C.S., Brastow, D.F., & Coss, D.C. 1990. Equipment, methods, and an automated data-entry station for PIT tagging. *Am. Fish. Soc. Symp.* 7: 335–340.

Punt, A.E. 2003. The performance of a size-structured stock assessment method in the face of spatial heterogeneity in growth. *Fish. Res.* 65: 391–409.

Ragonese, S. & Bianchini, M.L. 1992. Stima dei parametri di crescita di *Aristeus antennatus* nel Canale di Sicilia. *Oebalia* 17(s.1): 101–107.

Ragonese, S., Bianchini, M.L., & Gallucci, V.F. 1994. Growth and mortality of the red shrimp *Aristaeomorpha foliacea* in the Sicilian Channel (Mediterranean Sea). *Crustaceana* 67: 348–361.

Rahman, M.K. & Subramoniam, T. 1989. Molting and its control in the female sand lobster *Thenus orientalis* (Lund). *J. Exp. Mar. Biol. Ecol.* 128: 105–115.

Rudloe, A. 1983. Preliminary studies of the mariculture potential of the slipper lobster, *Scyllarides nodifer*. *Aquaculture* 34: 165–169.

Sastry, A.N. & French, D.P. 1977. Growth of American lobster, *Homarus americanus* Milne-Edwards, under controlled conditions. *Circ.–CSIRO, Div. Fish. Oceannogr. (Aust.)* No. 7: 11 (abstract).

Sheehy, M., Caputi, N., Chubb, C., & Belchier, M. 1998. Use of lipofuscin for resolving cohorts of western rock lobster (*Panulirus cygnus*). *Can. J. Fish. Aquat. Sci.* 55: 925–936.

Simpson, A.C. 1961. A contribution to the bionomics of the lobster (*Homarus vulgaris* Edw.) on the coast of North Wales. *Fish. Invest. (London), Ser. 2,* 23: 1–28.

Skewes, T.D., Pitcher, C.R., & Dennis, D.M. 1997. Growth of ornate rock lobsters, *Panulirus ornatus*, in Torres Strait, Australia. *Mar. Freshw. Res.* 48: 497–501.

Skinner, D.M. 1985. Moulting and regeneration. In: Bliss, D.E. (ed.), *The Biology of Crustacea: Integument, Pigments, and Hormonal Processes, Vol. 9*: pp. 44–146. New York: Academic Press.

Spanier, E. & Almog-Shtayer, G. 1992. Shelter preferences in the Mediterranean Slipper lobster: Effects of physical properties. *J. Exp. Mar. Biol. Ecol.* 164: 103–116.

Spanier, E. & Barshaw, D.E. 1993. Tag retention in the Mediterranean slipper lobster. *Isr. J. Zool.* 39: 29–33.

Spanier, E. & Lavalli K.L. 1998. Natural history of *Scyllarides latus* (Crustacea: Decapoda): A review of the contemporary biological knowledge of the Mediterranean slipper lobster. *J. Nat. Hist.* 32: 1769–1786.

Spanier, E., Tom, M., Pisanty, S., & Almog, G. 1988. Seasonality and shelter selection by the slipper lobster *S. latus* in the south eastern Mediterranean. *Mar. Ecol. Prog. Ser.* 42: 247–255.

Spanier, E., Tom, M., Pisanty, S., & Almog-Shtayer, G. 1990. Artificial reefs in the low productive marine environment of the Southeastern Mediterranean. *P.S.Z.N.I: Mar. Ecol.* 11: 61–75.

Spanier, E., Almog-Shtayer, G., & Fiedler, U. 1993. The Mediterranean slipper lobster *Scyllarides latus*: The known and the unknown. *BIOS* 1: 49–58.

Spanò, N., Ragonese, S., & Bianchini, M.L. 2003. An anomalous specimen of *Scyllarides latus* (Decapoda, Scyllaridae). *Crustaceana* 76: 885–890.

Stewart, J. 2003. Long-term recaptures of tagged scyllarid lobsters (*Ibacus peronii*) from the east coast of Australia. *Fish. Res.* 63: 261–264.

Stewart, J. & Kennelly, S.J. 1997. Fecundity and egg-size of the Balmain bug *Ibacus peronii* (Leach, 1815) off the east coast of Australia. *Crustaceana* 70: 191–197.

Stewart, J. & Kennelly, S.J. 2000. Growth of scyllarid lobsters *Ibacus peronii* and *I. chacei*. *Mar. Biol.* 136: 921–930.

Stewart, J., Kennelly, S.J., & Hoegh-Guldberg, O. 1997. Size at sexual maturity and the reproductive biology of two species of scyllarid lobsters from New South Wales and Victoria, Australia. *Crustaceana* 70: 344–367.

Templeman, W. 1948. Growth per moult in the American lobster. *Nfl. Dep. Nat. Resour. Res. Bull. (Fish.)* 18: 26–48.

Thomas, C.W., Carter, C.G., & Crear, B.J. 2003. Feed availability and its relationship to survival, growth, dominance and the antagonistic behaviour of the southern rock lobster, *Jasus edwardsii* in captivity. *Aquaculture* 215: 45–65.

Tolomei, A., Crear, B., & Johnston, D. 2003. Diet immersion time: Effects on growth, survival and feeding behaviour of juvenile southern rock lobster, *Jasus edwardsii*. *Aquaculture* 219: 303–316.

Tremblay, M.J. & Eagles, M.D. 1997. Molt timing and growth of the lobster, *Homarus americanus*, off northeastern Cape Breton Island, Nova Scotia. *J. Shellfish Res.* 16: 383–394.

Van Olst, J.C., Carlberg, J.M., & Hughes, J.T. 1980. Aquaculture. In: Cobb, J.S. & Phillips, B.F. (eds.), *The Biology and Management of Lobsters, Vol. II: Ecology and Management*: pp. 333–384. New York: Academic Press.

Verdoit, M., Pelletier, D., & Talidec, C. 1999. A growth model that incorporates individual variability for the Norway lobster population (*Nephrops norvegicus*, L. 1758) from the Bay of Biscay. *ICES J. Mar. Sci.* 56: 734–745.

von Bertalanffy, L. 1934. Untersuchungen über die Gesetzlichkeiten des Wachstums 1. Allgemeine Grundlagen der Theorie. *Roux. Arch. Entwicklungsmech. Org.* 131: 613–653.

Waddy, S.L., Aiken, D.E., & De Kleijn, D.P.V. 1995. Control of growth and reproduction. In: Factor, J.R. (ed.), *The Biology of the Lobster Homarus americanus*: pp. 217–266. New York: Academic Press.

Wickins, J.F., Beard, T.W., & Jones, E. 1986. Microtagging cultured lobsters, *Homarus gammarus* (L.), for stock enhancement trials. *Aquacult. Fish. Manag.* 17: 259–265.

Wilder, D.G. 1953. The growth rate of the American lobster (*Homarus americanus*). *J. Fish. Res. Board Can.* 10: 371–412.

Wolff, T. 1978. Maximum size of lobsters (*Homarus*) (Decapoda, Nephropidae). *Crustaceana* 34: 1–14.

Zoccarato, I., Benatti, G., Bianchini, M.L., Bocciglione, M., Conti, A., & Palmegiano, G.B. 1993. The effect of density and feeding level on performances and body composition in *Oncorhynchus mykiss*. *EAS Spec. Publ.* 18: 323–326.

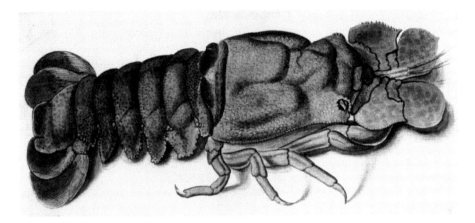

COLOR FIGURE 1.1 Original watercolor of "Squilla lata" by Gesner (1558, p. 1097), illustrating *Scyllarides latus*. The detail on this figure is accurate, including the coloration of the body (brown, lighter in some areas, darker in others) and the antennules (blue). (From Holthuis, L.B. 1996. *Zool. Verhand.* 70: 169–196; reprinted with permission.)

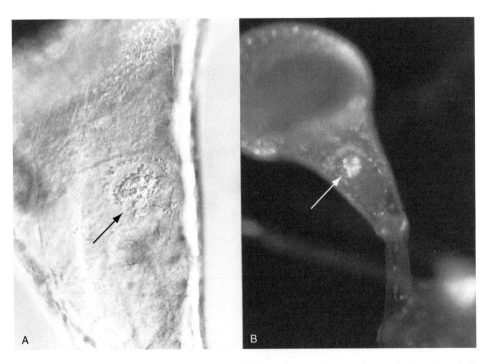

COLOR FIGURE 5.1 (A) Nomarski DIC microscopy and (B) fluorescence microscopy image of a first-stage phyllosoma of *Panulirus japonicus* in which the sinus gland/X-organ complex is clearly visible (arrows).

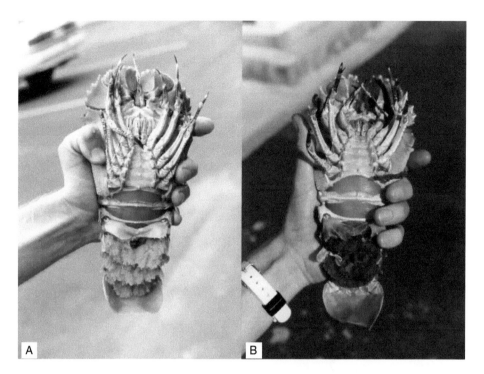

COLOR FIGURE 5.4 Different stages of egg development. (A) Newly spawned, orange-colored egg mass indicates early embryonic development, (B) prehatching-brown colored egg mass indicates appearance of eye pigment of embryo and reduction of yolk.

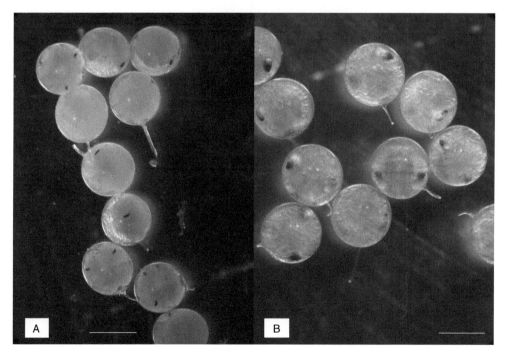

COLOR FIGURE 5.5 Microscopic images of egg development. (A) Newly spawned eggs, (B) pre-hatching eggs. Scale bars indicate 1 mm.

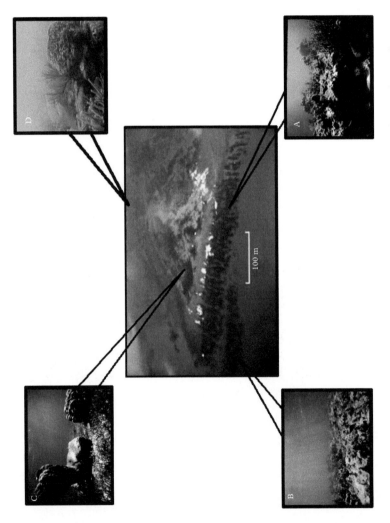

COLOR FIGURE 11.1 Looe Key Reef, Florida Keys, showing typical scyllarid habitats across the reef system. (A) High-relief spur-and-groove buttress zone, (B) low-relief spur-and-groove, (C) reef flat, and (D) patch-reef habitat.

COLOR FIGURE 14.1 The Galápagos slipper lobster *Scyllarides astori*. (Photograph by Alex Hearn.)

COLOR FIGURE 18.3 *Scyllarides latus* in an artificial reef at 20-m depth off Haifa, Israel, 2005. (Photo by S. Breitstein; used with permission.)

10

Directions for Future Research in Slipper Lobster Biology

Ehud Spanier and Kari L. Lavalli

CONTENTS

Abstract

Compared to clawed and spiny lobsters that have higher commercial values, the biological knowledge on scyllarid lobsters is limited at this time. However, the economic
importance of several large species of slipper lobsters has increased in recent years
due to shifts in fisheries from depleted spiny lobster stocks to autochthonous slipper
lobsters. Due to the large number of scyllarid species, the diverse habitats that they
occupy, and the different adaptations associated with these surroundings, the members
of this successful lobster family are also of special importance from the evolutionary
point of view. Considerable progress has been made in the recent years with respect
to some aspects of slipper lobster biology. This research has been often focused on a
few, relatively large species, mainly *Scyllarides* spp., *Ibacus* spp., and *Thenus* spp., but
has only examined limited life-history stages. Further in-depth studies are needed in
key biological areas such as taxonomy, phylogeny, and evolution; genetics, particularly
those studies that help to distinguish separate stocks within populations; morphology
and anatomy, physiology; larval, postlarval, juvenile, and adult ecology and behavior;
reproduction; and growth and its control. Future research in these directions should
be extended also to additional species. Improving our information base of these key
biological topics will also benefit fisheries management and help to preserve currently
overexploited populations, as well protect those populations that may become future
targets of fisheries.

10.1   Introduction

When considering future research in biology of scyllarid lobsters, one should refer to the considerable research done in the biology of the other two main families of lobsters, Nephropidae and Palinuridae, although even the biological knowledge on species from these family suffers from gaps in several areas. The relative wealth of information available on the biology of clawed and spiny lobsters, in comparison with that on slipper lobsters, probably stems from the higher economic importance of species within those families (see Spanier & Lavalli, Chapter 18). Yet, not only has the economic importance of slipper lobsters increased in recent years due to shifts in fisheries from depleted spiny lobster stocks to autochthonous slipper lobster species (see Hearn et al., Chapter 14; Radhakrishnan et al., Chapter 15; Spanier & Lavalli, Chapter 18 for examples), but scyllarids are also of special importance from the evolutionary point of view. Webber & Booth (Chapter 2) and Lavalli et al. (Chapter 7) point out that the family Scyllaridae contains a higher number of species than either the nephropids or the palinurids (see also Holthuis 1991). This diverse group of species displays varied early life history strategies and lives in highly diverse habitats and depths as postlarvae, juveniles, and adults — as such, they should display a variety of biological adaptations. Thus, Lavalli et al. (Chapter 7) further emphasizes that future studies on biological aspects of slipper lobsters are important, not only because they are exploited, but also because they are, in and of themselves, interesting animals that can provide evolutionary insights into other decapod species. This same argument was made by W.F. Herrnkind in 1977 at the first international workshop on lobster biology when he stated that the lobster is a very significant biological entity that is widely distributed, large in size, long lived, abundant, and ecologically consequential (Cobb & Phillips 1980). Slipper lobsters do not differ in any of those regards from clawed or spiny lobsters and may, in fact, provide unique opportunities to study evolutionary strategies (see Lavalli & Spanier, Chapter 1; Webber & Booth, Chapter 2; and Lavalli et al., Chapter 7 for further discussion of such strategies).

Phillips (2005), emphasizing the importance of biological knowledge to successful lobster fishery management in general, mentioned four biological disciplines of importance to this goal: larval ecology, juvenile biology, behavior, and genetics (for topics associated directly with lobster fisheries, see Spanier & Lavalli, Chapter 18). Reviewing the published biological papers in the proceedings of the last three International Conference and Workshop on Lobster Biology and Management — the 5th held in Queenstown New Zealand (*Marine and Freshwater Research* 48 (8), 2001, 655–1136), the 6th held in Key west, U.S.A. in 2000 (*Marine and Freshwater Research* 52 (8), 2001, 1033–1675), and the recent 7th held in Hobart, Tasmania, Australia in 2004 (*New Zealand Journal of Marine and Freshwater Research* 39 (2 + 3), 2005, 227–783), the majority deal with nephropid and palinurid species, and several additional disciplines of biological research can be identified. These include taxonomy, phylogeny, and evolution; morphology and anatomy; ecology (of adults and juveniles); physiology; reproduction; and growth. The following sections will use these biological topics to review the state of the knowledge in the family Scyllaridae.

10.2   Taxonomy, Phylogeny, and Evolution

Webber and Booth (Chapter 2) state that the Scyllaridae are a highly characteristic family of lobsters that are closely related to the Palinuridae and Synaxidae (together defined as the Achelata). These families share many characters, especially their unique larvae, which are exceptional in decapod crustaceans. The higher taxonomy of the family is also reasonably settled due, in large part, to the contributions of Lipcuis B. Holthuis and his colleagues (for details, see Webber & Booth, Chapter 2). However, in addition to the need for settling some relationships at lower taxonomic levels (e.g., *Thenus* spp.), the recent establishment of 13 new genera in the subfamily Scyllarinae, by far the largest and least known group of slipper lobsters, speaks to the need for new information on these and other less-studied species within the family. Additional species and perhaps other subfamilies will inevitably be discovered in the less explored parts of the oceans.

Webber and Booth (Chapter 2) also suggest that Scyllaridae have evolved more recently than the Palinuridae, and have done so along with the development of the major ocean basins to their present configuration. The present diversity of species (85 known to date) does not suggest the family, or any group within it, have resulted from reliction. They further propose that there is directional dispersal from the largest and most favorable areas into smaller and less favorable, more specialized niches and that slipper lobsters have undergone recent radiations into the various shallow shelf and deep water habitats where they are found today. To aid in exploration of these hypotheses, the use of molecular tools for the study of phylogeny of slipper lobsters, as has been done in spiny lobsters (e.g., Ovenden et al. 1997; Ptacek et al. 2001), may contribute to our understanding of the relationships between the various taxonomic groups of slipper lobsters, possible lineages within and between genera, species radiations, and other evolutionary trends. Also, further analysis of the evolution of life cycles (including migration strategies, if they exist, in at least some scyllarids; see Lavalli et al., Chapter 7), similar to the study of George (2005) in spiny lobsters will be needed when more information on ecology and behavior of separate ontogenetic phases of several scyllarid species become available for comparison.

10.3 Genetics

As Deiana et al. (Chapter 3) point out, the available information on genetics of Scyllaridae is very limited and is usually a marginal side product of studies of other decapods, especially those on spiny lobsters (e.g., Patek & Oakley 2003) and nephropids (Tam & Kornfield 1998). There are very few studies that focus solely on the genetics of slipper lobsters (e.g., Bianchini et al. 2003).

The available DNA data, genome size, and cytogenetic data are not enough to draw phylogenetic schemes among subfamilies within the Scyllaridae. The limited data also refers to very few species of Scyllaridae and there is clearly a need for more in-depth genetic studies on the many and diverse species within the family. These future genetic studies should include representatives from the various taxa to enable a better phylogenetic and evolutionary understanding of the family Scyllaridae as has been done, to a certain extent, in clawed (e.g., Jørstad et al. 2005) and spiny lobsters (e.g., Ovenden et al. 1997).

Genetic studies are important also from the standpoint of fisheries. Phillips (2005) pointed out that fishing may affect the genetic structure of a fished population (in his example, the western rock lobster, *Panulirus cygnus* George, 1962) through management schemes such as minimum size for both sexes, maximum size for females, protection of spawners, and different legal sizes within the season and for different localities. He suggests that such measurements may be encouraging the selection of early maturing, and thus slow growing populations as more energy is directed to reproduction rather than growth. If this is indeed the case, the genetic structure of fished (and often overexploited) populations of commercial species, such as *Scyllarides*, *Thenus* and *Ibacus* (see Spanier & Lavalli, Chapter 18) are of prime interest. Some of these commercially important species are distributed over large geographical ranges and are subject to exploitation in many countries in the same general region. As such, it is not clear if the lobsters being fished are from separate stocks within the same population, separate populations within the same metapopulation, or genetically dissimilar populations. Genetic studies may help to determine such questions and there is an urgent need for international scientific cooperation to enable such studies. The outcome of such genetic studies will be important for proper fisheries management of slipper lobster stocks and preservation of overfished populations or stocks within the population. The genetic information should also be important in cases of enhancement by laboratory-reared lobsters, to be considered where aquaculture of the species is available, or by restocking of adults where no success has been achieved in aquaculture (as was done in an experimental stock enhancement of adult Mediterranean slipper lobsters *Scyllarides latus* (Latreille, 1802) in Sicily by Bianchini et al. 1997, 1998; see Bianchini & Ragonese, Chapter 9 for further details). Restocking of a depleted region with considerable numbers of lobsters from another region (or country) can create "genetic contamination" if the lobster population in the source area is significantly genetically dissimilar from the autochthonous one. Similar genetic considerations can be raised regarding transport of lobsters (by ballast water of ships or man-made canals) to new biogeographic regions (e.g., Galil et al. 1989; Freitas & Castro 2005).

10.4   Larval Ecology and Behavior

Very limited knowledge is available on scyllarid larvae in comparison with information existing for such life history phases in clawed and spiny lobsters, in particular *Homarus americanus* H. Milne Edwards, 1837, and *Panulirus argus* (Latreille, 1804) (see Sekiguchi et al., Chapter 4 and Lavalli et al., Chapter 7). Information on larval ecology and behavior comes from both sampled wild specimens, with references to environmental conditions in the sampling area, and observations of laboratory cultured scyllarid larvae. Information from larval stages obtained in aquaculture also facilitates better identification of larval stages sampled in the wild. In both cases, the studies of scyllarid larvae are not nearly as advanced as those on larvae of nephropids and palinurids. There is limited success in sampling the larval, postlarval, and juvenile stages of Scyllaridae in their natural marine habitats. Also, only seven scyllarid species have been cultured through their full larval development (see Mikami & Kuballa, Chapter 5), which makes field identification of all stages of scyllarid phyllosomas difficult.

Sekiguchi et al. (Chapter 4) suggest that *Ibacus* and *Thenus* species have larger eggs and more advanced larval development at hatching; thus, their larvae are more likely to survive and be retained until settlement within coastal waters than are *Scyllarides* species. They presume that the higher fecundity of species belonging to this (commercially important) latter genus is an adaptation to offset high larval loss in oceanic waters. The longer larval durations (several months) typical of species of several scyllarids, such as *Scyllarides*, *Parribacus*, and *Arctides*, also increases opportunity for dispersal to other coastal regions.

According to Sekiguchi et al. (Chapter 4) and Johnston (Chapter 6), the few scyllarid phyllosomas studied appear to feed mainly on soft, fleshy foods and are likely to be preyed upon by a variety of pelagic fishes (as reported by Lyons 1970). The phyllosomas of quite a few scyllarid genera, in their later stages of development, can be found associated with gelatinous zooplankton, particularly jellyfish (Sekiguchi et al., Chapter 4 and references therein). The nature of this association is unknown at this time, but nematocysts in the gut suggest that some phyllosomas may consume the jellies (see Sekiguchi et al., Chapter 4; Mikami & Kuballa, Chapter 5, and Johnston, Chapter 6 for more information). Rearing studies typically use nonnatural food sources to feed larvae (Mikami & Kuballa, Chapter 5), although some studies indicate that live ctenophores in addition to chopped clam may be an excellent food source in culture (Radhakrishnan et al., Chapter 15).

Although there are recent advances in our understanding of larval transport and recruitment in Scyllaridae (e.g., Booth et al. 2005; Sekiguchi et al., Chapter 4), more research is needed in order to be able to address specific questions already dealt with in the larvae and postlarvae of some palinurid and nephropid species. These include questions on how postlarvae find the coast (e.g., Jeffs et al. 2005) and the particular effect of ocean processes on recruitment (e.g., Yeung et al. 2001).

There is obviously a need for more research in scyllarid larval ecology and behavior (see Lavalli et al., Chapter 7). Future research should address gaps in our knowledge of the early life-history stages of scyllarids on a variety of species from the different subfamilies. There should be close interactions between laboratory and field studies, particularly in species where aquaculture techniques have been perfected. Such studies can include larval response to light, thermoclines, pycnoclines, ability to swim in currents, etc. One of the problems of obtaining additional information on lobster larval ecology is the enormous cost of ship time needed to undertake sampling in the open sea (Phillips 2005). This obstacle can be partially overcome by utilizing physical oceanographic or other cruises (e.g., Fiedler & Spanier 1999) and in conducting similar experiments to those of Butler and colleagues in spiny lobsters (e.g., Eggleston et al. 1998; Acosta & Butler 1999; Butler & Herrnkind 2000). The increase in the knowledge of larval ecology of commercial species will undoubtedly benefit the fisheries management of these species, as we learn more about the recruitment process of the nisto into the benthic environment.

10.5   Juvenile Ecology and Behavior

Information is scantier when dealing with juveniles. In some species, such as *Scyllarides latus*, not a single juvenile has yet been found and it is speculated that these stages inhabit deep sea habitats (Spanier &

Lavalli 1998; Lavalli et al., Chapter 7). This lack of knowledge is in sharp contrast to the considerable number of publications accumulated on the biology of juvenile of clawed and spiny lobsters (e.g., Acosta et al. 1997; Butler et al. 1997; Incze et al. 1997; Wahle & Incze 1997; Butler & Herrnkind 2000 and references therein; Cowan et al. 2001; Diaz et al. 2005). The scanty and partial information available on juvenile ecology and behavior in Scyllaridae is reviewed by Lavalli et al. (Chapter 7) and includes the species of *Scyllarides*, *Ibacus*, and *Thenus*. The difficulties in detecting juvenile stages may stem from their cryptic appearance, their remote habitats, or both. Future development of proper methods to trap these stages ("collectors") and to observe them in their natural habitats, despite the limited accessibility (e.g., using Remotely Operated underwater Vehicles, see Spanier et al. 1994) may expand our knowledge on juvenile slipper lobsters. Also culturing slipper lobsters to their juvenile stage may enable observations of some facets of their biology and behavior, at least in captivity in naturalistic conditions. Since the juvenile stage in many species is identified as a "bottleneck" due to density-dependent mortality caused by shortage of shelter and food, it is important to know as much as possible about the settling postlarvae and early benthic juveniles to better understand recruitment processes. Thus, expanding the knowledge on juvenile stages of commercial species of slipper lobsters can also benefit the fishery management of these species.

10.6 Adult Behavior and Ecology

More is known about adult behavior of slipper lobsters than any other ontogenetic stage. This is because adults are captured more frequently and are often easy to keep under laboratory conditions. Nevertheless, most behavioral studies were conducted with limited numbers of commercially important species and represent the genera of only three subfamilies of Scyllaridae: *Scyllarides*, *Ibacus*, and *Thenus*. The majority of the studies were done with captive animals and only a few included observations and experimentation of slipper lobsters in their natural habitats. Lavalli et al. (Chapter 7) provides an extensive review of these behaviors, including feeding behavior, sheltering behavior and substrate preferences, mating behavior, intra- and interspecific interactions, diel-activity patterns, predators and antipredator behavior, daily and seasonal horizontal movement patterns, and swimming behaviors. Also the sensory basis of behavior in Scyllaridae is poorly known and knowledge on sensory biology of slipper lobster in essential sensory modalities such as chemoreception, mechanoreception, and vision is virtually absent compared to the information available regarding sensory modalities in clawed and spiny lobsters (see Lavalli et al., Chapter 7 for an expanded discussion). Future studies should be directed towards clarifying the sensory modalities that mediate behavior in slipper lobsters and to understand how these sensory modalities may differ from nephropid and palinurid lobsters. More emphasis should be directed into the study of behavior of a variety of slipper lobster species in their natural habitats. Since lobsters are nocturnally active, this presents a methodological difficulty. The use of underwater videos operating at very row-light levels or quiet, underwater Remotely Operated Vehicles (with brushless motors) with ultraviolet lights may, at least partially, solve this technical problem (see Spanier et al. 1994; Jury et al. 2001).

10.7 Morphology and Anatomy

Descriptions of morphological and anatomical features of lobsters, depending on the ontogenetic phase, are frequently by-products of taxonomic studies (e.g., Holthuis 1991; Webber & Booth, Chapter 2), of early life-history studies (e.g., Sekiguchi et al., Chapter 4 and references therein), and of behavioral or neuroethological studies (Lavalli et al., Chapter 7). Yet these studies, besides supplying limited information on selected species, life phases, and organs or level of organization, frequently fail to indicate the functions of the structure described. It is clear that there is a need for more specific, detailed studies referring to functional morphological and anatomical features of appendages and organ systems (e.g., the digestive system) of slipper lobsters. Focusing on a single system, such as the feeding apparatus and associated digestive system (Johnston, Chapter 6) can be a useful approach, although the great diversity of Scyllaridae and the great number of species, most of them unstudied, makes it extremely difficult to obtain a general picture of such systems, if one exists. It is clear that additional studies in a range of species with various adaptations are needed to understand evolutionary and ecological trends. Lavalli & Spanier (Chapter 1) indicate that no internal diagrams of scyllarid anatomy exist, but it is presumed that they follow the same internal body plan of other lobsters. Future studies should also aim to ascertain if this assumption is correct.

10.8   Physiology, Reproduction, and Growth

Sadly, few studies on basic physiological topics exist for slipper lobsters (e.g., Haond et al. 2001 and Faulkes 2004 are but two of the few). None of the articles published in the physiology sections of the Proceeding of the 5th, 6th, or 7th International Conference and Workshop on Lobster Biology and Management concerned slipper lobsters. Those present refer only to spiny and clawed lobsters where already much knowledge has been accumulated. This volume supplies some basic information on the adult and larval digestive systems and physiology (Johnston, Chapter 6) of a limited number of species; on the mineralization and biomechanics of the exoskeleton (Horne & Tarsitano, Chapter 8); on growth, molt increments, and reproduction maturity (Chapters 9, 14, 15, 16, 17); and provides some considerations of sensory modalities (Chapter 7). Using the example set by scientific achievements in the physiology of clawed and spiny lobsters (e.g., Waddy et al. 1995; Paterson et al. 1997; O'Grady et al. 2001; Kirubagaran et al. 2005), future research should aim at expanding knowledge in several key topics. These should include factors influencing growth and molt, with an emphasis on endocrinological control, reproductive hormones, and sexual pheromones; embryological development; blood physiology and responses to disease; respiratory physiology; reproductive physiology; muscle development, innervation, and control; and sensory modalities and their underlying neurophysiological control systems. The outcome of future physiological studies in Scyllaridae may have also practical implications in the aquaculture of slipper lobsters (e.g., reproductive physiology and embryology) and fisheries (e.g., problems of live transport of lobsters, see Taylor et al. 1997).

10.9   Conclusions

Considerable progress has been made in the recent years with respect to the biological knowledge of slipper lobsters. These developments are reflected in studies reviewed in the chapters of the present volume. Yet, because clawed and spiny lobsters are of significantly higher economic importance, the knowledge accumulated on the various key biological aspects (taxonomy, phylogeny and evolution; morphology and anatomy; larval, juvenile, and adults ecology and behavior; genetics; physiology; reproduction and growth) is notably more extensive. Where studies exist for slipper lobsters, they also often focus on a few, relatively large species that are of commercial interest (e.g., *Scyllarides* spp., *Ibacus* spp., and *Thenus* spp.). Slipper lobster researchers are challenged with the high number of species (85 recognized at present, see Webber & Booth, Chapter 2), the diverse habitats that they occupy, and the different adaptations associated with these surroundings. This diversity makes it difficult to describe a unified biological profile for the whole scyllarid family. However, it is possible that this evolutionary successful family of lobsters is actually characterized by flexibility that has made its numerous species adapt for living in many different habitats. Thus, it might be more profitable to conduct comparative studies of species that occupy similar habitats and have similar ecologies. Because populations are made up of individuals that are biologically successful based on their genetics and the expression of their genes (via their physiology, morphology, and behavior) as those genes interact with the environment, improved understanding of basic biology should ultimately prove profitable to management models for scyllarids (especially newer, individual-based, cradle-to-grave type models) and should help us preserve these species even when exploited.

References

Acosta, C.A. & Butler, M.J. 1999. Adaptive strategies that reduce predation of spiny lobster postlarvae during onshore transport. *Limnol. Oceanogr.* 44: 494–501.

Acosta, C.A., Matthews, T.R., & Butler, M.J., IV. 1997. Temporal patterns and transport processes in recruitment of spiny lobster, *Panulirus argus*, postlarvae to south Florida. *Mar. Biol.* 129: 79–85.

Bianchini, M.L., Ragonese, S., Greco, S., Chessa, L., & Biagi, F. 1997. Valutazione della fattibilità e potenzialità del ripopolamento attivo per la magnosa, *Scyllarides latus* (Crostacei Decapodi). *Final Rep. MiRAAF (Pesca Marittima)*: 1–145.

Bianchini, M.L., Greco, S., & Ragonese, S. 1998. Il progetto "Valutazione della fattibilità e potenzialità del ripopolamento attivo per la magnosa, *Scyllarides latus* (Crostacei Decapodi)": Sintesi e risultati. *Biol. Mar. Medit.* 5: 1277–1283.

Bianchini, M.L., Spanier, E., & Ragonese, S. 2003. Enzymatic variability of Mediterranean slipper lobsters, *Scyllarides latus*, from Sicilian waters. *Ann. Istrian Med. Stud.* 13: 43–50.

Booth, J.D., Webber, W.R., Sekiguchi, H., & Coutures, E. 2005. Diverse larval recruitment strategies within the Scyllaridae. *N. Z. J. Mar. Freshw. Res.* 39: 581–592.

Butler, M.J., IV & Herrnkind, W.F. 2000. Puerulus and juvenile ecology. In: Phillips, B.F. & Kittaka, J. (eds.), *Spiny Lobsters: Fisheries and Culture*, 2nd Edition: pp. 276–301. Oxford: Blackwell Scientific Press.

Butler, M.J., IV, Herrnkind, W.F., & Hunt, J.H. 1997. Factors affecting the recruitment of juvenile Caribbean spiny lobsters dwelling in macroalgae. *Bull. Mar. Sci.* 61: 3–19.

Cobb, J.C. & Phillips, B.F. 1980. Preface. In: Cobb, J.C. & Phillips, B.F. (eds.), *The Biology and Management of Lobsters*, Vol. 1: pp. xi–xiii. New York: Academic Press.

Cowan, D., Solow, A.R., & Beet, A. 2001. Patterns in abundance and growth of juvenile lobster, *Homarus americanus*. *Mar. Freshw. Res.* 52: 1095–1102.

Diaz, D., Zabala, M., Linares, C., Hreu, B., & Abelló, P. 2005. Increased predation of juvenile European spiny lobsters (*Palinurus elephas*) in a marine protected area. *N. Z. J. Mar. Freshw. Res.* 39: 447–453.

Eggleston, D.B., Lipcius, R.N., Marshal, & Ratchford, S.G. 1998. Spatiotemporal variation in postlarval recruitment of the Caribbean spiny lobsters in the central Bahamas: Lunar and seasonal periodicity, spatial coherence, and wind forcing. *Mar. Ecol. Prog. Ser.* 174: 33–49.

Faulkes, Z. 2004. Loss of escape responses and giant neurons in the tailflipping circuits of slipper lobsters, *Ibacus* spp. (Decapoda, Palinura, Scyllaridae). *Arthrop. Struct. Develop.* 33: 113–123.

Fiedler, U. & Spanier, E. 1999. Occurrence of *Scyllarus arctus* (Crustacea, Decapoda, Scyllaridae) in the eastern Mediterranean- preliminary results. *Ann. Istrian Med. Stud.* 17: 153–158.

Freitas, R. & Castro, M. 2005. Occurrence of *Panulirus argus* (Latreille, 1804) (Decapoda, Palinuridae) in the northwest islands of the Cape Verde Archipelago (Central-East Atlantic). *Crustaceana* 78: 1191–1201.

Galil, B., Pisanty, S., Spanier, E., & Tom, M. 1989. The Indo-Pacific lobster *Panulirus ornatus* (Fabricius, 1798) (Crustacea: Decapoda) a new Lessepsian migrant to the Eastern Mediterranean. *Israel J. Zool.* 35: 241–243.

George, R.W. 2005. Evolution of life cycles, including migration, in spiny lobsters (Palinuridae). *N. Z. J. Mar. Freshw. Res.* 39: 503–514.

Haond, C., Charmantier, G., Flik, G., & Wendelaar Bonga, S.E. 2001. Identification of respiratory and ion-transporting epithelia in the phyllosoma larvae of the slipper lobster *Scyllarus arctus*. *Cell Tissue Res.* 305: 445–455.

Holthuis, L.B. 1991. *Marine Lobsters of the World. An Annotated and Illustrated Catalogue of the Species of Interest to Fisheries Known to Date*. FAO Species Catalogue No. 125, Vol 13: pp. 1–292. Rome: Food and Agriculture Organization of the United Nations.

Incze, L.S., Wahle, R.A., & Cobb, J.S. 1997. Quantitative relationships between postlarval supply and benthic recruitment in the American lobster. *Mar. Freshw. Res.* 48: 729–743.

Jeffs, A.G., Mongomery, J.C., & Tindle, C.T. 2005. How do spiny lobster post-larvae find the coast?. *N. Z. J. Mar. Freshw. Res.* 39: 619–628.

Jørstad, K.E., Farestveit, E., Kelly, E., & Triantaphyllidis, C. 2005. Allozyme variation in European lobster (*Homarus gammarus*) throughout its distribution range. *N. Z. J. Mar. Freshw. Res.* 39: 515–526.

Jury, S.H., Howell, H., O'Grady, D.F., & Watson, W.H., III. 2001. Lobster trap video: *in situ* video surveillance of the behavior of *Homarus americanus* in and around traps. *Mar. Freshw. Res.* 52: 1125–1132.

Kirubagaran, R., Peter, D.M., Dharni, G., Vinithkumar, N.V., Sreeraj, G., & Ravindran, M. 2005. Changes in vertebrate-type steroids and 5-hydroxytryptamine during ovarian recrudescence in the Indian spiny lobster, *Panulirus homarus*. *N. Z. J. Mar. Freshw. Res.* 39: 527–537.

Lyons, W.G. 1970. Scyllarid lobsters (Crustacea, Decapoda). *FL Mar. Res. Lab., Mem. Hourglass Cruises* 1: 1–74.

O'Grady, D.F., Jury, S.H., & Waston, W.H. 2001. Use of treadmill to study the relationship between walking, ventilation and heart rate in the lobster *Homarus americanus*. *Mar. Freshw. Res.* 52: 1387–1394.

Ovenden, J.R., Booth, J.D., & Smolenski, A.J. 1997. Mitochondrial DNA of red and green rock lobsters (genus *Jasus*). *Mar. Freshw. Res.* 52: 1131–1136.

Ptacek, M.B., Sarver, S.K., Childress, M.J., & Herrenkind, W.F. 2001. Molecular phylogeny of the spiny lobster genus *Panulirus* (Decapoda: Palinuridae). *Mar. Freshw. Res.* 52: 1037–1047.

Patek, S.N. & Oakley, T.H., 2003. Comparative tests of evolutionary trade-offs in a palinurid lobster acoustic system. *Evolution* 57: 2082–2100.

Paterson, B.D., Grauf, S.G., & Smith, R.A. 1997. Haemolymph chemistry of tropical rock lobsters (*Panulirus ornatus*) brought to a mother ship from a catching dinghy in Torres Strait. *Mar. Freshw. Res.* 48: 835–838.

Phillip, B. 2005. Lobsters: The search for knowledge continues (and why we need to know!). *N. Z. J. Mar. Freshw. Res.* 39: 231–241.

Spanier, E. & Lavalli, K.L. 1998. Natural history of *Scyllarides latus* (Crustacea Decapoda): A review of the contemporary biological knowledge of the Mediterranean slipper lobster. *J. Nat. Hist.* 32: 1769–1786.

Spanier, E., Cobb, J.S., & Clancy, M. 1994. Impacts of remotely operated vehicles on the behavior of marine animals: An example using American lobsters. *Mar. Ecol. Prog. Ser.* 104: 257–266.

Tam, Y.K. & Kornfield, I. 1998. Phylogenetic relationships of clawed lobster genera (Decapoda, Nephropidae) based on mitochondrial 16S rRNA gene sequences. *J. Crust. Biol.* 18: 138–146.

Taylor, H.H., Paterson, B.D., Wong, R.J., & Wells, R.M.G. 1997. Physiology and live transport of lobsters: Report from a workshop. *Mar. Freshw. Res.* 48: 817–822.

Waddy, S.L., Aiken, D.E., & de kleijn, D.P.V. 1995. Control of growth and reproduction In: Factor, J.R. (ed.), *The Biology of the Lobster Homarus americanus*: pp. 217–266. New York: Academic Press.

Wahle, R.A. & Incze, L.S. 1997. Pre- and post-settlement processes in recruitment of the American lobster. *J. Exp. Mar. Biol. Ecol.* 217: 179–207.

Yeung, C., Jones, D.L., Criales, M.M., Jackson, T.L., & Richards, W.J. 2001. Influence of coastal eddies and counter-currents on the influx of spiny lobster, *Panulirus argus*, postlarvae into Florida Bay. *Mar. Freshw. Res.* 52: 1217–1232.

Part III

Fisheries of Slipper Lobsters

11

Observations on the Ecology of Scyllarides aequinoctialis, Scyllarides nodifer, and Parribacus antarcticus and a Description of the Florida Scyllarid Lobster Fishery

William C. Sharp, John H. Hunt, and William H. Teehan

CONTENTS

Abstract

We conducted a two-year study from 1987 through 1989 of the lobster populations within Looe Key National Marine Sanctuary, Florida, U.S.A. and collected information on the population structure, sex ratios, reproductive dynamics, and sheltering behavior of *Scyllarides aequinoctialis* (Lund, 1793), *S. nodifer* (Stimpson, 1866), and *Parribacus antarcticus* (Lund, 1793). We observed few ovigerous females of any scyllarid species, although we did document evidence of repetitive spawning without molting in *P. antarcticus*. *Scyllarides aequinoctialis* and *S. nodifer* tended to be encountered more frequently in different areas of the reef complex. *Scyllarides aequinoctialis* was primarily encountered sheltering within complex, high-relief, coral habitat, whereas *S. nodifer* appeared to be less specific in its sheltering preference and more commonly observed near unconsolidated sediments. Little evidence of gregarious behavior was observed in any species, as virtually all our observations were of individuals sheltering solitarily. *Scyllarides aequinoctialis* is apparently capable of a high degree of den fidelity, as a single individual was encountered sheltering in the same den on 23 of the 29 occasions that it was searched by divers during a 22-month period.

Scyllarid lobsters constitute only a minor fraction of Florida's lobster landings, and are landed primarily along the state's west coast as by-catch from the shrimp fishery.

In the Florida Keys and along the east coast of the state, they are occasionally harvested from spiny lobster traps and by divers. Given the dearth of information regarding scyllarid lobsters in Florida, the status of the fishery is unknown, and there remains need for further life-history information, especially regarding recruitment dynamics, growth rates, and reproductive biology before an effective assessment can be undertaken.

11.1   Introduction

Seven species of scyllarid lobsters representing four genera have been reported to occur in the coastal waters of Florida: *Scyllarides aequinoctialis* (Lund, 1793), *Scyllarides nodifer* (Stimpson, 1866), *Parribacus antarcticus* (Lund, 1793), *Scyllarus americanus* (S.I. Smith, 1869), *Scyllarus chacei* Holthuis, 1960, *Scyllarus depressus* (S.I. Smith, 1881), and *Arctides guineensis* (Spengler, 1799) (Williams 1984; Holthuis 1991). An eighth species, *Scyllarus faxoni* Bouvier, 1917, may also occur in the eastern Gulf of Mexico (Lyons 1970), but its presence there remains unconfirmed. Compared to the Caribbean spiny lobster (Palinuridae), *Panulirus argus* (Latreille, 1804), which supports one of the state's most valuable fisheries (Muller et al. 1997), scyllarid lobsters are of limited economic importance and are primarily landed as by-catch from other fisheries. Consequently, there has been comparatively limited amount of published information on these species.

Detailed taxonomic descriptions of adults for the earlier-mentioned species are available (Lyons 1970; Williams 1984), and there are numerous descriptions of the morphology and development of their phyllosomas (Robertson 1968, 1969a, 1969b, 1971; Lyons 1970; Johnson 1971; Sandifer 1971; Cline et al. 1978). However, the information available on the biology and ecology on any one of these species is fragmentary. Booth et al. (2005) reviewed much of what is known of scyllarid early life history and provided a thorough discussion of the larval recruitment strategies within the family (see also Sekiguchi et al., Chapter 4 for a similar review). Additionally, some effort has been directed at documenting the distribution of scyllarid phyllosomas in the coastal waters of the Florida Keys and understanding the transport and retention mechanisms as they relate to recruitment (Yeung & McGowan 1991; Yeung et al. 2000).

Information available on the biology and ecology of the postlarval stages of any species of scyllarid lobster is even more limited (see review by Booth et al. 2005, and references therein; see also Sekiguchi et al., Chapter 4). The swimming behavior of captive nistos has been described (e.g., Robertson 1968), and the postlarval duration is known for a few species that occur in Florida (Lyons 1970), but apparently there has been nothing published on their settlement habitat. Nistos are only occasionally captured on collectors designed to capture pueruli (Phillips & Booth 1994). We, too, have observed that even though the postlarvae of *Scyllarides nodifer* occur in the plankton of the nearshore waters of the Florida Keys (Yeung & McGowan 1991), and are caught in plankton nets around the interisland channels, they are rarely encountered on postlarval collectors that target *Panulirus argus* postlarvae by mimicking architecturally complex vegetative habitat favored by that species. As a result, we concur with Booth et al. (2005) that scyllarid settlement habitat is different from the palinurids.

Much of what has been documented on adult biology and ecology of scyllarids in Florida is based largely upon studies that relied upon examining individuals collected as by-catch of commercial fishing operations, on research cruises employing commercial fishing gears, or by observations of captive individuals held in aquaria. Lyons (1970) described the reproductive dynamics and other life-history parameters of several species of scyllarids from samples collected in research trawls along the Florida shelf in the eastern Gulf of Mexico. Some information exists on the feeding behavior of *S. nodifer*, *S. aequinoctialis*, and *Parribacus antarcticus* (Rudloe 1983; Lau 1988; Malcom 2003; see also Lavalli et al., Chapter 7), as does information on the sensory biology for the former two species (Malcom 2003; Weisbaum & Lavalli 2004; see also Lavalli et al., Chapter 7). Increased interest in a by-catch fishery for *Scyllarides nodifer* in the northern Gulf of Mexico during the 1980s resulted in some directed research on the species that included work on its physiology and biochemistry (Cline 1980; Cline & Hinton 1983; Hardwick & Cline 1984, 1985, 1986;

Cline & Hardwick 1985), reproduction (Cline et al. 1978; Hardwick & Cline 1990), behavior (Ogren 1977; Gilbert 1982; Malcom 2003), growth, and mariculture (Rudloe 1983).

This chapter is partitioned into two sections. First, we present previously unpublished information on scyllarid lobsters collected as part of a study documenting the population dynamics of spiny lobsters dwelling in Looe Key National Marine Sanctuary in the Florida Keys. We present observations of substrate preference and sheltering behavior of three reef-dwelling species: *S. aequinoctialis*, *S. nodifer*, and *P. antarcticus*. Although in many instances these data are quite limited, they either represent previously unknown aspects of the life history of these species or augment the scant published information. Second, we describe the Florida scyllarid lobster fishery by examining landings information collected by the National Marine Fisheries Service (NMFS), the Florida Marine Fisheries Information System (MFIS), Florida Fish & Wildlife Conservation Commission (FWC) fishery observers stationed on spiny lobster fishing vessels, and personal accounts provided to us by commercial fishers.

11.2 Observations on the Life History and Sheltering Behavior of Adult, Reef-Dwelling Scyllarid Lobsters

We conducted a two-year study from 1987 through 1989 of the lobster populations within Looe Key National Marine Sanctuary, Florida, U.S.A. The Sanctuary is located approximately 9 km south of Big Pine Key, Florida, in the Florida Keys archipelago, and encompasses approximately 18.2 km^2 of spur-and-groove coral-bank reef and associated habitats (Figure 11.1 shows the sanctuary and habitats within). Over the course of the study, these habitats were intensively sampled using SCUBA, and the sheltering behavior and habitat preference of lobsters dwelling there were documented. Deeper areas of the Sanctuary (~30 m), where sampling with SCUBA was less practical, were sampled using spiny lobster traps. Lobsters were captured at a predetermined portion of the sites to examine size structure and reproductive status (see Sharp et al. 1997 for sampling methodologies). Lobsters at the remaining sites were only visually assessed to avoid affecting their behavior. Although the study focused on the Caribbean spiny lobster, *Panulirus argus*, we also collected information on *Scyllarides aequinoctialis*, *S. nodifer*, and *Parribacus*

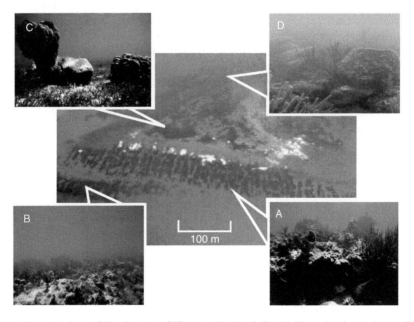

FIGURE 11.1 **(See color insert following page 242)** Looe Key Reef, Florida Keys, showing typical scyllarid habitats across the reef system. (A) High-relief spur-and-groove buttress zone, (B) low-relief spur-and-groove, (C) reef flat, and (D) patch-reef habitat.

antarcticus when these species were encountered. We present information on their population structure, sex ratios, reproductive dynamics, and sheltering behavior.

11.2.1   Population Structure

A total of 123 observations (including multiple sightings of the same individuals) of scyllarid lobsters were made during the two-year study. Of that total, 93 were of *Scyllarides aequinoctialis*, 26 were of *S. nodifer*, and four were of *Parribacus antarcticus*. From this total, we captured 34 *S. aequinoctialis,* 17 *S. nodifer*, and all four of the *P. antarcticus* to assess size structure and reproductive status. The *S. aequinoctialis* ranged in size from 63 to 117 mm carapace length (CL) (mean ±1, SE = 92.0 ± 2.0 mm CL). Of these, 13 were females (mean ±1, SE = 97.7 ± 2.9 mm CL; range = 83 to 117 mm CL) and 21 were males (mean ±1, SE = 88.7 ± 2.5 mm CL; range = 63 to 108 mm CL). The *S. nodifer* specimens ranged in size from 56 to 95 mm CL (mean ±1, SE = 78.9 ± 2.2 mm CL). Of these, nine were females (mean ±1, SE = 81.4±3.8 mm CL; range = 67 to 83 mm CL) and eight were males (mean ±1, SE = 76.1±1.7 mm CL; range = 63 to 108 mm CL). The *P. antarcticus* ranged in size from 60 to 83 mm CL (mean ±1, SE = 73.5 ± 5.2 mm CL). Of these, three were females (mean ±1, SE = 78.0 ± 3.6 mm CL; range = 71 to 83 mm CL) and one was male (60 mm CL).

11.2.2   Reproductive Dynamics

We observed few ovigerous females of any scyllarid species. Of the 13 captured female *Scyllarides aequinoctialis*, three were egg bearing. These ranged from 90 to 98 mm CL, and were captured during April and May. Two egg-bearing *S. nodifer* were observed and were 82 and 87 mm CL. Both were observed during July, which is within the reproductive season previously described for the species in the Gulf of Mexico (Lyons 1970; Hardwick & Cline 1990). No evidence of spermatophores was detected on the individuals of either species. Lyons (1970) reported the absence of observable spermatophores on ovigerous *S. nodifer*, and speculated that fertilization in this species is internal. *Scyllarides squammosus* (H. Milne Edwards, 1837) also appears to lack external spermatophores (DeMartini et al. 2005). In contrast, *Scyllarides latus* (Latreille, 1802) females have been observed bearing a spermatophore up to 10 days prior to extruding eggs (Martins 1985; Almog-Shtayer 1988; Spanier & Lavalli 1998), which suggests that different reproductive strategies may exist within the genus (Spanier & Lavalli 2006). We suggest that it is possible that the spermatophores of some species remain gelatinous and highly ephemeral, similar to those of the palinurid, *Jasus edwardsii* (Hutton, 1875) (Berry & Heydorn 1970).

Of the three female *Parribacus antarcticus* we encountered, two, both captured during July, bore evidence of reproductive activity. One (80 mm CL) clearly had recently spawned as evidenced by remnants of eggs on her pleopods, and possessed a hardened and eroded spermatophore located on the first abdominal somite. An external spermatophore located on the abdominal somite has been previously noted in this species (Lyons 1970). The other (71 mm CL) was bearing eggs and had a fresh spermatophore deposited atop a presumably eroded one (Figure 11.2), suggesting repetitive spawning without molting.

11.2.3   Habitat Preference and Sheltering Behavior

Scyllarides aequinoctialis and *S. nodifer* tended to be encountered more frequently in different areas of the reef complex. *Scyllarides aequinoctialis* clearly preferred shelters located within complex, high-relief, coral habitat to those in comparatively lower relief or unconsolidated sediments. Of the 91 individuals of *S. aequinoctialis* we observed, 23 were found sheltering within crevices along the shallow (<7 m water depth), spur-and-groove buttress zone (Figure 11.1A shows this shallow type of habitat). Fifty-four were found within a system of patch reefs lying inland of the buttress zone that consisted of either isolated coral mounds of varying structural complexity surrounded by sand or numerous smaller coral colonies located in large expanses of hard bottom (Figure 11.1D shows this patch reef habitat). In contrast, only nine individuals were encountered in the deeper (15 to 20 m water depth), low-relief, "drowned" spur-and-groove coral habitat lying seaward of the buttress zone (Figure 11.1B). Four lobsters were encountered in reef flat rubble areas (Figure 11.1C). The remaining individual was captured in a spiny lobster trap

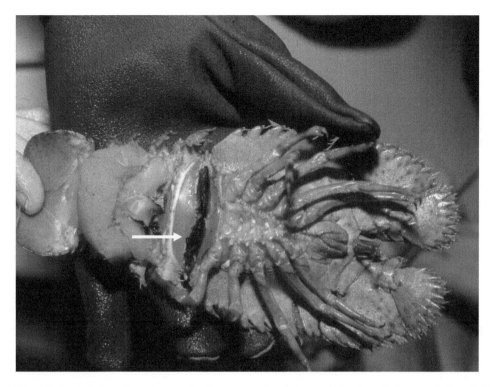

FIGURE 11.2 Ventral view of an ovigerous *Parribacus antarcticus* (71 mm CL) bearing a fresh spermatophore that had been deposited atop an eroded one.

placed at 30 m depth on unconsolidated sediments outside the seaward extent of the reef. We noted that this individual was one of the ovigerous individuals that we encountered, and that one of the other two ovigerous females was encountered in the deeper area of the reef seaward of the buttress zone. Although we acknowledge that our observations are limited, we speculate that ovigerous *S. aequinoctialis* may move to deeper waters to spawn.

Scyllarides nodifer, in contrast, appeared to be less specific in its sheltering preference than *S. aequinoctialis*. Ten (38%) of the 26 lobsters we encountered were observed on patch reefs and 14 (54%) were captured in spiny lobster traps set on unconsolidated sediments outside the seaward edge of the deep reef (30 m depth). One individual was encountered on a low-relief and another in high-relief spur-and-groove coral habitat. This distribution is consistent with the observations of Ogren (1977), who noted that *S. nodifer* may leave their daytime shelters in hard-bottom structure to forage on sediments at night. Hardwick & Cline (1990) noted both an absence of structure in the vicinity of their study population of *S. nodifer* in the northern Gulf of Mexico, and further reported that many of the individuals they observed were caked in mud. They speculated that *S. nodifer* there might bury in sediments during daylight hours (see Jones, Chapter 16 for a description of how scyllarids bury into substrates).

Although we only infrequently encountered *Parribacus antarcticus*, they seemed to prefer high-relief coral structure, similar to *S. aequinoctialis*. Of the four *P. antarcticus* that were encountered, three were found on the reef-buttress zone; the remaining individual was captured in a spiny lobster trap along the seaward edge of the reef. We noted that two of these four lobsters were encountered during night-time dives and were apparently foraging on the reef away from their shelters. We suspect that this species seeks particularly secretive shelters during the daytime, and that its presence often was undetected to a greater extent by divers conducting searches during the daytime than the other species.

Although gregarious behavior in *Scyllarides latus* has been well documented (Spanier et al. 1988, 1990; Spanier & Almog-Shtayer 1992), and reports exist of similar gregarious behavior in *S. nodifer* (Moe 1991), we observed virtually no instances of such behavior in either *S. aequinoctialis* or *S. nodifer*.

Of the 86 instances we observed *S. aequinoctialis* sheltering in a den, 84 were of a single individual, and the remaining two instances were of two individuals. *Scyllarides nodifer* was observed sheltering a total of 12 times, and in all instances the den was occupied by a solitary individual. *Parribacus antarcticus* was observed sheltered only once, and it was cohabiting with the obligate reef-dwelling palinurid lobster, *Panulirus guttatus* (Latreille, 1804). We note, however, that we have observed two *P. antarcticus* cohabiting within a shelter on other occasions.

Simultaneous occupancy of dens between scyllarid and palinurid lobsters was also rare. We observed *Panulirus argus* and *P. guttatus* sheltering within the habitats described earlier on 5097 occasions, and scyllarids on 99 occasions, but observed simultaneous cohabitation in a den by both a scyllarid and a palinurid lobster on only seven occasions. On those few occasions, each taxa usually occupied positions opposite one another within the den.

Both *Scyllarides aequinoctialis* and *S. nodifer* were rarely observed occupying the floor of dens, and *S. aequinoctialis* in particular showed a particular affinity for den ceilings, a sheltering behavior in marked contrast with *P. argus*, but similar to *P. guttatus* (see Sharp et al. 1997). Of the 88 *S. aequinoctialis* we observed in dens, 55 were on the ceiling, 18 were on the wall, and 15 were on the floor. Of our 12 observations of *S. nodifer* within a den, it was observed on the ceiling six times, the den wall five times, and once on the den floor. The lone *Parribacus antarcticus* we observed sheltering occupied the den wall.

Scyllarides aequinoctialis is apparently capable of a high degree of den fidelity. During the course of one year, divers searched one lobster den located on a patch reef on 18 occasions, usually at two-week intervals, and observed one *S. aequinoctialis*, which we were confident was the same individual, on 17 of those occasions. In all, the den was searched by divers a total of 29 times over 22 months, and the lobster was observed on 23 of those occasions. We captured this individual at the end of the study and found it was a male measuring 108 mm CL.

11.3   Florida's Scyllarid Lobster Fishery

Scyllarid lobsters constitute only a minor fraction of Florida's lobster landings, which are dominated by *Panulirus argus*. This spiny lobster supports an intensive trap and dive fishery along the state's southeastern coast, especially along the Florida Keys archipelago. Since 1980, the *P. argus* fishery has landed an average of nearly 2722 tons, with an average ex-vessel value estimated to be US$21 million. In contrast, annual landings of scyllarid lobsters over the same time period have averaged 17 tons, with an annual value of US$82,000 (National Marine Fisheries Service, General Canvas).

The vast majority of Florida's scyllarid landings occur along the state's west coast (NMFS) (Figure 11.3 shows landings by region), and are landed primarily in trawls as by-catch of the shrimp fishery (Hardwick & Cline 1990; Moe 1991) (Figure 11.4A shows landings by gear type for the west coast of Florida). Although landings of scyllarid lobsters are not identified by species, landings along the west coast north of the Florida Keys are undoubtedly composed largely of *Scyllarides nodifer*, as it is the only species of scyllarid lobster that commonly occurs there and attains a size sufficient to be exploited for the seafood industry (Lyons 1970). In the Florida Keys, scyllarid lobsters are landed primarily as by-catch in spiny lobster traps (Figure 11.4B shows landings by gear type for the Florida Keys; Figure 11.5 shows the typical spiny lobster trap used in the Florida Keys) and the species making up this by-catch consist of *S. nodifer*, *S. aequinoctialis*, and *Parribacus antarcticus* (FWC, unpublished data). However, these species are infrequently encountered in this gear and account for only a small proportion of total landings. For example, during one spiny lobster fishing season, fishery observers encountered only 155 scyllarid lobsters in the approximately 21,000 spiny lobster traps they sampled (Matthews et al. 2005). Landings of scyllarid lobsters along the east coast are landed in approximately equal proportions by spiny lobster traps and by divers (Figure 11.4C shows landings by gear type for the east coast of Florida).

A part-time fishery developed for *S. nodifer* in the Gulf of Mexico during the 1980s (Moe 1991), and landings increased rapidly during the early part of the decade (Figure 11.3 shows landings by coast and by year). Hardwick & Cline (1990) noted that shrimpers operating along the west coast of the state directed

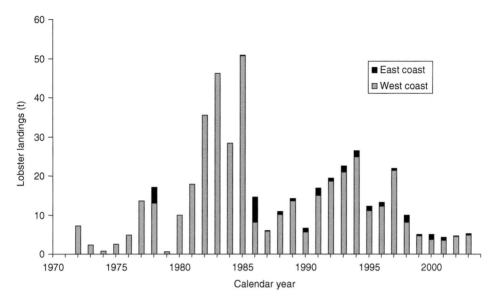

FIGURE 11.3 Landings of slipper lobsters from the east and west coast of Florida from 1972 to 2003 reported by the National Marine Fisheries Service (NMFS).

fishing effort toward *S. nodifer* during the spring and summer in their "off-season". They provided a detailed description of one such fishing operation and noted that fishers modified their shrimp trawls to withstand the habitat in which they had frequently encountered *S. nodifer*, and then concentrated their fishing effort in those specific areas. The authors also noted that ovigerous females were commonly encountered during these time periods (spring and summer), and observed females with clipped pleopods (i.e., eggs removed) in seafood markets. They suggested that the fishery was exploiting reproductively active lobsters that were migrating into warmer, shallower waters.

Landings of scyllarid lobsters along the west coast peaked in 1985, and then decreased rapidly during the next several years. While landings again progressively increased during the 1990s, they never approached those of the peak landings years (Figure 11.3). The reduction in landings may be related to regulatory changes implemented during 1987 that prohibited both the possession of ovigerous *S. nodifer* and the removal of eggs by clipping their pleopods. Additionally, since 1990, the state has required turtle-excluding devices (TEDS) on shrimp trawls that may have also reduced the efficiency with which the gear captures lobsters. Annual scyllarid lobster landings statewide have remained below 5 tons since 1999 as a decline in the Gulf of Mexico shrimp fishery has resulted in fewer fishing trips targeting shrimp (Figure 11.6 shows the decline in the number of shrimping trips by year).

In addition to being landed for the seafood market, there also exists a small live-market fishery for scyllarid lobsters for the marine aquarium trade. Since 1990, the MFIS has recorded landings of scyllarid lobsters for this fishery (Figure 11.7A to Figure 11.7C show the aquarium market landings by gear type). As with the landings marketed for the seafood industry, most of these landings are by-catch from the west coast shrimp fishery. Divers collect the majority of scyllarids sold to the aquarium market from the remainder of the state regions, but annual landings by these collectors have typically been no greater than a few hundred lobsters annually. In the Florida Keys, divers land *S. nodifer*, *S. aequinoctialis*, and *P. antarcticus*. *Scyllarus americanus* landed as by-catch from the bait shrimp fishery is also occasionally marketed for the aquarium trade. Of these species, *S. nodifer* is the most popular with aquarists, but the demand for scyllarid lobster in general is limited (K. Nedimyer & F. Young, personal communication).

There is no information available regarding a recreational fishery for scyllarid lobsters. However, *P. argus* supports an intensive recreational dive fishery that typically attracts approximately 60,000 people statewide during the opening month of the lobster fishing season (Sharp et al. 2005), and some recreational

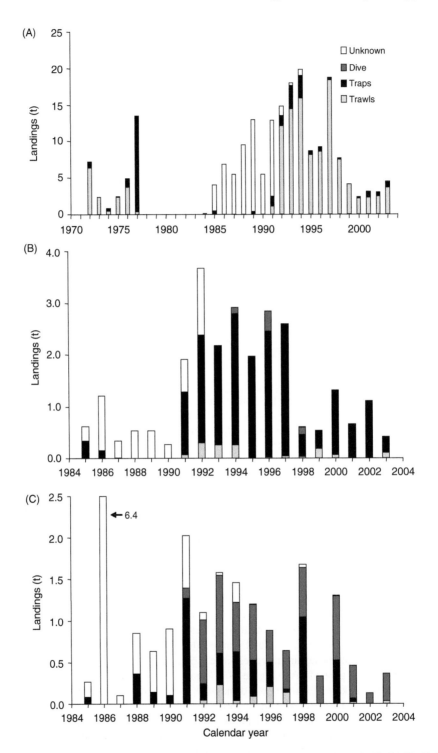

FIGURE 11.4   Gear-specific slipper lobster landings, 1972 to 2003. (A) the west coast of Florida, (B) the Florida Keys, and (C) the east coast of Florida. West coast landings from 1972 to 1976 are reported by the National Marine Fisheries Service and include landings from the Florida Keys. All landings from 1984 to 2003 are reported by the Florida Marine Fisheries Information System.

FIGURE 11.5   Two typical wood-slat lobster traps used in the Florida Caribbean spiny lobster (*Panulirus argus*) fishery.

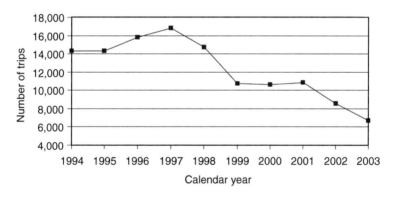

FIGURE 11.6   Number of fishing trips reporting food shrimp landed on the west coast of Florida reported by the Florida Marine Fisheries Information System, 1994 to 2003.

fishers will harvest scyllarid lobsters if they observe them. An intensive creel survey of recreational spiny lobster fishers conducted at the height of the *P. argus* fishing season in the Florida Keys, which are the most popular fishing grounds for the species, clearly indicated that scyllarids are not targeted by these fishers, as few observations of scyllarids were noted in their lobster catches. However, a popular regional outdoors magazine, *Florida Sportsman*, has occasionally run articles noting that scyllarid lobsters could be an attractive alternative to spiny lobsters for many recreational lobster fishers, especially for those living along the northern Gulf of Mexico. Nevertheless, because of their exceedingly cryptic sheltering behavior, scyllarid lobsters are unlikely to become a substantial recreational fishery in Florida. Yet, the status of this stock is completely unknown and there remains a need for further life history information, especially that regarding recruitment dynamics, growth rates, behavior, and reproductive biology before an effective assessment can be undertaken.

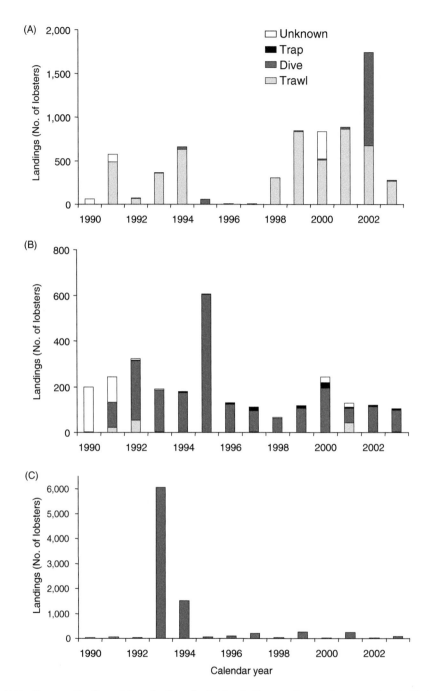

FIGURE 11.7 Gear-specific slipper lobster landings, landed for the live aquarium market reported to the Florida Marine Fisheries Information System 1990 to 2003. (A) the west coast of Florida, (B) the Florida Keys, and (C) the east coast of Florida.

Acknowledgments

We thank Rick Beaver, Barbara Blonder, Carolyn Schultz, Doug Heatwole, Brenda Hedin, and Tom Matthews for their assistance in collecting the data at Looe Key. The National Oceanic and Atmospheric

Administration, National Marine Sanctuary Program provided support for this research under contract 50-DGNC-6-00093. We thank Billy Causey, the former Looe Key National Marine Sanctuary manager, for his recognition that spiny lobsters are an excellent species for evaluating Marine Sanctuary management and for his support of this research. We also thank Forrest Young and Ken Nedimyer for sharing with us their experience in the marine aquarium trade. Finally, we thank Linda Simurra-Sharp and two anonymous reviewers for their cogent comments that greatly improved this chapter.

References

Almog-Shtayer, G. 1988. Behavioural-Ecological Aspects of Mediterranean Slipper Lobsters in the Past and of the Slipper Lobster Scyllarides latus in the Present. M.A. thesis: University of Haifa, Israel. 165 pp.

Berry, P.F. & Heydorn, A.E.F. 1970. A comparison of the spermatophoric masses and mechanisms of fertilization in southern African lobsters (Palinuridae). *Invest. Rept. Oceanogr. Res. Inst. South Africa* 25: 1–18.

Booth, J.D., Webber, W.R., Sekiguchi, H., & Coutures, E. 2005. Diverse larval recruitment strategies within the Scyllaridae. *N. Z. J. Mar. Freshw. Res.* 39: 581–592.

Cline, G.B. 1980. Chemical composition and cells of the hemolymph of the slipper lobster *Scyllarides nodifer*. *J. Ala. Acad. Sci.* 51: 1783.

Cline, G.B. & Hardwick, C.W., Jr. 1985. Phosphoglucomutase and phosphohexose isomerase isozymes of slipper lobsters *Scyllarides nodifer* from the Gulf of Mexico and *Scyllarides aequinoctialis* from the eastern Caribbean Sea. In: Herrnkind, W.F. (ed.), *Proceedings of a Workshop on Florida Spiny Lobster Research and Management, 1982*. Florida State University Marine Laboratory Technical paper, No. 32, p. 3.

Cline, G.B. & Hinton, J. 1983. A comparison of phosphglucomutuase and phosphohexose isomerase isozymes of muscle extracts of slipper lobsters, *Scyllarides nodifer* and *Scyllarides aequinoctialis*. *The Iso. Bull.* 16: 78.

Cline, G.B., Chilcutt, D., & Lindsay, R.A. 1978. Histological studies and rearing of phyllosomes of the slipper lobster *Scyllarides nodifer*. *J. Ala. Acad. Sci.* 49: 73.

DeMartini, E.E., McCracken, M.L., Moffitt, R.B., & Wetherall, J.A. 2005. Relative pleopod length as an indicator of size at sexual maturity in slipper (*Scyllarides squammosus*) and spiny Hawaiian (*Panulirus marginatus*) lobsters. *Fish. Bull.* 103: 23–33.

Gilbert, K.M. 1982. Behavioral observations on the slipper lobster Scyllarides nodifer (Stimpson), with notes on Scyllarus chacei (Holthuis) and Scyllarus depressus (Smith) (Decapoda: Palinuriodea). M.Sc. thesis: University of Alabama, Tuscaloosa, Alabama.

Hardwick, C.W., Jr. & Cline, C.B. 1984. Genetic characterization of a population of *Scyllarides nodifer* using isoelectric focusing (IEF) on some gene products *J. Ala. Acad. Sci.* 55: 140.

Hardwick, C.W., Jr. & Cline, C.B. 1985. Differences in isoelectric focusing and electrophoretic mobility of isozymes and proteins in tissue extracts of the slipper lobster, *Scyllarides nodifer* (Stimpson). *The Iso. Bull.* 18: 53.

Hardwick, C.W., Jr. & Cline, C.B. 1986. Isozymic analysis of the slipper lobster *Scyllarides nodifer* (Stimpson) from the northern Gulf of Mexico. *The Iso. Bull.* 19: 34.

Hardwick, C.W., Jr. & Cline, C.B. 1990. Reproductive status, sex ratios and morphometrics of the slipper lobster *Scyllarides nodifer* (Stimpson) (Decapoda: Scyllaridae) in the northeastern Gulf of Mexico. *Northeast Gulf Sci.* 11: 131–136.

Holthuis, L.B. 1991. *Marine Lobsters of the World. An Annotated and Illustrated Catalogue of the Species of Interest to Fisheries Known to Date*. FAO Species Catalogue No. 125, Vol. 13: pp. 1–292. Rome: Food and Agriculture Organization of the United Nations.

Johnson, M.W. 1971. The phyllosoma larvae of slipper lobsters from the Hawaiian Islands and adjacent areas (Decapoda, Scyllaridae). *Crustaceana* 20: 77–103.

Lyons, W.G. 1970. Scyllarid lobsters (Crustacea: Decapoda). Florida Marine Research Laboratory. *Mem. Hourglass Cruises* 1: 1–74.

Lau, C.J. 1988. Dietary comparison of two slow-moving crustacean (Decapoda: Scyllaridae) predators by a modified index of relative importance. *Proceedings of the 6th International Coral Reef Symposium, Vol. 2*. Townsville, Australia, 8–12 August 1988.

Malcom, C. 2003. Description of the Setae on the Pereiopods of the Mediterranean Slipper Lobster, *Scyllarides latus*, the Ridged Slipper Lobster, *S. nodifer*, and the Spanish Slipper lobster, *S. aequinoctialis*. M.Sc. thesis: Southwest Texas State University, San Marcos, Texas.

Martins, H.R. 1985. Biological studies of the exploited stock of the Mediterranean locust lobster *Scyllarides latus* (Latreille, 1803) (Decapoda: Scyllaridae) in the Azores. *J. Crust. Biol.* 5: 294–305.

Matthews, T.R., Cox, C., & Eaken, D. 2005. Bycatch in Florida's spiny lobster trap fishery. *Proceedings of the 47th Gulf and Caribbean Fisheries Institute* November 1994: pp. 66–78. Margarita, Venezuela.

Moe, M.A., Jr. 1991. *Lobsters: Florida, Bahamas, the Caribbean*. Plantation, FL: Green Turtle Publications. 510 pp.

Muller, R.G., Hunt, J.H., Matthews, T.R., & Sharp, W.C. 1997. Evaluation of effort reduction in the Florida Keys spiny lobster, *Panulirus argus*, fishery using an age structured population analysis. *Mar. Freshw. Res.* 48: 1045–1058.

Ogren, L.H. 1977. Concealment behaviour of the Spanish lobster, *Scyllarides nodifer* (Stimpson), with observations on its diel activity. *Northeast Gulf Sci.* 1: 115–116.

Phillips, B.F. & Booth, J.D. 1994. Design, use, and effectiveness of collectors for catching the puerulus stage of spiny lobsters. *Rev. Fish. Sci.* 2: 255–289.

Robertson, P.B. 1968. The complete larval development of the sand lobster, *Scyllarus americanus* (Smith), (Decapoda, Scyllaridae) in the laboratory, with notes on larvae from the plankton. *Bull. Mar. Sci.* 18: 294–342.

Robertson, P.B. 1969a. Biological investigations of the deep sea. No. 48. Phyllosoma larvae of a scyllarid lobster, *Arctides guineensis*, from the western Atlantic. *Mar. Biol.* 4: 143–151.

Robertson, P.B. 1969b. The early larval development of the scyllarid lobster *Scyllarides aequinoctialis* (Lund) in the laboratory, with a revision of the larval characters of the genus. *Deep-Sea Res.* 16: 557–586.

Robertson, P.B. 1971. The larvae and postlarvae of the scyllarid lobster *Scyllarus depressus* (Smith). *Bull. Mar. Sci.* 21: 841–865.

Rudloe, A. 1983. Preliminary studies of the mariculture potential of the slipper lobster, *Scyllarides nodifer.* *Aquaculture* 34: 165–169.

Sandifer, P.A. 1971. The first two phyllosomas of the sand lobster, *Scyllarus depressus* (Smith) (Decapods, Scyllaridae). *J. Eli. Mitch. Sci. Soc.* 87: 183–187.

Sharp, W.C., Hunt, J.H., & Lyons, W.G. 1997. Life history of the spotted spiny lobster, *Panulirus guttatus*, an obligate reef-dweller. *J. Mar. Freshw. Res.* 48: 687–698.

Sharp, W.C., Bertelsen, R.D., & Leeworthy, V.R. 2005. Long-term trends in the recreational lobster fishery of Florida, United States: landings, effort, and implications for management. *N. Z. J. Mar. Freshw. Res.* 39: 733–747.

Spanier, E. & Almog-Shtayer, G. 1992. Shelter preferences in the Mediterranean slipper lobster: Effects of physical properties. *J. Exp. Mar. Biol. Ecol.* 164: 103–116.

Spanier, E. & Lavalli, K.L. 1998. Natural history of *Scyllarides latus* (Crustacea: Decapoda): A review of the contemporary biological knowledge of the Mediterranean slipper lobster. *J. Nat. Hist.* 32: 1769–1786.

Spanier, E. & Lavalli, K.L. 2006. *Scyllarides* spp. In: Phillips, B.F. (ed.), *Lobsters: Biology, Management, Aquaculture, and Fisheries*: pp. 462–496. Oxford, London: Blackwell Publishing.

Spanier, E., Tom, M., Pisanty, S., & Almog-Shtayer, G. 1988. Seasonality and shelter selection by the slipper lobster *S. latus* in the south eastern Mediterranean. *Mar. Ecol. Prog. Ser.* 42: 247–255.

Spanier, E., Tom, M., Pisanty, S., & Almog-Shtayer, G. 1990. Artificial reefs in the low productive marine environment of the Southeastern Mediterranean. *P.S.Z.N.I. Mar. Ecol.* 11: 61–75.

Weisbaum, D. & Lavalli, K.L. 2004. Morphology and distribution of antennular setae of scyllarid lobsters (*Scyllarides aequinoctialis*, *S. latus*, and *S. nodifer*) with comments on their possible function. *Invert. Biol.* 123: 324–342.

Williams, A.B. 1984. *Shrimps, Lobsters and Crabs of the Atlantic Coast of the Eastern United Stated, Maine to Florida*: pp. 1–550. Washington, D.C.: Smithsonian Institution Press.

Yeung, C. & McGowan, M.F. 1991. Differences in inshore-offshore and vertical distribution of phyllosoma larvae of *Panulirus*, *Scyllarus* and *Scyllarides* in the Florida Keys in May–June, 1989. *Bull. Mar. Sci.* 49: 699–714.

Yeung, C., Criales, M.M., & Lee, T.N. 2000. Unusual level abundance of *Scyllarides nodifer* and *Albunea* sp. during an intrusion of low-salinity Mississippi flood water in the Florida Keys in September 1993: Insight into larval transport from upstream. *J. Geophys. Res. (C Oceans)* 105: 28741–28758.

12

The Northwestern Hawaiian Islands Lobster Fishery: A Targeted Slipper Lobster Fishery

Gerard T. DiNardo and Robert B. Moffitt

CONTENTS

Abstract

Resource surveys conducted in the late 1970s detected large populations of lobsters in the Northwestern Hawaiian Islands, prompting the development of a distant-water, multispecies, trap fishery. The fishery primarily targeted the Hawaiian spiny lobster, *Panulirus marginatus* (Quoy & Gaimard, 1825), and the scaly slipper lobster, *Scyllarides squammosus* (H. Milne Edwards, 1837), but three other lobster species were caught in low abundance: the green spiny lobster, *Panulirus penicillatus* (Olivier, 1791), the ridgeback slipper lobster, *Scyllarides haanii* (De Haan, 1841), and the Chinese slipper lobster, *Parribacus antarcticus* (Lund, 1793). In this chapter, the population structure and demographic parameters of the scaly slipper lobster in the Northwestern Hawaiian Islands are reviewed, and fishery statistics from the lobster trap fishery are presented. Geo-referenced estimates of abundance are provided and factors contributing to their increase are proffered. Recognizing the importance of spatial structure in Northwestern

Hawaiian Islands lobster populations, management implications and data requirements to advance stock assessments are discussed, as well as the importance of Marine Protected Areas as a management tool.

12.1   Introduction

Slipper or shovel-nosed lobsters of the family Scyllaridae are found throughout the world's tropical and temperate oceans. Although they are often a desirable incidental catch in a commercial fishery, they are generally considered too small and scarce to warrant targeted harvesting (Nishikiori & Sekiguchi 2001; Freitas & Santos 2002; Vance et al. 2004). More recently, however, a few species in a few locales have become major target species for small-to-moderate-scale fisheries (Coutures & Chauvet 2003; Molina et al. 2004; Haddy et al. 2005; see also Hearn et al., Chapter 14, Radhakrishnan et al., Chapter 15, and Haddy et al., Chapter 17) — this is the case in Hawaii.

Scyllaridae are represented in Hawaii by nine species (Table 12.1). Of these nine species, only the two *Scyllarides* spp. are of commercial importance. In the inhabited main Hawaiian Islands (MHI), the scaly slipper lobster, *Scyllarides squammosus* (H. Milne Edwards, 1837), and, to a lesser extent, the ridgeback slipper lobster, *Scyllarides haanii* (De Haan, 1841), along with two species of spiny lobster, the green spiny lobster, *Panulirus penicillatus* (Olivier, 1791), and the endemic Hawaiian spiny lobster, *Panulirus marginatus* (Quoy & Gaimard, 1825), are mainly harvested by recreational divers for subsistence uses and only incidentally harvested as a portion of commercial fish trapping operations (Hawaii State Division of Aquatic Resources, unpublished data). Other species may occasionally be collected for subsistence use or for the aquarium trade. The annual commercial catch of slipper lobsters in the MHI is generally less than 1000 kg (Hawaii State Division of Aquatic Resources, unpublished data).

In the uninhabited and remote Northwestern Hawaiian Islands (NWHI), resource surveys conducted in the late 1970s detected large populations of spiny and slipper lobsters, prompting the development of a distant-water, multispecies, trap fishery. The Hawaiian spiny lobster and scaly slipper lobster were primarily targeted by the fishery, but three other species, the green spiny lobster, ridgeback slipper lobster, and Chinese slipper lobster, *Parribacus antarcticus* (Lund, 1793), were caught in low abundance. In this chapter, we review the population structure and life-history characteristics of the scaly slipper lobster in the NWHI, and document the commercial lobster fishery that has been operating in this area for over 20 years.

TABLE 12.1

List of Hawaiian Slipper Lobster Species (PIFSC Unpublished Data)

Scyllarides haanii (De Haan, 1841)
Scyllarides squammosus (H. Milne Edwards, 1837)
Parribacus antarcticus (Lund, 1793)
Arctides regalis Holthuis 1963
Galearctus[a] *aurora* (Holthuis, 1982)
Chelarctus[a] *cultrifer* (Ortmann, 1897)
Eduarctus[a] *modestus* (Holthuis, 1960)
Biarctus[a] *vitiensis* (Dana, 1852)
Petrarctus[a] *demani* (Holthuis, 1946)[b]

[a] Recently separated from *Scyllarus* (Holthuis, 2002).
[b] Only phyllosomes and nistos collected in Hawaii to date.

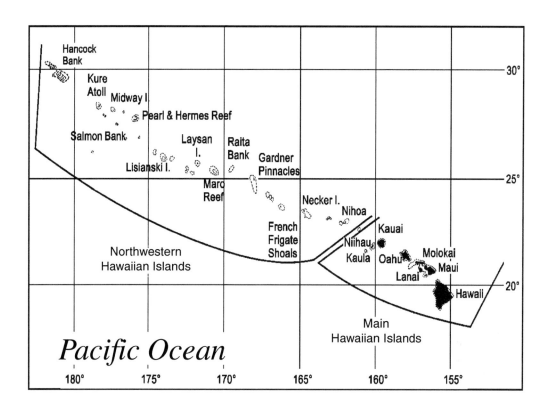

FIGURE 12.1 The Hawaiian Archipelago.

12.2 Geographic Setting

The NWHI ecosystem is highly fragmented and consists of a series of isolated islands, banks, atolls, islets, and reefs (hereafter referred to as banks) that extend 1500 nautical miles west-northwest of the main Hawaiian Islands (Figure 12.1 illustrates this ecosystem). The NWHI lie near the center of the Subtropical Gyre, and a weak geostrophic current flows along the NWHI from northwest to southeast. The bank summits are composed largely of flat, low-to-moderate relief terraces between depths of 20 to 50 m (Parrish & Boland 2004). Higher relief features, including isolated coral colonies, pits, and overhangs, are sparsely scattered throughout. These terraces are dominated, at least seasonally, by algal meadows. The summits and adjacent slopes are the primary lobster habitat. The amount of lobster habitat varies between banks, ranging from 4 to 153 hectares, and lobster concentrations at any particular bank are generally patchy.

12.3 Life-History Characteristics and Population Structure

Knowledge about a population's vital rates (i.e. growth, mortality, etc.), composition, and structure are essential to good resource management. For example, whether an animal lives 20 years or eight years, it provides insight into proper use of the resource. Also, whether the animals represent a single homogeneous population or a group of spatially structured populations influences assessment approach methodologies and harvesting strategies.

12.3.1 Distribution

There is considerable overlap in the range of depths at which Hawaiian spiny lobster and scaly slipper lobster are found in the NWHI. Hawaiian spiny lobsters are found mainly at depths ranging from 10 to 80 m, with highest concentrations occurring between 30 and 50 m (Uchida & Tagami 1984). Scaly slipper lobsters inhabit deeper waters, ranging from 30 to 120 m, with the highest concentrations occurring between 50 and 70 m. Ridgeback slipper lobsters are generally found even deeper, ranging from 60 to 200 m, with the greatest concentrations between 100 and 150 m. The reason for the apparent separation is unknown, but may be a result of competition for available habitat. In recent years at Maro Reef, and possibly other banks at the northwestern end of the NWHI, scaly slipper lobster concentrations in shallow areas (30 to 50 m) have increased as the concentrations of Hawaiian spiny lobster have decreased.

12.3.2 Age and Growth

Growth estimates are difficult to obtain for crustaceans. Crustaceans lack persistent hard parts that are frequently used to determine age in other organisms (e.g., otoliths in fishes). Growth estimates can be obtained, however, from length-frequency analyses, tag-and-release studies, or a combination thereof (see Bianchini & Ragonese, Chapter 9 for further information on growth analyses). The understanding of age and growth for the common slipper lobster in the NWHI is poor. Tagging studies, using passive integrated transponder (PIT) tags (Biomark), were initiated in September 2004 for slipper lobsters in the NWHI and the results are forthcoming. Until these results are finalized, the best growth information available is a preliminary growth curve generated for the common slipper lobster in the NWHI by modal progression analysis of tail width (TW) (Polovina & Moffitt 1989). This growth curve was based on TW frequency distributions obtained on research cruises conducted in only one month (between June and August) for each of three consecutive years, 1986 to 1988, and TW modes were not readily apparent or easy to track with such infrequent sampling. That being said, results indicated that slipper lobsters grew to 56 mm (the minimum legal size at the time) TWs in 3.3 years after postlarval settlement.

12.3.3 Reproduction

Egg-bearing, female scaly slipper lobsters are found throughout the year, with peak abundance in the late spring and early summer (May to July) (DiNardo & Wetherall 1999). In contrast to the Hawaiian spiny lobster, spermatophores are rarely observed on females and are generally gelatinous, suggesting that fertilization occurs rapidly after sperm transfer. Fecundity is strongly positively related to body size, with the median-sized female of 60 mm TW bearing approximately 90,000 eggs (DeMartini & Williams 2001). Eggs average 0.67 mm in diameter and are not related to body size. The egg numbers and diameter are similar to those observed in the congeners *Scyllarides latus* (Latreille, 1802) (Martins 1985; Almog-Shtayer 1988; Spanier & Lavalli 1998) and *Scyllarides astori* Holthuis, 1960 (see Hearn et al., Chapter 14). In these species, there are ~100,000 to 150,000 eggs per female and egg diameter ranges from 0.6 to 0.7 mm.

Scaly slipper lobster phyllosomas remain pelagic for three to six months prior to transforming to benthic juveniles (PIFSC, unpublished data; see Sekiguchi et al., Chapter 4 for more information on larval and postlarval biology). Nursery habitat for this species has not been identified. Small juveniles with carapace lengths less than 30 mm are rarely encountered in our traps (PIFSC, unpublished data).

Size at maturity for females, as measured by the presence of eggs on the pleopods, showed a general decline over time (Polovina et al. 1988; DeMartini & Kleiber 1998), suggesting that it is positively related to population density in a manner similar to that reported for the Hawaiian spiny lobster (Polovina 1989). This relationship, however, was very weak, due in part to the large number of nonegg-bearing, though mature, females in the samples. A more rigorous method of indicating size at maturity was developed and verified histologically for both slipper and spiny lobster using relative pleopod length (DeMartini et al. 2005). Use of this method on females collected in 2000 and 2001 resulted in an estimate of female maturity of 47.6 mm TW, or about three years of age using the growth curve mentioned earlier. The pleopod measurement has become a standard measurement on our research cruises, which will allow for monitoring of changes in size at maturity with changes in population abundance over time.

12.3.4   Recruitment

Little is known about slipper lobster recruitment in the NWHI. As mentioned above, nursery habitat has not been identified. Because of the protracted pelagic larval phase of scaly slipper lobsters (three to six months) and Hawaiian spiny lobsters (11 to 12 months), recruitment to a bank is likely dependent, in part, on lobster reproduction at surrounding banks (see Sekiguchi et al., Chapter 4 for hypotheses concerning larval dispersal and nisto recruitment). Populations inhabiting discrete banks are connected by the dispersal of larvae between banks. This results in banks acting as either recruitment sources, sinks, or both. Historical research survey data collected prior to any significant increase in anthropogenic activities in the NWHI suggest that the region between Necker Island and Maro Reef is a major sink area for Hawaiian spiny lobster larvae (Uchida et al. 1980). Banks northwest of Maro Reef are considered recruitment sources, receiving only sporadic recruitment. Populations with low or sporadic recruitment are less resilient and very susceptible to depletion and overfishing.

12.3.5   Population Structure

Recent advances in our understanding of the spatial structure of NWHI lobster populations and dynamics of larval transport indicate that lobster populations in the NWHI constitute a metapopulation. A metapopulation is a group of populations inhabiting discrete patches of suitable habitat that are connected by the dispersal of individuals between habitat patches (Hanski 1991). The population structure and spawning strategy of NWHI lobsters supports the definition of a metapopulation reasonably well. Genetic studies indicate that a single homogeneous population of spiny lobster occurs in the NWHI (Shaklee & Samollow 1984; Seeb et al. 1990), adding additional support to the metapopulation notion. Genetic studies on scaly slipper lobsters have only recently been conducted and the results are forthcoming (see Deiana et al., Chapter 3 for additional genetic studies on other species of slipper lobsters). Dependence among local populations on lobster in the NWHI implies that it is conceivable for a bank to undergo population declines even though the bank experiences little or no anthropogenic activities (e.g., fishing).

12.4   History of the Fishery

A fishery is a system of interactions within and among the populations of fish being harvested, the populations of fishermen, the ecosystem, and fisheries policy. Typical to most developing fisheries, the NWHI lobster fishery started out small and expanded rapidly. As a result, a suite of input and output management controls was applied to the NWHI lobster fishery in an effort to halt the expansion while, at the same time, ensuring sustainability. In this section, the history of the NWHI lobster fishery is described, as well as adopted management measures.

12.4.1   Fishing Gears, Practices, and Management

In the mid-1970s, research cruises conducted by the Pacific Islands Fisheries Science Center (PIFSC) and the State of Hawaii identified commercial concentrations of trappable Hawaiian spiny lobster. Within a very short time, two vessels that had already been fishing for deep-slope snappers and groupers in the area added lobster trapping to their fishing operations, landing a mix of fresh fish and live Hawaiian spiny lobster for local markets. Trips were short (~10 days) and trapping effort low (50 to 100 trap hauls per vessel fishing day). The California two-chambered wire trap was used (hereafter referred to as the California trap; Figure 12.2). The basic California trap is rectangular in shape and is constructed with heavy-gauge galvanized wire. It measures 92 cm in length, by 71.5 cm in width, and 41 cm in height, and has an open mesh 4.7 cm long by 9.7 cm wide. A number of field alterations were applied to the basic design in attempts to modify the fishing characteristics and catch rates of this trap. This included sheathing the trap with finer mesh, enlarging the entrance cones, coating the wire with plastic of various colors, and reducing the mesh size of the bait container (Clarke & Sumida, unpublished report). The cumulative effects of these modifications on commercial catch rates are unknown. The California traps were effective at catching Hawaiian spiny lobster (~80% of the catch by number), but relatively ineffective at catching

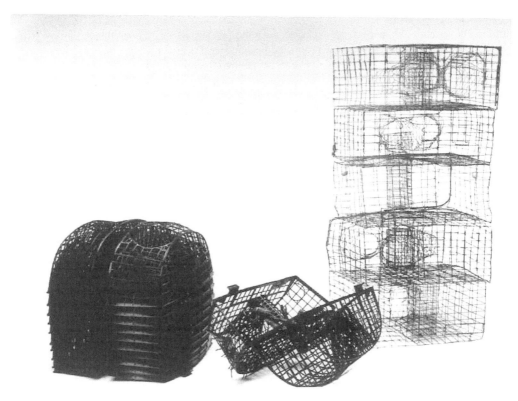

FIGURE 12.2 The rectangular California wire trap (on the right) shown stacked one atop the other and next to the stackable, black plastic Fathoms Plus® trap (on the left). (Photo by John L. Tarantino from Fathoms Plus, Inc.; used with permission.)

scaly slipper lobster (~9% of the catch) (Moffitt et al., in press). The fishery at this time was largely unregulated; the only restrictions were State of Hawaii regulations prohibiting the sale of lobsters under one pound in weight (0.45 kg) and the prohibition of selling egg-bearing females.

Between 1981 and 1986, the NWHI lobster fishery developed rapidly from a modest home-market operation targeting live Hawaiian spiny lobster into a multispecies operation exporting live lobster and frozen tails internationally (Clarke et al. 1992). Associated with the fisheries growth were transitions in commercial fishing gear, vessel size, fleet participation, and number of traps fished. Trips lengthened to between 40 and 60 days and concentrated solely on lobster trapping. The daily number of trap hauls increased to about 400 for each vessel (Clarke & Sumida, unpublished report). The 1984 to 1985 period marked the transition from the use of the California trap to the Fathoms Plus® black polyethylene plastic trap (hereafter referred to as black plastic trap; see Figure 12.2 for a comparison of the two trap types and Figure 12.3 for a detailed view of the Fathoms Plus® trap). Intensity of fishing effort, catch composition, and catch rates were strongly influenced by this change in gear type. In addition, deck-space constraints, which had restricted the number of California traps carried by each vessel, eased with the introduction of the compact black plastic trap, which could be broken down and nested for transport (see Figure 12.3). Boats could now easily carry 800 to 1200 traps, and the average number of traps fished daily increased from 400 traps to 1000 traps. The resulting increase in trap-carrying capacity, in turn, contributed to an immediate rise in fishing effort and fishing mortality (Gates & Samples 1986). By 1985, only one year after its introduction, 90% of all traps fished were of the black plastic design (Clarke & Sumida, unpublished report).

The basic shape of the black plastic trap has changed little since its introduction. It is oval, slightly tapered towards the top, and is domed at the center. Plate steel or cast lead ballast is bolted, wired, or banded to the base of each trap half. In late 1985, a significant modification was made to the basic design of the black plastic trap. Eager to market a trap for the deepwater shrimp fisheries developing in Hawaii

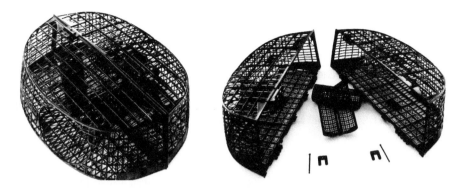

FIGURE 12.3 Fathom Plus® black plastic trap as fished in its closed position (left) and in its open position (right). (Photo by John L. Tarantino from Fathoms Plus, Inc.; used with permission.)

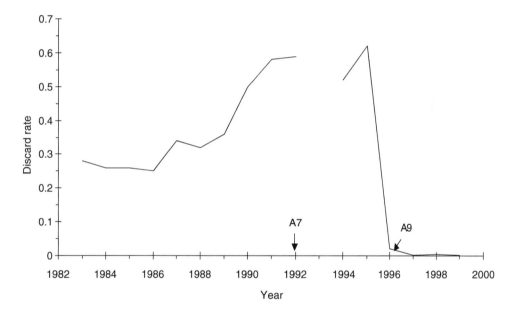

FIGURE 12.4 Annual estimates of the reported discard rate for lobsters from the NWHI lobster trap fishery. A7 and A9 show when Amendments 7 and 9 were implemented, respectively.

and along the west coast of the United States, the manufacturer of the black plastic trap added vertical rungs into the existing trap mesh. (Raymond Clarke, personal communication). As a result, the trap mesh of the current black plastic trap is between 56 and 74% less than the early version of the same trap and 80 to 89% less that of the California wire trap.

The effect of gear changes and reduced mesh size on lobster populations cannot be accurately assessed because historical size structure data was not available from the NWHI commercial lobster fishery. However, problems with high catch rates of small, immature lobsters and associated discard mortality were identified as early as 1980 (MacDonald & Stimson 1980). The requirement that black plastic traps include escape vents was adopted in 1987 and implemented in 1988. From 1983 through 1995, the lobster (Hawaiian spiny lobsters and scaly slipper lobsters combined) discard rate (the reported ratio of lobsters discarded to total lobsters caught) generally increased, rising from 0.28 in 1983 to 0.62 in 1995. After 1995, the discard rate decreased significantly as a result of a relaxation of the minimum legal-size requirement in favor of an optional retain-all policy (Figure 12.4 illustrates the rise and fall of the discard rate from 1983 to 1996). Gear alterations (introduction of the black plastic trap and reduction in trap mesh size) and spatiotemporal shifts in fishing effort contributed to the reported increase in discard rate between 1983 and 1995.

The change to the black plastic trap greatly magnified the importance of scaly slipper lobster in the fishery. The percentage of scaly slipper lobster in the catch increased from less than one tenth that of Hawaiian spiny lobster to about equal numbers (DiNardo & Marshall 2001). In response to the increasing catch, NWHI lobster fishermen actively developed the market for frozen, scaly slipper lobster tails. By 1987, scaly slipper lobster landings nearly equaled that of Hawaiian spiny lobster, although the price remained much lower (US$10.16/lb vs. US$13.00/lb, respectively) (Dollar & Landgraf 1992).

12.4.2 Fishery Management

The Western Pacific Fisheries Management Council (WPFMC) is responsible for developing fishery regulations in U.S. Federal waters of the Pacific Islands Region. The WPFMC adopted a Federal Crustacean Fishery Management Plan (FMP) in 1983 to manage the rapidly growing NWHI lobster fishery. A history of management measures affecting this fishery is found in Table 12.2.

Several approaches have been used since 1983 to model NWHI lobster populations and establish levels of sustainable yield. From 1985 to 1987, maximum sustainable yield (MSY) for the NWHI was estimated using surplus production methods (Polovina et al. 1988). After 1988, a discrete population model was fit to the NWHI commercial data (pooled across all banks) to estimate population size and biological parameters (Haight & Polovina 1993). The model expresses the number of exploitable lobsters (all species combined) in a given month as a function of the number of exploitable lobsters in the previous month, adjusted for natural mortality, fishing mortality, and recruitment. In 1992, catch quotas were adopted as a management tool (Amendment 7 of the Crustaceans FMP), and the discrete population model of Haight & Polovina (1993) was used to estimate the exploitable population of lobsters in the NWHI. In 1996, a constant harvest rate strategy that allows only a 10% risk of overfishing in a given year and the retention of all lobsters caught was adopted to establish annual catch quotas (Amendment 9 of the Crustaceans FMP; DiNardo & Wetherall 1999). Spatial management commenced in 1998, requiring estimates of exploitable population in four management areas, and the discrete population model was used to compute exploitable population estimates for each management area. The fishery was closed in 2000 as a precautionary measure because of increasing uncertainty in the population models used to assess stock status. In December 2000, President William Jefferson Clinton, through Executive Order (EO) 13178 and later through EO 13196, established the Northwestern Hawaiian Islands Coral Reef Ecosystem Reserve, which may prohibit commercial lobster fishing in the NWHI indefinitely.

12.4.3 Commercial Catch Statistics

The total reported catch and landings of all species of lobster peaked in 1985 at approximately 2,736,000 and 2,031,000 lobsters, respectively, and generally declined from 1986 to 1995 (Table 12.3 provides a breakdown of the landings by spiny and slipper lobsters; Figure 12.5 shows catch and effort for all species combined). Fishing effort peaked in 1986 at approximately 1,290,000 trap hauls and declined to 834,000 trap hauls in 1988 before increasing to 1,180,000 trap hauls in 1990. After 1990, fishing effort generally declined.

The fishery initially targeted Hawaiian spiny lobster, but by 1985, gear modifications and improved markets led to an increase in scaly slipper lobster landings. Catches of scaly slipper lobster remained high from 1985 to 1987, fell into a general decline from 1988 to 1996, and increased significantly from 1997 to 1999 (Figure 12.6 shows the catch and effort for spiny and slipper lobsters separated). Slipper lobsters comprised 45% of the catch in 1985, 11% in 1995, and ~63% in 1999 (see Table 12.3 for more information).

The proportion of fishing effort and reported catch at each bank in the NWHI has varied both spatially and temporally. Although as many as 16 banks in the NWHI were fished on an annual basis, the majority of fishing effort has been directed at four banks: Maro Reef, Gardner Pinnacles, St. Rogatien, and Necker Island (Figure 12.7 shows the relative effort among banks). Between 1984 and 1989, most of the fishing effort was directed at Maro Reef. After 1989, fishing effort decreased at Maro Reef and increased significantly at Gardner Pinnacles and Necker Island. In 1996 and 1997, the majority of fishing

TABLE 12.2

NWHI Lobster Fishery Regulations (FMP, Amendments, and Major Administrative Actions), 1983–2000

Year	FMP Amendment/Actions	Primary Regulatory Effect[a]
1983	FMP takes effect (3/8/83)	Permits, logbooks, legal sizes, and trap dimensions.
	Amendment 1	Adopt State of Hawaii measures in EEZ around main Hawaiian Islands.
	Amendment 2	Limit the size of trap entrances.
1985	Amendment 3	Redefine minimum size by tail width and set smaller minimum size.
1986	Amendment 4	Prohibits capture of slipper lobster in refugia.
1987	Amendment 5	Minimum size for slipper lobster; require escape vents in traps.
1990	Amendment 6	Defines overfishing as Spawning Stock Biomass per Recruit.
1991	Amendment 7	Limited access system; adjustable annual harvest quota; closed season.
1993	Emergency closure	Fishery closed for entire year.
1994	Emergency closure	Fishery closed following in-season revision (decrease) in harvest quota.
	Amendment 8	Revision to limited access system (removal of two-year, use–lose requirement).
1995	Commercial fishery closed: Conducted experimental fishery	One vessel operates under experimental (exploratory) permit (with observer).
1996	Amendment 9	Revised annual harvest guideline procedure; elimination of in-season adjustment to quota; removal of minimum size and condition restrictions. A regulatory "framework" procedure was implemented to ease annual regulatory changes.
1998	Spatial management adopted	Council established bank-specific harvest guidelines for the 1998 lobster season to prevent the potential risk of overexploiting the lobster population at Necker Island.
2000	Fishery closed	Fishery closed due to increasing uncertainty in the population models used to assess stock status.
2000	Executive order	President Clinton, through Executive Order 13178, establishes the NWHI Coral Reef Ecosystem Reserve which may prohibit commercial lobster fishing for at least 10 years.

[a] This table should not be viewed as a substitute for the detailed federal regulations implementing these and additional measures in the fishery.

Source: Adapted from Kawamoto, K.E. & Pooley, S.G. 2000. Annual report of the 1998 Western Pacific lobster fishery (with preliminary 1999 data). Honolulu Laboratory Southwest Fisheries Science Center, National Marine Fisheries Service, NOAA, Honolulu, Hawaii. Southwest Fisheries Science Center Administrative Report H-00-02, 38 pp.

effort was directed at Necker Island. Spatial management commenced in 1998, redistributing fishing effort throughout the archipelago.

In general, the observed spatiotemporal shifts in fishing effort between banks are attributed to declines in spiny lobster catch rates; as Hawaiian spiny lobsters were fished down and catch rates at a particular bank fell below some minimum economic threshold, fishing effort shifted to more productive banks. By the mid-1990s, fishing was generally limited to Necker Island where relatively higher concentrations of spiny lobsters were found. With the adoption of spatial management in 1998, fishing effort was redistributed throughout the NWHI, and the major target of the fishery changed to scaly slipper lobster.

Commercial lobster landings per unit effort (pooled across all banks) declined from 2.75 lobsters/trap haul in 1983 to 0.98 lobsters/trap haul in 1987, then increased to 1.26 lobsters/trap haul in 1988, before declining to an average of 0.63 lobsters/trap haul between 1991 and 1995 (Figure 12.8 shows the annual LPUE from 1982 to 2000). Landings per unit effort (LPUE) increased to an average of 1.68 lobsters/trap

TABLE 12.3

Summary of Catch and Effort Data from Federal Logbooks for the NWHI Lobster Fishery, 1983 to 2000

Year	No. Vessels	No. Trips	No. Banks	Trap Hauls	Spiny Lobster Reported Catch				Slipper Lobster Reported Catch				Reported Landings	Reported Discards	Total Reported Catch
					Mature	Immature	Berried	Total	Mature	Immature	Berried	Total			
1983	4	19	3	64,000	158,000	51,000	10,000	218,000	18,000	6200	1700	26,000	176,000	68,000	244,000
1984	13	41	7	371,000	677,000	239,000	75,000	991,000	271,000	9000	8000	288,000	948,000	331,000	1,279,000
1985	17	66	13	1,040,000	1,002,000	355,000	132,000	1,489,000	1,029,000	96,000	121,000	1,245,000	2,031,000	705,000	2,736,000
1986	16	62	16	1,290,000	843,000	298,000	153,000	1,294,000	1,005,000	55,000	121,000	1,181,000	1,848,000	627,000	2,475,000
1987	11	40	12	805,000	393,000	233,000	101,000	727,000	395,000	36,000	43,000	474,000	788,000	414,000	1,202,000
1988	9	29	13	834,000	888,000	279,000	115,000	1,282,000	168,000	69,000	41,000	278,000	1,056,000	504,000	1,560,000
1989	11	33	13	1,070,000	944,000	369,000	169,000	1,482,000	216,000	69,000	49,000	334,000	1,160,000	655,000	1,815,000
1990	14	45	14	1,180,000	591,000	464,000	181,000	1,236,000	184,000	56,000	67,000	307,000	775,000	769,000	1,544,000
1991	9	21	5	297,000	132,000	192,000	29,000	353,000	35,000	8700	6000	49,700	167,000	236,000	403,000
1992	12	28	9	685,000	248,000	278,000	82,000	608,000	163,000	48,000	29,000	240,000	411,000	437,000	848,000
1993[a]	—	—	—	—	—	—	—	—	—	—	—	—	—	—	—
1994	5	5	5	168,000	85,000	61,000	39,000	185,000	46,000	28,000	11,000	84,000	131,000	139,000	270,000
1995[b]	1	1	3	64,000	35,000	34,000	21,000	90,000	3300	7400	c	11,500	38,300	61,000	99,300
1996	5	5	2	115,000	123,000	—	42,000	165,000	18,000	—	4000	22,000	187,000	2000	189,000
1997	9	9	4	178,000	140,000	—	36,000	176,000	121,000	—	13,000	134,000	310,000	c	310,000
1998	5	9	12	171,000	69,000	—	13,000	82,000	113,000	—	16,000	129,000	211,000	c	211,000
1999	6	6	13	236,000	73,000	—	13,000	86,000	134,000	—	12,000	146,000	232,000	c	232,000
2000[a]	—	—	—	—	—	—	—	—	—	—	—	—	—	—	—

[a] Fishery closed.
[b] Experimental fishery.
[c] Fewer than 1000 lobsters.

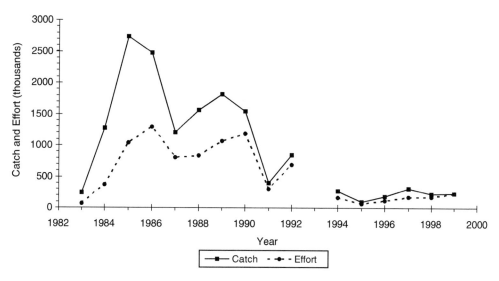

FIGURE 12.5   Annual metrics of reported catch and fishing effort (trap hauls) in the NWHI commercial lobster fishery.

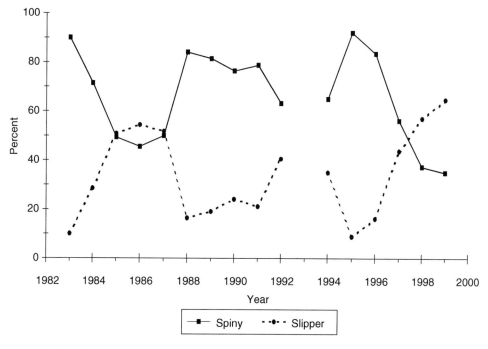

FIGURE 12.6   Annual percentages on mature spiny and slipper lobster catches (see text for legal definition).

haul between 1996 and 1997 prior to declining to 1.0 lobster/trap in 1999. This sudden increase in reported LPUE during the 1996 and 1997 fishing seasons resulted from changes in management policies and fishing strategies, and not significant increases in the population. The 1996 and 1997 commercial fisheries operated under Amendment 9 of the Crustaceans FMP, which allowed all lobsters caught and decked to be landed (eliminated regulatory discards). Also, most of the fishing effort in 1996 and 1997 was directed at Necker Island, the most productive bank. In addition, areas with higher concentrations of scaly slipper lobster were specifically targeted by some participants during the 1997 to 1999 commercial fishery, representing a change in fishing strategy. In previous years, minimum size limits were imposed

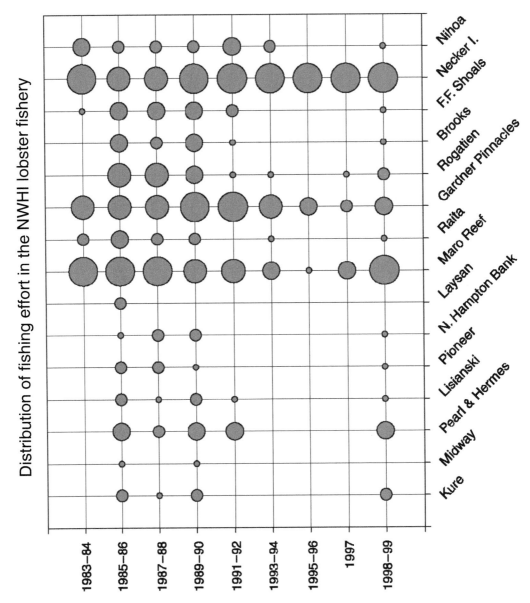

FIGURE 12.7   Spatial distribution of commercial fishing effort in the NWHI lobster trap fishery. In each time step, a circle's diameter represents the relative magnitude of reported fishing effort; the larger the circle the greater the reported fishing effort.

and fishing occurred on several banks, including less productive banks. The drop in LPUE during the 1998 and 1999 fishing seasons resulted from the adoption of spatial management that redistributed fishing effort throughout the NWHI. Reevaluating the 1996 to 1997 and 1998 to 1999 LPUEs by assuming historical minimum legal sizes results in hypothetical average LPUEs of 1.21 and 0.91, respectively.

The reported LPUE time series from banks, where at least five years of commercial fishing data are available, all exhibit similar declining trends (Figure 12.9 shows catch and effort for 13 banks). For many of the banks, a 50% drop in LPUE was reported between 1983 and 1987. Insufficient data make it difficult to assess specific causes for the observed declines in NWHI lobster LPUEs, but fishing mortality is a likely contributor. Significant increases in lobster LPUEs were observed at some banks in 1997 and 1998, and resulted from a switch in target species — Hawaiian spiny lobster to scaly slipper lobster.

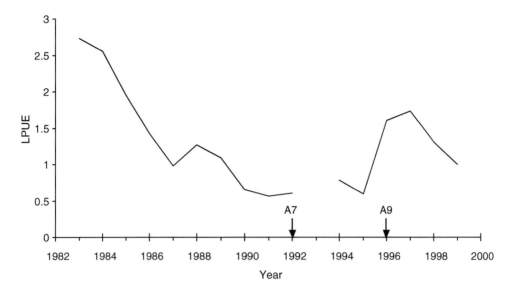

FIGURE 12.8 Annual commercial LPUE for the NWHI lobster trap fishery, 1983 to 1999. A7 and A9 show when Amendments 7 and 9 were implemented, respectively.

12.5 Estimates of Abundance

Estimates of population abundance, either absolute or relative, are necessary for understanding changes in population number, composition, and other vital rates for estimating harvest rates and as a basis for effective resource management. Abundance estimates for fished species are derived using either fishery-dependent or fishery-independent data, and generally expressed as relative abundance (catch per unit of fishing effort). For NWHI lobsters, relative abundance estimates are derived from both fishery-dependent data, as well as fishery-independent data. In this section, the sources of data to facilitate abundance estimation are described, as well as potential factors affecting lobster abundance in the NWHI.

12.5.1 PIFSC Annual NWHI Lobster Survey

A fishery-independent lobster resource survey is conducted annually by the PIFSC and data are available from 1984 to 1989 and 1991 to 2004. The survey is currently configured to monitor local populations of lobster in the NWHI, but has also been used: (1) to evaluate the performance of commercial and research survey gear, (2) to calibrate gear types, (3) as a platform for short-term experiments (e.g., studies of handling mortality), and (4) for the collection of biological and oceanographic data. The survey uses a fixed-site design stratified by depth, and at each site shallow (<20 fathoms or 36.6 m) and deep (≥20–50 fathoms or 36.6–91.4 m) stations are sampled. Ten strings of eight traps each are set at the shallow station and two to four strings of 20 traps each are set at the deep station. Traps are fished overnight and baited with 1.5 to 2.0 pounds (0.68 to 0.91 kg) of cut-up, previously frozen, mackerel. Data on species, carapace length and TW, sex, and reproductive condition (berried or unberried) are collected for each lobster caught, as well as the latitude and longitude of the traps recorded at the string level. The geographical extent of the trap survey has generally been limited to Necker Island and Maro Reef, with infrequent trips to adjacent banks.

Between 1984 and 1991 a variety of gears and gear configurations were used in the resource survey. California traps were used from 1985 to 1991. Black plastic traps without escape vents were first used in 1984, and since 1992 have been used exclusively in the survey. While trap-comparison studies were conducted to provide a conversion formula for the California trap and black plastic trap catch per unit effort (CPUE) in terms of lobster per trap haul, the studies were incomplete. Thus, in computing a CPUE

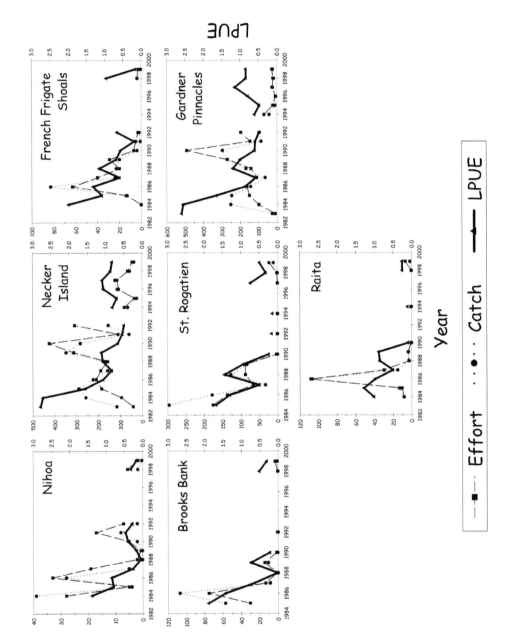

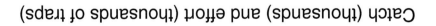

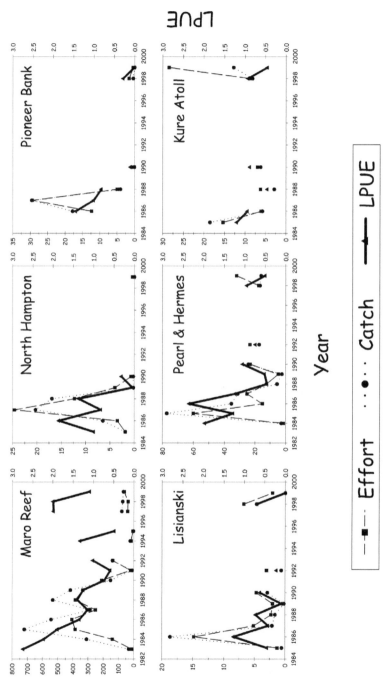

FIGURE 12.9 Bank-specific annual metrics of LPUE, fishing effort (trap hauls), and landings from the NWHI lobster fishery.

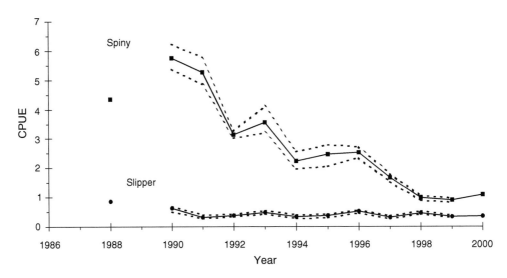

FIGURE 12.10 Annual metrics of spiny and slipper lobster CPUEs from research surveys at Necker Island.

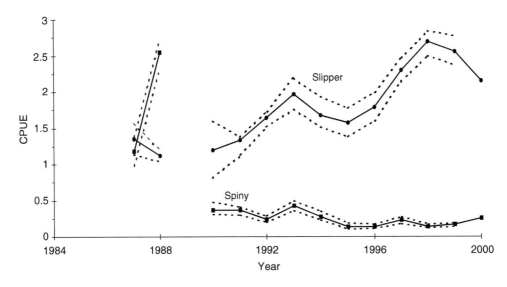

FIGURE 12.11 Annual metrics of spiny and slipper lobster CPUEs from research surveys at Maro Reef.

time series from research survey data, we are limited to years in which black plastic traps were fished in significant numbers at both shallow and deep stations (>50% of the total traps fished). For Maro Reef this corresponds to years 1987 to 2004 and for Necker Island years 1988 to 2004.

Since 1990, Necker Island Hawaiian spiny lobster research CPUEs have generally decreased (Figure 12.10 shows CPUEs for spiny and slipper lobsters separately). Significant drops in CPUE were observed in 1992, 1994, and 1998. Scaly slipper lobster research CPUEs have remained at relatively low levels at Necker Island between 1988 and 2004.

Hawaiian spiny lobster research CPUEs at Maro Reef declined significantly after 1988 and have since remained low (Figure 12.11). Scaly slipper lobster CPUEs at Maro Reef have generally been increasing, with significant increases occurring after 1991. These changes suggest a switch in species dominance at Maro Reef, with an increase in scaly slipper lobster as Hawaiian spiny lobster were fished down and habitat became available to the slipper lobsters.

12.5.2   Factors Affecting Abundance

Research to date has identified a dynamic change in the spatial and temporal structure of the NWHI lobster population. Prior to large-scale commercial fishing, Hawaiian spiny lobsters were the dominant species in many regions of the NWHI. Throughout the 1980s, Hawaiian spiny lobster density declined, and in 1990 a collapse of many local Hawaiian spiny lobster populations was observed. In subsequent years, scaly slipper lobster density increased, and presently this species is the dominant species of lobster in the NWHI.

The observed decline of Hawaiian spiny lobster density in the NWHI has generally been attributed to large-scale, ocean-climate events that significantly altered oceanic productivity and lobster recruitment. The recruitment of Hawaiian spiny lobster was found to correlate with the strength of the subtropical countercurrent, suggesting that ocean circulation patterns affect the transport and survival of lobster larvae during their pelagic larval cycles. A concurrent decline in Hawaiian spiny lobster density at Laysan Island, a bank in the NWHI experiencing very little commercial fishing, provided additional support for the ocean-climate hypothesis (Polovina & Mitchum 1992; Polovina et al. 1995). However, recent improvements in our understanding of the spatial structure of NWHI lobster populations, the dynamics of larval transport, and the commercial fishery indicates that the decline of spiny lobster and observed switch in species dominance resulted from a combination of oceanographic, biotic (e.g., habitat and competition), and anthropogenic (e.g., fishing) factors.

12.6   Management Implications and Data Requirements for Stock Assessments

Treating Hawaiian spiny and scaly slipper lobster in the NWHI as metapopulations is consistent with available data and represents a departure from the status quo. Given the connectivity among local populations of lobsters in the NWHI, overfishing or depletion of local populations could result in catastrophic effects to the population as a whole (e.g., reduction in average recruitment or recruitment failure), particularly when a large number of local populations, or the most productive populations, are overfished. Also, when spatial correlation among local populations is highly bank specific, relationships between population size and fishing can become decoupled, masking the true effect of fishing. The decline of Hawaiian spiny lobsters at Laysan Island may provide an example of this decoupling. A major component of an effective management plan for spatially structured populations, such as NWHI lobsters, is the establishment of refugia or no-take marine protected areas (MPAs). Setting aside refuges provides a buffer against the risk of overfishing the system (Guenette et al. 1998). While it is unclear when the NWHI lobster fishery will reopen, future management should consider the establishment of no-take MPAs.

Recognizing the importance of spatial structure changes the data requirements to advance NWHI lobster population models and stock assessments. While the discrete population model relied solely on commercial catch and effort data as input, spatially structured models require data (both biological and fishery related) with greater spatial resolution. Because of life history differences between Hawaiian spiny and scaly slipper lobsters (e.g., growth, natural mortality, and recruitment), subsequent models will also need to be species specific. The increased data requirements will require modifications to the existing data collection and monitoring programs to facilitate model development and effective resource management.

Acknowledgments

The authors are indebted to the numerous scientists who participated on the various lobster resource surveys to the Northwestern Hawaiian Islands; without their dedication and persistence there would be no data. The authors would also like to thank the crew of the NOAA vessels *Townsend Cromwell* and *Oscar Elton Sette* for providing logistical support and assisting with sampling operations. Finally, we would like to acknowledge two anonymous reviewers whose comments strengthened the manuscript.

References

Almog-Shtayer, G. 1988. Behavioural-ecological Aspects of Mediterranean Slipper Lobsters in the Past and of the Slipper Lobster Scyllarides latus in the Present. M.A. thesis: University of Haifa, Israel. 165 pp.

Clarke, R.P., Yoshimoto, S.S., & Pooley, S.G. 1992. A bioeconomic analysis of the Northwestern Hawaiian Islands lobster fishery. *Mar. Resour. Econ.* 7: 115–140.

Coutures, E. & Chauvet, C. 2003. Study of an original lobster fishery in New Caledonia (Crustacea: Palinuridae & Scyllaridae). *Atoll Res.* 504: 1–7.

DeMartini, E.E. & Kleiber, P. 1998. Estimated body size at sexual maturity of slipper lobster *Scyllarides squamosus* at Maro Reef and Necker Island (Northwestern Hawaiian Islands), 1986–1997, Honolulu Laboratory, Southwest Fisheries Science Center, National Marine Fisheries Service, NOAA, Honolulu, HI. *Southwest Fish. Sci. Cent. Admin. Rep. H-98-02*, 14 pp.

DeMartini, E.E. & Williams, H.A. 2001. Fecundity and egg size of *Scyllarides squammosus* (Decapoda: Scyllaridae) at Maro Reef, Northwestern Hawaiian Islands. *J. Crust. Biol.* 21: 891–896.

DeMartini, E.E., McCracken, M.L., Moffitt, R.B., & Wetherall, J.A. 2005. Relative pleopod length as an indicator of size at sexual maturity in slipper (*Scyllarides squamosus*) and spiny Hawaiian (*Panulirus marginatus*) lobsters. *Fish. Bull.* 103: 23–33.

DiNardo, G.T. & Marshall, R. 2001. Status of lobster stocks in the Northwestern Hawaiian Islands, 1998–2000, Honolulu Laboratory, Southwest Fisheries Science Center, National Marine Fisheries Service, NOAA, Honolulu, HI. *Southwest Fish. Sci. Cent. Admin. Rep. H-01-04*, 47 pp.

DiNardo, G.T. & Wetherall, J.A. 1999. Accounting for uncertainty in the development of harvest strategies for the Northwestern Hawaiian Islands lobster-trap fishery. *ICES J. Mar. Sci.* 56: 943–951.

Dollar, R.A. & Landgraf, K.C. 1992. Annual report of the 1991 Western Pacific lobster fishery, Honolulu Laboratory Southwest Fisheries Science Center, National Marine Fisheries Service, NOAA, Honolulu, Hawaii 96822-2396. *Southwest Fisheries Science Center Administrative Report H-92-10*, 26 pp.

Freitas, A.E.T.S. & Santos, M.C.F. 2002. The "sapata" lobster *Scyllarides brasiliensis* Rathbun (1906) (Crustacea: Decapoda: Scyllaridae) studies in Pernambuco and Alagoas states coast — Brazil. *Bol. Tec. Cient. CEPENE* 10: 123–143.

Gates, P.D. & Samples, K.C. 1986. Dynamics of fleet composition and vessel fishing patterns in the Northwestern Hawaiian Islands commercial lobster fishery: 1983–1986, Honolulu Laboratory Southwest Fisheries Science Center, National Marine Fisheries Service, NOAA, Honolulu, Hawaii. *Southwest Fisheries Science Center Administrative Report H-86-17C*, 32 pp.

Guenette, S., Lauck, T., & Clark, C. 1998. Marine reserves: From Beverton and Holt to the present. *Rev. Fish Biol. Fish.* 8: 251–272.

Haddy, J.A., Courtney, A.J., & Roy, D.P. 2005. Aspects of the reproductive biology and growth of Balmain bugs (*Ibacus* spp.) (Scyllaridae). *J. Crust. Biol.* 25: 263–273.

Haight, W.R & Polovina, J.J. 1993. Status of lobster stocks in the Northwestern Hawaiian Islands, 1992, Honolulu Laboratory Southwest Fisheries Science Center, National Marine Fisheries Service, NOAA, Honolulu, HI. *Southwest Fish. Sci. Cent. Admin. Rep. H-93-0*, 23 pp.

Hanski, I. 1991. Single-species metapopulation dynamics: Concepts, models, and observations. *Biol. J. Linn. Soc.* 42: 17–38.

Holthuis, L.B. 2002. The Indo-Pacific scyllarine lobsters (Crustacea, Decapoda, Scyllaridae). *Zoosystema* 24: 499–683.

Kawamoto, K.E. & Pooley, S.G. 2000. Annual report of the 1998 western Pacific lobster fishery (with preliminary 1999 data). Honolulu Laboratory Southwest Fisheries Science Center, National Marine Fisheries Service, NOAA, Honolulu, Hawaii. *Southwest Fish. Sci. Cen. Admin. Rep. H-00-02*, 38 pp.

MacDonald, C.D. & Stimson, J.S. 1980. Population biology of spiny lobsters in the lagoon at Kure Atoll — preliminary findings and progress to date. In: Grigg, R.W. & Pfund, R.T. (eds.), *Proceedings of the Symposium on Status of Resource Investigations in the Northwestern Hawaiian Islands. April 24–25, 1980*: pp. 161–174. Honolulu: University of Hawaii, UNIHI-SEAGRANT-MR-80-04.

Martins, H.R. 1985. Biological studies of the exploited stock of the Mediterranean locust lobster *Scyllarides latus* (Latreille, 1803) (Decapoda: Scyllaridae) in the Azores. *J. Crust. Biol.* 5: 294–305.

Molina. L., Chasiluisa, C., Murillo, J.C., Moreno, J., Nicolaides, F., Barreno, J.C., Vera, M., & Bautil, B. 2004. Pesca Blanca y pesquerías que duran todo el año, 2003. In: Murillo, J.C. (ed.), *Evaluación de las Pesquerías en la Reserva Marina de Galápagos. Informe Compendio 2003*: pp. 103–139. Santa Cruz, Galápagos: Fundación Charles Darwin y Parque Nacional Galápagos.

Moffitt, R.B., Johnson, J., & DiNardo, G. Spatiotemporal analysis of lobster trap catches: Impacts of trap fishing on community structure. *Atoll Res. Bull.* 543: (in press).

Nishikiori, K. & Sekiguchi, H. 2001. Spiny lobster fishery in Ogasawara (Bonin) Islands, Japan. *Bull. Jap. Soc. Fish. Oceanogr.* 65: 94–102.

Parrish, F.A. & Boland, R.C. 2004. Habitat and reef-fish assemblages of banks in the Northwestern Hawaiian Islands. *Mar. Biol.* 144: 1065–1073.

Polovina, J.J. 1989. Density dependence in spiny lobster, *Panulirus marginatus*, in the Northwestern Hawaiian Islands. *Can. J. Fish. Aquat. Sci.* 46: 660–665.

Polovina, J.J. & Mitchum, G.T. 1992. Variability in spiny lobster *Panulirus marginatus* recruitment and sea level in the Northwestern Hawaiian Islands. *Fish. Bull.* 90: 483–493.

Polovina, J.J. & Moffitt, R.B. 1989. Status of lobster stocks in the Northwestern Hawaiian Islands, 1988, Honolulu Laboratory Southwest Fisheries Science Center, National Marine Fisheries Service, NOAA, Honolulu, HI. *Southwest Fish. Sci. Cent. Admin. Rep. H-89-3*, 10 pp.

Polovina, J.J., Moffitt, R.B., & Clarke, R.P. 1988. Status of stocks of lobsters in the Northwestern Hawaiian Islands, 1987, Honolulu Laboratory Southwest Fisheries Science Center, National Marine Fisheries Service, NOAA, Honolulu, HI. *Southwest Fish. Sci. Cent. Admin. Rep. H-88-3*, 8 pp.

Polovina, J.J., Haight, W.R., Moffitt, R.B., & Parrish, F.A. 1995. The role of benthic habitat, oceanography, and fishing on the population dynamics of the spiny lobster (*Panulirus marginatus*) in the Hawaiian Archipelago. *Crustaceana* 68: 203–212.

Seeb, L.W., Seeb, J.E., & Polovina, J.J. 1990. Genetic variation in highly exploited spiny lobster *Panulirus marginatus* populations from the Hawaiian archipelago. *Fish. Bull.* 88: 713–718.

Shaklee, J.B. & Samollow, P.B. 1984. Genetic variation and population structure in a spiny lobster, *Panulirus marginatus*, in the Hawaiian archipelago. *Fish. Bull.* 82: 693–702.

Spanier, E. & Lavalli K.L. 1998. Natural history of *Scyllarides latus* (Crustacea: Decapoda): A review of the contemporary biological knowledge of the Mediterranean slipper lobster. *J. Nat. Hist.* 32: 1769–1786.

Uchida, R.N. & Tagami, D.T. 1984. Biology, distribution, population structure, and pre-exploitation abundance of spiny lobster, *Panulirus marginatus* (Quoy and Gaimard, 1825), in the Northwestern Hawaiian Islands. In: Grigg, R. & Tanoue, K.Y. (eds.), *Proceedings of the Second Symposium on Resource Investigations in the Northwestern Hawaiian Islands, May 25–27, 1983*: pp. 157–198. Honolulu: University of Hawaii. Vol. 1.

Uchida, R.N., Uchiyama, J.H., Tagami, D.T., & Shiota, P.M. 1980. Biology, distribution, and estimates of apparent abundance of the spiny lobster, *Panulirus marginatus* (Quoy and Gaimard), in waters of the Northwestern Hawaiian Islands: Part 2, Size distribution, legal to sublegal ratio, sex ratio, reproductive cycle, and morphometric characteristics. In: Grigg, R.W & Pfund, R.T. (eds.), *Proceedings of the Symposium on Status of Resource Investigations in the Northwestern Hawaiian Islands, April 24–25, 1980*: pp. 131–142. Honolulu: University of Hawaii.

Vance, D., Smit, N., & Turnbull, C. 2004. Bugs. In: *National Oceans Office. Description of Key Species Groups in the Northern Planning Area*: pp. 275–280. Hobart, Australia: National Oceans Office.

13

The Biology of the Mediterranean Scyllarids

Daniela Pessani and Marco Mura

CONTENTS

Abstract

The five species of scyllarids actually living in the Mediterranean are discussed with regard to adult distribution, ecology, reproductive biology, larval development, and laboratory rearing. The native species, *Scyllarides latus* (Latreille, 1802) and *Scyllarus arctus* (Linnaeus, 1758), have been studied extensively and, therefore, many aspects of their biology are well known. The pygmy slipper lobster, *Scyllarus pygmaeus* (Bate, 1888), formerly considered to be a juvenile of *S. arctus*, needs, however, further study. Only one specimen of each of the two exotic species (*Acantharctus posteli* (Forest, 1963) and *Scyllarus caparti* Holthuis, 1952) has been collected to date in the Mediterranean. Updated data relevant to the distribution and reproductive periods of these species are given, along with the identification key of the adults. With regard to the phyllosoma (larval) and the postlarval stage ("pseudibacus" or "nisto"), data are incomplete, scarce, or even entirely absent (e.g., *S. caparti*). All past and current information relevant to the distribution of the larvae in the plankton, as well as attempts at rearing, are reviewed in this chapter. Since phyllosoma stage I larvae are known for at least four of the five Mediterranean species, this chapter provides an identification key for them. This chapter outlines the biology and ecology of the five Mediterranean species. General biology of scyllarids and specific information, such as genetics and behavior, of the Mediterranean species are covered in other chapters of this book (see Deiana et al., Chapter 3 for information on genetics, Sekiguchi et al., Chapter 4 on larval ecology, and Lavalli et al., Chapter 7, on behavior) and are not included in the present review.

13.1 Introduction

Five species of the family Scyllaridae (Crustacea, Decapoda) live in the Mediterranean Sea. Only two of them, *Scyllarides latus* (Latreille, 1802) and *Scyllarus arctus* (Linnaeus, 1758), are historically considered "true" (native) Mediterranean scyllarids — actually, the English vernacular name for *S. latus* is the "Mediterranean locust lobster". Thanks to the work of Forest & Holthuis (1960), *S. pygmaeus* (Bate, 1888) is also considered a Mediterranean species, but it is not endemic as it is also found in the Atlantic Ocean. Within the past 25 years, two additional species of scyllarids (considered exotics) have been found in Mediterranean waters: the Atlantic *Acantharctus posteli* (Forest, 1963) and *Scyllarus caparti* Holthuis, 1952.

Only two species (*S. latus* and *S. arctus*) have been and still are commercially fished in considerable amounts. They have even been used as ornaments (on a display board, in a restaurant in Cyprus, Eastern Mediterranean — Lewinsohn & Holthuis 1986). Today only *S. latus* is included in the European Economic Community (EEC) Habitats Directive (see Section 13.2), while all three nonexotic species are included in the Annex II of the Italian Aree Specialmente Protette in Mediterraneo (ASPIM) Directive.

Information and data available in literature are used and compiled together with unpublished data from our research and observations. With regard to the morphology of adults and larvae, only a short description of the characters useful for specific identification is given.

13.2 Autochthonous Species: *Scyllarides latus*

Vernacular names of this species include: Mediterranean slipper (or locust) lobster (English); grande cigale (French); magnosa (Italian); cavaco (carrasca) or lagosta-da-pedra (Portuguese) and cigarra (Spanish). For a complete list of local names, see Palombi & Santarelli (1961) and Spanier & Lavalli (1998). This species is the subject of intensive fishing because of its large size and tasty flesh (Holthuis 1991; Spanier et al. 1993; Bianchini et al. 1997; Spanier & Lavalli 1998), and, as a result, has become rare along the European coast of the western Mediterranean and in its Atlantic Ocean range of distribution (Martins 1985a; Spanier 1991). It is still quite common and fished, mainly by self-contained underwater breathing apparatus (SCUBA) and trammel nets, in the eastern Mediterranean along the coasts of Israel, Cyprus, and Turkey, along the southern coast of Crete (Greece), and along the North African coast. A reasonable population of the species has been discovered off the coast of Albania (Relini et al. 1999), most likely because the fishery is not yet as developed as it is in the rest of the Mediterranean. The species is subject to the 92/43/EEC Council Directive dated May 21, 1992, dealing with the preservation of natural and seminatural habitats, as well as of wild flora and fauna (Habitats Directive; Annex V: Animal and plant species of Community interest), and, thus, exploitation may be subject to management measures.

In Italy, this species is subject to uncontrolled fishing by trammel nets and underwater fishing (Relini et al. 1999), even though as early as 1976, it was already evident that it was becoming very rare (Orsi Relini et al. 1976). As a result of their rarity, Orsi Relini et al. (1976) refrained from collecting specimens of *Scyllarides latus*, even for scientific purposes. Fishing in Sardinia and any holding of animals have been forbidden for a period of three years beginning in 2000 (Froglia 2001). A feasibility study on stock enhancement in Italy was carried out by Bianchini et al. (1998) and considered the biological and economic aspects typical for such a project. The authors deemed that "the restocking might produce feasible results; still, further studies and a thorough cost/benefits analysis are required before actual implementation".

Several aspects of the biology of *S. latus* have been studied in the Azores Islands (Martins 1985a, 1985b) and in the southeastern Mediterranean in Israel (Spanier 1987; Almog-Shtayer 1988; Spanier 1989, 1991; Spanier et al. 1988, 1990, 1991, 1993; Spanier & Almog-Shtayer 1992; Spanier & Barshaw 1993; Spanier & Lavalli 1998, 2006; see also Lavalli et al., Chapter 7 and Bianchini & Ragonese, Chapter 9). The following is a summary of the data available on the species with some new additional information.

13.2.1 Short Description of the Adult

The rostrum is prominent and t-shaped. The carapace is quadrangular, with lateral granular borders; the gastric region bears three high and conspicuous tubercles. The thoracic sternum is not very large, with two conical tubercles on each segment. Abdominal segments two to four have an obtuse carina, consisting of a row of blunt tubercles. The first abdominal segment bears dorsally a circular spot and two lateral triangular spots. Abdominal segments are heavily crenate dorsolaterally. The fourth antennal segment bears an anterolateral tooth that is hooked and twisted out of the plane of the segment. The carpus of first pereiopod bears a distinct dorsal groove. The color of the dorsal surface is rusty red, while that of the ventral surface is deep yellow (according to Holthuis 1991; Spanier & Lavalli 1998, Figure 13.1).

The maximum total length (TL) is 450 mm, but specimens are usually not longer than 300 mm. In the eastern Mediterranean, males average 95.1 mm carapace length (CL) and reach a maximum of 115 mm CL, while females average 99.6 mm CL and reach a maximum of 140 mm CL. Comparing Atlantic specimens (as per Martins 1985a) with southeastern Mediterranean *Scyllarides latus*, Spanier & Lavalli (1998) stated that the former seemed to be heavier.

In terms of the growth rate, Bianchini et al. (2003) maintained in the laboratory specimens of *Scyllarides latus* and observed 59 cases of molting. On average, freshly molted specimens showed an incremental increase in weight of about 12%; the increment was higher in juveniles (24%) than in adults (6%). In addition, the authors tagged and released 115 adults of *S. latus*; a total of 29 specimens were recaptured (some multiple times) (see Bianchini & Ragonese, Chapter 9 for further information). One tagged female was caught by a trammel net after 1575 days at large, about 5 km away from the place of release. The specimen was fully viable and showed an increase of 28% in carapace length and 100% in body weight.

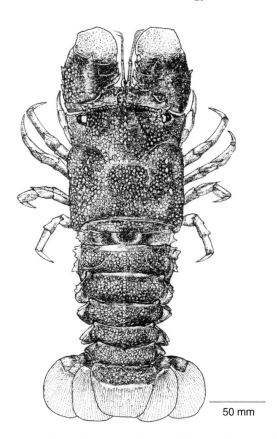

50 mm

FIGURE 13.1   *Scyllarides latus* (adult): dorsal view. From Holthuis 1991 used with permission.

13.2.2   Distribution and Ecology of the Adult

The species is present throughout the Mediterranean basin, except in the Northern and Central Adriatic (Holthuis & Gottlieb 1958; Holthuis 1991; Spanier & Lavalli 1998; Relini et al. 1999). The findings of the species in the Mediterranean, subsequent to 1991, are shown in Figure 13.2. *Scyllarides latus* is found also in the central eastern Atlantic Ocean, from the coast of Portugal (in the vicinity of Lisbon) to Senegal, Madeira, the Selvagens Islands, and the Cape Verde Islands (Holthuis 1991; Spanier & Lavalli 1998). It is generally nocturnal and typically dwells inshore on rocky substrates of infra- (*aka* shallow subtidal, dominated by barrens and seaweeds) and circalittoral (*aka* deeper subtidal, dominated by reef animals and sponge gardens) zones, except during its long-term movements (see Lavalli et al., Chapter 7 for more details on slipper lobster movement patterns). It is found in caves (Gili & Macpherson 1987; Manconi & Pessani 2003) and under boulders in the upper infralittoral zone, at 9 m depth. *Scyllarides latus* is found at depths ranging from 2 to 50 m, but occasionally is found as deep as 400 m (Lewinsohn & Holthuis 1986). However, in the eastern and southern waters of Sicily, it has been found in fishing gear operating at depths of 10 to 800 m (Pipitone & Arculeo 2003). In laboratory studies, *S. latus* seems to prefer rough substrata (Spanier & Lavalli 1998) similar to its natural habitat, but this preference may change with time in captivity (Barshaw & Spanier 1994; see also Lavalli et al., Chapter 7).

   Scyllarides latus feeds mostly on bivalves and gastropods (Spanier 1987; Spanier et al. 1988, 1993). Weak intraspecific interactions over food were observed in captivity by Barshaw & Spanier (1994) and Chessa et al. (1998).

13.2.3   Reproductive Biology

Many phases of the reproductive process have been observed in the laboratory, since sexually mature specimens of *Scyllarides latus* easily copulate, spawn, and brood in captivity (Spanier & Lavalli 1998).

FIGURE 13.2 Locations of the autochthonous and exotic species of scyllarids in the Mediterranean, subsequent to 1991. Note sources of location data are: *Scyllarides latus* — (1) Grippa 1993; (2) Manconi & Pessani 2003; (3) Pipitone & Arculeo 2003. *Scyllarus arctus* — (4) Relini et al. 1995; Rostagno 1996; Modena et al. 2001; (5) Sturani 1992; (6) Grippa 1993; (7) Manconi & Pessani 2003; (8) Rinelli et al. 1998; (9) Pipitone & Tumbiolo 1993. *Scyllarus pygmaeus* — (10) De Domenico et al. 2003; (11) Manconi & Pessani 2003; (12) Pipitone & Tumbiolo 1993; (13) Koukouras et al. 1993. *Acantharctus posteli* — (14) Garcia Raso 1982. *Scyllarus caparti* — (15) Froglia 1979.

Males produce gelatinous spermatophores which, in the laboratory, females carry for some days (6 to 10) before extruding eggs (Spanier & Lavalli 1998). In laboratory settings, fertilization of the entire egg mass often appears to be incomplete. Approximately 100,000 eggs (Almog-Shtayer 1988; Spanier et al. 1993) of ~0.5 mm diameter are spawned in the spring or early summer, just after mating, and are carried by females for between four to six weeks in Israel (Almog-Shtayer 1988) or six to eight weeks in the Azores (Martins 1985a). Reproductive peaks apparently vary according to the specific location and may be associated with local environmental conditions, such as water temperatures (Spanier et al. 1988). The egg mass is sometimes kept for a long time even if it has not been fertilized (Bianchini & Ragonese 2003). Berried females take care of the eggs, actively cleaning them with their modified, chelate fifth pereiopods. The lobsters are sensitive to handling and lose almost all of their eggs when frequently manipulated (Bianchini & Ragonese 2003), so care is required when holding berried females in laboratory settings.

Sex ratios appear close to unity, but are biased more toward males in the Azores Islands (Martins 1985a) and off Sicily (Bianchini et al. 1996), and more toward females in the eastern Mediterranean (Spanier et al. 1988). In the Atlantic, the ratio is 1.2 (Martins 1985a), while in Mediterranean waters off the coast of Israel, it is 1.08 (Spanier & Lavalli 1998).

13.2.4 Embryonic Development

The developmental phases *in ovo* can be defined by means of a six-stage scale, based on either egg color (Martins 1985a, 1985b) or egg dimension (Bianchini & Ragonese 2003) (see Table 13.1). The embryonic development is not synchronous (Bianchini & Ragonese 2003): the embryos develop through a nauplius, a prenaupliosoma, and an early-naupliosoma phase (a well-defined larva with evident eyes and nailed extremities), up to a naupliosoma. A few hours after hatching (maximum five, Martins 1985b), the free-swimming naupliosoma becomes a stage I phyllosoma, which reaches the postlarval nisto stage after several months and many molts (Spanier & Lavalli 1998; Bianchini & Ragonese 2003). Prenaupliosoma and naupliosoma stages have been illustrated by Bianchini & Ragonese (2003), and the naupliosoma was described and drawn by Martins (1985b) (Figure 13.3).

TABLE 13.1

Recognition of the Six-Stage Scale of *in ovo* Development Stages in *Scyllarides latus*

Stage[a]	Stage[b]	Egg Color	Egg Diameter [μm]	Description
1	0	Light yellow	600–650	Weak and scarce filaments; nonfecundated, or at the beginning of segmentation
2	1	Orange yellow	660–720	Well-developed filaments; eyes not visible at all
3	2	Yellow orange	740–760	Eyes hardly visible, embryo very difficult to see even at microscope; nauplius
4	3	Bright orange	770–790	Visible eyes, feathered antennas and rounded limbs; prenaupliosoma
5	4	Reddish orange	810–840	Evident eyes and chromatophores, well-defined larva, nailed extremities; naupliosoma
6	5	Brownish orange	865	Ready to hatch, limbs visible through the shell; advanced naupliosoma

[a] See Martins (1985a).
[b] See Bianchini & Ragonese (2003).

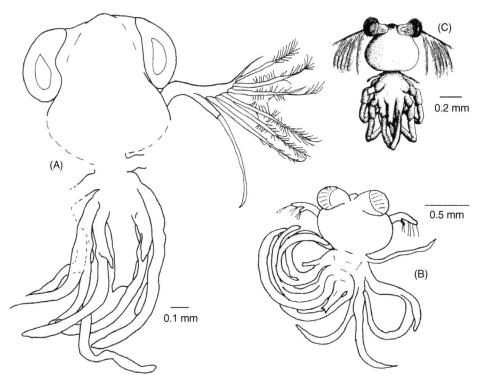

FIGURE 13.3 *Scyllarides latus*: (A) prenaupliosoma; (B) naupliosoma; (C) naupliosoma. (From Bianchini & Ragonese 2003 (A, B) and Martins 1985b (C); used with permission.)

13.2.5 Larval Development

13.2.5.1 Brief History

Many attempts have been made to obtain all stages (probably 11) of the phyllosomas and the postlarval, or nisto stage (*aka* pseudibacus stage) of *Scyllarides latus*. Sampling of planktonic specimens has been

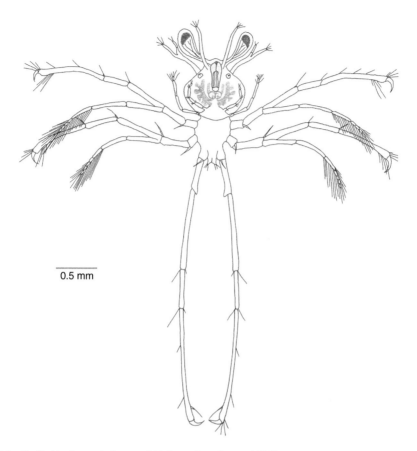

FIGURE 13.4 *Scyllarides latus*: phyllosoma I. Redrawn from Santucci 1925.

attempted (Martins 1985b; Fiedler & Spanier 1996), as has culture of larvae obtained from ovigerous females (Martins 1985b). Unfortunately, neither field sampling nor laboratory culture has been successful. The phases of the larval development in *Scyllarides* were described by Robertson (1969) and Crosnier (1972) (see also Sekiguchi et al., Chapter 4); therefore, the information presented herein is mostly based on planktonic specimens.

Bouvier (1913) remarked upon the correspondence between *Pseudibacus veranyi* Guérin Ménville, 1855 and the nisto stage of *S. latus*. As a result, *Pseudibacus* lost its value as a genus, and instead became a noun indicating the postlarval stage of *Scyllarides* spp.

13.2.5.2 Morphological Notes

Among the Mediterranean scyllarid species, the phyllosoma of *Scyllarides latus* can be recognized easily by its antennae, clearly biramous in the early stages, and the pear-shaped cephalic shield. The pereiopods bear a coxal spine only in the early stages. The stage I phyllosoma (Figure 13.4) has been described and illustrated based on either planktonic (Santucci 1925, 1928) or laboratory-reared (Martins 1985b; Glavic et al. 2001) specimens. Comparisons between identical stages of the two types of specimens show some discrepancies with regard to the presence of setae or spines on the antennule and on the antennal exopod and endopod, the articulation (or its absence) of the antennal endopod, and the length of the antennule compared with one of the antenna (Table 13.2). We assume that a few details have been either misinterpreted or not detected by all of the prior authors.

Stage II phyllosoma planktonic specimens have been sampled, described, and illustrated by Fiedler & Spanier (1996); in this stage, the shape of the cephalic shield is not pear-shaped, but rhomboidal. Stage III and IV phyllosomas have not been described. A stage V phyllosoma caught off the Balearic Islands

TABLE 13.2

Scyllarides latus: Comparison Between Planktonic and Laboratory-Reared Phyllosoma I Observed by Various Researchers

Characters	Santucci (1925)	Santucci (1928)	Martins (1985b)	Glavic et al. (2001)
TL [mm]	1.7–1.8	1.7–1.8	1.8	1.59–1.76
A1	4 Apical setae	4 Apical setae	Like Santucci (1928)	3 Setae + 1 spine
A2 endopod	Not articulate, with apical setae	Not articulate, with thin apical setae	Like Santucci (1928)	Articulate, with 3 apical spines
A2 exopod	2 Apical spines	Thin apical setae	Like Santucci (1928)	2 Apical spines
A1 vs. A2	A1 as long as A2	A1 shorter than A2	A1 longer than A2	A1 longer than A2

Abbreviations used: A1 = Antennule; A2 = antenna; TL = Total length.

and described and illustrated by Stephensen (1923) as *Thenus orientalis* (Lund, 1793) is a larva of *S. latus* because of its location, size (TL = 7 mm), and the presence of setose exopodites on the fourth pereiopods. Stage VI and VII phyllosoma have not been described. Some non-Mediterranean stage VIII phyllosomas have been described and illustrated by Maigret (1978), and have a TL of 17.5 to 19.7 mm. Their cephalic shield is oval and as wide as the thorax. Stage IX phyllosomas have not been described. As with stage V, stage X phyllosomas identified as *T. orientalis* by Santucci (1926) could be larvae of *S. latus* because of their location (off Sicily), their abdominal length, the absence of gill buds on maxillipeds and pereiopods, and a TL of 30 mm. Stage XI phyllosomas have been collected off the Azores Islands and drawn and described by Martins (1985b). Their TL is 48 mm, all of the pereiopods bear gill buds, the pleopods are biramous, and the uropods extend beyond the telson.

Nistos differ from small, adult specimens of *S. latus* with regard to their carapace (slightly dorsally convex, with a smooth surface that is free from tubercles), their carapace lateral margins (with some teeth before the hepatic groove and 11 to 12 teeth after it), and the dorsal carinae of their abdominal segments (with spine-like projections) (Bouvier 1913). The size of the nisto specimens has not been reported (Spanier & Lavalli 1998).

No live juveniles of *S. latus* have ever been sampled. The smallest juvenile (male) of 64 mm CL (200 g) was sampled by Almog-Shtayer (1988), while the smallest specimens caught by Bianchini et al. (1996) weighed 100 and 103 g. Spanier & Lavalli (1998) reported two small exuviae found on the Israeli Mediterranean coast, one of 38 mm CL at 15 m depth and one of 47.5 mm CL found washed ashore. They also reported a small preserved male of 34.3 mm CL located at the Museum of Zoology of the University of Florence. It had been collected by a scientific trawl in Italian waters, possibly at depth greater than 400 m.

13.2.6 Larval Rearing

No published information is available regarding the methods used in the rearing attempts of the species (Glavic et al. 2001; Bianchini & Ragonese 2003). According to Martins (1985b), stage I phyllosomas that hatched in the laboratory from ovigerous females died without molting to stage II, probably because the water temperature in the laboratory (21 to 24°C) was too high compared to their natural environment. (For more information on larval rearing, see Mikami & Kuballa, Chapter 5.)

13.2.7 Geographical and Seasonal Distribution of Phyllosoma Stages

Information on collection locations and times of the year is sparse. The specimens were collected as follows: stage I phyllosomas were collected in July, August, and September off Ganzirri (Sicily), Naples, and Messina, respectively (Santucci 1925, 1928). Stage II phyllosomas were collected in the eastern Mediterranean (no month indicated — Fiedler & Spanier 1996), while stage V phyllosomas were collected east of the Balearic Islands in August (Stephensen 1923). Stage X phyllosomas were collected in January, always off Ganzirri (Santucci 1926).

13.3   Autochthonous Species: *Scyllarus arctus*

The vernacular names for *Scyllarus arctus* include: lesser slipper lobster, small European lobster, or locust lobster (English); petite cigale (French); magnosella (Italian); cigarra-do-mar (galé) or lagosta-da-pedra (Portuguese); and santiaguiño (Spanish). For the complete list of the local names, see Palombi & Santarelli (1961). Although this species is edible and used as food, there is no special or dedicated fishery; it is typically collected as a "by-catch" in fisheries using nets, trawls, dredges, and seines (Relini et al. 1999), or via SCUBA. In some places, the divers seem to have decimated the populations (Relini et al. 1999). The lobsters are sold at local markets, but their small size and irregular catch make them unprofitable (Relini et al. 1999).

The species seems to be quite suitable for rearing either as isolated individuals or in groups, and shows a wide range of tolerance for salinity variation. In the laboratory, at temperatures of 20°C, it can tolerate salinities ranging from 29.5 to 65.0‰ (Vilotte 1982).

13.3.1   Short Description of the Adult

The rostrum does not protrude. The carapace is rectangular with crenate borders and three acute teeth before the cervical groove. In the thoracic sternum, the anterior margin has a wide triangular groove while the last segment bears an anteroposteriorly, medial, flattened tubercle. The second pleuron is acutely pointed distally and directed posteriorly. Abdominal segments two to five bear branching, narrow grooves only on the exposed parts of the segments; a complete transverse groove is present only on the first segment. The fourth antennal segment has a single oblique median carina, bears two teeth on its outer margin and three to four teeth on its inner margin (according to Holthius 1991, Figure 13.5). The carapace is colored reddish brown, with scales and spines white colored distally. The abdominal segments are a pale blue with orange stripe. Males are smaller than females, being, on average, 78 mm TL compared to 138 mm TL. The maximum TL reported is 160 mm (Relini et al. 1999).

13.3.2   Distribution and Ecology of the Adult

In the eastern Atlantic, *Scyllarus arctus* is found from the south coast of the British Islands (Plymouth) to the Azores, Madeira, and Canary Islands. It is found in the Mediterranean in the infra- and circalittoral zones, mainly on rocky substrates with holes and hiding places, at depths from five down to 50 m. The species has been collected also between 50 and 100 m (Orsi Relini & Costa 1981), between 100 and 200 m (Relini et al. 1999), and even at depths greater than 200 m (Naples, Italy — Moncharmont 1979). It is possible to find the species also in *Posidonia oceanica* (Linneaus) Delile, 1813 meadows, muddy substrates (Holthuis 1991), and caves (Bianchi et al. 1986; Manconi & Pessani 2003). Specimens of *S. arctus* have been observed, although rarely, under boulders of the upper infralittoral at depths from 5 to 12 m (eastern and western Ligurian Riviera — Sturani 1992; Rostagno 1996). According to Holthuis (1991), the species is distributed in the Eastern Atlantic Ocean as well as in the entire Mediterranean; the reports of the species in the Mediterranean, subsequent to 1991, are presented in Figure 13.2.

The tendency to seek a shelter was confirmed by Modena et al. (2001) who fished for specimens of *S. arctus* around the wreck of the vessel *Haven*, off Varazze (Ligurian Sea). The collecting area was characterized by lumps of tar and tar residues that made the seabed less uniform and full of crevices and shelter. As is the case with most other lobsters, this species is mainly nocturnal in its activity (Udekem d'Acoz 1999).

13.3.3   Reproductive Biology

In the Mediterranean, ovigerous females (70 mm minimum TL) are present all year, with reproductive peaks in March/April and July/September (Relini et al. 1999). Studies on gonadal structure and gametogenesis (Cau et al. 1988) confirm these as the main reproductive periods. Numerical variations of oogonia and oocytes were correlated and showed two consecutive laying periods for the females of this species. The proliferation of oogonia and maturation of oocytes are cyclical with two respective

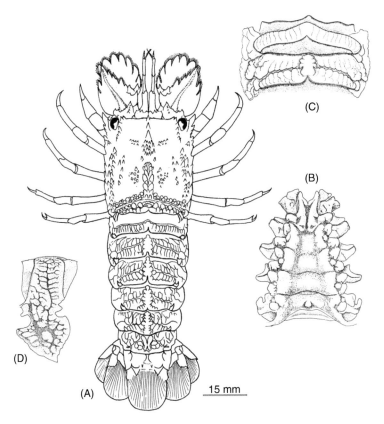

FIGURE 13.5 *Scyllarus arctus* (adult): (A) dorsal view; (B) sternum; (C) dorsal view of the first two abdominal segments; (D) lateral view of the 2nd pleuron. (From Holthuis 1991; used with permission.)

peaks, one in May and December and one in March and July/August. On the other hand, spermatogenesis is a continuous process, with two periods of massive spermatogonial division in December and March that produces sperm for fertilization of the eggs extruded just after March and July/August (Cau et al. 1988).

Females (TL varying from 100 to 130 mm) produce between 30,000 and 70,000 eggs, which are extruded and packed together in a dense cluster prior to being cemented onto the pleopods (Vilotte 1982). Egg diameter ranges from 0.4 to 0.5 mm. Egg color is, at first, a golden yellow, but then becomes orange (Vilotte 1982). Ovigerous females have been observed to use their large, spade-like antennae either to bury into mud, sand, or gravel, or to hide among and under stones (Mura et al. 1984).

The sex ratio (males/females) is close to unity (1.06) (Mura et al. 1984).

13.3.4 Larval Development

13.3.4.1 Brief History

In the laboratory, Vilotte (1982) has observed, but not described in detail, few mobile larvae, with legs not yet outstretched; these probably were naupliosoma specimens, which died within a few hours. With regards to descriptions of the phyllosoma larvae, the situation is probably more confusing than that for *Scyllarides latus.*

According to Holthuis (1977), in 1827 Risso described and illustrated the phyllosoma of *Scyllarus arctus* as *Chrysoma mediterraneum.* Dohrn (as cited in Stephensen 1923) obtained a phyllosoma from the egg. Stephensen (1923) also collected a great number of phyllosoma specimens from the field, which he attributed to *S. arctus.* Many subsequent researchers (Robertson 1969; Barnich 1996; Gonzáles-Gordillo &

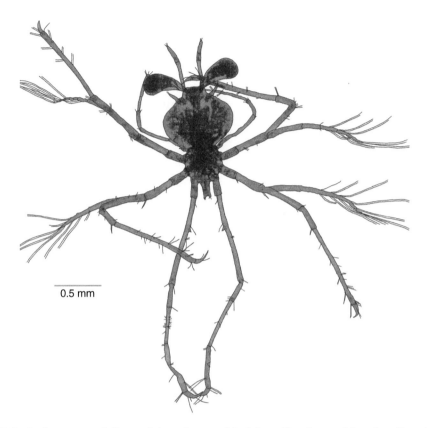

FIGURE 13.6 *Scyllarus arctus*: phyllosoma I, drawn from an original photo. (Photo by one of the authors (Pessani), redrawn by G. Rappini.)

Rodríguez 2000) have expressed doubts concerning the correct identification of these specimens. It is probable that some of them were phyllosoma stages of other Mediterranean *Scyllarus* species, mainly *Scyllarus pygmaeus*. As a consequence, other authors (Santucci 1925; Gurney 1942; Kurian 1954; Williamson 1983; Fiedler & Spanier 1999) who have dealt with Mediterranean *Scyllarus* phyllosomas, based on the work by Stephensen (1923), could have identified larvae of both *S. artcus* and *S. pygmaeus* as *S. arctus* phyllosomas.

Earlier authors thought that the complete larval development of the *Scyllarus* species consisted of eight (Phillips & McWilliams 1986), nine (Robertson 1971), or 13 stages (Barnich 1996). In plankton samples collected in the Eastern Mediterranean, Fiedler & Spanier (1999) recorded the third to the twelfth phyllosoma stages of *S. arctus*: the stage XII phyllosoma bore gill buds on the legs, a character peculiar to the final phyllosoma stage before the molt to the nisto (Robertson 1969). Most of their phyllosomas were found in eddies. Thus, they suggested that this physical, but ephemeral, oceanographic structure can serve as a developmental habitat for this species.

In the laboratory, Vilotte (1982) observed the first stage of *S. arctus* hatched from ovigerous females; many phyllosomas molted to stage II, but only some to stage III. Pessani et al. (1999) succeeded in obtaining the complete larval development of this species. From stage I (Figure 13.6), the phyllosomas progressed through 16 molts before metamorphosizing into the nisto stage. The problem of the presence of additional instars during the laboratory rearing and their comparison with the planktonic stages was already pointed out by Vilotte (1982), and is currently under further investigation by Pessani (unpublished data, 2001; see also Mikami & Kuballa, Chapter 5 for a discussion of the production of additional instars with laboratory culture).

Bouvier (1913) noted the resemblance between *Nisto laevis* (Sarato, 1885) and *Nisto asper* (Sarato, 1885) and the natant stage of *S. arctus*. As a consequence, *Nisto* (originally described as a subgenus of the genus *Arctus*) has lost its generic (or subgeneric) status and has become the noun indicating the

TABLE 13.3

Size of Phyllosomal Stages of *Scyllarus arctus* and Phyllosoma Identified as *S. arctus* by Various Authors, but Most Likely *S. pygmaeus* (See Text, Section 13.3.4.2, for Further Explanation)

Stage	Pessani et al. (1999); Pessani, Unpublished Data, 2001	TL [mm]		Fiedler and Spanier (1999)
		Stephensen (1923)	Santucci (1925)	
1	1.1–1.4	1.5	1.0–1.5	—
2	1.2–1.8	1.5–2.0	1.5–2.5	—
3	1.6–2.2	2.5–3.0	2.5–4.0	3.3–3.4
4	2.3–2.8	3.0–5.0	4.0–6.0	4.3–4.9
5	2.9–3.6	4.0–9.0	6.0–7.5	6.1
6	3.6–4.3	7.0–17.0	7.5–9.0	7.0–7.9
7	4.4–5.1	9.0–20.0	9.0–14.0	—
8	4.6–5.6	18.0–23.0	14.0–22.0	10.1–12.3
9	6.0–6.75	17.0–28.0	19.0–26.5	12.8–13.7
10	7.8–8.0			15.5–17.0
11	8.8–9.7			19.0
12	9.8–10.2			21.6–24.4
13	11.6–12.3			
14	13.6–14-6			
15	14.5–15.8			
16	16.5–18.3			

Abbreviations used: TL = total length.

postphyllosomal (natant) or postlarval stage of *Scyllarus* spp. These two species are now synonomous with *S. arctus* (Holthuis 1991). Bouvier (1913, 1917) emphasized that the two species of *Nisto* might be subsequent stages (the first, *N. asper*, probably a natant stage; the second, *N. laevis*, probably a reptant stage) of the same species, *S. arctus*, during the phase of metamorphosis from a planktonic larva to a benthic adult. Vilotte (1982) suggested that the nisto stage has only one molt, taking into account Fedele's (1926) observations. Fedele obtained a nisto in the laboratory from a final-stage phyllosoma of *S. arctus* sampled from the plankton. He observed the molting of other nisto specimens and their transformation to juveniles. It is, therefore, highly doubtful that the nisto specimens described and identified by Sarato as *N. asper* might be from another scyllarid species, such as *S. pygmaeus*.

13.3.4.2 Morphological Notes

In Table 13.3 the TL of *S. arctus* instars cultured by Pessani et al. (1999, unpublished data 2001) is compared with the TL of the phyllosoma stages identified as *S. arctus* by other investigators. Table 13.4 summarizes the characters that enable one to distinguish the 16 cultured phyllosoma instars, and compares these characters with those provided by previous authors (Santucci 1928; Stephensen 1923; Fiedler & Spanier 1999) describing field-caught or laboratory-observed specimens. As can be seen, there is no agreement with respect to the appearance of the characters in planktonic specimens among the various authors (see Table 13.4). Furthermore, the specimens described by Santucci (1928) seem to be smaller and more precocious than those described by Stephensen (1923). Likewise, the Eastern Mediterranean specimens described by Fiedler & Spanier (1999) were also small. As such, these specimens may be phyllosomas of *S. pygmaeus*, rather than *S. arctus*.

Besides the historic and doubtful descriptions mentioned above, the nisto of *S. arctus* has been described recently based on living specimens collected by Garcia Raso (1982). The main distinctive features include the carapace (very depressed, with two mid-dorsal teeth and a crenate carina between the mid-dorsal line and the lateral margins), the last thoracic sternite (bearing two strong, lateral, posteriorly directed spines), and the second abdominal pleuron (with smooth and distally pointed margins). The specimens obtained in the laboratory show the same features, but with the first four abdominal pleurons having notched margins

TABLE 13.4

Characteristics that Separate Each Stage of *Scyllarus arctus* from Those Preceding

Stage as per Pessani et al. (1999), Pessani Unpublished Data, 2001	Characteristics	Stage as per Stephensen (1923)	Stage as per Santucci (1925)	Stage as per Fiedler & Spanier (1999)
1	Eyestalk unsegmented	1	1	—
2	P4 bud shorter than abdomen	2	2	—
3	P5 bud	3	3	—
4	P4 exopod naked	4	3	—
5	A2 a small pointed process on outer side	4	3	—
6	P4 exopod setose	5	4	—
	Uropods as simple buds	4 or 5	3	8
7	A1 peduncle 2-segmented	5	4	—
8	P5 bud longer than half the abdomen	7 (?)	5	—
9	A1 peduncle 3-segmented	6	5	—
	pleopods as simple buds	6	4	8
10	Abdomen straight	6	5	9
11	P5 endopod 2-segmented	7	6	—
12	Pleopods as bilobed buds	7	6	10
	uropods as biramous buds	6	5	10
13	P5 reaching the tip of the abdominal spine	—	–	—
14	Pleopods as elongated buds	9	7	12
15	Pleopods biramous buds	8	8	—
16	Gill buds at the base of mxp2–3 and P1–P4	—	9	12

As observed by Pessani et al. (1999), Pessani (unpublished data 2001) compared with different stages of phyllosomae identified as *S. arctus* by different authors (Stephensen 1923, Santucci 1925, Fiedler & Spanier 1999). Abbreviations used: A1 = antennule; A2 = antenna; mxp 2–3 = maxillipeds of the 2nd and 3rd pair; P1–P5 = pereiopods of the 1st to 5th pair.

(Pessani, unpublished data 2001). Nisto specimens of *S. arctus* differ from juveniles mainly in their softer exoskeleton and the lack of almost all pigmentation (Vilotte 1982). Reported sizes for nistos range from 17.0 to 22.0 mm TL (Vilotte 1982), 20 to 21 mm TL (Garcia Raso 1982), and 10.3 to 20.0 TL mm (Pessani, unpublished data 2001).

13.3.5 Larval Rearing

Vilotte (1982) reported the different conditions used in laboratory rearing of *Scyllarus arctus* phyllosomas. She concluded that rearing en masse was unsatisfactory, since the phyllosomas tended to aggregate, twisted their long legs, and then died, probably because they did not succeed in disentangling from each other. A temperature of 23 to 24°C (during the day) and 21 to 22°C (during the night) with a photoperiod (L:D) of 12 : 12 h seemed to be most suitable for the molt from stage I to II. Pessani et al. (1999) reared the phyllosomas en masse at 20 ± 1°C, in a 80 l aquarium, using a closed system with natural seawater. The larvae were fed at first on *Artemia salina* Leach, 1819 nauplii, then after ~80 days on whisked fish meat and beef meat. The phyllosomas molted to the nisto stage after 192 days (Pessani et al. 1999; Pessani, unpublished 2001).

13.3.6 Geographical and Seasonal Distribution of Phyllosoma and Nisto Stages

The information regarding the geographical and seasonal distribution of the phyllosoma and nisto stages of *S. arctus* in the Mediterranean seems less relevant in light of the earlier-mentioned, possible misinterpretation of species. Nevertheless, the presence of the larvae in a given area and in a given season could be compared to detailed information about the distribution and the reproductive period of the adults. This comparison may lead to a better understanding of the biology of the two species.

The phyllosomas of *S. arctus* have been collected throughout the Mediterranean, from the Straits of Gibraltar to Marmara Sea, from Ligurian Sea to Egyptian (Stephensen 1923) and Lebanese (Lakkis & Zeidane 1988) coastal waters, in the Adriatic Sea, and not far off the Mediterranean coast of Israel (Galil 1993). Using neuston nets to perform plankton hauls, Fiedler & Spanier (1999) found *S. arctus* phyllosomas about 260 km off the Israel coast in March. In contrast, Stephensen (1923) collected phyllosoma stages I through VIII (especially I through IV) in July and August, while sampling only few phyllosomas (stages V through VIII) in the winter months. In the Adriatic Sea, nearly all of the eight phyllosoma stages were found in August (Kurian 1956). Nisto specimens were found off Nice, France at unspecified times (Vilotte 1982), in the Ionian Sea in December (Stephensen 1923), and at Malaga at unspecified times (Garcia Raso 1982). Those found by Stephensen (1923) and Garcia Raso (1982) were identified as *S. arctus*.

13.4   Autochthonous Species: *Scyllarus pygmaeus*

The vernacular names for this species include: pygmy locust lobster (English); cigale naine (French); magnosella (Italian); cigarra-do-mar (galé) or lagosta-da-pedra (Portuguese); and cigarra enana (Spanish). For the complete list of the local names, see Fischer et al. (1987).

Scyllarus pygmaeus is quite rare in the Mediterranean and this explains its late acknowledgment as an autochthonous species. According to Holthuis (1977), in 1827 Risso described *S. arctus* and *S. cicada*, based on small-sized specimens collected near Nice, France. These small scyllarids were probably juvenile forms of *S. arctus*, but it is also possible that they were adults of *S. pygmaeus*, a species not yet identified at that time. Bate (1888, cited in Holthuis 1977) described and illustrated the species under the name of *Arctus pygmaeus* after a specimen collected off the Canary Islands. This species was considered to be endemic to the Atlantic, and small specimens of scyllarids present in the Mediterranean were identified as young *S. arctus*, when actually they were adults of *S. pygmaeus* (Forest & Holthuis 1960). Caroli (see Forest & Holthuis 1960) had already hypothesized (but not published) that the small Mediterranean specimens identified as *S. arctus* might actually be *S. pygmaeus* instead. Moreover, De Man (see Forest & Holthuis 1960) noted morphological differences among the Mediterranean specimens identified as *S. arctus*, but did not conclude that those differences resulted from a misidentification of some of the specimens, which really were adults of *S. pygmaeus* rather than juveniles of *S. arctus* (Forest & Holthuis 1960). According to Moncharmont (1979), a young female *S. pygmaeus* collected at Naples in 1878 (or 1879) was kept at the Natural History Museum in Leiden (Holland).

13.4.1   Short Description of the Adult

The rostrum is slightly pronounced. The carapace is rectangular with three distinct, sharp teeth in the median line before the cervical groove. The anterior margin of the sternum has a U-shaped incision medially, and a median, conical tubercle is present on the last segment. The second pleuron of the abdominal segments is distally rounded and directed ventrally. The smooth anterior part of the tergites of abdominal segments two to six show a distinct branching groove on each half. The fourth antennal segment bears a single median carina, two teeth on the outer margin, and three to four teeth on the inner margin (according to Holthuis 1991, Figure 13.7). The color is pale brownish or pinkish, with a pale blue and an orange stripe on abdominal segments.

Males are smaller, reaching a maximum TL of 54 mm, while females reach a maximum TL of 65 mm. Forest & Holthuis (1960) provided the seven morphological characters which enable one to distinguish *S. pygmaeus* specimens from the juveniles of *S. arctus*. Cau et al. (1978) added another character concerning the tergite of the first abdominal segment: in *S. pygmaeus*, a white median stripe is followed, on each side of the tergite, by a black zone, a white stripe, and a reddish zone whereas in *S. arctus*, the tergite is red-orange, bearing in the middle two small, hole-like areas, covered by hairs, and surrounded by a black zone.

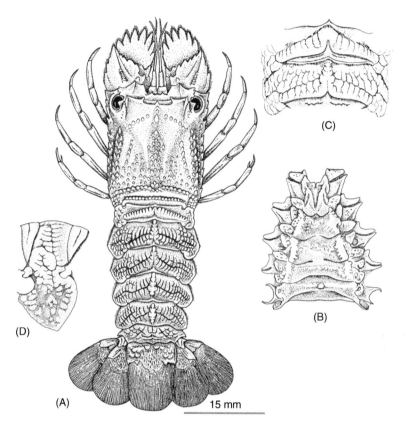

FIGURE 13.7 *Scyllarus pygmaeus* (adult): (A) dorsal view; (B) sternum; (C) dorsal view of the first two abdominal segments; (D) lateral view of the 2nd pleuron. (From Holthuis 1991; used with permission.)

13.4.2 Distribution and Ecology of the Adult

Scyllarus pygmaeus is found in the Eastern Atlantic along the Madeira, Canary, and Cape Verde Islands at depths from 20 to 1200 m. In the Mediterranean, *S. pygmaeus* can be found from western Mediterranean (Melilla, Spanish Morocco) as far as the Israeli coast and Cyprus. Holthuis (1991) stated that it had not yet been reported along the North African coast east of Morocco. Figure 13.2 shows the reports of the species in the Mediterranean, subsequent to 1991. The species lives on sandy and coralline substrates, and in *Caulerpa prolifera* (Forsskål) Lamouroux, 1809 and *Posidonia oceanica* (Linnaeus) Delile, 1813 beds, often buried in the substratum or hidden among *P. oceanica* rhizomes. Its depth distribution ranges between 5 and 146 m. In Sardinia waters, where the adults were found in abundance from 5 to 50 m (Cau et al. 1978), *S. pygmaeus* seemed less frequent in late spring and summer when the lush *Posidonia* leaves prevent the trawl nets from working properly; as a consequence *S. pygmaeus* specimens are caught less frequently or not at all (Cau et al. 1978). Recently, the species has also been reported in caves (Manconi & Pessani 2003).

S. *pygmaeus* feeds on small living bivalves and dead fish (Cau et al. 1978). As with other scyllarids, it has no offensive adaptations and thus it is also necrophagous and detritivorous.

13.4.3 Reproductive Biology

Ovigerous females are present during the whole year, with the reproductive peak between May and July. In Sardinia waters, there is a reproductive low in August and September (Mura & Pessani 1994). In the Eastern Mediterranean, the reproductive period lasts longer: some ovigerous females have been observed in December, at depths from 27 to 82 m (Lewinsohn & Holthuis 1986).

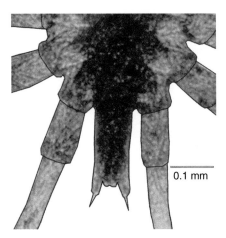

FIGURE 13.8 *Scyllarus pygmaeus*: telson of the phyllosoma I. (Photo by one of the authors (Pessani), redrawn by G. Rappini.)

The minimum reproductive TL is 23 mm for females (Forest & Holthuis 1960) and 20 mm for males. In the laboratory, the eggs are carried for 60 to 70 days before hatching (Cau et al. 1982).

The sex ratio (males/females) is 0.94. This ratio may be influenced by the behavior of males, which tend both to hide and to move more than females. Consequently, male have lower probability of detection (Cau et al. 1982).

13.4.4   Larval Development

13.4.4.1   Brief History

The larval development of *Scyllarus pygmaeus* is not well known. Bate (1888, cited by Thiriot 1974) illustrated the naupliosoma of *Arctus* (= *Scyllarus*) *pygmaeus* taken from the egg. In the laboratory, a stage I phyllosoma was obtained by Thiriot (1974), who did not describe the larva, and by one of the author of this chapter (Mura). The larvae died immediately after hatching. As regards the nisto, Bouvier (1917, p. 115) stated that the specimen described by Bate (1888) as *Arctus immaturus* is nothing but the second stage of a *S. pygmaeus* nisto.

13.4.4.2   Morphological Notes

Mura & Pessani (1994) summarized some morphological characteristics of the small stage I phyllosoma. The few features indicated by the authors were insufficient to distinguish the stage I phyllosoma of *S. pygmaeus* from that of other species. Using comparable characteristics, the current reexamination of the few specimens available shows that the cephalic shield is almost round and the terminal spines of the abdomen go beyond the joint between the first and second segment of P4 (Figure 13.8).

13.5   Exotic Species: *Acantharctus posteli*

Holthuis (2002) has assigned the species *Scyllarus posteli* (Forest, 1963) to the genus *Acantharctus*; it is characterized by a sharp median spine in the middle of the last segment of the thoracic sternum (*Acanthus* = thorn; *arctus* = junior synonym of *Scyllarus*). In the Mediterranean, only a single specimen of the species has been reported (Garcia Raso 1982).

13.5.1   Short Description of the Adult

The rostrum is absent in this species. The carapace is subquadrate with lateral margins tridentate and diverging anteriorly. The dorsal median carina is rugose with pregastric and cardiac teeth; the lateral

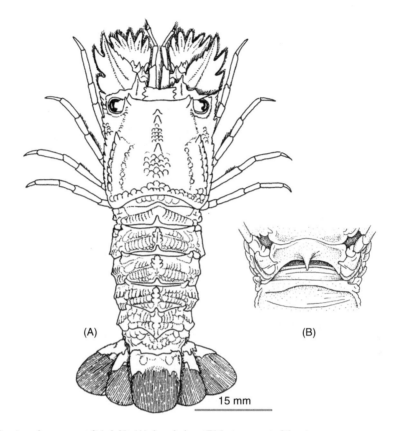

FIGURE 13.9 *Acantharctus posteli* (adult): (A) dorsal view; (B) last segment of the sternum.

carinae are coarsely rugose. The anterior margin of the thoracic sternum is concave and the last sternal segment bears a prominent median spine. In the male, the posterolateral angles of the sternum are tooth-like. The median carina on abdominal segments two through five is obtuse, and the second pleuron is triangular with its distal angle acutely pointed. No color has been recorded for this species (according to Galil et al. 2002) (Figure 13.9).

In the Atlantic Ocean, males range in TL from 30 to 65 mm; females range in TL from 48 to 52 mm (Anadón 1981). The only recorded specimen in the Mediterranean, which was not sexed, had a TL of 48.65 mm (Garcia Raso 1982).

13.5.2 Distribution and Ecology of the Adult

Along the eastern Atlantic coast (from Senegal to the Congo and Cadiz, Atlantic Spain), this tropical and subtropical species lives on either sandy bottoms with algae (*Caulerpa* spp.) or muddy bottoms, at depths between 0–5 and 70 m (Garcia Raso 1982; Galil et al. 2002). Along the Mauritanian coast, *Acantharctus posteli* dwells in rocky, sandy, and detritic substrates, between 39 and 60 m (Anadón 1981). The single Mediterranean specimen was collected in Malaga, Spain in the infralittoral zone (depth not indicated) (Garcia Raso 1982) (Figure 13.2).

13.5.3 Reproductive Biology

There is very little information available about the reproductive biology of this species. In the Atlantic, ovigerous females have been collected in March, April, May, July, November, and December (Anadón 1981). The minimum TL of sexually mature females is 30 mm (Anadón 1981).

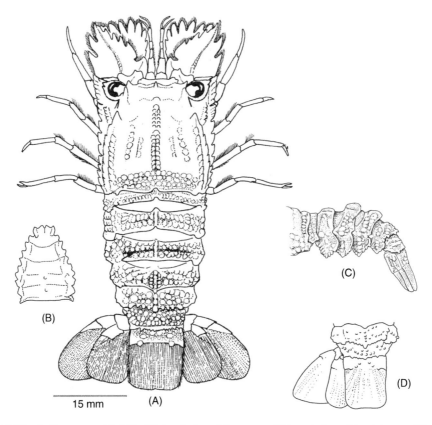

15 mm   (A)

FIGURE 13.10 *Scyllarus caparti* (adult): (A) dorsal view; (B) sternum; (C) lateral view of the abdomen; (D) telson and left uropods.

13.5.4   Larval Development

13.5.4.1   Morphological Notes

A stage I phyllosoma has been reared in the laboratory from an ovigerous female (Gonzáles-Gordillo & Rodríguez 2000); otherwise no attempts at rearing have been described. The body measurements of the stage I phyllosoma were: TL of 1.173 ± 0.05 mm; carapace shield length (CSL) of 0.682 ± 0.028 mm; and carapace shield width (CSW) of 0.705 ± 0.034 mm. On the distal half of the antennule margin, a short plumose seta is visible. The abdomen is shorter than the first proximal segment of the fourth pereiopod.

13.6   Exotic Species: *Scyllarus caparti*

13.6.1   Short Description of the Adult

The rostrum is small. The carapace is subquadrate with slightly diverging, dentate lateral margins. The median carina of the carapace bears three rounded teeth and two rows of tubercles. The anterior margin of the thoracic sternum has two rounded lobes separated by a short fissure; the last segment bears an obtuse median tubercle. A prominent median carina is present on abdominal segments two to four, and the second pleuron is triangular. The color is brownish-grey (according to Galil et al. 2002) (Figure 13.10).

In the Mediterranean, an ovigerous female was recorded with a carapace length 10 mm CL (Froglia 1979), corresponding to an approximate TL of 27 mm.

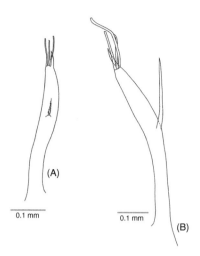

0.1 mm

0.1 mm

(A)

(B)

FIGURE 13.11 Antennule of the phyllosoma I of (A) *Acantharctus posteli* and (B) *Scyllarus arctus.* ((A) modified from Gonzáles Gordillo & Rodriguez (2000); (B) used with permission and drawn from an original photo of one of the authors (Pessani) by G. Rappini.)

13.6.2 Distribution and Ecology

The species is found in the tropical and subtropical Eastern Atlantic, typically on sandy and muddy bottoms at depths ranging from 30 to 60 m (Galil et al. 2002). Along the Mauritanian coast, it was found on rocky, sandy, and detritic substrates at depths between 39 and 80 m (Anadón 1981). In the Mediterranean, there has only been one record (an ovigerous female) from the central Adriatic Sea (Froglia 1979) (Figure 13.2).

13.6.3 Reproductive Biology and Larval Development

The reproductive biology is unknown. Unfertilized eggs have been recorded as having a diameter of 0.34 mm (Zariquiey Alvarez 1968). The larval stages have not yet been described.

13.7 Key for the Identification of the Species

Abbreviations used: A1 = antennule; CSL = cephalic shield length; CSW = cephalic shield width; mxp3 = maxillipeds of the 3rd pair; P4 = pereiopods of the 4th pair; TL = total length.

13.7.1 Adults

An identification key, based on the available information about the morphology of the adults (Zariquiey Alvarez 1968; Holthuis 1991; Galil et al. 2002) not associated with color, is provided below and should be of use for the identification of live and preserved specimens.

1. Mxp3 exopodite with a flagellum . . . . *Scyllarides latus*
   Mxp3 exopodite without a flagellum. 2
2. Prominent spine in the middle of the last thoracic sternite . . . . *Acantharctus posteli*
   No median spine on last thoracic sternite. . . . . (*Scyllarus* spp.) 3
3. Abdominal somites two to four bearing dorsally a prominent median carina . . *Scyllarus caparti*
   Not as above 4
4. Pleuron of the 2nd abdominal somite pointed and directed posteriorly . . . . *Scyllarus arctus*
   Pleuron of the 2nd abdominal somite rounded and directed ventrally . . . . . . *Scyllarus pygmaeus*

13.7.2 Phyllosoma I

The proposed identification key only concerns the stage I phyllosoma, because this stage is the only one identified for four of the five species of Mediterranean scyllarids.

1. Biramous antenna . . . . *Scyllarides latus*
   Uniramous antenna, triangle-shaped 2
2. Abdomen longer than the first proximal segment of P4; TL/CSW = 1.87 . *Scyllarus pygmaeus*
   Abdomen shorter than the first proximal segment of P4; TL/CSW = 1.5 to 1.7 3
3. Short seta on the margin of A1, about 1/6 of A1 length (apical setae excluded) (Figure 13.11A);
   CSW/CSL = 1.03 to 1.04 . . . .*Acantharctus posteli*
   Long seta on the margin of A1, roughly 1/3 of A1 length (apical setae excluded) (Figure 13.11B);
   CSW/CSL = 1.2 to 1.3 . . . .*Scyllarus arctus*

13.8   Conclusions

The present review points out that there is a great need to obtain information about the distribution and status of the populations of adult scyllarids in the Mediterranean. In recent decades, observations of autochthonous species have been made primarily in the waters off Spain, Italy, Greece, and Israel, but nothing is known about their presence along other Mediterranean coasts. Since this information is essential for the implementation of effective protective measures throughout the Mediterranean, we recommend the use of modern techniques for sampling and censusing populations and the larvae from which the adult populations come.

With regard to the lack of information on the larval development for all Mediterranean species, it is also evident that there is a need to define effective rearing techniques (see Mikami & Kuballa, Chapter 5 for one such successful technique) that will facilitate accurate descriptions of larval morphology. This is the only possible way, aside from genetic analyses, to enable proper identification of the planktonic larvae. We hope that international circum-Mediterranean projects will be launched to enable coordinated actions of investigation and intervention with respect to Mediterranean scyllarids.

Acknowledgments

The authors want to thank Giuseppe C. Rappini of C.S.T. Centro Servizi Tecnici for his technical support and advice during the revision of the text.

References

Almog-Shtayer, G. 1988. Behavioural-ecological Aspects of Mediterranean Slipper Lobsters in the Past and of the Slipper Lobster Scyllarides latus in the Present. M.A. thesis: University of Haifa, Israel. 165 pp.

Anadón, R. 1981. Crustáceos Decápodos (excl. Paguridea) recogidos durante la campaña "Atlor VII" en las costas noroccidentales de África (Noviembre 1975). *Res. Exp. Cient. (suppl. Inv. Pesq.)* 9: 151–159.

Barnich, R. 1996. *The Larvae of the Crustacea Decapoda (excl. Brachyura) in the Plankton of the French Mediterranean Coast.* Göttingen: Cuvillier Verlag. 124 pp.

Barshaw, D. & Spanier, E. 1994. The undiscovered lobster — a first look at the social behavior of the Mediterranean slipper lobster, *Scyllarides latus* Decapoda, Scyllaridae. *Crustaceana* 67: 187–197.

Bate, C.S. 1888. Report on the Crustacea Macrura collected by *H.M.S. Challenger* during the years 1873–76. *Report Scientific Results Voyage Challenger (Zool.)* 24: 1–942.

Bianchi, C.N., Cevasco, M.P., Diviacco, G., & Morri, C. 1986. Primi risultati di una ricerca ecologica sulla grotta sottomarina di Bergeggi (Savona). *Boll. Mus. Ist. Biol. Univ. Genova* 52: 267–293.

Bianchini, M.L. & Ragonese, S. 2003. *In ovo* embryonic development of the Mediterranean slipper lobster, *Scyllarides latus*. *The Lobster Newsletter* 16(1): 10–12.

Bianchini, M.L., Bono, G., & Ragonese, S. 2003. Long term recaptures and growth of slipper lobsters, *Scyllarides latus*, in the Strait of Sicily (Mediterranean Sea). *Crustaceana* 74: 673–680.

Bianchini, M.L., Greco, S., & Ragonese, S. 1998. Il progetto valutazione della fattibilità e potenzialità del ripopolamento attivo per la magnosa, *Scyllarides latus* (Crostacei Decapodi): Sintesi e risultati. *Biol. Mar. Medit.* 5: 1277–1283.

Bianchini, M.L., Chessa, L., Greco, S., Ragonese, S., & Scarpeli, G. 1996. Morphometric aspects of slipper lobsters, *Scyllarides latus*. In: *Abstracts of the 6th Colloquium Crustacea Decapoda Mediterranea*, Florence September 12–15, 1996, pp. 11–12.

Bianchini, M.L., Napolitano, L., Palmegiano, G.B., Ragonese, S., & Sicuro, B. 1997. Osservazioni preliminari sulla composizione delle carni in magnosa, *Scyllarides latus* (Crustacea Decapoda). *Biol. Mar. Medit.* 4: 306–308.

Bouvier, E.L. 1913. Sur le genres *Pseudibacus* et *Nisto*, et le stade natante des Crustacés decapodes macroures de la famille des *Scyllarides*. *C. R. Hebd. Acad. Sci. Paris* 156: 1643–1648.

Bouvier, E.L. 1917. Crustacés décapodes (Macroures marcheurs) provenant des campagnes des yachts Hirondelle et Princesse-Alice (1885–1915). *Rés. Camp. Scient. Monaco* 50: 1–104.

Cau, A., Deiana, A.M., & Mura, M. 1978. Sulla frequenza e bionomia di *Scyllarus pygmaeus* (Bate) in acque neritiche sarde. *Natura Milano* 69: 118–124.

Cau, A., Deiana, A.M., & Mura, M. 1982. Aspetti biologici di *Scyllarus pygmaeus* (Bate, 1888) (Crustacea, Malacostraca). *Quad. Civ. Staz. Idrobiol. Milano* 10: 29–36.

Cau, A., Davini, M.A., Deiana, A.M., & Salvatori, S. 1988. Gonadal structure and gametogenesis in *Scyllarus arctus* (L.) (Crustacea, Decapoda). *Boll. Zool.* 5: 299–306.

Chessa, L.A., Pais, A., Serra, S., Scardi, M., & Ligios, L. 1998. Preferenze alimentari della magnosa, *Scyllarides latus* (Latreille, 1803), in cattività. *Biol. Mar. Medit.* 5: 504–507.

Crosnier, A. 1972. Naupliosoma, phyllosomes et pseudibacus de *Scyllarides herklotsi* (Herklots) (Crustacea, Decapoda, Scyllaridae) récoltés par l'Ombago dans le sud du Golfe de Guinée. *Cah. O.R.S.T.O.M., sér. Océanogr.* 10: 139–149.

De Domenico, F., Cosentino, A., & Spanò, N. 2003. Fauna carcinologica di due ambienti insulari: Archipelago toscano e Isole Pontine. *Biol. Mar. Medit.* 10: 550–554.

Fedele, M. 1926. La metamorfosi dal phyllosoma dello *Scyllarus arctus*. *Boll. Soc. Nat. Napoli* 37: 215–223.

Fiedler, U. & Spanier, E. 1996. Stage II phyllosoma of the Mediterranean slipper lobster *Scyllarides latus*. *The Lobster Newsletter* 9(1): 12.

Fiedler, U. & Spanier, E. 1999. Occurrence of larvae of *Scyllarus arctus* (Crustacea, Decapoda, Scyllaridae) in the Eastern Mediterranean — preliminary results. *Ann. Istrian Med. Stud.* 17: 153–158.

Fischer, W., Bauchot, M.-L., & Schneider, M. (eds.) 1987. *Fiches FAO d'identification des Espèces pour les Besoins de la Pêche, révision 1*, Vol. 1: pp. 293–319. Méditerranée et Mer Noire (Zone de Pêche 37).

Forest, J. & Holthuis, L.B. 1960. The occurrence of *Scyllarus pygmaeus* (Bate) in the Mediterranean. *Crustaceana* 1: 156–163.

Froglia, C. 1979. Segnalazione di alcuni Crostacei Decapodi nuovi per la fauna adriatica. *Quad. Lab. Tecnol. Pesca.* 2: 191–196.

Froglia, C. 2001. Gli invertebrati marini. Schede di riconoscimento e brevi note biologiche per le specie soggette a normativa comunitaria o nazionale. In: Gramitto, M.E. (ed.), *La Gestione della Pesca Marittima in Italia. Fondamenti Tecnico-biologici e Normativa Vigente*: pp. 150–194. Roma: Consiglio Nazionale delle Ricerche.

Galil, B. 1993. The composition and diversity of planktonic larval Decapoda off the Mediterranean coast of Israel. *Final Reports on Research Projects Dealing with the Effects of Pollutants on Marine Communities and Organisms*: pp. 131–151. Athens: UNEP.

Galil, B., Froglia, C., & Nöel, P. 2002. Crustaceans: Decapods and stomatopods. In: Briand, F. (ed.), *CIESM Atlas of Exotic Species in the Mediterranean*: pp. 1–192. Monaco: CIESM Publishers.

Garcia Raso, J.E. 1982. Familia *Scyllaridae* Latreille 1825 (*Crustacea, Decapoda*) en la Región Sur-mediterránea española. *Bol. Asoc. esp. Entom.* 6: 73–78.

Gili, J. & Macpherson, E. 1987. Crustáceos decápodos capturados en cuevas submarinas del litoral Balear. *Inv. Pesq.* 51 (Suppl.): 285–291.

Glavic, N., Kozul, V., Tutman, P., Glamuzina, B., & Skaramuca, B. 2001. Morphological characteristics of Mediterranean slipper lobster, *Scyllarides latus* (Latreille, 1802) (Decapoda: Scyllaridae) stage I phyllosoma. *Rapp. Comm. Intern. Expl. Scient. Mer Médit.* 36: 271.

Gonzáles-Gordillo, J.I. & Rodríguez, A. 2000. First larval stage of *Scyllarus posteli* Forest, 1963 and *Processa macrodactyla* Holthuis, 1952 hatched in the laboratory (Crustacea, Decapoda). *Ophelia* 53: 91–99.

Grippa, G. 1993. Notes on Decapod fauna of "Arcipelago Toscano". *Bios* 1: 223–239.

Gurney, R. 1942. *Larvae of Decapod Crustacea*: pp. 230–236. London: Ray Society.

Holthuis, L.B. 1977. The Mediterranean decapod and stomatopod Crustacea in A. Risso's published works and manuscripts. *Ann. Mus. Hist. Nat. Nice* 5: 37–88.

Holthuis, L.B. 1991. *Marine Lobsters of the World. An Annotated and Illustrated Catalogue of the Species of Interest to Fisheries Known to Date*. FAO Species Catalogue No. 125, Vol. 13: pp. 1–292. Rome: Food and Agriculture Organization of the United Nations.

Holthuis, L.B. 2002. The Indo-Pacific scyllarine lobsters (Crustacea, Decapoda, Scyllaridae). *Zoosystema* 24: 499–683.

Holthuis, L.B. & Gottlieb, F. 1958. An annotated list of the Decapod Crustacea of the Mediterranean coast of Israel, with an appendix listing the Decapoda of the Eastern Mediterranean. *Bull. Res. Counc. Israel* 7B: 1–126.

Koukouras, A., Dounas, C., & Eleftheriou, A. 1993. Crustacea Decapoda from the cruises of "Calypso" 1955, 1960, in the Greek waters. *Bios* 1: 193–200.

Kurian, C.V. 1956. Larvae of decapod Crustacea from the Adriatic. *Acta Adriat.* 6: 1–108.

Lakkis, S. & Zeidane, R. 1988. Le meroplankton des eaux Libanaises: Larves de Crustacés Décapodes. *Rapp. P.-V. Réun. Comm. Intern. Expl. Scient. Mer Médit.* 31: 238.

Lewinsohn, C. & Holthuis, L.B. 1986. The Crustacea Decapoda of Cyprus. *Zool. Verh.* 230: 1–64.

Maigret J. 1978. Contribution à l'ètude des langoustes de la côte occidentale de l'Afrique (Crustacés, Decapodes, Palinuridae). 5. Les larges phyllosomes de Scyllaridae et de Palinuridae reçueillies au large des côtes du Sahara. *Bull. Ist. Fondam. Afrique Noire* 40: 36–80.

Manconi, R. & Pessani, D. 2003. Biologia ed ecologia. Il popolamento: Crostacei Decapodi. In: Cicogna, F., Bianchi, C.N., Ferrari, G., & Forti, P. (eds.), *Grotte Marine. Cinquant'anni di Ricerca in Italia*: pp. 187–193. Roma: Ministero dell'Ambiente e della Tutela del Territorio.

Martins, H.R. 1985a. Biological studies on the exploited stock of the Mediterranean locust lobster *Scyllarides latus* (Latrielle, 1803) (Decapoda, Scyllaridae) in the Azores. *J. Crust. Biol.* 5: 294–305.

Martins, H.R. 1985b. Some observations on the naupliosoma and phyllosoma larvae of the Mediterranean locust lobster, *Scyllarides latus* (Latreille, 1803), from the Azores. *International Council for the Exploration of the Sea C.M.K.*, 52 Shellfish Committee, 13 pp.

Modena, M., Mori, M., & Vacchi, M. 2001. Note su alcuni crostacei malacostraci raccolti in aree adiacenti alla m/c Haven (Mar Ligure). *Biol. Mar. Medit.* 8: 675–679.

Moncharmont, U. 1979. Notizie biologiche e faunistiche sui Crostacei Decapodi del Golfo di Napoli. *Annu. Ist. Mus. Zool. Univ. Napoli* 23: 33–132.

Mura, M. & Pessani, D. 1994. Descrizione del primo stadio larvale e notizie sul periodo riproduttivo di alcune specie di Decapodi. *Biol. Mar. Medit.* 1: 391–392.

Mura, M., Cau, A., & Deiana, A.M. 1984. Il genere *Scyllarus* Fabr., 1775 nel Mediterraneo centro occidentale. *Rend. Seminari Fac. Sc. Univ. Cagliari* 54 (Suppl.): 267–274.

Orsi Relini, L. & Costa, M.R. 1981. I Decapodi litorali di Portofino, III: Specie raccolte con vari attrezzi da pesca. *Quad. Lab. Tecnol. Pesca* 3 (1 suppl.): 165–174.

Orsi Relini, L., Arata, P., & Costa, M.R. 1976. I Crostacei Decapodi litorali di Portofino: II Raccolte subacquee e casi di foresia. *Boll. Mus. Ist. Biol. Univ. Genova* 44: 81–92.

Palombi, A. & Santarelli, M. 1961. *Gli Animali Commestibili dei Mari d'Italia. Descrizione — Biologia — Pesca — Valore Economico e Nomi Italiani Dialettali e Stranieri dei Pesci — Tunicati — Echinodermi — Molluschi — Crostacei ad Uso dei Pescatori di Professione, Dilettanti e Subacquei* (ed. 2): pp. 1–437. Milano, Italy: Hoepli Editore.

Pessani, D., Pisa, G., & Gattelli, R. 1999. The complete larval development of *Scyllarus arctus* (Decapoda, Scyllaridae) in the laboratory. *Abstracts of the 7th Colloquium Crustacea Decapoda Mediterranea*, Lisbon, September 6–9, 1999. pp. 143–144.

Phillips, B.F. & McWilliam, P.S. 1986. Phyllosoma and nisto stages of *Scyllarus martensii* Pfeffer (Decapoda, Scyllaridae) from the Gulf of Carpentaria, Australia. *Crustaceana* 51: 133–154.

Pipitone, C. & Arculeo, M. 2003. The marine crustacea decapoda of Sicily (central Mediterranean Sea): A checklist with remarks on their distribution. *Ital. J. Zool.* 70: 69–78.

Pipitone, C. & Tumbiolo, M.L. 1993. Decapod and stomatopod crustaceans from the trawlable bottoms of the Sicilian Channel (central Mediterranean Sea). *Crustaceana* 65: 358–364.

Relini, G., Relini, M., & Torchia, G. 1995. La barriera artificiale di Loano. *Biol. Mar. Medit.* 2: 21–64.

Relini, G., Bertrand, J., & Zamboni, A. 1999. Synthesis of the knowledge on bottom fishery resources in central Mediterranean (Italy and Corsica). *Biol. Mar. Medit.* 6 (suppl.): 593–600.

Rinelli, P., Spanò, N., Giordano, D., Perdichizzi, F., & Greco, S. 1998. Organismi bentonici e fauna demersale in un'area del Tirreno Meridionale. *Biol. Mar. Medit.* 5: 80–89.

Robertson, P.B. 1969. The larval development of the scyllarid lobster, *Scyllarides aequinoctialis* (Lund) in the laboratory, with a revision of the larval characters of the genus. *Deep Sea Res.* 16: 557–586.

Robertson, P.B. 1971. The larvae and postlarva of the scyllarid lobster *Scyllarus depressus* (Smith). *Bull. Mar. Sci.* 21: 841–865.

Rostagno, M. 1996. I Crostacei Decapodi del Piano Meso ed Infralitorale Inferiore del Mar Ligure. Thesis: Università degli Studi di Torino, Torino, Italy.

Santucci, R. 1925. Contributo allo studio dello sviluppo post-embrionale degli "Scyllaridea" del Mediterraneo. II. *Scyllarus arctus* (L.). III. *Scyllarides latus* Latr. *Mem. R. Com. Talassogr. Ital.* 121: 16 pp.

Santucci, R. 1926. Fillosomi di Scillaridi esotici nel mediterraneo. *Monit. Zool. Ital.* 27: 19–23.

Santucci, R. 1928. Il primo stadio post-embrionale di *Scyllarides latus* Latreille. *Mem. R. Com. Talassogr. Ital.* 144: 1–7.

Spanier, E. 1987. Mollusca as food for the slipper lobster *Scyllarides latus* in the coastal waters of Israel. *Levantina* 68: 713–716.

Spanier, E. 1989. How to increase the fisheries yield in low productive marine environments. *Ocean 89*: 297–301.

Spanier, E. 1991. Artificial reefs to insure protection of the adult Mediterranean slipper lobster, *Scyllarides latus* (Latreille, 1803). In: Boudouresque, C.F., Avon, M., & Gravez, V. (eds.), *Les Espéces Marines á Protéger en Méditerranée*: pp. 179–185. France: GIS Posidonia Publ.

Spanier, E. & Almog-Shtayer, G. 1992. Shelter preferences in the Mediterranean slipper lobster: Effects of physical properties. *J. Exp. Mar. Biol. Ecol.* 164: 103–116.

Spanier, E. & Barshaw, D.E. 1993. Tag retention in the Mediterranean slipper lobster. *Isr. J. Zool.* 39: 29–33.

Spanier, E. & Lavalli, K.L. 1998. Natural history of *Scyllarides latus* (Crustacea: Decapoda): A review of the contemporary biological knowledge of the Mediterranean slipper lobster. *J. Nat. Hist.* 32: 1769–1786.

Spanier, E. & Lavalli, K.L. 2006. *Scyllarides* spp. In: Phillips, B.F. (ed.), *Lobsters: Biology, Management, Aquaculture, and Fisheries*: pp. 462–496. Oxford, London: Blackwell Publishing.

Spanier, E., Almog Shtayer, G., & Fiedler, U. 1993. The Mediterranean slipper lobster *Scyllarides latus*: the known and the unknown. *Bios* 1: 49–58.

Spanier, E., Tom, M., Pisanty, S., & Almog, G. 1988. Seasonality and shelter selection by the slipper lobster *Scyllarides latus* in the southeastern Mediterranean. *Mar. Ecol. Prog. Ser.* 42: 247–255.

Spanier, E., Tom, M., Pisanty, S., & Almog-Shtayer, G. 1990. Artificial reefs in the low productive marine environment of the Southeastern Mediterranean. *P.S.Z.N.I. Mar. Ecol.* 11: 61–75.

Spanier, E., Weihs, D., & Almog-Shtayer, G. 1991. Swimming of the Mediterranean slipper lobster. *J. Exp. Mar. Biol. Ecol.* 145: 15–31.

Stephensen, K. 1923. Decapoda Macrura (excl. Sergestidae). *Rep. Danish Oceanogr. Exped. 1908-10 Medit. & Adj. Seas* 2: 1–252.

Sturani, M. 1992. Studio Bionomico dei Fondali a Massi in Liguria Orientale: Analisi della Fauna Bentonica Vagile. Thesis: Università degli Studi di Torino, Torino, Italy.

Thiriot, A. 1974. Larves de Décapodes Macrura et Anomura, espèces européennes; Caractères morphologiques et observations écologiques. *Thalassia Jugosl.* 10: 341–378.

Udekem d'Acoz, C.d'. 1999. Inventaire et distribution des crustacés décapodes de l'Atlantique nord-oriental, de la Méditerranée et des eaux continentales adjacentes au nord de 25°N. *Patrimoines naturels (M.N.H.N./ S.P.N.)* 40: 1–383.

Vilotte, O. 1982. *Recherches sur la Biologie et Elevage de la Cigale de Mer Scyllarus arctus (L., 1758) (Crustacea, Decapoda, Scyllaridae).* Ph.D. thesis: Universitè des Sciences et Techniques du Languedoc, Montpellier, France.

Williamson, D.I. 1983. Crustacea Decapoda: Larvae. VIII. Nephropidea, Palinuridea, and Eryonidea. *Fich. Ident. Zoopl.* 167/168 : 1–8.

Zariquiey Alvarez, R. 1968. Crustáceos Decápodos Ibéricos. *Inv. Pesq.* 32: 1–510.

14

Biology and Fishery of the Galápagos Slipper Lobster

Alex Hearn, Veronica Toral-Granda, Camilo Martinez, and Gunther Reck

CONTENTS

Abstract

Until recently, the Galápagos slipper lobster, *Scyllarides astori* Holthuis, 1960, received little attention regarding its biology, natural history, and fishery. However, as the main fishery resources (i.e., spiny lobster and sea cucumber) have been depleted, there is increasing pressure to expand the fishery of *S. astori* from a mainly local resource, to include export to continental Ecuador. Biological studies have focused on the distribution, abundance, diet, and population dynamics of this species within the Galápagos Marine Reserve (GMR). *Scyllarides astori* is a nocturnal forager, preferring rocky habitats that provide shelter in the form of crevices and caves. It is found throughout the GMR, but in higher numbers in cooler waters, with a depth distribution from the immediate subtidal zone to at least 40 m. Preliminary growth data suggest slow growth and an asymptotic carapace length of 17.53 cm for males and 16.38 cm for females. Reproductive studies suggest that onset of sexual maturity occurs over a narrow size range (20 to 25 cm TL), reproductive activity is more pronounced in the warmer months, and fecundity is related to total length. Fishing is carried out by divers who catch *S. astori* by hand mainly at night, from small (<6 m) fiberglass or wooden vessels with onboard compressors that provide air to the divers by means of a 50 m, plastic hose. Current annual landings exceed 12 metric tons live weight. Other than a ban on the landing of ovigerous females, there are no regulations specific to the capture of *S. astori*. However, there is a preliminary (and not yet enforced) zonation scheme within the GMR that prohibits extractive activities in 18% of the coastal waters. Additionally, the Galápagos Five-Year Fishing Calendar identifies the need to incorporate specific regulations for 2004, including closed seasons and a minimum landing size.

14.1 Introduction

Of the 13 known species in its genus (Holthuis 1991), the Galápagos slipper lobster, *Scyllarides astori* Holthuis, 1960, is one of the largest, generally reaching a total length (TL) of 30 cm, although the maximum recorded size is 37.8 cm (Hearn 2004). It is the only native species of slipper lobster occurring in the Galápagos Archipelago (Hickman & Zimmerman 2000) (Figure 14.1). It is found on rocky substrates, preferring vertical walls with pockmarks, caves, and crevices (Barr 1968; Martínez et al. 2002), where it shelters during the day and forages nocturnally.

Although the genus *Scyllarides* is widespread throughout the Indo-Pacific area, the geographical range of *S. astori* is unclear. Holthuis & Loesch (1967) and Hickman & Zimmerman (2000) suggest that it may be endemic to the Galápagos Islands, while Reck (1983) notes the occurrence of larvae around the Cocos Islands & Johnson (1971) indicates the possible existence of breeding populations elsewhere in the eastern Pacific. According to Holthuis (1991) and Gotshall (1998), *S. astori* is also found in the Gulf of California.

FIGURE 14.1   **(See color insert following page 242)** The Galápagos slipper lobster *Scyllarides astori*. (Photograph by Alex Hearn.)

The Galápagos Marine Reserve (GMR), created in 1998, lies in the Eastern Pacific, approximately 1000 km due west off the coast of Ecuador. The reserve is one of the largest in the world and as a UNESCO World Heritage Site, covers a total area of 138,000 km² (Heylings et al. 2002). It has an extraordinary range of biological communities, high levels of endemism, and an abundance of charismatic species, making it one of the world's top marine protected areas (Bustamante et al. 2002).

Traditionally, local fishing activity around the Galápagos Islands was focused on the endemic bacalao, or grouper, *Mycteroperca olfax* Gill, 1862 and the red and green spiny lobsters, *Panulirus penicillatus* (Olivier, 1791) and *Panulirus gracilis* Streets, 1871, respectively (Reck 1983). In recent years, the sea cucumber *Isostichopus fuscus* (Ludwig, 1875) (Murillo et al. 2002) has become the most important source of income for the local fishing sector, providing a gross annual income of around US$3 million (Murillo et al. 2002) and fueling a large rise in the number of registered commercial vessels (Murillo et al. 2004). However, since 2000, the sea cucumber population has shown signs of progressive and severe depletion (Hearn et al. 2005a; Toral-Granda 2005a). At the same time, spiny lobster landings have decreased steadily from a maximum of 85 metric tons of tail in 2000 to 25.7 metric tons in 2005, stimulating concern as to the sustainability of the fisheries as a whole within the GMR (Hearn et al. 2005b).

The slipper lobster was historically a minor resource associated with the spiny lobster fishery (Reck 1983), and is currently mainly limited to local consumption. However, over the last few years, partly in response to declining revenue from the sea cucumber and spiny lobster fisheries, landings have increased (Martínez et al. 2002; Molina et al. 2004) and there is mounting pressure to allow large-scale export to the continent.

This chapter summarizes the current state of knowledge regarding the biology and fishery for this species in Galápagos, based on research projects and the Participatory Fisheries Monitoring Program at the Charles Darwin Research Station.

14.2 Brief History of Research

The first biological studies on *Scyllarides astori* in Galápagos were carried out by Barr (1968), who collected basic information on habitat, reproduction, behavior, and diet from a small number of individuals. Reck (1983) carried out a study of the spiny and slipper lobsters of Galápagos in 1976 to 1977 and 1979 onboard a commercial fishing vessel; he determined biometric relationships, reproductive aspects, population structure; and assessed fishing activity. For the next two decades, little attention was given to the lobster populations of Galápagos. Martínez (2000) expanded on Barr's study of habitat and diet, and commenced a four-year study of the size structure and relative abundance of *S. astori* in the different islands within the Archipelago (Martínez et al. 2002), including a mark-recapture survey, which was expanded upon (Hearn et al. 2003; Hearn 2004) to evaluate growth parameters and determine optimal levels of exploitation. Basic studies on the reproductive biology of the three species of lobsters were undertaken by Toral-Granda et al. (2002) and Martínez et al. (2002) from 1998 to 2000.

From 1997 to 2003, the Fisheries Monitoring Program at the Charles Darwin Research Station accepted responsibility for the collection and compilation of all fisheries data from the local fishing sector. During this period, observers, both aboard fishing vessels and at landing points, collected information on the geographical distribution of the fishing effort directed at *S. astori*. In addition, biological and economic data were obtained for calculations of catch-per-unit-effort and cost-benefit analysis of fishing activity (Murillo et al. 2003; Molina et al. 2004). The data collected have provided some baseline information on the geography and relative abundance of the species within the GMR.

14.3 Distribution and Abundance

The monitoring program described above, combined with baseline ecological studies of the coastal subtidal zone around the GMR, provides information regarding the spatial distribution of *S. astori* and its relative abundance in different locations around the islands. Work relating to depth distribution is limited to the depth range accessible by divers.

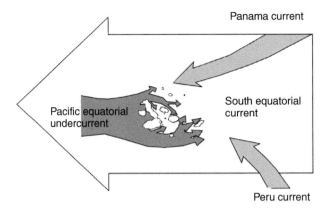

FIGURE 14.2   The Galápagos Archipelago. The Pacific Equatorial Undercurrent (also known as the Cromwell Current) collides with the Galápagos platform, producing upwelling of cool, nutrient-rich waters to the surface. The Panama Current brings tropical waters from the coast of Central America. The Peru Current is the extension of cold waters which originate in equatorial upwelling zones between Galápagos and the coasts of continental Ecuador and Peru. The South-Equatorial Current is the main surface current, which flows to the west, and whose direction and intensity throughout the year are influenced by the Panama and Peru current systems. Illustration by: Mats Vedin. (From Danulat, E. & Edgar, G.J. (eds.) 2002 *Reserva Marina de Galápagos*. Linea Base de la Biodiversidad. Fundación Charles Darwin/Servicio Parque Nacional Galápagos, Santa Cruz, Galápagos, Ecuador. 484 pp.; used with permission.)

14.3.1   Geographical Distribution

The Galápagos Marine Reserve covers an area of 138,000 km^2 (Heylings et al. 2002), including 15 islands and over 100 islets, rocks, and atolls of volcanic origin (Snell et al. 1996). As an oceanic archipelago, the marine conditions within the reserve are greatly affected by large-scale ocean processes, namely the South Equatorial Current, the Humboldt Current from Peru, and the Panama Current from Central America. The Cromwell Current from the west brings cold water to the archipelago, forming areas of upwelling and high productivity on the western coasts of many of the islands (Figure 14.2) and in the western part of the archipelago in general (Banks 2002). As a result of these currents, the archipelago can be divided into a number of biogeographical zones on the basis of their physical attributes and marine biological communities. Edgar et al. (2004) proposed the following zones: a warm, far northern zone; a cool, western zone; a mixed warm, northern zone; and a mixed central and southern zone.

During the Stanford Oceanographic Expedition (Barr 1968), seven sites were sampled for lobsters, out of which *Scyllarides astori* were found at four: the waters off the islands of Floreana, Fernandina, and Isabela (two sites). No *S. astori* were found in the waters off Genovesa, Santa Cruz, and Plazas islands. From 2000 to 2004, a comprehensive study of the distribution of both spiny and slipper lobsters was carried out by the Charles Darwin Foundation, comprised of surveys around most of the major islands. Dive teams made up of scientists and local fishers carried out periodic, nocturnal 50 m band transect surveys with a 1 m band width at 42 sites (Figure 14.3). Abundance was estimated as the number of lobsters observed by a diver per hour. In contrast with similar studies on spiny lobsters, where catch rates by fishers were significantly higher, there was no difference in the catch rate of *S. astori* between fishers and scientists (Hearn, unpublished data); this has been attributed to the behavior of *S. astori*, which is generally found in the open after dark and whose escape movement is sluggish. Generally, *S. astori* was found to be solitary, unlike the red spiny lobster that is often found in groups of up to 20 individuals in caves (Toral-Granda et al. 2002) or a congener of *S. astori*, the Mediterranean slipper lobster, *Scyllarides latus* (Latreille, 1802) that is also found in groups (Spanier & Lavalli 1998; Lavalli & Spanier 2001; see Lavalli et al., Chapter 7 for a description of slipper lobster gregarious behavior).

The highest abundance (individuals observed per diver hour) of *S. astori* was found in the cooler, western part of the archipelago (Fernandina and western Isabela), and in Santiago and Rábida. In the north and far north (Wolf, Pinta, Genovesa, and Marchena), however, where sea surface temperatures are highest, *S. astori* was rare (less than one individual observed per diver hour). The islands of Española,

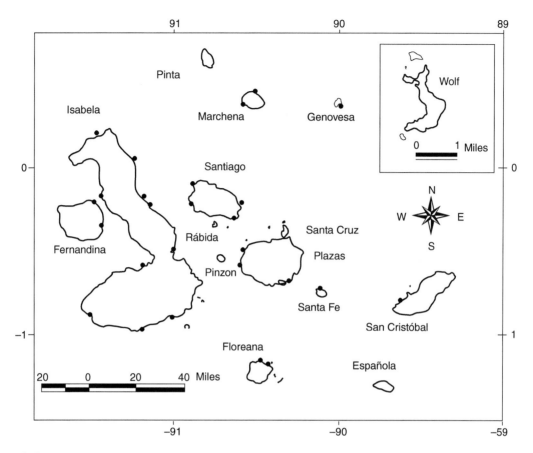

FIGURE 14.3 Sample sites (black circles) around different islands for 2000 to 2004 lobster research program.

Floreana, San Cristóbal, and Santa Cruz all showed similar levels of abundance, ranging between two to four individuals observed per diver hour (Figure 14.4). These numbers suggest that the overall population size of the slipper lobster may be an order of magnitude lower than that of the red spiny lobster, which exceeds 10 individuals per diver hour in five major islands (Hearn 2004).

14.3.2 Depth Distribution

The depth range of *Scyllarides astori* is poorly known (Holthuis 1991), although Reck (1983) observed individuals from shallow depths to at least 40 m, while Gotshall (1998) stated that depth could range up to 90 m. According to Martínez et al. (2002) and Hearn (2004), it is less abundant in the immediate subtidal zone, the preferred habitat for the red spiny lobster *Panulirus penicillatus*, where approximately five individuals were observed per diver hour. In contrast, the observed abundance at depths greater than 10 m was more than double (Figure 14.5).

14.4 Population Structure and Morphometrics

The size and sex structure of the population of *Scyllarides astori* in the GMR has been estimated from both fisheries landings and diver surveys; however, these methods both involve a sampling bias against juveniles, so there is a considerable amount of uncertainty in the population structure displayed by smaller individuals. For morphometric measurements, TL was taken to be the distance between the rostrum and

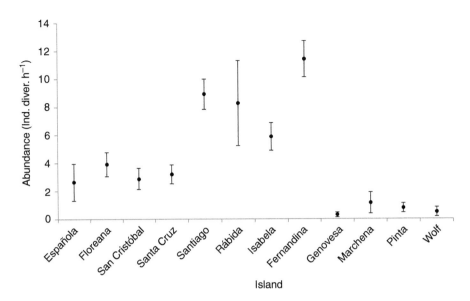

FIGURE 14.4   Mean (±SE) relative abundance of the slipper lobster *Scyllarides astori* by island during diving surveys 2000 to 2004, expressed as the number of individuals observed per diver hour. (From Hearn, A. 2006. *J. Exp. Mar. Biol. Ecol.* 328: 87–97; used with permission.)

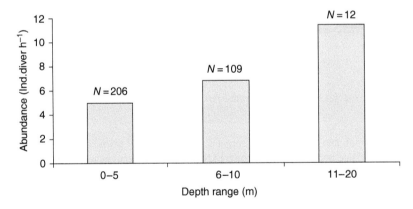

FIGURE 14.5   Relative abundance of slipper lobsters *Scyllarides astori* at different depth ranges throughout the Galápagos Islands 2000 to 2004, showing number of surveys (*N*) carried out at each range.

the posterior edge of the telson, whereas carapace length (CL) was measured as the distance between the rostrum and the posterior edge of the carapace.

14.4.1   Sex Ratio and Sexual Dimorphism

The sex of *Scyllarides astori* may be determined easily by inspection of the abdomen (as in other congeners, see Holthuis 1991). Females display large fronded pleopods, whereas males display reduced pleopods on only the first abdominal segment. The last pair of walking legs on females is chelate, presumably to aid in grooming of eggs. Overall sex ratios did not differ significantly from unity ($\chi^2 = 0.8$, df $= 7$) for those islands where more than 10 specimens were obtained (Hearn 2004). This appears to be consistent with other species — reported sex ratios for *Scyllarides nodifer* (Stimpson 1866) were approximately 1:1 (Hardwick & Cline 1990), as were those for *S. latus* (Spanier & Lavalli 1998).

The relationship between TL and CL was found to be significantly different between the two sexes (ANCOVA, $P = 0.000$; df = 436; Figure 14.6). This was consistent with an earlier study by Reck (1983) that also showed males to have a slightly larger CLs than females of a same given TL.

14.4.2 Length–Weight Relationship

The length–weight relationship for *S. astori* displays a significant difference between the sexes, with males being heavier than females of the same length (Figure 14.7). The difference becomes more pronounced at

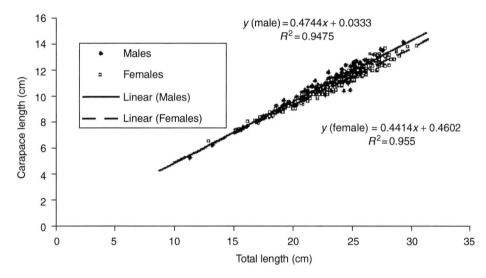

FIGURE 14.6 Relationship between TL and CL (separate sexes) of *Scyllarides astori*, obtained in 2000 to 2004. (From Hearn, A. 2004. *Evaluación de las Poblaciones de Langostas en la Reserva Marina de Galápagos. Informe Final 2002–2004.* Santa Cruz, Galápagos, Ecuador: Fundación Charles Darwin y Dirección Parque Nacional Galápagos. 96 pp.; used with permission.)

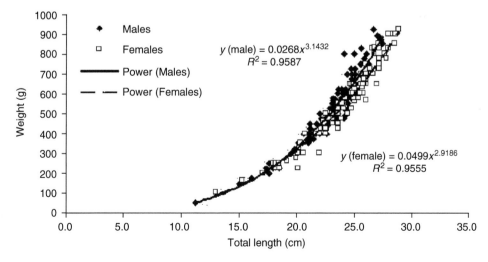

FIGURE 14.7 Length–weight relationships (separate sexes) for *Scyllarides astori*, obtained in 2000 to 2004. (From Hearn, A. 2004. *Evaluación de las Poblaciones de Langostas en la Reserva Marina de Galápagos. Informe Final 2002–2004.* Santa Cruz, Galápagos, Ecuador: Fundación Charles Darwin y Dirección Parque Nacional Galápagos. 96 pp.; used with permission.)

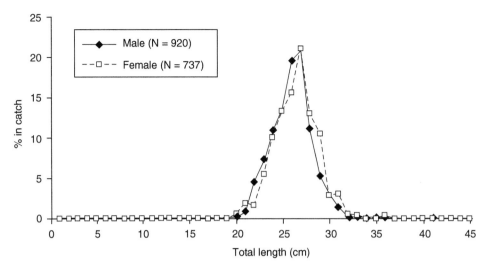

FIGURE 14.8   Average size structure of *Scyllarides astori* from 2001 to 2003, by sex, captured in the artisanal fleet.

larger sizes, after sexual maturity is attained. This may be explained by differential use of energy — once sexually mature, the female may divert more energy toward egg production, whereas the male may direct more energy toward growth in order to have an advantage of size over competitors for females, is the case as with other decapod crustaceans (Hartnoll 1974).

14.4.3   Size Frequency Distribution

The size structure of *Scyllarides astori* captured by the local artisanal fleet is characterized by having a narrow, unimodal distribution that has not displayed significant variations during the period 1997 to 2003 (Hearn 2004; Molina et al. 2004). Few individuals smaller than 20 cm TL or larger than 30 cm TL are landed (Figure 14.8). The narrow size range was also noticed in the 1970s by Reck (1983), so it is unlikely to be a consequence of fishing pressure and probably reflects the difficulties in catching juveniles (which may inhabit a different depth range or display cryptic behavior), as well as slow growth at sizes greater than 26 cm TL (Hearn 2004).

14.5   Reproductive Biology

There is insufficient information on the reproductive biology of *Scyllarides astori* in Galápagos, although prior studies have been able to provide indications on the seasonality of reproduction, the fecundity and duration of incubation, and the size of females at onset of sexual maturity. However, the copulation event has not been observed either in the wild or in the laboratory and it is not known whether migrations related to reproduction take place, as is the case with some spiny lobster species (Phillips et al. 2000).

14.5.1   Seasonality of Reproduction

The occurrence of ovigerous females of *Scyllarides astori* is a relatively rare event — out of 1142 females obtained from sampling throughout the archipelago in 2000 to 2004, only 13.1% were with egg masses (Martínez et al. 2002). During this period, there were only 19 months in which representative samples ($N \geq 10$ females) were obtained. Of these months, those displaying the highest percentages of ovigerous females (15 to 30%) were consistently the warm months, from December to April–May. In contrast, generally few to no ovigerous females were recorded during the cool season from May–June to November (Figure 14.9).

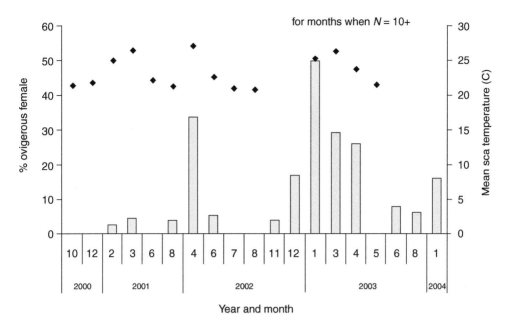

FIGURE 14.9 Percentage of ovigerous females in samples of slipper lobster (*Scyllarides astori*) for months where the number of females was equal to or greater than 10 (columns). Mean monthly sea surface temperature (from AVHRR data provided by NASA-JPL) when available (diamonds).

Based on these data, it is proposed that *S. astori* reproduces at most once a year, during the warm season. Due to the low numbers of ovigerous females obtained during sampling, it is assumed that they may engage in more cryptic behavior or migrate to deeper waters (Hearn 2004). Other species of the genus are also thought to reproduce seasonally, in the warm months, such as *S. latus* in the Azores (Martins 1985) and in the Mediterranean (Spanier & Lavalli 1998); and *S. nodifer* in the Gulf of Mexico (Hardwick & Cline 1990)

14.5.2 Size of Sexual Maturity

Estimations of size at onset of sexual maturity (SOM) have been calculated for females (Hearn 2004) using log-transformed normalized proportions of ovigerous females in 2 cm TL size classes. The smallest ovigerous female obtained over four years of sampling (2000 to 2004) was 20.1 cm TL. The onset of sexual maturity appeared to occur over a narrow size range, possibly corresponding to a single, pubertal molt, with 25% maturity occurring at 21 cm TL and 95% maturity occurring at 23 cm TL (Figure 14.10). SOM, defined as the size at which 50% of females have egg masses, corresponds to 22.1 cm TL (10.2 cm CL). *Scyllarides squammosus* (H. Milne Edwards, 1837), found in the Hawaiian islands, attains sexual maturity at a size of 6.6 to 6.7 cm CL, and maximum lengths of around 8 cm CL, although significant variation was found between reefs, suggesting that this plasticity may also apply for *S. astori* in Galápagos, where such a range of biophysical conditions exist between islands.

14.5.3 Fecundity

The only study carried out to date on the fecundity of *Scyllarides astori* was based on the analysis of the egg masses of 19 females (Toral-Granda et al. 2002). Brood size ranged from 87,000 to 360,000 eggs, and was found to be size dependent ($r = 0.784$, df = 18, $P \leq 0.05$) (Figure 14.11). As in other lobster species (nephropids and palinurids), the size of the egg clutch increases with the size of the female (Aiken & Waddy 1979). This is particularly important in stocks that are fished, as fishing pressure can result in a decrease in the size of onset of maturity (Polovina 1989; Chubb 1994). A population consisting of

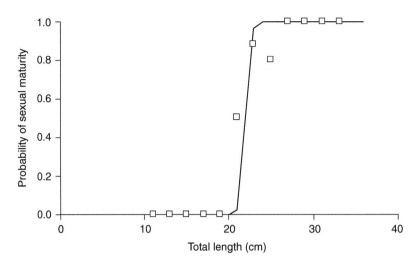

FIGURE 14.10   Sexual maturity (normalized data points) of female slipper lobsters, *Scyllarides astori*, in the Galápagos Archipelago from individuals obtained during 2000 to 2004 sampling period (*N* = 1142). (From Hearn, A. 2004. *Evaluación de las Poblaciones de Langostas en la Reserva Marina de Galápagos. Informe Final 2002–2004*. Santa Cruz, Galápagos, Ecuador: Fundación Charles Darwin y Dirección Parque Nacional Galápagos. 96 pp.; used with permission.)

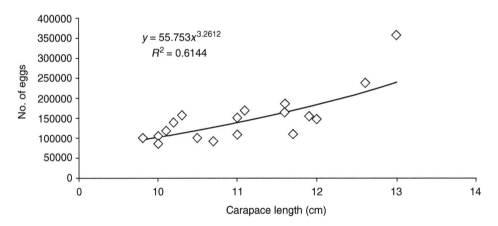

FIGURE 14.11   Total number of eggs in relation to carapace length (CL) for *Scyllarides astori* in the Galápagos islands.

smaller females will produce fewer eggs than a population with larger females present. Mean brood size was 147,000 eggs (±14,600 SE) (Martínez et al. 2002). This can be compared with the brood size of *S. latus*, which ranges from 151 to 356,000 eggs according to Martins (1985).

14.6   Feeding Ecology

In the field, *Scyllarides astori* has been observed preying upon the white sea urchin, *Tripneustes depressus* A. Agassiz, 1863. Additionally, in laboratory experiments, *S. astori*'s preferred food item was the white sea urchin (Martínez 2000). However, analysis of stomach contents showed a varied diet. Over a 14-month period in 1998 to 1999, the stomachs of 314 individuals were analyzed, out of which 49% (*n* = 154) were empty. The remains of 10 taxa were observed in those individuals whose stomachs were full or partially full (Figure 14.12). Besides digestive material, which made up 35% of stomach contents, bivalve remains were the most common, represented by the families Arcidae, Mytilidae, Isognomonidae, and Pinnidae.

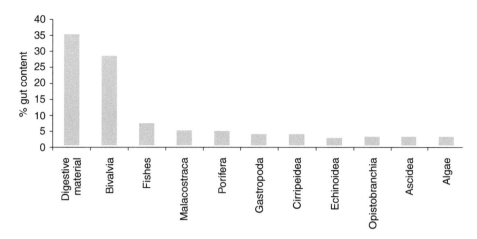

FIGURE 14.12 Overall gut content of *Scyllarides astori* (N = 160) examined in 1998 to 1999. (From Martínez, C.E. 2000. *Ecología trófica de Panulirus gracilis, P. penicillatus y Scyllarides astori (Decapoda, Palinura) en sitios de pesca de langosta en las islas Galápagos*. Tesis de Licenciatura: Universidad del Azuay, Ecuador. 102 pp. and Danulat, E. & Edgar, G.J. (eds.) 2002. *Reserva Marina de Galápagos*. Linea Base de la Biodiversidad. Fundación Charles Darwin / Servicio Parque Nacional Galápagos, Santa Cruz, Galápagos, Ecuador. 484 pp.; used with permission.)

An analysis of variance (ANOVA) test did not show evidence of significant differences between diet and sex (ANOVA; $F = 0.022$; df $= 1$; $P = 0.881$) (Martínez 2000). Bivalves were also the preferred diet of the Mediterranean slipper lobster *S. latus* (Spanier et al. 1993).

14.7   Growth

As with other lobsters, the growth process in *Scyllarides astori* depends on the frequency of molting and the size increment per molt. Growth in lobsters is notoriously difficult to measure, with different methods having different drawbacks — observations in captivity generally involve artificial temperature and feeding conditions, while observations in the wild are rare and tagging results may often depend on the type of tag, with varying mortality and tag-shedding rates (Spanier & Barshaw 1993).

14.7.1   Mark-Recapture Program

From 2000 to 2004, a mark-recapture program was carried out at the Charles Darwin Foundation in order to evaluate the populations of the commercially important lobsters occurring in the Galápagos Archipelago (Martínez et al. 2002; Hearn et al. 2003; Hearn 2004, 2006). Slipper lobsters were obtained by divers during the abundance surveys described in Section 14.3.1. Standard T-bar tags (made by Hallprint, Australia) were inserted ventrally into the musculature on the first abdominal segment, at a point halfway between the midline and the right-hand edge (Figure 14.13). Total length of each individual was measured as the distance between the rostrum and the posterior edge of the telson, whereas CL was measured as the distance between the rostrum and the posterior edge of the carapace. All individuals of *Scyllarides astori* caught were sexed, measured, weighed, and tagged before being released into the sea at their site of capture.

When caught by the local fishers, the marked slipper lobsters were reported and measured in exchange for a small reward, to ensure that information would be obtained on change in size over time. A total of 1926 *S. astori* were tagged and released back into the wild out of which 116 (6%) were recaptured and reported by the local fishers (17 in 2001, 44 in 2002, 47 in 2003, and eight in 2004). Of these, 59 individuals (35 males and 24 females) were landed whole, thus providing complete data. The total length of the tagged individuals ranged from 9.8 to 37.8 cm, although 86.5% of the specimens ranged from 20.1 to 28 cm TL (Hearn 2004).

FIGURE 14.13 Tagging specimen of *Scyllarides astori* ventrally into the musculature of the first abdominal segment. (Photograph by Hamish Saunders, used with permission.)

14.7.2 Molt Frequency

Molting frequency and seasonality in scyllarid lobsters varies among species and possibly among populations of the same species — *Scyllarides latus* displays a seasonal winter molt according to Spanier & Lavalli (1998), but molts throughout the year according to Bianchini et al. (2001). The molt process for *Scyllarides astori* has not been observed either in the wild or in the laboratory. During the 2000 to 2004 study, only two recently molted individuals were observed (Hearn 2004). In a study in 1979 (Reck 1983), no recently molted individuals were found in February/March or May/June periods, whereas in August, 3.2% of males and 1.5% of females were found to have molted recently, and in November, these values increased to 5.2 and 2.8%, respectively. These results indicate some degree of seasonality, with molting following after the reproductive season (Section 14.5.1). It is not known whether *S. astori* migrates to deeper water to molt, or behaves in a more cryptic fashion during this vulnerable period.

Molt periodicity was estimated by grouping the recaptured lobsters according to the time spent at large, into 120 day intervals. The proportion of individuals that had molted in each time interval was converted into a probability curve, with the assumption that there was no seasonality to molting and no relationship between the size of individuals (for the size range observed) and their molt frequency. As suggested by Reck (1983), the first assumption is probably untrue, so the results must be treated as approximations.

The results (Figure 14.14) suggest a low molt frequency, as there was only a 7 to 14% chance of an individual molting within four months of being tagged, and an 80% chance of molting within 420 days of tagging. An annual molt (Figure 14.14) would display significantly higher values. Thus, it is assumed that *S. astori*, within the 20.1 to 28 cm size range, molts every 18 to 24 months (Hearn 2004, 2006).

14.7.3 Determination of Growth Parameters

For growth analysis, data obtained during the mark-recapture program, from 13 males and seven females (individuals which had molted) were used (Table 14.1). Growth was assumed to comply with the von

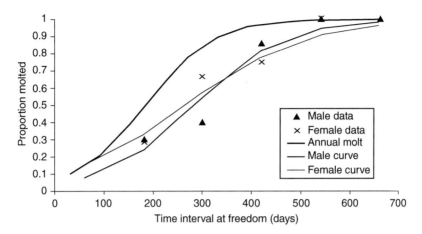

FIGURE 14.14 Probability of molting at different time intervals from tagging to recapture, compared with theoretical probability curve for annual molts with no seasonality. (From Hearn, A. 2004. *Evaluación de las Poblaciones de Langostas en la Reserva Marina de Galápagos. Informe Final 2002–2004*. Santa Cruz, Galápagos, Ecuador: Fundación Charles Darwin y Dirección Parque Nacional Galápagos. 96 pp. and Hearn, A. 2006. *J. Exp. Mar. Biol. Ecol.* 328: 87–97; used with permission.)

TABLE 14.1

Carapace Length at Tagging (CL_1) and at Recapture (CL_2), Number of Days from CL_1 to CL_2 (dT), and Annual Size Increment (dL yr^{-1}) for *Scyllarides astori* Used in Growth Analysis (where dT is Larger than 150 Days and dL is Greater than Zero)

Sex	CL_1	CL_2	Days at Liberty	dL/dT	% increase
Males	10.1	10.8	224	1.14	6.9
	12.0	13.2	242	1.81	10.0
	9.8	10.5	268	0.99	7.1
	11.2	11.8	299	0.73	5.4
	10.0	11.6	318	1.84	16.0
	9.3	10.3	397	0.92	10.8
	10.2	11.0	399	0.78	7.8
	11.4	13.0	404	1.45	14.0
	9.8	11.0	420	1.04	12.2
	10.4	11.6	420	1.04	11.5
	10.2	11.1	420	0.78	8.8
	11.0	11.5	610	0.3	4.5
	8.1	9.9	1127	0.58	22.2
Females	7.7	8.5	212	1.38	10.4
	10.0	10.4	268	0.6	4.0
	10.9	12.0	395	1.02	10.1
	11.2	12.0	420	0.7	7.1
	10.6	11.4	429	0.68	7.5
	11.1	12.1	489	0.75	9.0
	11.8	13.6	619	1.06	15.3

Source: Adapted from Hearn, A. 2006. *J. Exp. Mar. Biol. Ecol.* 328: 87–97; with permission.

Bertalanffy (1934) growth equation:

$$L_{(t)} = L_\infty \times [1 - \exp^{(-k(t-to))}]$$

where L_t is total length at age t, L_∞ is the asymptotic total length, k is the rate at which L_∞ is approached, and t_0 is the age of a slipper lobster at total length zero if growth always conforms to the equation (see Bianchini & Ragonese, Chapter 9 for an expression of this equation for carapace length).

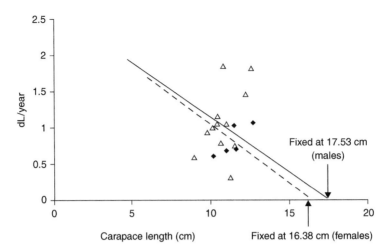

FIGURE 14.15 "Forced" Gulland & Holt plot for *Scyllarides astori* with fixed L_∞ at 17.53 cm CL for males (triangles and continuous line, $N = 13$); 16.38 cm CL for females (diamonds and broken line, $N = 7$). (From Hearn, A. 2006. *J. Exp. Mar. Biol. Ecol.* 328: 87–97; used with permission.)

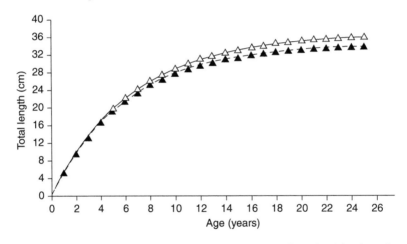

FIGURE 14.16 Length at age for *Scyllarides astori* with growth parameters for males (triangles and continuous line, L_∞=36.9 cm, K= 0.153 yr^{-1}) and females (diamonds and broken line, L_∞=34.5 cm, K= 0.162 yr^{-1}). For both sexes, the value of t_0 was assumed to be zero. (From Hearn, A. 2006. *J. Exp. Mar. Biol. Ecol.* 328: 87–97; used with permission.)

The parameter L_∞ was estimated for both sexes using the Powell-Wetherall analysis (Powell 1979; Wetherall et al. 1987; Sparre & Venema 1992) of the size distribution of individuals obtained during the 2000 to 2004 mark-recapture program. L_∞ was 36.9 cm for males and 34.5 cm for females (Hearn 2006). These values were then transformed to carapace length (CL$_\infty$). The growth data were used in a "forced" Gulland & Holt plot (Gulland & Holt 1959; Pauly 1984; Gayanilo & Pauly 1997) in order to determine the growth parameter k, as a function of the slope of the linear regression of the CL against the change in size over time (Figure 14.15).

The forced Gulland and Holt plots with fixed CL$_\infty$ of 16.38 cm for males and 17.53 cm for females gave estimates of the growth constant k of 0.124 for males and 0.145 for females. Assuming that the parameter t_0 is equal to zero, the consequent length-at-age curves display similar growth patterns for both sexes, whereby the age of full recruitment corresponding to TL 24 to 26 cm is seven to eight years (Figure 14.16) (Hearn 2006).

The results of the tagging experiments, carried out to determine the growth parameters of *S. astori*, displayed a slow-growing species in relation to other scyllarid species, such as *Ibacus peronii* Leach, 1815 and *I. chacei* Brown & Holthuis, 1998 (Stewart & Kenelly 2000), both of which display rapid growth rates for at least the first four or five years. There is little information available on the growth of other species of the genus *Scyllarides* under natural conditions, but *S. nodifer* was reported as growing from larval phase to 30 cm total length in 16 to 18 months in the laboratory (Rudloe 1983). Long-term recaptures of tagged *S. latus* from the Mediterranean (Bianchini et al. 2001) suggested slow growth rates similar to *S. astori*. These results must be considered preliminary and, as shown in Figure 14.15, are highly variable. Additionally, the narrow size range for which data exist imply the extrapolation of the growth curves for smaller sizes, which may not necessarily grow according to the von Bertalanffy growth function.

14.8 Fisheries Management

Since the creation of the GMR in 1998, fishing activity in Galápagos is limited to the local fishing sector, which is organized into four fishing cooperatives — one each on the islands of Santa Cruz and Isabela, and two on the island of San Cristobal. As a multiuse marine reserve, human activities in the GMR are regulated by means of a local Participatory Management Board (PMB), which is made up of the five main stakeholders: the Galápagos National Park Service (GNPS), the Charles Darwin Foundation (CDF), the Fishing Sector (an amalgamation of the four cooperatives), the Tourism Sector, and the Naturalist Guides Sector. Agreements are reached by consensus and ratified by the Inter-Institutional Management Authority (IMA), the ultimate decision-making body of the GMR. In the event that no consensus is reached within the PMB, the issues are transferred to the IMA which decides by majority vote. The IMA is made up of seven voting members: the Ministers of Environment, Tourism, Defence, and Commerce and Fisheries, as well as the Galápagos Fishing and Tourism Sectors, and the representative of the national environmentalist organization, CEDENMA (Ecuadorian Committee for the Defense of the Environment). The GNPS acts as secretary to the Minister of Environment, who chairs the IMA, and the CDF provides technical advice at IMA meetings.

One of the major human uses of the GMR is fishing activity, currently restricted by law to the local artisanal fleet, which is made up of approximately 1000 fishermen operating from small fiberglass and wooden vessels with outboard engines (Bustamante et al. 2000; Toral-Granda et al. 2002; Murillo et al. 2004). Fishing activity is regulated through the GMR Management Plan and a Five-Year Fishing Calendar (2002 to 2006), which incorporates a series of indicators for each major fishery. Currently, in accordance with the Five-Year Fishing Calendar, there is no closed season or minimum landing size for *Scyllarides astori*. However, there is a ban on the capture and landing of ovigerous females, and the Calendar recognized the need to incorporate a closed season and a minimum landing size by the end of 2004. As of July 2006, this is still pending.

There exists a preliminary zoning scheme for activities within the GMR, which was decided by consensus within the PMB in 2000. The entire marine area was divided into zones and subzones. The coastal area (zone 2) was divided into four subzones. Subzone 2.1 makes up 8% of the coastal area, and is set aside for comparison and protection. Subzone 2.2 makes up 10% of the coastal area, and mainly covers the tourist visitor sites and breeding sites of endemic birds, such as the flightless cormorant and waved albatross, whereas Subzone 2.3 makes up the largest proportion of the coastal area (77%) and is the only subzone in which extractive activities such as fishing are permitted. A final subzone, 2.4, covering nursery grounds and protected areas, actually consists of the areas immediate to the human settlements, making up 5% of the coastline (Figure 14.17). According to the Fishing Calendar, fishing activities, including the capture of *S. astori* may only be carried out in subzone 2.3. However, due to the lack of physical markers, and the limited patrolling capability of the GNPS and Ecuadorian Navy, there is, as yet, little adherence to the zonation scheme by the fishers (Hearn 2004). Additionally, the Fishing Calendar stressed the need to identify nursery areas for *S. astori* and incorporate them into the zonation for this species from 2004. This has not been carried out yet.

Scyllarides astori is defined as a product for local consumption; its export is limited to 10 lb of lobster tail or 12 individuals per person each time that person leaves the islands, subject to a permit obtained from the GNPS. Since 2002, approximately 3 metric tons of tails (corresponding to 8.4 metric tons live weight) are exported annually to continental Ecuador (Galápagos National Park Archives). The Fishing

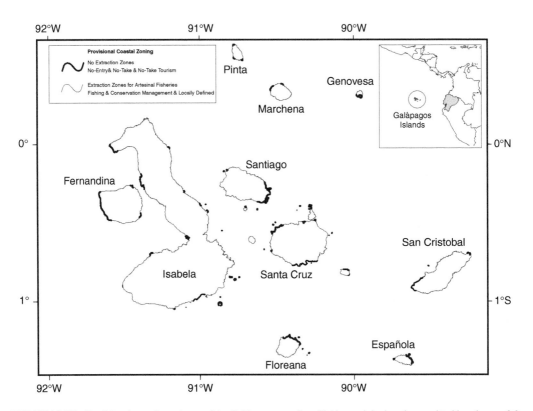

FIGURE 14.17   Provisional zonation scheme of the Galápagos coastline. Fishing activity is only permitted in sub-zone 2.3.

Calendar also stipulates that a marketing study be carried out from 2003. This has not been discussed yet within the framework of the PMB, although sufficient information exists in order to carry this out, and the commissions have been created.

14.8.1   Fishery Catches

According to Holthuis (1991), there was no traditional fishery in the Galápagos targeting *Scyllarides astori*. Rather, this species was considered to be associated with the spiny lobster (Reck 1983), and used for food when caught. There are reports of interest in this species from the aquarium trade (e.g., Holthuis 1991), although it was considered to be caught too infrequently to be of economic importance. In recent years, however, catches of *S. astori* have increased, partly as a response to decreasing revenue from spiny lobster catches (Murillo et al. 2004), even though it is thought that the total catch of *S. astori* contributes <15% of the entire lobster catch for Galápagos (Bustamante et al. 2000).

   Scyllarides astori is caught by hand, usually at night, by divers operating a surface supply, or "hooka," system from small fiberglass or wooden vessels (up to 6 m length). As a complementary resource, not as important as the sea cucumber and spiny lobster resources, each of which has defined annual fishing seasons, the fishing effort directed at *S. astori* varies on a monthly basis (Figure 14.18). Approximately 2 metric tons of whole *S. astori* are reported by the fisheries observer program each year. The lowest reported catches correspond to the sea cucumber fishing season, which has been carried out over a 60-day period from May to July since its opening in 1999 until 2003. During this period, very little effort is focused on resources other than the sea cucumber. From September to December, the spiny lobster fishing season, *S. astori* is caught alongside the red and green spiny lobsters. Because fishers are required to report all spiny lobster catches, any *S. astori* also tend to be reported at the same time during this period. In January and February, many fishermen take advantage of the good weather and the lack of more lucrative fisheries

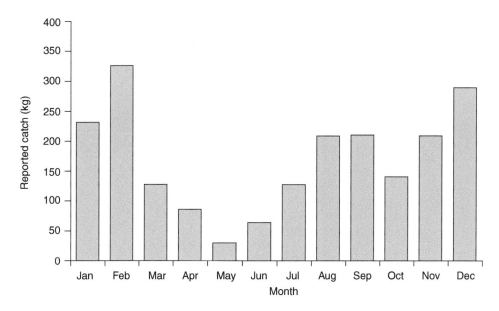

FIGURE 14.18 Mean monthly reported catch of *Scyllarides astori* 1997 to 2001.

(spiny lobster and sea cucumber) to spend several days or weeks targeting the slipper lobster in more remote locations such as western Isabela (Murillo et al. 2003).

As a minor resource, fishers are not obliged to report their catches of *S. astori*. However, it is possible to estimate the amount of *S. astori* removed during the spiny lobster fishing season by extrapolating the reported catch from fishing trips with observers to a value for the entire fleet over the four month period. It is estimated that ~70% of the total volume of *S. astori* landed is caught during the spiny lobster season (Molina et al. 2004). For the remaining months of the year, since 2002, it has been possible to obtain the number of fishing trips targeting *S. astori* from the Port Authorities and thereby estimate a catch for the entire year (Murillo et al. 2003, Molina et al. 2004). In 2002, the total catch was estimated to be 13.8 metric tons (Murillo et al. 2003), whereas for 2003, the value decreased slightly to 12.8 metric tons (Molina et al. 2004) live weight. Large quantities of *S. astori* have been removed from San Cristóbal, Floreana, Santa Cruz, and southern Isabela (Figure 14.19), corresponding to the areas closest to human settlements, which would have the lowest transport costs for the fishers.

The catch-per-unit-effort (CPUE), expressed in terms of live weight removed per diver day, was 19 kg/diver day in 2002, and declined to 10.7 kg /diver day in 2003 (Murillo et al. 2003; Molina et al. 2004). Values are not available for all islands due to the low number of reported catches, but Table 14.2 summarizes CPUE data in 2002 and 2003 for those islands where values are available.

The mean size of individuals landed has varied slightly from 1997 to 2003, but there is no significant trend, although the mean values for 1998 and 1999 were higher than in other years (Figure 14.20).

In comparison with the spiny lobster (US$10 per lb tail), the slipper lobster has a relatively low value, and does not exceed US$3 per whole individual or US$2 per lb of tail (Murillo et al. 2003). For this reason and its relatively low abundance in comparison with the spiny lobsters, the expectations that *S. astori* may be a lucrative alternative resource seem unfounded.

14.9 Discussion

As one of the minor fishing resources, little attention has been paid to *Scyllarides astori* in comparison with other, more lucrative resources, such as the spiny lobster or sea cucumber. There is a clear need, as identified in the Fishing Calendar, to improve the current regulations surrounding the extraction of this species.

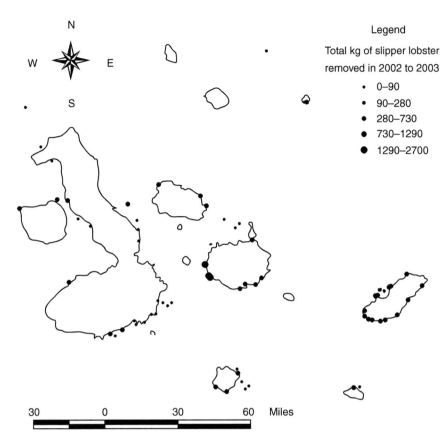

FIGURE 14.19   Estimated live weight of *Scyllarides astori* removed from different sites in the period 2002 to 2003. Note: Darwin and Wolf islands not shown.

TABLE 14.2

Catch-Per-Unit-Effort (Expressed in Kilograms Live Weight Per Diver Day) of *Scyllarides astori* by Island in 2002 and 2003

	2002		2003	
Region	**N (trips)**	**CPUE (kg. diver/day)**	**N (trips)**	**CPUE (kg. diver/day)**
Floreana	11	32.2		
N & W San Cristóbal	18	38.8		
S & E San Cristóbal	21	27.1	26	8.6
Santa Cruz	34	10.3	21	13.7
Santiago	6	7.4		
Total/Average	90	19.0	47	10.7

In 2003, data is presented for San Cristóbal as a single region. Source: Murillo et al. 2003; Molina et al. 2004.

The current no-take areas within the zonation scheme of the GMR were designed more to provide protection for nesting marine birds than as refuges for exploited marine resources. As a result, there exists a need to incorporate some islets and vertical walls, the preferred habitat of *S. astori*, within the no-take areas.

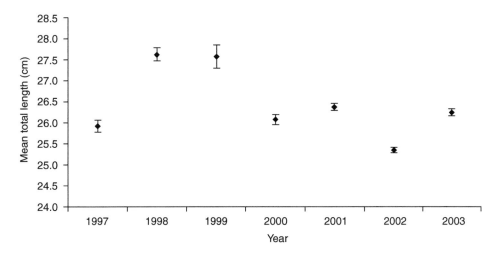

FIGURE 14.20 Mean size of landings of *Scyllarides astori* within the Galápagos Marine Reserve, 1997 to 2003.

A fishing season must be defined in accordance with the Fishing Calendar, although minimum landing size and Total Allowable Catches (TAC) are also important measures that must be considered in order to account for *S. astori*'s slow growth. A key aspect of the management of *S. astori* as a resource is the monitoring of fishing activity. A mandatory certification system, such as that in use for the spiny lobster and sea cucumber, should be adopted for this species, and for all the minor resources within the GMR. This should include information on effort, location of fishing sites, and biological information such as size, molt state, and ovigerous state of representative samples from the catch. The mark-recapture program of the Charles Darwin Foundation is expected to continue. Preliminary results from recaptured tagged individuals suggest that slipper lobsters tend to remain in a general area, although focused movement studies have not yet been carried out.

The existing knowledge of the biology of *S. astori* is still patchy (Hearn 2004, 2006). Very little is known about its early life history, including larval development and transport, settlement, and early benthic growth and ecology. Finally, the ecological role of the Galápagos slipper lobster has yet to be quantified. As a resource facing increasing pressure, it is necessary to understand its role as both predator and prey, and the impacts that changes in its population may have on the remaining benthic communities within the GMR.

Acknowledgments

The authors would like to thank the dive team at the marine department of the Charles Darwin Foundation, the Galápagos National Park Service, and the local Fishing Sector for their collaboration. Much of the work described in this chapter was funded through the US Agency for International Development, the Government of Ecuador through a loan from the Inter-American Development Bank, and the World Wildlife Foundation. Thanks also to the Pew Charitable Trusts Marine Conservation Fellowship program and the Pew Collaborative Initiative Fund Award. This document is contribution number 1029 of the Charles Darwin Foundation.

References

Aiken, D.E. & Waddy, S.L. 1979. Maturity and reproduction in the American lobster. *Can. Tech. Rep. Fish. Aquat. Sci.* 932: 59–71.

Banks, S. 2002. Ambiente físico. In: Danulat, E. & Edgar, G.J. (eds.), *Reserva Marina de Galápagos*. Santa Cruz, Galápagos, Ecuador: pp. 22–35. Linea Base de la Biodiversidad Fundación Charles Darwin y Servicio Parque Nacional de Galápagos.

Barr, L. 1968. Some aspects of the life history, ecology and behaviour of the lobsters of the Galápagos islands. *Stanford Oceanogr. Exp.* 17: 254–262.

Bianchini, M.L., Bono, G., & Ragonese, S. 2001. Long-term recaptures and growth of slipper lobsters, *Scyllarides latus*, in the Strait of Sicily (Mediterranean Sea). *Crustaceana* 74: 673–680.

Bustamante, R.H., Reck, G.K., Ruttenberg, B.I., & Polovina, J. 2000. The Galápagos spiny lobster fishery. In: Phillips, B.F. & Kittaka, J. (eds.), *Spiny Lobsters: Fisheries and Culture*: pp. 210–220. Oxford: Blackwell Scientific Press.

Bustamante, R.H., Wellington, G.M., Branch, G.M., Edgar, G.J., Martínez, P., Rivera, F., Smith, F., & Witman, J. 2002. Outstanding marine features of Galápagos. In: Bensted-Smith, R. (ed.), *A Biodiversity Vision for the Galápagos Islands*: pp. 60–71. Puerto Ayora, Galápagos: Charles Darwin Foundation and World Wildlife Fund.

Chubb, C.F. 1994. Reproductive biology: Issues for management. In: Phillips, B.F., Cobb, J.S., & Kittaka, J. (eds.), *Spiny Lobster Management*: pp. 181–212. London: Blackwell Scientific.

Danulat, E. & Edgar, G.J. (eds.) 2002. *Reserva Marina de Galápagos*. Linea Base de la Biodiversidad. Fundación Charles Darwin / Servicio Parque Nacional Galápagos, Santa Cruz, Galápagos, Ecuador. 484 pp.

Edgar G.J., Banks, S., Fariña, J.M., Calvopiña, M., & Martínez, C. 2004. Regional biogeography of shallow reef fish and macro-invertebrate communities in the Galapagos archipelago. *J. Biogeogr.* 31: 1107–1124.

Gayanilo, F.C., Jr. & Pauly, D. (eds.) 1997. *FAO-ICLARM Stock Assessment Tools. (FiSAT) Reference Manual*. FAO Computerised Information Series (Fisheries). No. 8, Rome. 262 pp.

Gotshall, D.W. 1998. *Sea of Cortez Marine Animals. A Guide to the Common Fishes and Invertebrates from Baja California to Panama*. Monterey, CA: Sea Challengers. 140 pp.

Gulland, J.A. & Holt, S.J. 1959. Estimation of growth parameters for data at unequal time intervals. *J. Cons. Perm. Int. Explor. Mer.* 25: 47–49.

Hardwick, C.W. & Cline, G.B. 1990. Reproductive status, sex ratios and morphometrics of the slipper lobster *Scyllarides nodifer* (Stimpson) (Decapoda: Scyllaridae) in the northeastern Gulf of Mexico. *Northeast Gulf Sci.* 11: 131–136.

Hartnoll, R.G. 1974. Variations in growth pattern of some secondary characteristics in crabs (Decapoda, Brachyura). *Crustaceana* 27: 131–136.

Hearn, A. 2004. *Evaluación de las Poblaciones de Langostas en la Reserva Marina de Galápagos. Informe Final 2002–2004*. Santa Cruz, Galápagos, Ecuador: Fundación Charles Darwin y Dirección Parque Nacional Galápagos. 96 pp.

Hearn, A. 2006. Life history of the slipper lobster *Scyllarides astori* Holthuis 1960, in the Galápagos Islands, Ecuador. *J. Exp. Mar. Biol. Ecol.* 328: 87–97.

Hearn, A., Pinillos, F., & Sonnenholzner, J. 2003. *Evaluación de las Poblaciones de Langostas en la Reserva Marina de Galápagos. Informe Anual 2002*. Puerto Ayora, Galápagos: Estación Científica Charles Darwin y Servicio Parque Nacional Galápagos. 42 pp.

Hearn, A., Martínez, P., Toral-Granda, M.V., Murillo, J.C., & Polovina, J. 2005a. Population dynamics of the exploited sea cucumber *Isotichopus fuscus* in the Western Galapagos Islands, Ecuador. *Fish. Oceanogr.* 14: 377–385.

Hearn, A., Castrejón, M., Reyes, H., Nicolaides, F., Moreno, J., & Toral-Granda M.V. 2005b. Evaluación de la pesquería de langosta espinosa (*Panulirus penicillatus* y *P. gracilis*) en la Reserva Marina de Galápagos. In: Hearn, A. (ed.), *Evaluación de las Pesquerías en la Reserva Marina de Galápagos. Informe Compendio 2004*: pp. 63–98. Fundación Charles Darwin, Santa Cruz, Galapagos, Ecuador.

Heylings, P., Bensted-Smith, R., & Altamirano, M. 2002. Zonificación e historia de la Reserva Marina de Galápagos. In: Danulat, E. & Edgar, G. (eds.), *Reserva Marina de Galápagos, Línea Base de la Biodiversidad*: pp. 10–21. Santa Cruz, Galápagos, Ecuador: Fundación Charles Darwin y Servicio Parque Nacional Galápagos.

Hickman, C.P. & Zimmerman, T.L. 2000. *A Field Guide to Crustaceans of Galapagos. An Illustrated Guidebook to the Common Barnacles, Shrimps, Lobsters and Crabs of the Galapagos Islands*. Lexington, VA: Sugar Spring Press. 156 pp.

Holthuis, L.B. 1991. *Marine Lobsters of the World. An Annotated and Illustrated Catalogue of the Species of Interest to Fisheries Known to Date.* FAO Species Catalogue No. 125, Vol 13: pp. 1–292. Rome: Food and Agriculture Organization of the United Nations.

Holthuis, L.B. & Loesch, H. 1967. The lobsters of the Galapagos Islands (Decapoda, Palinuridae). *Crustaceana* 12: 214–222.

Johnson, M.W. 1971. The palinurid and scyllarid lobster larvae of the tropical eastern Pacific and their distribution as related to the prevailing hydrography. *Bull. Scripps. Inst. Oceanogr.* 19: 1–36.

Lavalli, K.L. & Spanier, E. 2001. Does gregariousness function as an antipredator mechanism in the Mediterranean slipper lobster *Scyllarides latus*? *Mar. Freshwater Res.* 52: 1133–1143.

Martínez, C.E. 2000. *Ecología trófica de Panulirus gracilis, P. penicillatus y Scyllarides astori (Decapoda, Palinura) en sitios de pesca de langosta en las islas Galápagos.* Tesis de Licenciatura: Universidad del Azuay, Ecuador. 102 pp.

Martínez, C.E., Toral, V., & Edgar, G. 2002. Langostino. In: Danulat, E. & Edgar, G.J. (eds.), *Reserva Marina de Galápagos, Línea Base de la Biodiversidad*: pp. 216–232. Santa Cruz, Galápagos, Ecuador: Fundación Charles Darwin y Servicio Parque Nacional Galápagos.

Martins, H.R. 1985. Biological studies of the exploited stock of the Mediterranean locust lobster *Scyllarides latus* (Latreille, 1802) (Decapoda: Scyllaridae) in the Azores. *J. Crust. Biol.* 5: 294–305.

Molina, L., Chasiluisa, C., Murillo, J.C., Moreno, J., Nicolaides, F., Barreno, J.C., Vera, M., & Bautil, B. 2004. Pesca Blanca y pesquerías que duran todo el año, 2003. In: Murillo, J.C. (ed.), *Evaluación de las Pesquerías en la Reserva Marina de Galápagos. Informe Compendio 2003*: pp. 103–139. Santa Cruz, Galápagos: Fundación Charles Darwin y Parque Nacional Galápagos.

Murillo, J.C., Martínez, P., Toral-Granda, M.V., & Hearn, A. 2002. Pepino de Mar. In: Danulat, E. & Edgar, G.J. (eds.), *Reserva Marina de Galápagos, Línea Base de la Biodiversidad*: pp. 176–197. Santa Cruz, Galápagos, Ecuador: Fundación Charles Darwin y Sevicio Parque Nacional Galápagos.

Murillo, J.C., Chasiluisa, C., Molina, L., Moreno, J., Andrade, R., Bautil, B., Nicolaides, F., Espinoza, E., Chalén, L., & Barreno, J.C. 2003. Pesca blanca y pesquerías que duran todo el año en Galápagos, 2002. In: Murillo, J.C. (ed.), *Evaluación de las pesquerías en la Reserva Marina de Galápagos. Informe Compendio 2002*: pp. 97–124. Santa Cruz, Galápagos, Ecuador: Fundación Charles Darwin y Servicio Parque Nacional Galápagos.

Murillo, J., Nicolaides, F., Reyes, H., Moreno, J., Molina, L., Chasiluisa, C., Bautil, B., Villalta, M., García, L., & Ronquillo, J. 2004. Estado pesquero y biológico de las dos especies de langosta espinosa en el año 2003. Análisis comparativo con las pesquerías 1997–2002. In: Murillo, J.C. (ed.), *Evaluación de las pesquerías en la Reserva Marina de Galápagos. Informe Compendio 2003*: pp. 50–102. Santa Cruz, Galápagos, Ecuador: Fundación Charles Darwin y Dirección Parque Nacional Galápagos.

Pauly, D. 1984. Length converted catch curves: A powerful tool for fisheries research. *Fishbyte* 2: 18–19.

Phillips, B.F., Chubb, C.F., & Melville-Smith, R. 2000. The status of Australia's rock lobster fisheries. In: Phillips, B.F. & Kittaka, J. (eds.), *Spiny Lobsters, Fisheries and Culture.* Second Edition: pp. 45–77. Oxford, U.K.: Fishing News Books, Blackwell Scientific.

Polovina, J. 1989. Density-dependence in spiny lobster, *Panulirus marginatus*, in the Northwestern Hawaiian Islands. *Can. J. Fish. Aquat. Sci.* 46: 660–665.

Powell, D.G. 1979. Estimation of mortality and growth parameters from the length-frequency in the catch. *Rapports et Proces-Verbaux des Réunions du Conseil International pour l'Exploration de la Mer* 175: 167–169.

Reck, G.K. 1983. *The Coastal Fisheries in the Galapagos Islands, Ecuador. Description and Consequences for Management in the Context of Marine Environmental Protection and Regional Development.* Dissertation zur Erlangung des Doktorgrades. Christian-Albrechts-Universität zu Kiel. Kiel, Alemania. 231 pp.

Rudloe, A. 1983. Preliminary studies of the mariculture potential of the slipper lobster *Scyllarides nodifer*. *Aquaculture* 34: 165–169.

Snell, H.M., Stone, P.A., & Snell, H.L. 1996. Special paper: A summary of geographical characteristics of the Galapagos Islands. *J. Biogeog.* 23: 619–624.

Spanier, E. & Barshaw, D.E. 1993. Tag retention in the Mediterranean slipper lobster. *Israel J. Zool.* 3: 29–33.

Spanier, E. & Lavalli, K.L. 1998. Natural history of *Scyllarides latus* (Crustacea: Decapoda): A review of the contemporary biological knowledge of the Mediterranean slipper lobster. *J. Nat. Hist.* 32: 1769–1786.

Spanier, E., Almog-Shtayer, G., & Fiedler, U. 1993. The Mediterranean slipper lobster *Scyllarides latus*, the known and the unknown. *Proceedings of the Fourth Colloquium Crustacea Decapoda Mediterranea, Thessaloniki, Greece, April 25–28, 1989*. pp. 49–58.

Sparre, P. & Venema, S.C. 1992. *Introduction to Tropical Fish Stock Assessment. Part 1. Manual*. FAO Fish. Tech. Pap. No. 306/1 FAO, Rome. 376 pp.

Stewart, J. & Kennelly, S.J. 2000. Growth of the scyllarid lobsters *Ibacus peronii* and *I. chacei*. *Mar. Biol.* 136: 921–930.

Toral-Granda, M.V. 2005. Requiem for the Galapagos sea cucumber fishery? *SPC Beche-de-Mer Information Bulletin* 21: 5–8.

Toral-Granda, M.V., Espinoza, E., Hearn, A., & Martínez, C. 2002. Langostas espinosas. In: Danulat, E. & Edgar, G.J. (eds.), *Reserva Marina de Galápagos, Línea Base de la Biodiversidad. Fundación*: pp. 190–215. Santa Cruz, Galápagos, Ecuador: Charles Darwin y Servicio Parque Nacional Galápagos.

von Bertalanffy, L. 1934. Untersuchungen über die Gesetzlichkeiten des Wachstums 1. Allgemeine Grundlagen der Theorie. *Roux. Arch. Entwicklungsmech. Org.* 131: 613–653.

Wetherall, J.A., Polovina, J., & Ralston, R. 1987. Estimating growth and mortality in steady state fish stocks from length-frequency data. *ICLARM Conf. Proc.* 13: 53–74.

15

Biology and Fishery of the Slipper Lobster, Thenus orientalis, *in India*

Edakkepravan V. Radhakrishnan, Mary K. Manisseri, and Vinay D. Deshmukh

CONTENTS

Abstract

Commercially exploited, edible lobsters in India belong to the families Palinuridae and Scyllaridae, the latter representing only one species, *Thenus orientalis* (Lund, 1793). The introduction of mechanized trawlers for fishing and the attractive prices that export of frozen rock lobster tail fetched in the early 1970s resulted in the exploitation of lobster resources on a commercial scale. *Thenus orientalis* found along both the east and west coasts of India formed and continues to sustain fisheries of importance along the northwest and southeast coasts, where their landings are mainly as by-catch of trawlers. India has a multispecies lobster fishery: *T. orientalis* dominates the fishery in Gujarat and northern Tamil Nadu, *Panulirus polyphagus* Herbst, 1793 dominates in Maharashtra, *Puerulus sewelli* Ramadan, 1938 dominates in Kerala and *Panulirus ornatus* Fabricius, 1798 and *P. homarus* Linnaeus, 1758 dominates in southern Tamil Nadu. The total landings of lobsters by trawlers in the country improved from 800 metric tons in 1968 to 4075 metric tons in 1985, fluctuated around 2200 metric tons for more than a decade, and then declined to 1438 metric tons in 2004. The northwest region, accounted for more than 50% of the catch, followed by the southwest and southeast regions. In the waters off Mumbai, the fishery for *T. orientalis* showed a phase of development during 1978 to 1982 with an average annual catch of 160 metric tons, a more or less stable phase during 1983 to 1987 with an average annual landing of 261 metric tons, and a phase of decline during 1988 to 1994, the catch being just 2 metric tons each in 1994 and 1996. Mortality estimates and stock assessment of the fishery for the period 1980 to 1985 showed that the exploitation ratio of males was 0.85 ($Z = 4.59$ and $M = 0.69$), while that for females was 0.82 ($Z = 3.84$ and $M = 0.69$). The E_{max} was only 0.55 for both the sexes. This was a classic example of recruitment overfishing of *T. orientalis* resulting in rapid decline of the fishery and annihilation of the stock. In the state of Gujarat where the maximum landings of *T. orientalis* are reported, size groups of 60 to 80 mm carapace length formed the mainstay of the fishery in 2004. Analysis of gut contents showed that bottom sediment was more frequent in the guts of smaller-sized lobsters, whereas mollusks were more common in the guts of larger animals. Analysis of the length frequency of *T. orientalis* collected at Mumbai during 1980 to 1986 showed that the males and females had an asymptotic total length (L_∞) of 257.3 ± 7.7 mm and 256.4 ± 9.5 mm, respectively, while the growth coefficients (K) were 0.53 ± 0.04 and 0.46 ± 0.04, respectively. Higher incidence of mature and berried females was noticed during October to March, indicating a specific spawning season for the species. Captive breeding and larval rearing of *T. orientalis* was achieved under laboratory conditions; in addition, both *T. orientalis* and *Petrarctus* (formerly *Scyllarus*) *rugosus* (H. Milne Edwards, 1837) attained sexual maturity, mated, and spawned under captive conditions. The spermatophoric mass adhered to the postventral sternite and anterior abdominal region of female in the form of a white, longitudinal, jellylike mass. Oviposition was completed within six to eight hour and the spermatophore was lost in about 12 h after mating. For *T. orientalis* the incubation period was 35 days. The phyllosoma larvae metamorphosed through four stages and molted into the nisto stage in 26 days, with a survival rate of 22%. The nisto, a nonfeeding stage, molted into juvenile in four days. Fishery regulations for each maritime state have to be formulated and implemented considering the resource status of each state. Currently comanagement of the resource through participatory approach is practiced, and in consideration of the socioeconomic impact of the lobster fishery, enforcement of regulations currently considers the socioeconomic and livelihood issues of various stakeholders.

15.1   Introduction

The edible lobsters exploited commercially in India belong to the families Palinuridae and Scyllaridae. While the former is represented by the littoral species *Panulirus polyphagus* (Herbst, 1793), *Panulirus homarus* (Linnaeus, 1758), and *Panulirus ornatus* (Fabricius, 1798), as well as the deep-sea species, *Puerulus sewelli* Ramadan, 1938, the latter is represented by only one species of commercial importance, the sand lobster, *Thenus orientalis* (Lunda, 1793). Other, noncommercially important slipper lobsters recorded from Indian waters include *Scyllarides elisabethae* (Ortmann, 1894) from Vizhinjam, *Scyllarides tridacnophaga* Holthuis, 1967 from Mandapam, *Parribacus antarcticus* (Lund, 1793) from Mandapam and Minicoy, *Scyllarus rubens* Alcock & Anderson, 1894, *Scyllarus batei* Holthuis, 1946, *Eduarctus* (formerly *Scyllarus*) *martensii* (Pfeffer, 1881) from the west coast, *Petrarctus* (formerly *Scyllarus*) *rugosus* (H. Milne Edwards, 1837) from Chennai, *Biarctus* (formerly *Scyllarus*) *sordidus* (Stimpson, 1860) from both coasts, and *Scyllarus tutiensis* Srikrishnadhas, Rahman & Anandasekaran, 1991 from Tuticorin. Although rich in species diversity, only one or two species of scyllarids dominate the landings along the different coastal states of the country.

Commercial exploitation of lobsters in India began in the 1950s. Initially, lobsters were caught by bottom-set gill nets, bag nets, baited traps, and manually operated dragnets. Lobsters gained attention

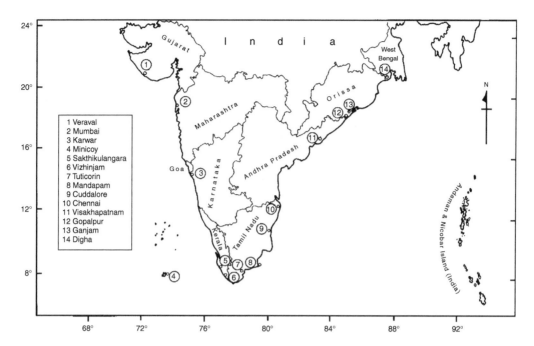

FIGURE 15.1 Major fishing centers (1 and 10) and areas of occurrence of *Thenus orientalis* along the Indian coast.

in the early 1970s when export of frozen rock lobster tail was initiated. The advent of trawlers as an efficient gear for fishing also helped in increasing the exploitation of lobsters. Exploitation has risen in the recent past due to heavy demand and attractive prices for live lobsters in the international market. Although lobsters are one of the most valuable and highly priced seafood items in great demand both in the domestic market and export industry, they support only minor fisheries and constitute <1% of the edible crustaceans landed in India (Anonymous 2005). However, among crustaceans, lobsters are second only to shrimps in the export earnings.

While *T. orientalis* is landed along both the east and west coasts of India, it is only along the northwest and southeast coasts of the country that the species forms fisheries of importance (Figure 15.1). Being a bottom-dwelling species, *T. orientalis* is caught mostly by trawlers and is landed as by-catch. Landing reports in India on this species date from the present to nearly six decades ago (Alikunhi 1948; Prasad & Tampi 1960; Chhapgar & Deshmukh 1964; Hossain 1975, 1978; Hossain et al. 1975; Radhakrishnan et al. 1995). Reviews of the status of lobster fisheries in India include those by Kagwade et al. (1991), Radhakrishnan (1995), Radhakrishnan & Manisseri (2003), and Radhakrishnan et al. (2005).

In addition, a number of studies on the biology of this species have been conducted. Prasad & Tampi (1957) and Kagwade & Kabli (1991) described the development of the eggs and larvae of *T. orientalis*. Kizhakudan et al. (2004a) reared larvae to settlement in the laboratory, and reported the techniques for such breeding and rearing programs. Reproductive biology, age and growth, conversion factors, and morphometry of *T. orientalis* from Mumbai (formerly called "Bombay"), where the species supported a lucrative fishery, were studied by Kagwade & Kabli (1996a, 1996b) and Kabli & Kagwade (1996). The decline and collapse of the fishery for the species from Mumbai was described by Deshmukh (2001). Subramanian (2004) reported on the fishery for the species off Chennai (formerly called "Madras") along the southeast coast of India.

15.2 Fishery Distribution

Thenus orientalis has been reported from different regions along the coast of India: Digha, Gopalpur, the Ganjam coast, the Andhra coast, the Chennai coast, and along different maritime states on the west coast, such as Kerala, Maharashtra, and Gujarat (Figure 15.1). However, fisheries of considerable magnitude have been reported only from the states of Gujarat and Maharashtra on the northwest coast (Kagwade et al. 1991). At Chennai in Tamil Nadu, the species constitutes a small fishery and is landed as by-catch by trawlers (Subramanian 2004). Operations of bottom-set gill nets landing small numbers of *T. orientalis*, along with cephalopods, elasmobranchs, sciaenids, and threadfin breams at several villages in the northern Tamil Nadu coast also have been reported (Kizhakudan et al. 2004b). Kerala State, on the southwest coast supports a fishery of small magnitude. *Thenus orientalis* is landed year-round (except during 15th June to 30th July when trawling is banned) as by-catch of trawlers from a depth of about 20 to 60 m at Sakthikulangara.

A lucrative fishery for *T. orientalis* once occurred along the coast of Maharashtra, but it failed to withstand the unusually heavy exploitation in the 1980s, even though the lobster was not a target fishery, and the stock became almost entirely depleted in the early 1990s. At present, Gujarat reports maximum landings of the species, although the stock shows depletion at an alarming rate when compared to landings of the previous decades (Radhakrishnan et al. 2005). As such, the multispecies lobster fishery in India shows an interesting pattern of distribution, with *T. orientalis* dominating the fishery in Gujarat, *Panulirus polyphagus* dominating in Maharashtra, *Puerulus sewelli* in Kerala, *P. ornatus* and *P. homarus* along the southern region of Tamil Nadu, and *T. orientalis* and *P. homarus* along the northern part of Tamil Nadu. Other species of minor importance landed are the spiny lobsters, *Panulirus longipes* (A. Milne Edwards, 1868), *Panulirus versicolor* (Latreille, 1804), *Panulirus penicillatus* (Olivier, 1791), *Linuparus somniosus* Berry & George, 1972, *Palinustus waguensis* Kubo, 1963, and the nephropid lobster *Nephropsis stewarti* Wood-Mason, 1872 (Radhakrishnan & Manisseri 2003).

15.3   Biology

Thenus orientalis has a depressed body, a trapezoidal-shaped carapace, broad and flattened antennae, and small brown eyes. The body color is brownish-black with reddish-brown granules. The species is a bottom-dwelling form that generally occurs on soft bottoms with sand, mud, shells, or gravel (Kizhakudan et al. 2004a). Sexes are distinguished by the position of the gonopores at the base of the third pereiopod for females and fifth pereiopod for males. The planktonic phyllosoma larval phase is not as long as that in the spiny lobsters (Kizhakudan et al. 2004a). *Thenus orientalis* generally buries into soft substrates with its eyes and antennules visible during day time. It actively swims for nocturnal feeding, which is how it is captured by trawlers as by-catch. In India, the species is landed mostly from 20 to 50 m depth. (For more detailed information on the ecology and behavior of *Thenus* spp. see Jones, Chapter 16).

15.3.1   Size-Frequency Distribution

The carapace length (CL) of *T. orientalis* landed at Veraval in 2004 ranged from 36 to 95 mm and 41 to 100 mm in males and females, respectively. Lobsters belonging to the size group 60 to 80 mm CL form the mainstay of the fishery. Females predominate in a 1:1.7 ratio, suggesting that more females are present in that particular fishing ground. Kagwade & Kabli (1996a) who studied the fishery for *T. orientalis* in the waters off Mumbai during 1985 to 1986 reported the occurrence of males in the size range of 41 to 277 mm total length (TL) and females in the range of 55 to 275 mm TL. At Sakthikulangara, the fishery was represented by *T. orientalis* in the size range of 111 to 115 mm TL and 101 to 180 mm TL for males and females, respectively, with lobsters in the size range of 141 to 155 mm TL forming the mainstay of the fishery during 2004 to 2005. Subramanian (2004), in his study of the fishery for species in the waters off Chennai during 1982 to 1999, observed that females were slightly larger, ranging in size from 71 to 275 mm TL, while males measured from 71 to 265 mm TL. Mean sizes were 153.6 mm TL and 156.9 mm TL for males and females, respectively. Males belonging to the size range 91 to 220 mm TL and females in the size range 91 to 240 mm TL comprise most of the catch. Sexes were almost equally distributed, ruling out the possibility of sexual segregation. The author suggested recruitment into the fishery at around 110 mm TL, at the age of 1.5 years. Kizhakudan et al. (2004b), in their study on *T. orientalis* landed in

TABLE 15.1

Morphometric Relationship of Male and Female *Thenus orientalis* at Mumbai

Body Dimensions	Sex	Relationships	
Carapace and total length	M	$CL = -0.4047 + 0.4361 \times TL$	$(r = 0.96)$
	F	$CL = 4.8053 + 0.3832 \times TL$	$(r = 0.98)$
Carapace length and total weight	M	$TW = 0.002077 \times CL^{2.678}$	$(r = 0.91)$
	F	$TW = 0.000348 \times CL^{3.114}$	$(r = 0.93)$
Total length and total weight	M	$TW = 0.0001964 \times TL^{2.7013}$	$(r = 0.92)$
	F	$TW = 0.000115 \times TL^{2.7938}$	$(r = 0.96)$
Carapace length and abdominal weight	M	$AW = 0.003051 \times CL^{2.3703}$	$(r = 0.91)$
	F	$AW = 0.00004064 \times CL^{2.3032}$	$(r = 0.93)$
Total length and abdominal weight	M	$AW = 0.0001252 \times CL^{2.6105}$	$(r = 0.91)$
	F	$AW = 0.00004064 \times CL^{2.8234}$	$(r = 0.96)$
Abdominal weight and total weight	M	$AW = 5.6541 + 0.3732 \times TW$	$(r = 0.94)$
	F	$AW = 0.41 + 0.4365 \times TW$	$(r = 0.97)$

small numbers by bottom-set gill nets along the Chennai–Cuddalore stretch in Tamil Nadu, reported modal size classes of 41 to 65 mm CL and 41 to 75 mm CL for males and females, respectively.

15.3.2 Dimensional Relationship

Kabli & Kagwade (1996) studied the morphological relationship between the various characters of *Thenus orientalis*. A total of 3660 specimens (all nonovigerous) of size range 52 to 227 mm TL were measured (Table 15.1). The relationship between the carapace length (length between the median spine on the anterior margin and the posterior margin of carapace) and total length (length between the median spine on the anterior margin of the carapace and the posterior margin of telson) of the two sexes indicated that, for any given carapace size up to about 42.5 mm CL, the total length of males was greater than that for a female of the same carapace length, but above 42.5 mm CL, it was smaller than that of females (Kabli & Kagwade 1996). The relationship between the CL and total weight showed that for all sizes, males were marginally heavier than females. Abdomens of males were slightly heavier than those of females up to about 200 mm TL, but beyond that the females had heavier abdomens. Differences in the abdominal weight at different carapace lengths was negligible between the two sexes. Instead of grouping by weight or sex, processors group-market lobsters in one lot for processing and freezing according to their commercial grades.

These observations differ greatly from those on spiny lobsters *Jasus lalandii* (H. Milne Edwards, 1837) (Fielder 1964), *Panulirus homarus* (Heydorn 1969), and *P. polyphagus* (Kagwade 1987) wherein the carapace size in males is shorter than that of females until maturity and thereafter elongates. The carapace length of males in spiny lobsters increases with maturity in order to accommodate the highly convoluted vas deferens, while the abdominal length in females increases with the total length in order to carry a large number of eggs on the pleopods. In comparison, the vas deferens in *T. orientalis* is less convoluted and fewer eggs are carried on the pleopods. Possibly these differences in reproductive organs and strategies could result in the lack of pronounced differences in various morphometric relationships between TL and CL for the two sexes of this species. Moreover, unlike the spiny lobsters where males grow to a much larger size than the females, the two sexes of *T. orientalis* grow to almost the same size, the largest male recorded being 277 mm TL and the largest female being 275 mm TL. In the Mediterranean slipper lobster, *Scyllarides latus* (Latreille, 1802), however, Martins (1985) in the Azores Islands and Almog-Shtayer (1988) and Spanier & Lavalli (1998) in Israel found that females are larger than males (see Bianchini & Ragonese, Chapter 9 for further information on growth and size relationships in *Scyllarides* spp.). Jones (1990), in a study on the morphological characteristics of *Thenus* species from northeastern Australia, also stated that the morphometric differentiation between sexes was related to characteristics associated with egg bearing.

15.3.3 Food and Feeding

The food materials observed in the gut of *T. orientalis* obtained from commercial catches were mostly amorphous and unrecognizable. Kabli (1989) gave a detailed account of the food and feeding of the species from Mumbai waters. She observed high incidences of empty guts (74%), especially in the smaller-sized males and females. When food was present, it did not appear to differ with regard to season or to size of the lobster. However, bottom sediment was more frequent in the guts of smaller-sized lobsters and mollusks were more common in the guts of larger animals. Overall, mollusks ranked high (27.7%) followed by bottom sediments (24.1%), fishes (22.9%), crustaceans (10.7%), polychaetes (4.2%), and miscellaneous food items (10.4%) like sponges, sipunculid worms, foraminiferans, etc. It is interesting to note that mollusks were reported as the preferred food of the Mediterranean slipper lobster *Scyllarides latus* by Spanier (1987) and also in the gut contents of the Galápagos slipper lobster, *S. astori* Holthuis, 1960 by Martínez (2000) (see Hearn et al., Chapter 14). Molluskan diet items included shells, together with viscous tissue and gastric glands of gastropods and pelecypods. The shells formed an essential constituent of the diet and may be used to replenish calcium and other minerals essential during ecdysis or, alternatively, could be used as an aid in grinding foodstuff in the cardiac stomach. Cephalopods, that are abundant along the northwest coast of India, were identified in the gut by the presence of suckers, jaws, radula, pen, eye lenses, arms or tentacles, and sometimes body parts of cuttlefish (*Sepiella* spp.) and squid (*Loligo duvaucelii* D'Orbigny, 1835).

The slipper lobster, *T. orientalis*, is a slow-moving crustacean compared to fast-swimming squids and cuttlefishes. Therefore, the likelihood of these lobsters actively capturing such mollusks is low, and their presence in the gut is probably due to consumption of animals that have recently mated, spawned, and spent their energy reserves, such that they are quiescent at the bottom, or to those that have died and are available on the substrate. Likewise, fish remnants suggest opportunistic scavenging rather than active predation. Furthermore, the idea that *T. orientalis* is a scavenger is based on finding fish eggs and planktonic organisms, such as arrow worms (*Sagitta* spp.) and ostracods (*Cypris* spp.) in their guts — items that could have been engulfed by this bottom feeder only when such prey had begun to decay after dying. On the bottom, slipper lobsters also feed on polychaetes and crustaceans while browsing. The presence of polychaetes in the gut was evident from the iridescent setae, jaws, and electrae, while crustaceans, such as crabs and prawns, were conspicuous by the presence of their shells, eggs, and appendages. Occasionally, entire shrimp and shell fragments of barnacles were also seen in the gut. The animals also feed on sponges and sipunculid worms present on sandy bottoms. Mud and sand formed an important constituent of the food of these burrowing animals, further suggesting that they may feed on detritus. Thus, based on the stomach contents, it could be inferred that slipper lobster is an opportunistic, omnivorous, benthic feeder. It burrows in soft and sandy mud (see Jones, Chapter 16 for further information), engulfs sediments consisting of sand and mud, and then preys on organisms that it encounters in this way.

15.3.4 Age and Growth

The size of the smallest specimen of *T. orientalis* collected from commercial landings at Mumbai by Kabli & Kagwade (1996) was 41 mm TL; it was assumed that at 33 mm TL, the nisto transformed into a juvenile. This almost corresponds with the size of first juvenile obtained in a hatchery experiment (Kizhakudan et al. 2004a). To examine the age and growth of commercially landed *T. orientalis*, Kabli & Kagwade (1996) used modal progression analysis where modal lengths obtained from length-frequency distributions of commercially landed *T. orientalis* were plotted at monthly intervals, and the trend in progression of modes through time was drawn by an eye-fitted line. This line, when extrapolated with reference to intermodal slopes, intersected the time axis to indicate the growth of broods in successive months and also the time of origin of the brood. The lengths at each time interval, given as age in months, were read along each of the curves and averaged to obtain growth of the lobster at successive time intervals. Accordingly, the males attained total lengths of 109 mm, 166 mm, and 213 mm and females attained total lengths of 110 mm, 165 mm, 203 mm at the end of one, two, and three years, respectively (Kabli & Kagwade 1996). In the case of females, the growth curves showed that these lobsters reached 231 mm and 242 mm TL in 4 and 4.5 years. Using the growth increments at half-yearly intervals, the estimated von Bertalanffy (1934) growth functions (VBGF) for the two sexes were:

$$\text{Males}: L_t = 368 \times [1 - e^{0.1279(t+0.7368)}]$$

$$\text{Females}: L_t = 300 \times [1 - e^{0.1690(t+0.70)}]$$

The length frequency of *T. orientalis* collected at Mumbai during 1980 to 1986 was analyzed using fish stock assessment tools (FiSAT) computer software. The Bhattacharya and Faben's method was employed to obtain VBGF parameters in which t_0 was assumed to be zero. The males and females had asymptotic total length ($L\infty$) of 257.3 ± 7.7 mm and 256.4 ± 9.5 mm, respectively, while the growth coefficients (K) were 0.53 ± 0.04 and 0.46 ± 0.04, respectively.

15.3.5 Sex Ratio, Size at Maturity, and Fecundity

The sex ratio of *T. orientalis* showed males (<130 mm TL) dominating numerically at smaller sizes, the male to female proportion being more or less equal at middle sizes (111 to 170 mm TL), and females outnumbering males above 230 mm TL. Month-wise distribution in the catches showed that sexes were not markedly segregated from each other at any size or month, and each sex numerically dominated the other at one time or another, with the result that the annual ratio of males and females remained 1:1.1 (Kagwade & Kabli 1996b). According to Subramanian (2004), the sex ratio of *T. orientalis* from Chennai waters did not significantly deviate from 1:1. However, the ratio was slightly weighted in favor of females, with average annual ratios of 52.7% and 55.8% during 1995–1996 and 1996–1997, respectively, and with increasingly wider margins among the larger-size groups, especially those above 220 mm TL.

The total length of the smallest ovigerous female collected from Mumbai waters was 98 mm TL, but the cumulative percentage of maturing, mature, berried, and spent specimens collected during September to April showed that 50% of the females mature at 107 mm TL (Kagwade & Kabli 1996b) (Figure 15.2). The pattern of distribution of ova in the ovaries, presence of semispent ovaries even when the females were in berry, and two ovigerous peaks in the same spawning season suggests that the species has an extended breeding season from September to April. In addition, about four to five broods recruit into the fishery every year. Juvenile lobsters measuring 40 to 50 mm TL appeared in commercial catches during April to June, suggesting an abbreviated larval life for the species, unlike that of the spiny lobsters and *Scyllarides* spp. (see Sekiguchi et al., Chapter 4 and Pessani & Mura, Chapter 13 for information about larval periods in various species). Such abbreviated larval development was confirmed via larval

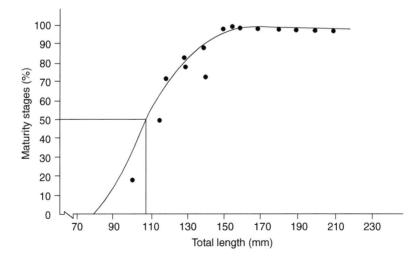

FIGURE 15.2 Size at 50% maturity for female *Thenus orientalis* from Mumbai waters. (Reprinted from Kagwade, P.V. & Kabli, L.M. 1996b. *Indian J. Fish.* 43: 13–25; with permission.)

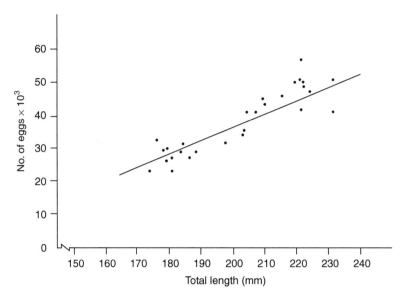

FIGURE 15.3 Linear regression line for fecundity with total length of female *Thenus orientalis* from Mumbai waters. (Reprinted from Kagwade, P.V. & Kabli, L.M. 1996b. *Indian J. Fish.* 43: 13–25; with permission.)

rearing (Kizhakudan et al. 2004a; see Section 15.5 for more information). The smallest maturing or mature female *T. orientalis* recorded by Subramanian (2004) from Chennai waters measured 93.5 mm TL. However, substantial numbers of such females belonged to 101 to 110 mm TL size group or above.

Kagwade & Kabli (1996b) estimated fecundity of the species from the number of eggs attached to the pleopods of ovigerous females. The fecundity of 32 females, ranging in size from 157 to 238 mm TL, showed a linear relationship between total length and the number of eggs (Figure 15.3). It is represented as: Fecundity (in thousand eggs) = −46.66 + 0.4164× TL. From this relationship, it is estimated that *T. orientalis* of 165 mm TL produces 22,050 eggs, while a specimen of 240 mm TL produces 53,280 eggs. Females above 130 mm TL were mostly mature.

15.3.6 Maturation

Kizhakudan et al. (2004a) during their experimental studies on breeding and larval rearing of *Thenus orientalis* and *Petarctus rugosus* in the laboratory observed that the animals attained sexual maturity under captive conditions and mated in the holding tanks. Two females of *T. orientalis* attained sexual maturity at 75.0 mm and 70.5 mm CL, extruded eggs under captive condition, and released viable phyllosoma larvae. On the other hand, *P. rugosus* attained maturity and spawned viable eggs at a carapace length ranging from 17.0 to 25.2 mm. Both species mated during intermolt and produced fertilized eggs. The spermatophoric mass was attached to the postventral sternite and anterior abdominal region of the female in the form of a white-colored, longitudinal, jellylike mass. Mating generally occurred at night and extrusion of the eggs started within five to seven hours. Oviposition was completed within six to eight hours and the spermatophore was lost about 12 hours postmating.

The reproductive biology of *T. orientalis* from Mumbai waters was studied by Kagwade & Kabli (1996b). There are no external indicators of maturation stages for male *T. orientalis*, and internal changes in the size of testes are not recognizable. However, in mature individuals, sperm are observed throughout the year. In females, ovigerous setae appear at about 60 mm TL, when the animals are still immature, and additional setae emerge with increasing size of the female. However, these setae have no correlation with maturity (Kagwade & Kabli 1996b). Therefore, determination of maturity depends mainly on the ovarian condition. Based on the size of the ovary and ova during the process of maturation, five stages were recognized by the authors: immature, early-maturing, late-maturing, mature, and spent. Females

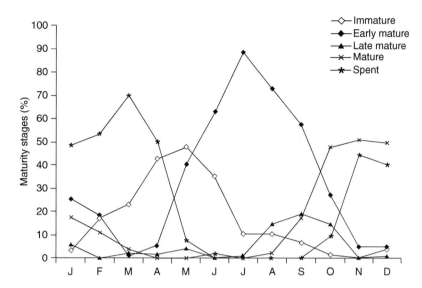

FIGURE 15.4 Monthly distribution of maturity stages of female *Thenus orientalis* from Mumbai during 1985 to 1986.

in an ovigerous condition also displayed semispent ovaries, wherein part of the ova had already been spawned.

15.4 Spawning Season

In a study on *Thenus orientalis* from Chennai on the southeast coast, Subramanian (2004) reported seasonal abundance of females in "berry". His data showed two annual spawning periods during June to August and February to March, the latter of which revealed a sharper peak. Females in berry were totally absent during the northeast monsoon months of October to December. Kabli & Kagwade (1996) studied the spawning season of *T. orientalis* at Mumbai on the northwest coast based on the percentage of different maturity stages in the landings during 1985 to 1986. Lobsters with immature ovaries appeared throughout the year, with a significant increase in frequency during February to June. Early-maturing stages were most frequent during May to September, late-maturing stages were most common during August to October, and mature and spent females predominated during September to April (Figure 15.4). Ovigerous females also appeared during September to April, but were most frequent from November to January. Thus, the different maturity stages of *T. orientalis* began to appear in landings from May and progressed from immature stage through early- and late-maturing stages to fully mature stage by September when berried females started appearing in landings. The number of berried females increased in the following months, but dwindled to negligible numbers by April. Substantially higher incidences of mature females during October to December, berried females during October to March, and spent individuals during November to April suggested that there was a definite spawning period for the species. Kabli & Kagwade (1996) concluded that the species spawned during September to April, but had two peaks: the major in November and the minor in March. Hossain (1978) also suggested that *T. orientalis* from the east coast of India (Andhra coast) spawned more than once a year. Branford (1982) reported that there was only a single spawning period for *T. orientalis* occurring off the Tokar delta in the Red Sea. This well-defined, single spawning period was restricted to three to four months and is rarely exhibited by other tropical species.

Kagwade and Kabli (1996a) reported that commercial catches of *T. orientalis* at Mumbai consisted mainly of three-year old males and 4.5-year old females. *Thenus orientalis* is a slow-growing species and the females appeared to grow more slowly than males after attaining the sexual maturity. The sex ratio in the catches also indicated that sexes might not be segregated at any definite size or month. These results suggest that, during the peak spawning period, females remained in the fishing grounds and did not

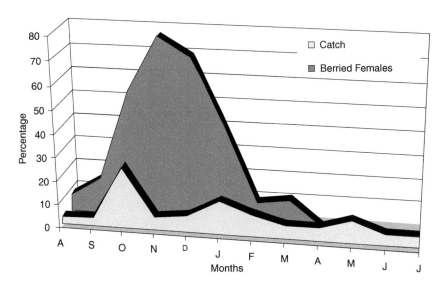

FIGURE 15.5   Average monthly catch of *Thenus orientalis* and occurrence of egg-bearing females at Mumbai during 1980 to 1986.

undertake breeding migrations to either shallower or deeper waters. Thus, they were equally vulnerable to heavy trawling pressure as males.

The fecundity of the sand lobster is generally low (20,000 to 50,000 eggs) compared to 0.14 to 4.72 million eggs for the spiny rock lobster *Panulirus polyphagus* (Kagwade 1988), but is on the same order of magnitude as the Mediterranean slipper lobster, *Scyllarides latus* (Almog-Shtayer 1988). Thus, at Mumbai, the combination of slow growth, low fecundity, and spawning restricted to a short period that coincides with heavy trawling activities (see Figure 15.5 for landings of berried females), may have resulted in *T. orientalis* suffering overfishing. In the absence of effective management measures, the exploitation continued unabated, particularly during the spawning period, and over a number of years this exploitation led to almost total extermination of the resource.

15.5   Captive Breeding and Seed Production

Breeding and larval rearing of the slipper lobsters *Thenus orientalis* and *Petrarctus rugosus* have been achieved in captivity at the Central Marine Fisheries Research Institute (CMFRI), Chennai (Kizhakudan et al. 2004a). Juveniles and subadults of *T. orientalis* (30 to 60 mm CL), collected from the gill-net fishery along the south Chennai coast were maintained in the laboratory in rectangular fiberglass reinforced plastic (FRP) tanks with continuous recirculation of water and fed *ad libitum* on fresh clam *Meretrix casta* Chemnitz, 1782. They attained sexual maturity, mated in the holding tanks, and produced fertilized eggs. The incubation period for *T. orientalis* was about 35 days.

Thenus orientalis phyllosoma larvae metamorphosed through four stages and molted into the postlarval nisto stage, completing the larval phase in 26 days posthatching (Kizhakudan et al. 2004a). The duration of each stage from phyllosoma I to IV ranged from five to seven days. Newly hatched larvae were reared in a "clear water system with no algal medium" and fed on chopped clam meat and live ctenophores collected from the sea. Floating plastic basins with bottom netting were used to hold larvae from phyllosoma stage III onward. On the last day of phyllosoma stage IV, the larvae stopped feeding, underwent considerable changes in appearance with the abdomen turning cylindrical and the tail opaque, and swam actively. When provided with substratum-like pieces of net or oyster shells, they clung to these materials. Molting to nisto stage occurred during midnight. Kizhakudan et al. (2004a) observed the larvae becoming rigid, with the setae on the walking legs beating vigorously, and the nisto breaking out from the old shell with jerking movements. The nisto were transparent initially but turned brownish later. The survival rate from phyllosoma I to the nisto was 22%. The nisto, a nonfeeding stage, molted into a juvenile after four days. The juveniles had a hard exoskeleton that resembled the adults in all respects.

In the case of laboratory-held female *P. rugosus*, the incubation period ranged from 23 to 35 days (Kizhakudan et al. 2004a). The phyllosoma larvae advanced through eight stages and metamorphosed into the nisto stage. The larvae were stocked at the rate of 10 larvae per 500 ml seawater, in a "clear water system." However, *Nannochloropsis* sp. was added at the time of feeding with *Artemia* nauplii during the first three phyllosoma instars. From phyllosoma IV onward, the larvae were fed with chopped meat of *M. casta*. The duration of each stage from phyllosoma I to VIII ranged from two to six days, with the total number of days taken for the phyllosomas to settle as nistos being 32 days posthatching. The survival rate from phyllosoma I to nisto was 1%.

Mikami & Greenwood (1997) reported successful rearing of newly hatched phyllosoma larvae of two species of *Thenus*, *T. orientalis* and a *Thenus* sp., to the juvenile stage, under laboratory conditions. Phyllosoma larvae of the two species of *Thenus* developed through four phyllosomal instars, the development being equivalent to the four larval stages of *T. orientalis* previously described from planktonic material. The authors studied the influence of different food regimes on the development of phyllosoma of *T. orientalis* and reported that the duration ranged from 27 to 45 days. Certain food regimes resulted in poor survival, prolonged duration of instars, and poor molt increments resulting in about 50% of phyllosomas developing an extra instar (5th instar) before metamorphosing to the nisto stage.

15.6   Production Trends

Lobsters are landed by both mechanized trawlers and indigenous gears, such as bottom-set gill nets, trammel nets, and traps. However, the mechanized sector accounts for >75% of the landings of lobster, comprising mainly *Panulirus polyphagus*, *Thenus orientalis*, and *Puerulus sewelli* (Radhakrishnan & Manisseri 2003). The total catch of lobsters in the country improved steadily from 800 metric tons in 1968, to 2000 metric tons in 1973, and 3000 metric tons in 1975. The landings showed a declining trend thereafter, only totaling 700 metric tons in 1980, but increased again to reach a peak of 4075 metric tons in 1985. The landings by trawlers then fluctuated around 2200 metric tons for more than a decade, and then declined further from 2387 metric tons in 2000 to 1389 metric tons, 1364 metric tons, 1245 metric tons, and 1438 metric tons, during 2001, 2002, 2003, and 2004, respectively (Figure 15.6). A regional breakdown of the catch during the five-year period from 1999 to 2003 showed that the northwest region contributed 60% of the landings, the southwest 27%, while the southeast contributed only 12%. Statewise, during 1999 to 2003, Gujarat recorded the maximum percent of the catch (34%), followed by Maharashtra

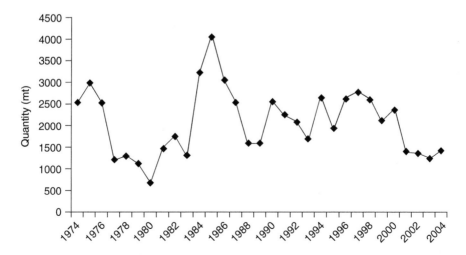

FIGURE 15.6   Total annual lobster landings (metric tons) in India, during 1974 to 2004.

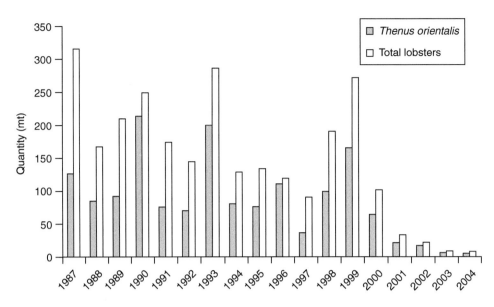

FIGURE 15.7 Comparison of annual total landings of *Thenus orientalis* and total catch of all lobster species (metric tons) at Veraval during 1987–2004.

(26%), Kerala (25%), and Tamil Nadu (11%). The northwest region recorded maximum landings in all the years from the 1970s, with the primary species being *T. orientalis* and *P. polyphagus*.

The fishery and biological characteristics of *T. orientalis* and *P. polyphagus* were examined at Veraval in Gujarat and Mumbai in Maharashtra. According to Kagwade et al. (1991), at Veraval the average annual catch of *T. orientalis* was 148.3 metric tons at a catch rate of 1.7 kg per unit effort (trawler), which formed 54.9% of the total lobster landings during the six-year period from 1980 to 1985. Subsequently, landings showed wide fluctuations, decreasing from 126 metric tons in 1987 to 85 metric tons in 1988, and reaching a peak of 215 metric tons in 1990 (Figure 15.7). During the following five-year period, landings ranged from 72 metric tons (1992) to 200 metric tons (1993). Although the catch further declined to 36 metric tons in 1997 and showed an improvement to 166 metric tons in 1999, the average annual landing during 1991 to 2000 was only 97.7 metric tons (Radhakrishnan et al. 2005). There was a steady decline in the landings of the species at Veraval in subsequent years, with landings recorded as 22, 18, 7, and 6 metric tons during the years 2001, 2002, 2003, and 2004, respectively. The species is landed at this center as by-catch by trawlers engaged in multiday fishing. There used to be no fishing during the southwest monsoon season, from June to August in certain years (Kagwade et al. 1991). The premonsoon period witnessed a poor fishery, while the postmonsoon period had a better fishery for the species. The peak landings were reported to occur during December to January.

Exploitation of *T. orientalis* from the waters off Mumbai was initiated in the late 1970s. Kagwade et al. (1991) reported average annual landings of 184.9 metric tons of *T. orientalis* at a catch rate of 4.4 kg per unit effort (trawler); this formed 46% of the total lobster landings at this center during an eight-year period from 1978 to 1985. A steep decline (50 metric tons) occurred in the catch in 1989 and, while the following year showed a slight improvement, the catch continued to decline in subsequent years, being a meager 7.5 metric tons in 1992 (Figure 15.8). Landings further declined to 2 metric tons in both 1994 and 1996, and the fishery for this species then collapsed entirely. From the year 1997 onward, the species is represented only in low numbers in the landings of other target species. The depletion of the fishery for *T. orientalis* is also reflected in the total landings of lobsters at Mumbai: the average annual landing declined from 348 metric tons and 462 metric tons during 1981–1985 and 1986–1990, respectively, to 129 metric tons and 131 metric tons during 1991–1995 and 1996–2000, respectively.

According to Deshmukh (2001), the slipper lobster fishery along the coast of Maharashtra began in 1978 with a meager catch of 1.5 metric tons and reached a maximum of 375 metric tons at a catch rate of

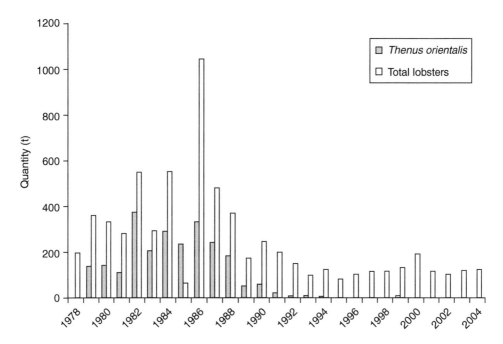

FIGURE 15.8 Comparison of annual total landings of *Thenus orientalis* and total catch of all lobster species (metric tons) at Mumbai during 1978–2004.

8.3 kg per boat trip in 1982. Although a second peak was noticed in 1986 (334 metric tons), the fishery showed a declining trend, and finally collapsed in the 1990s. Thus, the fishery spanning from 1978 to 1994, landed a total of 2401 metric tons, at a catch rate of 2.5 kg per boat trip. The author segregated the fishery into three phases: the 1st being a phase of development, during 1978 to 1982, with an average annual catch of 160 metric tons and catch rate of 4.2 kg per boat trip. During the second phase (1983 to 1987) the fishery was more or less stable with an average annual catch of 261 metric tons and a catch rate of 5.7 kg per boat trip. The third phase, during 1988 to 1994, was a declining phase when the catches steadily fell from 183 metric tons in 1988 to 2 metric tons in 1994. The catch rate also declined from 3.6 kg per boat trip in 1988 to 0.04 kg per boat trip in 1994. Monthly percentage of catch, pooled for the first phase of the fishery, showed that *T. orientalis* was landed throughout the year, with peak landings in June and December in the first phase, June and October in the second phase, and October and January in the third phase. Courtney (2002), in a similar study on Moreton Bay Bugs in Queensland, reported a significant decline in landings over the five-year period from 1997 to 2001, but attributed the decline to reduction in fishing effort.

Thenus orientalis currently supports a fishery of small magnitude and is landed as by-catch of trawlers at Sakthikulangara in Kerala State, along the southwest coast. The species is landed in low numbers along with cephalopods, shrimps, and fishes. *Thenus orientalis* is also landed as by-catch of trawlers at Chennai, the only center along the southeast coast from where a fishery, though of small magnitude, is reported for the species. Subramanian (2004) reported the abundance and catch trend of *T. orientalis* along this coast, based on studies conducted during April 1982 through March 1999. The average annual landing was around 10 metric tons during 1982 to 1985, increasing to 20 metric tons during 1986 to 1990, with only one decrease to 6.7 metric tons during April 1988 to March 1989. There was a significant increase in the abundance of the species in the following years (1991 to 1995), with the annual landings rising above 60 metric tons and reaching a maximum of 115 metric tons in 1994 to 1995. Subsequently, landings decreased and about 20 metric tons of lobsters were landed annually from 1996 to 1998. Studies conducted in subsequent years have shown that the species continues to be landed as by-catch of trawlers, with the average annual landings ranging from 25 to 35 metric tons. The season for the fishery in this region

TABLE 15.2

Stock Parameters of *Thenus orientalis* Landed at Mumbai During 1980 to 1985

Parameter	Sex	1980	1981	1982	1983	1984	1985	Pooled
Z	M	4.46	4.85	3.47	3.95	3.58	3.51	4.59
	F	2.92	3.30	2.67	2.90	3.44	2.32	3.84
M	M	0.69	0.69	0.69	0.69	0.69	0.69	0.69
	F	0.69	0.69	0.69	0.69	0.69	0.69	0.69
E	M	0.85	0.86	0.80	0.83	0.81	0.80	0.85
	F	0.76	0.79	0.74	0.76	0.80	0.70	0.82

Z, total mortality coefficient; M, natural mortality; E, exploitation ratio.

spanned over September to February, with peak landings during October to November, and coincided with the northeast monsoon months. Abundance of smaller-sized populations of the species measuring <110 mm TL marked recruitment into the fishery during May–July and January–February (Subramanian 2004). *Thenus orientalis* is landed in very small quantities by bottom-set gill nets as well, along the coastal stretch of Northern Tamil Nadu (Kizhakudan et al. 2004b). Peak landings of 5 metric tons were reported during September to November 2003. Fishermen ventured into deeper waters (40–60 m) during December to February and caught comparatively larger specimens, with about 40% being berried females (Kizhakudan et al. 2004b).

15.7 Mortality and Exploitation Rates

Analysis of mortality and exploitation rates for male and female *Thenus orientalis* from Mumbai waters for the period 1980 to 1985 is presented in Table 15.2. The total mortality coefficient (Z), natural mortality coefficient (M), exploitation rate (U), and the E_{max} exploitation ratio that gives maximum relative yield per recruit ('Y'/R) were estimated. During this entire five-year period, the exploitation ratio of males was 0.85 ($Z = 4.59$ and $M = 0.69$), while that for females was 0.82 ($Z = 3.84$ and $M = 0.69$). The E_{max} was only 0.55 for both sexes. These values clearly indicated that the species was grossly overexploited. The exploitation ratio at which the relative biomass per recruit ('B'/R) was reduced to 50% was 0.31 for males and 0.33 for females. At exploitation ratios of 0.85 for males and 0.82 for females, the relative biomass per recruit was reduced to 3.3% and 4.8% of the virgin biomass, respectively. Owing to such reduction in biomass, the stock could not generate enough new recruits and, as a result, the fishery collapsed. In 1980 to 1986, the species occurred in all months throughout the year, but the maximum abundance was during October to March when the region landed 62% of its annual catch (Deshmukh 2001). Because of its single, well-defined breeding period and its slow growth and low fecundity (Kagwade & Kabli 1996b; see Section 15.5 above), large-scale removal of the females might have been detrimental to the recruitment process, resulting in the rapid decline of the fishery and annihilation of the stock in the waters off Mumbai. This is a classic example of recruitment overfishing, which is not precluded by growth overfishing.

15.8 Management

The maritime states in India manage their own fisheries and fishing regulations, but often these are not strictly enforced. A typical management problem occurs when lobsters are caught incidentally in shrimp and fish trawls. In this situation, it may not be practical to enforce management measures such as closure of fishery during breeding season, minimum legal size (MLS) for fishing, or gear restrictions for fishing. The Ministry of Commerce and Industry of the Government of India enforced MLS for the export of four species of lobsters including *Thenus orientalis* by banning export of whole lobsters weighing < 150 g.

Fishing regulations for each maritime state have to be formulated and implemented considering the resource status of each state. As no stock of *T. orientalis* is left in the sea off Mumbai, total conservation of the remaining residual population is the only possible option and requires a full ban on landing of the species with the return of lobsters captured by the fishery to the sea until the stock is revived. Since success has been achieved in captive breeding and hatchery production, sea ranching of hatchery-produced juveniles may also be considered. When conceiving the project, revival of the depleted stock should be the objective, rather than the economic gain. A participatory management program has already been initiated with the financial support of the Ministry of Commerce under which fishermen are educated on the long-term impact of destructive fishing methods and fishing of egg-bearing lobsters (Radhakrishnan et al. 2005). As lobster fishing is a socioeconomic activity, implementation of any regulatory measure needs to consider the socioeconomic aspects and livelihood issues of the various stakeholders associated with this activity in order for the measure to be successfully implemented.

Acknowledgments

We thank Dr. Mohan Joseph Modayil, Director of the Central Marine Fisheries Research Institute, for the facilities provided and encouragement. We also gratefully acknowledge receipt of data on landing of lobsters by trawlers from the Fishery Resources Assessment Division of the Institute. The support from Mr. R.Thangaraja and Ms. K.V. Sajitha is also gratefully acknowledged.

References

Almog-Shtayer, G. 1988. Behavioural-ecological aspects of Mediterranean slipper lobsters in the past and of the slipper lobster Scyllarides latus in the present. M.A. thesis: University of Haifa, Israel. 165 pp.

Alikunhi, K.H. 1948. Note on the metamorphosis of phyllosoma larvae from the Madras plankton. *Proceedings of the 35th Indian Science Congress, Pt. III*: Abstract 193.

Anonymous 2005. *Annual Report 2004–2005*. Central Marine Fisheries Research Institute, Cochin, India.

Branford, J.R. 1982. Notes on scyllarid lobster *Thenus orientalis* (Lund, 1793) off Tokar Delta (Red Sea). *Crustaceana* 38: 21–224.

Chhapgar, B.F. & Deshmukh, S.K. 1964. Further records of lobsters from Bombay. *Bombay. Nat. Hist. Soc.* 61: 203–207.

Courtney, A.J. 2002. The status of Queensland's Moreton Bay bug *(Thenus* spp.) and Balmain bug (*Ibacus* spp.) stocks. *Southern Fisheries Centre Deception Bay, Information Series* Q102100: 1–18. Queensland Govt. Dept. of Primary Industries.

Deshmukh, V.D. 2001. Collapse of sand lobster fishery in Bombay waters. *Indian. J. Fish.* 48: 71–76.

Fielder, D.R. 1964. The spiny lobster *Jasus lalandii* (H. Milne-Edwards), in South Australia, II- Reproduction. *Aust. J. Mar. Freshw. Res.* 15: 133–144.

Heydorn, A.E.F. 1969. The rock lobster of the South African west coast *Jasus lalandi* (H. Milne-Edwards) 2. Population studies, behavior, reproduction, moulting, growth and migration. *S. Afr. Div. Sea. Fish. Invest. Rep.* 71: 1–32.

Hossain, M.A. 1975. On the squat-lobster, *Thenus orientalis* (Lund) off Visakapatnam (Bay of Bengal). *Curr. Sci.* 44: 161–162.

Hossain, M.A. 1978. Few words about the sand lobster, *Thenus orientalis* (Lund) (Decapoda: Scyllaridae) from Andhra Coast. *Seaf. Export J.* 10: 43–46.

Hossain, M.A., Shyamsundari, K., & Rao, K.H. 1975. On the landing of sand lobster *Thenus orientalis* (Lund). *Seaf. Export J.* 7: 1–6.

Jones, C. M. 1990. Morphological characteristics of Bay Lobsters, *Thenus* Leach species (Decapoda, Scyllaridae), from north-eastern Australia. *Crustaceana* 59: 265–275.

Kabli, L.M. 1989. *Biology of sand lobster, Thenus orientalis (Lund)*. Ph.D. thesis: University of Bombay, Bombay, India. 378 pp.

Kabli, L.M. & Kagwade, P.V. 1996. Morphometry and conversion factors in the sand lobster *Thenus orientalis* (Lund) from Bombay waters. *Indian J. Fish.* 43: 249–254.

Kagwade, P.V. 1987. Age and growth of spiny lobster *Panulirus polyphagus* (Herbst) of Bombay waters. *Indian J. Fish.* 34: 389–398.

Kagwade, P.V. 1988. Fecundity in the spiny lobster *Panulirus polyphagus* (Herbst). *J. Mar. Biol. Assoc. India* 30: 114–120.

Kagwade, P.V. & Kabli, L.M. 1991. Embryonic development of larvae on the pleopods of the spiny lobster *Panulirus polyphagus* (Herbst) and the sand lobster *Thenus orientalis* (Lund) from Bombay waters. *Indian J. Fish.* 38: 73–82.

Kagwade, P.V. & Kabli, L.M. 1996a. Age and growth of the sand lobster *Thenus orientalis* (Lund) from Bombay waters. *Indian J. Fish.* 43: 241–247.

Kagwade, P.V. & Kabli, L.M. 1996b. Reproductive biology of the sand lobster *Thenus orientalis* (Lund) from Bombay waters. *Indian J. Fish.* 43(1): 13–25.

Kagwade, P.V., Manickaraja, M., Deshmukh, V.D., Rajamani, M., Radhakrishnan, E.V., Suresh, V., Kathirvel, M., & Rao, G.S. 1991. Magnitude of lobster resources of India. *J. Mar. Biol. Assoc. India* 33: 150–158.

Kizhakudan, J.K., Thirumilu, P., Rajapackiam, S., & Manibal, C. 2004a. Captive breeding and seed production of scyllarid lobsters — opening new vistas in crustacean aquaculture. *Mar. Fish. Infor. Serv. T & E Ser.* 181: 1–4.

Kizhakudan, J.K., Thirumilu, P., & Manibal, C. 2004b. Fishery of the sand lobster *Thenus orientalis* (Lund) by bottom-set gillnets along Tamilnadu coast. *Mar. Fish. Infor. Serv. T & E Ser.* 181: 6–7.

Martínez, C.E. 2000. Ecología trófica de Panulirus gracilis, P. penicillatus y Scyllarides astori (decapoda, Palinura) en sitios de pesca de langosta en las islas Galápagos. Tesis de Licenciatura: Universidad del Azuay, Ecudator. 102 pps.

Martins, H.R. 1985. Biological studies on the exploited stock of the Mediterranean locust lobster *Scyllarides latus* (Latreille, 1803) (Decapoda, Scyllaridae) in the Azores. *J. Crust. Biol.* 5: 294–305.

Mikami, S. & Greenwood, J.G. 1997. Complete development and comparative morphology of larval *Thenus orientalis* and *Thenus* sp. (Decapoda: Scyllaridae) reared in the laboratory. *J. Crust. Biol.* 17: 289–308.

Prasad, R.R. & Tampi, P.R.S. 1957. On the phyllosoma of Mandapam. *Proc. Natl. Inst. Sci. India* 23B: 48–67.

Prasad, R.R. & Tampi, P.R.S. 1960. Phyllosoma of scyllarid lobsters from Arabian Sea. *J. Mar. Biol. Assoc. India* 2: 241–249.

Radhakrishnan, E.V. 1995. Lobster fisheries of India. *Lobster Newslett.* 8(1): 12–13.

Radhakrishnan, E.V. & Manisseri, M.K. 2003. Lobsters. In: Mohan J.M. & Jayaprakash, A.A. (eds.), *Status of Exploited Marine Fishery Resources of India*: pp. 195–202. Cochin, India: Central Marine Fisheries Research Institute.

Radhakrishnan, E.V., Kasinathan, C., & Ramamoorthy, S. 1995. Two new records of scyllarids from the Indian coast. *Lobster Newslett.* 8(1): 9.

Radhakrishnan, E.V., Deshmukh, V.D., Manisseri, M.K., Rajamani, M., Kizhakudan, J.K., & Thangaraja, R. 2005. Status of the major lobster fisheries in India. *N. Z. J. Mar. Freshw. Res.* 39: 723–732.

Spanier, E. 1987. Mollusca as food for the slipper lobster *Scyllarides latus* in the coastal waters of Israel. *Levantina* 68: 713–716.

Spanier, E. & Lavalli, K.L. 1998. Natural history of *Scyllarides latus* (Crustacea: Decapoda): A review of the contemporary biological knowledge of the Mediterranean slipper lobster. *J. Nat. Hist.* 32: 1769–1786.

Srikrishnadhas, B., Kaleemur Rahman, M.D., & Anandasekharan, A.S.M. 1991. A new species of scyllarid lobster *Scyllarus tutieneis* (Scyllarida: Decapoda) from the Tuticorin Bay in the Gulf of Mannar. *J. Mar. Biol. Assoc. India* 33: 418–421.

Subramanian, V.T. 2004. Fishery of sand lobster *Thenus orientalis* (Lund) along Chennai coast. *Indian J. Fish.* 51: 111–115.

Thiagarajan, R., Krishnapillai, S., Jasmine, S., & Lipton, A.P. 1998. On the capture of a live South African cape locust lobster at Vizhinjam. *Mar. Fish. Infor. Serv. T & E Ser.* 158: 18–19.

von Bertalanffy, L. 1934. Untersuchungen über die Gesetzlichkeiten des Wachstums 1. Allgemeine Grundlagen der Theorie. *Roux. Arch. Entwicklungsmech. Org.* 131: 613–653.

16

Biology and Fishery of the Bay Lobster, Thenus *spp.*

Clive M. Jones

CONTENTS

Abstract

Bay lobsters of the genus *Thenus* are a common and valuable by-catch of the shrimp trawl fisheries of northern Australia. Until recently, this was considered a monospecific genus, but two species are now recognized from Australia, *Thenus orientalis* (Lund, 1793) and *Thenus indicus* Leach, 1816, and additional species are likely to be described from other regions. The economic importance of the Australian species was at odds with the meager biological information available until detailed studies were made in the 1980s and 1990s, which generated important information for fisheries management, and revealed the specialized nature of these lobsters. *Thenus* spp. inhabit the soft, sedimentary mud and sand of the continental shelf, particularly in inter-reef areas along the tropical coastline of Australia. Their morphology and behavior share much in common with other scyllarids, but also have unique features which reflect successful adaptation to their environment. Most notable are their ability to swim, often long distances, and the capacity to bury into the sediment. *Thenus indicus* is generally the smaller of the two species (maximum size ~65 mm carapace length, CL) and inhabits shallow, inshore waters, 10 to 30 m deep in areas characterized by fine sand and silt. *Thenus orientalis* grows to a larger size (maximum size ~95 mm CL) and inhabits waters of 30 to 60 m depth, where sediments are characterized by medium to coarse sands. For both species, sexual dimorphism is very subtle. Mating is believed to involve a brief encounter, and the spermatophoric mass is likely to be very short lived. Mean fecundity is 12,455 eggs for *T. indicus* and 32,230 for *T. orientalis*. Reproductive seasonality is marked by two spawning peaks corresponding to spring and midsummer. Growth is quite rapid: for both species a CL of 40 mm (~40 g total weight) is reached within the first 12 months, and approximately 80% of maximum size is reached by two years of age. Maximum age for *T. indicus* is between two and four years and between four and eight years for *T. orientalis*. Both species are nocturnally active with clear peaks in activity at dusk and just prior to dawn. During daylight hours, *Thenus* spp. bury themselves in the sediment with only the eyes and antennules exposed. During periods of activity, *Thenus* spp. occasionally leap from the sediment into a swimming mode effected by contractions of the abdomen and controlled by movement of the antennae. Such locomotion swimming may be sustained for periods up to 40 min, enabling lobsters to move distances of several hundreds of meters. The fishery for *Thenus* in Australia is managed and regulations include a total ban on the taking of berried females and a minimum size of 75 mm CL. CPUE has been stable over the past two decades, although total catch has diminished with reductions in fishing effort. For Queensland, a maximum catch of 755 metric tons was recorded in 1997, and more recently, the annual catch has been around 400 metric tons.

16.1   Introduction

Of the 20 genera constituting the Scyllaridae, *Thenus* is possibly the most economically significant. Although localized commercial catches of other genera including *Ibacus*, *Scyllarides*, *Parribacus*, and *Evibacus* have also been reported from trawl and trap fisheries throughout the Indian and Pacific Oceans, and for *Scyllarides* in the Mediterranean, they are generally negligible (Holthuis 1991).

Thenus spp. contribute to many of the demersal trawl fisheries which operate along the tropical coasts of the Indian Ocean and the Western Pacific region (Ben-Tuvia 1968; Prasad & Tampi 1968; Isarankura 1971; Shindo 1973; Shirota & Ratanachote 1973; Hossain 1974; Hossain et al. 1975; Saeger et al. 1976; Mutagyera 1979; Pauly 1979; Branford 1980; Ivanov & Krylov 1980; Jones 1984; 1990a; Courtney 2002). Its presence in such trawl catches has often received special comment in the literature, as it represents the only lobster of commercial importance among the great diversity of other crustaceans and fish. Although catch rates are usually low relative to other commercially important species, there are some seasons and localities within existing trawl fisheries where considerably higher catch rates of *Thenus* spp. have been attained.

The type locality of *Thenus indicus* Leach, 1816 is the East Indies and China; however, this species has been recorded from shallow coastal waters throughout the Indo-West Pacific. In summarizing over 200 years of recorded sightings, Holthuis (1946) noted that *T. indicus* was common in the western Indo-Pacific region, but entirely lacking in the Oceania region. The recording of *T. indicus* from the Kermadec Islands (Chilton 1910) north of New Zealand is likely to be erroneous. Biological survey information and fisheries statistics (Matilda & Hill 1981; Seefried 1983; Jones 1984: Kailola et al. 1993; Courtney 2002) from the east Australian region indicate that populations of *T. indicus* from northeastern Australia are particularly abundant.

Within Australia, *Thenus* spp. are distributed along the subtropical and tropical coasts from northern New South Wales to Shark Bay in Western Australia (George & Griffin 1972; Holthuis 1991). Throughout this range, these lobsters contribute to the incidental by-catch of trawl fisheries that fish for penaeid shrimp and, in some localities, scallops (*Amusium* spp.). Taiwanese demersal fish trawlers operating in the shallow waters between northern Australia and the Indonesia/New Guinea archipelago also take considerable quantities of *Thenus* spp. (Liu 1976).

The commercial potential of *T. indicus* in Australia was first recognized by the fishermen of Moreton Bay in southern Queensland, when catches of this lobster were made in 1888 during an exploratory trawl survey (Fison 1888). This discovery gave rise to a common name widely applied to this species, "Moreton Bay Bug." The name "Bay Lobster" has recently gained more extensive popularity, and is used herein in reference to lobsters of the genus *Thenus*. The history, development, and status of the fishery for bay lobsters in Queensland are documented by Jones (1984) and Courtney (2002).

The genus *Thenus* has been considered monospecific until quite recently. The history of the nomenclature is reviewed by Webber & Booth (Chapter 2). In reviewing the literature referring to this species (in particular Boone 1937; Barnard 1950; Ben-Tuvia 1968; Hossain 1978a, 1978b; Hossain 1979; Branford 1980; Fischer & Bianchi 1984; Jefferies et al. 1984; Davie 2002), it is apparent that morphological variability exists between specimens from different localities, although such variability was not recognized within localities. Via collections of *T. indicus* made in the Gulf of Carpentaria in 1983 and in eastern Queensland waters in 1984 and 1985, Jones (1988) observed morphological variability sufficient to differentiate two distinct morphs, which were later suggested to be distinct species (Jones 1990b) and which were subsequently assigned the names *Thenus orientalis* (Lund, 1793) and *T. indicus* (Davie 2002). Further revision of the genus has been made through examination of material from throughout the known distribution, and several species are likely to be recognized (Davie 2002). Although further naming changes are likely, the names *T. indicus* and *T. orientalis* will be used for this discussion concerning the Australian species.

A comprehensive study of the two Australian species was undertaken by Jones (1988) based on trawled samples of populations throughout northeastern Australia and the bulk of the following information is based on the findings from that study. It is likely that the biology of the other *Thenus* species is very similar to that described here.

16.2 Morphology

16.2.1 External Morphology

Among the specific characteristics of *Thenus* species, the position of the orbits on the lateral extremes of the carapace is sufficient to distinguish this genus from all other scyllarid genera (Figure 16.1). Several morphological characteristics common to all *Thenus* species are worth noting: the body is dorsoventrally compressed, with the carapace being more broad than long. The anterolateral borders of carapace are broadly expanded upward, and stalked eyes lie within deep orbits protected by several robust teeth. The carapace and abdomen are sculptured with fine granules and tubercles among a patchy tomentum, and the carapace and first five abdominal segments have an elevated median dorsal ridge, which has three forward

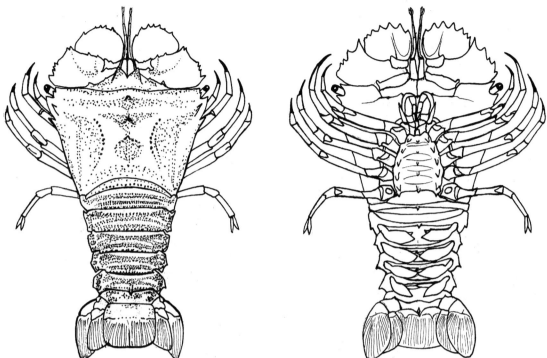

FIGURE 16.1   Dorsal (left) and ventral (right) views of *Thenus* spp. to illustrate gross morphological features.

and upwardly directed teeth. A rostral process consisting of two short, sharp spines is separated by a deep V-shaped sinus. The second and fourth segments of the antennae are broadly flattened and expanded, and bear long curving teeth on their anterolateral margins. The antennules rise from beneath the rostral process and extend just beyond the anterior margins of the antennae. The abdominal segments each have their lateral margins expanded downward concealing the pleopods. The telson is longer than broad, calcified, and stiff proximally, but thin and soft distally; the uropods are well developed and expanded beyond the posterior margin of the telson.

Ventrally, four pairs of pleopods arise from abdominal segments two to five and decrease in size posteriorly. The exopods are larger than the endopods. The five pairs of pereiopods are relatively long and slender, and the first four are equipped with sharp dactyls, the tips of which are ensheathed in a keratin sheath. The dactyls of the posterior three pairs of pereiopods bear two rows of coarse hairs each. The maxillipeds are relatively large and well developed and their branchiostegites are smooth and relatively featureless. Figure 16.1 illustrates many of the morphological characteristics of *Thenus*.

16.2.2   Specific Characteristics

Morphological characteristics that may be used to distinguish between the two Australian species of *Thenus* are subtle. The most conspicuous of these distinguishing characteristics are presented in Table 16.1. Although fishers recognize the differences and the species are fished from different fisheries, the species are not distinguished in the marketplace. The most obvious of the differences are that *Thenus indicus*, known to fishers as the mud bug, tends to be brown, whereas *Thenus orientalis* (known as the reef bug) is more reddish in body coloration. The former has no pigment spotting on the pereiopods or telson, but the latter displays distinct spotting on these appendages.

16.2.3   Morphometrics

Jones (1990b) conducted a broad morphometric study of *Thenus* spp. across their range in northeastern Australia representing each sex, two species, and five geographical locations. An analysis of variability of

TABLE 16.1

Morphological Characteristics Used to Distinguish *T. indicus* and *T. orientalis* (Jones 1990b)

	Thenus indicus	*Thenus orientalis*
Common name	Mud bug	Reef bug
Body color	Dark brown	Light red/brown
Telson/uropods	Yellow	Pink/red
Pleopods	Yellow	Pink/red
Pigmentation	Pigment spotting absent on pereiopods and telson	Conspicuous dark pigment spotting on pereiopods and proximal portion of telson
Pereiopods	Slender	Robust
Dorsal profile	Slightly concave, slender	Slightly convex, robust
Rostral processes	Sharp, directed anteriorly and upward	Blunt and more robust, more elevated
Second antennal segment	Several sharp anteromarginal teeth, the most conspicuous on the anterolateral border directed forward and outward, five teeth	Anteromarginal teeth larger and broader, anterolateral tooth less pronounced and curved slightly backward, four teeth

morphometry was made by comparing each of six body dimensions (total length, carapace length (CL), carapace width, 2nd pereiopod length, 2nd abdominal segment width, and telson length) and total weight between samples using linear regression and analysis of covariance. Carapace length was shown to be the most accurate and precise index of size out of the morphometric dimensions examined. Based on several hundred measurements of this index for both sexes of *T. indicus* and *T. orientalis* from five locations, it was evident that *T. orientalis* grows to a larger size than *T. indicus*, and that females of both species attain a larger size than males.

The comparative morphometry of the two *Thenus* species indicates great similarity, supporting the hypothesis that they may have only recently speciated from a common ancestral stock (Jones 1988). The morphometric distinctions that were apparent were consistent with *T. orientalis* being heavier and more robust in body form than *T. indicus*. This disparity is likely to be strongly influenced by environmental factors. *Thenus orientalis* inhabits inter-reef areas characterized by coarse, coralline sediments in which a more robust morphology would be advantageous for burying into the sediment. *Thenus indicus* inhabits inshore areas characterized by fine, silty mud, for which a heavier, stronger exoskeleton would not be helpful.

The relationship of CL and total weight for each species (sexes combined) is as follows:

Thenus indicus: Total weight (g) $= 0.0006 \times CL^{3.0031}$
Thenus orientalis: Total weight (g) $= 0.0006 \times CL^{2.9929}$

Morphometric differentiation between sexes was related most noticeably to characteristics associated with egg bearing. Females carry the brood of eggs beneath the abdomen, attached to the pleopods. A comparatively longer and wider abdomen of the female facilitates the accommodation of the brood, and a relatively longer telson assists in its physical protection. These increased abdominal dimensions explain the relatively greater weight of females. Thus, the greater mean size of females may be associated with egg-carrying capacity and the maximization of reproductive efficiency.

The absence of clear allometric changes in body morphometry attendant to sexual maturity of male *Thenus* spp. and scyllarids in general is exceptional among lobster species. Allometric growth of the pereiopods of male palinurids and of the chelae of male nephropids has significance to both attainment of physical maturity and to behavioral maturity. Courtship behavior in these families involves the display of enlarged appendages (Atema & Cobb 1980). The absence of equivalent morphology and behavior in *Thenus* spp. may be related to their low density and solitary existence, for which courtship is of no value (see Lavalli et al., Chapter 7).

16.2.4   Protein Specificity to Clarify Morphological Variability

An electrophoretic analysis of general proteins in the abdominal musculature of *T. indicus* and *T. orientalis* conducted by Jones (1988) used polyacrylamide gel electrophoresis as per Keenan and Shaklee (1985). The results revealed clear discontinuities in protein make up, representing fixed genetic differences between the two species (Jones 1990b; Davie 2002). Investigation of interpopulation variability within each species will necessitate more exhaustive procedures involving molecular techniques (see Deiana et al., Chapter 3 for examples of such techniques that are being used on some scyllarid species already).

16.3   Population Structure

Populations of *Thenus indicus* and *T. orientalis* have been examined to describe population characteristics, including natural abundance and its spatial and temporal distribution as estimated from catch figures and catch per unit effort (CPUE) information and size and sex composition. Methods are described by Jones (1993).

16.3.1   Spatial Distribution

Populations studied by Jones (1993) represented areas of 12,000 and 9000 km^2 for *T. indicus and T. orientalis* respectively, and for both species the spatial distribution of lobsters was highly aggregated. For *T. indicus* the mean catch per hour and mean catch per hectare over all sampling stations and using standard trawl fishery gear, was 16.8 and 1.83, respectively. For *T. orientalis* mean catch per hectare was 2.01. Although actual abundance of lobsters is likely to be higher than the catch figures suggest, because the fishing method does not catch every lobster, the estimates indicate that the populations sampled represent a minimum of 2.2 million and 1.8 million individuals for each species.

Analysis of variance of catch per hour between stations indicates significant variability, supporting the finding that lobsters are patchy or aggregated in their distribution. It also suggests that the characteristics of the sampling areas are differentially suited to the habitation of *Thenus* spp. Characteristics most likely to influence lobster abundance are the nature of sediment and the water depth. Temperature and salinity were homogeneous among sampling stations and therefore are not responsible for the variability.

Correlation analyses of catch rate and the type of sediment indicate a significant association between catch and the prevalence of particular particle sizes. The results suggest that sediments characterized by having >70% of particles <0.25 mm are most preferred for *T. indicus* and sediments characterized by having >60% of particles between 0.25 and 1.0 mm are preferred by *T. orientalis*. A strong relationship also exists for depth and catch, such that *T. indicus* is most abundant at depths between 10 and 30 m and *T. orientalis* is most abundant at depths between 40 and 50 m. This strongly suggests a niche separation between these two species. The results are supported by anecdotal information from commercial fishers, that *T. indicus* is an inhabitant of relatively shallow, inshore waters, with muddy sediment (and thus the common name "mud bug"), and that *T. orientalis* is typically abundant in offshore, deeper water with coarse sandy sediments.

Localized distribution patterns within populations of lobsters have received little attention in the literature. A major contributing factor is likely to be the complexity of habitats, particularly for palinurids, and the difficulty of describing (mathematically) the distribution of individuals. Nevertheless, *in situ* observations of *Panulirus* species (Herrnkind et al. 1975; Goni et al. 2001; Negrete-Soto et al. 2002; Butler 2003) have indicated that the distribution of palinurids can be nonrandom, as has been described here for *Thenus* spp. Nonrandom distributions have also been described for other scyllarid species (Lyons 1970; Spanier & Lavalli 1998), although the contributing factors are not clear (Spanier & Almog-Shtayer 1992).

Aggregated or contagious distributions are typical of many benthic species (Elliot 1977) and are usually attributed to the influence of environmental factors. The degree of aggregation and dispersion of a species in a particular area are likely to be a reflection of the habitat preference of the species and the relative availability of that habitat type.

Although sampling techniques vary widely, the catch densities of many tropical palinurids (Peacock 1974; Kanciruk 1980; Lara et al. 1998) are of the same order of magnitude (0 to 10 individuals/ha) as those for *Thenus* spp. In contrast, the density of temperate lobsters inhabiting uniform depositional substrates (e.g., *Nephrops* spp.) or more complex rock/sediment substrates (e.g., *Homarus* spp.) (Cooper & Uzmann 1980; Tuck et al. 1997) is often much greater, on the order of hundreds or thousands of individuals per hectare. Such latitudinal differences are typical of many faunal groups and can be attributed to differences in environmental productivity (Morgan 1980) and complexity.

Although sediment characteristics and depth have a significant influence on the distribution of *Thenus* spp., biological factors may also be important. There is no evidence of behaviorally stimulated aggregation for reproductive or migratory purposes, but the influence of other species (prey, predator, or competitors) on that of *Thenus* spp. is likely to be significant.

16.3.2 Temporal Distribution

Analysis of dispersion values throughout the year suggests significant variability in the degree of aggregation of *T. indicus* lobsters through time. Two periods of increased abundance are evident: the first in autumn (March/April) and the second in spring (October) (Jones 1988). The catch in each of these two periods is significantly greater than that in the months immediately before or after them. Recruitment pulses of postlarval individuals are likely to be responsible for these periodic increases in catch rates. In contrast, *T. orientalis* is more uniformly distributed through time. Seasonal variability in abundance of *Thenus* spp. was similar to that reported for other tropical lobster species (Lyons et al. 1981; Spanier et al. 1988).

16.3.3 Sex Ratio and Size Structure of Populations

Sex ratio of *T. indicus* is typically constant at 1:1 throughout the year. *Thenus orientalis* populations, however, have a preponderance of males, with a sex ratio of 0.57 (Jones 1988). Although sex ratio was reasonably stable within each population of *Thenus* spp. examined, sex ratios documented in the literature for *T. indicus* (sic) from other localities (Shirota & Ratanachote 1973; Branford 1980) indicate wide geographic variability.

Among the larger lobster species which support commercial fishing, *Thenus* is one of a very few genera which inhabit tropical, sedimentary (soft) substrates. *Panulirus polyphagus* (Herbst, 1793) is moderately abundant on soft substrates in northern India (Radhakrishnan et al. 2005). In the case of *Thenus* spp., this habitat requirement has led to several specialized adaptations. In other respects, the population parameters of *Thenus* spp. are similar to those of other lobsters, particularly tropical species.

The size structure of *Thenus* spp. populations is characterized by the distinctiveness of the 0+ year class and the broad overlap of subsequent year classes. This pattern is similar to other tropical lobsters (Lyons et al. 1981; Negrete-Soto et al. 2002; Butler 2003) and to many tropical species in general that have protracted recruitment and variable growth rates. For both species, primary recruitment of juveniles to the population occurs in midsummer, although low level and perhaps secondary recruitment appears to occur throughout the year. Examination of the population structures of the two *Thenus* species indicates fundamental similarities that are to be expected between two such closely related species.

Thenus orientalis are typically larger than *T. indicus* as evidenced by comparisons of mean carapace length (CL) between the two species. The difference in mean size indicates that *T. orientalis* are between 27 and 28% larger (in CL) than *T. indicus*. This represents a difference in mean total weight of approximately 113% in males and 90% in females between the species.

The systematic affinity of the Scyllaridae and Palinuridae (Webber & Booth, Chapter 2) is strongly reflected in the similarities in population structure between *Thenus* spp. and various palinurid species. Some modification of typical palinurid population characteristics that enable *Thenus* spp. to effectively inhabit the soft benthos are evident, and add support to the hypothesis that the scyllarids are a recent evolutionary derivative of the palinurid stock (see Webber & Booth, Chapter 2).

16.4 Reproduction

16.4.1 Morphology of Reproductive Organs

Both *Thenus indicus* and *T. orientalis* are dioecious. Although lobsters have been found with more than two genital apertures (Jones 1988) in which both ovarian and testicular tissue is apparent, there is no evidence

TABLE 16.2

Sexually Dimorphic Characteristics of *Thenus* spp. (Jones 1988)

	Characteristics	
	Male	**Female**
Genital aperture	Situated on the coxae of 5th legs	Situated on coxae of 3rd legs; half the diameter of that of males
Pleopods	Relatively small	Relatively large
Pleopod setation (mature lobsters)	No ovigerous setae	Endopodites possess long ovigerous setae
Dactyl of 5th pereiopods	Pointed with small fixed claw at tip[a]	Distally bulbous with concave depression ventrally; tip with tiny fixed claw[a]
Setal fringe of telson	Setae relatively short	Setae relatively long

[a]This claw represents a vestigial chela, present and functional in other Scyllarid genera (females only).

of intersexuality, as has been reported for *Nephrops* and other lobsters (Farmer 1972). Several macroscopic morphological characteristics can be used to distinguish the sexes (Table 16.2). The differences are subtle, a feature uncharacteristic of the Decapoda in general.

Morphometric differences between sexes are apparent, although these too can be subtle. These differences are all related to egg-carrying capacity. Relative to males, females are characterized by having larger pleopods, equipped with ovigerous setae, and a greater growth rate of total length, abdominal width, and telson length (relative to CL). These characteristics all contribute to maximizing egg-carrying capacity.

The morphological nature of the fifth pereiopod in *Thenus* spp. is uncharacteristic of the Scyllaridae. In females of all other scyllarid genera (and all palinurids), the dactyl of this leg is equipped with a small, functional chela which is used in the manipulation and cleaning of eggs and the release of sperm from the spermatophore, as described for palinurids (Aiken & Waddy 1980). In *Scyllarides* spp., these chela bear short, stiff, brushlike setae that may aid in all of the earlier-mentioned functions (Spanier & Lavalli 2006; see Lavalli et al., Chapter 7 for a detailed description and figures). In females of *Thenus* spp., however, this chela is greatly reduced in size and is fixed and the dactyl is swollen distally to form a subchela. Of equal significance is the presence of this same tiny, fixed subchela on the male fifth pereiopod. The function, if any, of this fixed subchela in male and female *Thenus* spp. is unknown.

The lack of sexually dimorphic characteristics is also atypical of lobsters and decapod crustaceans in general. Courtship and copulation among nephropids and palinurids involves quite complex ritualistic behavior (Lipcius & Herrnkind 1987; MacDiarmid & Kittaka 2000) in which the chelae (of nephropids) or legs and antennae (of palinurids) play an important role. By virtue of low natural density, social interactions in *Thenus* spp. are likely to be uncommon. Development of courtship behavior would confer little advantage, and the absence of appropriate morphology in *Thenus* spp. is evidence of mating simplicity. A sexual attractant mechanism would be of greater significance (Atema & Engstrom 1971). While the operation of a chemical attractant (pheromone) was not apparent in the observations made and is unlikely in a low density population, the use of sound is a possibility (Atema & Cobb 1980); however, sound production has only been reported for nephropids and some species of palinurids (Phillips et al. 1980).

Internally, the reproductive systems of both male and female lobsters are of the general decapod form (Farmer 1975; Phillips et al. 1980; Rupert & Barnes 1994). The genital apertures of both sexes consist of a simple structure, with no evidence of a seminal receptacle (thelycum) in females (Jones 1988) as described for nephropids (Farmer 1974), or of an intromittent organ in males (Fielder 1964; Farmer 1974).

Jones (1988) has described the general morphology of *Thenus* spp. gonads. Ovaries consist of two simple unconvoluted tubes joined by a transverse bridge at a point approximately one third of their length from the anterior extremity. Posterior to the bridge, each ovary has a thin straight oviduct which passes ventrolaterally to the genital apertures on the coxae of the third pereiopods. Lobsters sometimes have

TABLE 16.3

Size (mm CL) of *T. indicus* and *T. orientalis* at which Physiological Maturity Occurs

Species	Smallest Mature	Largest Immature	Size at 50% Maturity
Thenus indicus	42.0	58.0	52.0
Thenus orientalis	46.0	70.0	58.0

As indicated by smallest lobster with mature ovaries, largest lobster with immature ovaries, and size at which 50% of female lobsters are mature. From Jones (1988).

additional genital apertures on the coxae of the fourth pereiopods. These apertures are also connected to the ovaries by oviducts, and presumably are functional. Supernumerary genital apertures have been described for several nephropids (Farmer 1972). Ova are roughly spherical in shape with a mean diameter of 1.12 mm (range 0.90 to 1.49 mm). They are bright orange and semiopaque when first laid and remain so for most of the incubatory period; however, several days prior to hatching they become brown and transparent.

Testes are also in the form of paired tubes connected by a transverse bridge. Each testis is composed of a tightly coiled and continuous tube with many sacculi. The vasa deferentia are connected to the testes at a point posterior to the transverse bridge. Each is proximally very thin and convoluted, widening distally. The structure is similar to that described by Matthews (1954) for three other scyllarid genera. The only macroscopic feature indicating maturity in males is the swollen and opaque nature of the genital aperture integument.

16.4.2   Gonad (Physiological) Maturity

Since there is variance associated with size at first physiological maturity, the expression of size at which 50% of the population is mature has become conventional (George & Morgan 1979; Somerton 1980). The size at 50% maturity for female *Thenus* spp. is presented in Table 16.3. *Thenus orientalis* matures at a larger size than *T. indicus*.

16.4.3   Morphological (Physical) Maturity

For the telson length/carapace length relationship of females, a discontinuity is apparent, indicating a morphological transformation consistent with a relative increase in telson length growth rate that is likely to be associated with maturity. No other morphometric discontinuities have been found (Jones 1988).

By virtue of indistinguishable (macroscopically) maturation processes of testes and the absence of morphometric discontinuities of body parameters examined, it is not possible to estimate size at maturity for male *Thenus* spp. However, because growth of male and female *Thenus* spp. is equivalent, it is assumed that maturation also occurs at a similar rate and that maturity of males is reached at a similar size to that of females. Hossain's (1978a) estimates of size at first maturity for male and female *T. indicus* differed by less than three mm CL.

Because physical maturity occurred at a considerably smaller size than physiological maturity, size at functional maturity for 50% of individuals is estimated to be 52.0 mm and 58.0 mm CL for *T. indicus* and *T. orientalis* respectively. These estimates are consistent with those made by Kneipp (1974) and Hossain (1978a, 1978b).

16.4.4   Mating/Fertilization/Oviposition

Neither precopulatory courtship behavior nor mating within either species of *Thenus* has been directly witnessed in over 2000 h of remote video observation of captive populations (Jones 1988). This suggests that mating involves a very brief encounter.

Unlike many palinurids, male *Thenus* spp. do not appear to deposit a persistent spermatophoric mass in the process of mating. In many palinurids, the well-developed claw on the female's fifth pereiopod is used to pinch and scrape away the tough coating on the spermatophoric mass deposited by the male (Aiken & Waddy 1980). The small size of the chela and its fixed nature in *Thenus* spp. suggests that if a spermatophoric mass is deposited, it is likely to be soft and short-lived, similar to that of *Jasus* (MacDiarmid & Kittaka 2000). Such soft, nonpersistent masses have been observed by Kizhakudan et al. (2004) for *T. orientalis* and *Petarctus rugosus* (H. Milne Edwards, 1837), and females oviposited within eight h postmating and lost the spermatophore within 12 h.

Beneath the female's fifth pereiopod subchela there is a hollowed depression that may be involved in transporting the eggs from the genital aperture to the pleopods. Fertilization is presumed to occur soon after mating when the eggs are transported from the genital apertures to the pleopods or a short time afterwards when the eggs have been deposited on the pleopods. Unpublished data for *Thenus* spp. (unpublished observations by D.A. Ritz cited in Phillips et al. (1980)) suggest fertilization and oviposition occur within six h of mating; more recently, Kizhakudan et al. (2004) noted that ovipositioning occurred within five to seven h postmating and was completed by eight h postmating.

In the remaining scyllarid genera, the male pereiopods all terminate with a similar, sharp dactyl (Holthuis 1991). The presence of a small, fixed subchela on the fifth pereiopod of male *Thenus* spp., with unknown function, is anomalous.

16.4.5 Fecundity

A summary of the total weight of the egg mass (brood weight) and fecundity estimates for *Thenus* spp. is given in Table 16.4. The rate of increase with size of each parameter is significantly greater for *T. orientalis* than for *T. indicus*. Yet, since *T. indicus* mature at a smaller size, the fecundity of this species and brood weight are greater than that of *T. orientalis* until approximately 65 mm CL.

The relationship of fecundity and size (CL) is described by the following functions (see Radhakrishnan et al., Chapter 15 for alternative functions based on TL for *T. orientalis*).

> *T. indicus*: Fecundity $= 658.7 \times$ CL (mm) $- 26,329$
> *T. orientalis*: Fecundity $= 1273.2 \times$ CL (mm) $- 67,049$

The fecundity estimates for both species are of a magnitude greater than those for nephropid lobsters (Aiken & Waddy 1980) and significantly less than those of palinurids (MacDiarmid & Kittaka 2000). There is a considerable attrition of eggs over the incubation period (Jones 1988).

Although individual fecundity estimates were on the order of five to 50,000 (see also Radhakrishnan et al., Chapter 15), the likelihood of two or more spawnings per season indicates that the functional fecundity per female may be considerably higher. The release of eggs on more than one occasion can be attributed to several causes. A limited capacity to hold eggs on the pleopods and to store sufficient sperm to ensure maximal fertilization are possible explanations for lower fecundity. A comparison of reproductive features (Table 16.5) indicates that this strategy is in contrast to that typical of palinurids and nephropids.

16.4.6 Reproductive Seasonality

Immature lobsters of both species are most prevalent in mid- to late summer (January to March) when recruitment of postlarval juveniles to the population is at a maximum. The presence of immature lobsters

TABLE 16.4

Summary of Data Gathered from Fecundity Estimates of *Thenus* spp. (From Jones [1988])

Species	N	CL (mm)		Total Egg Mass (g)		Estimated Fecundity	
		Range	Mean	Range	Mean	Range	Mean
Thenus indicus	44	46.7–70.5	58.8	4.1–17.8	10.0	3686–25,134	12,455
Thenus orientalis	32	61.8–95.5	78.0	10.7–38.4	25.9	5579–54,746	32,230

TABLE 16.5

A Comparison of Characteristics Significant to the Reproductive Strategies of Various Spiny and Clawed Lobster Genera in Comparison with *Thenus* spp.

Characteristic	*Thenus*	*Jasus*	*Sagmariasus verreauxi*	*Panulirus*	*Nephrops*
Body size	Medium	Large	Large	Large	Small
Age at maturity (years)	1	3–11	6–7	2–6	2–3
Spawnings per season	2+	1	1	1–5	1
Fecundity (1000s)	10–30	200–2000	1000–2000	100–1000	0.5–4
Egg size (mm)	1.12	0.49	0.3–0.5	0.35–0.46	1.5
Incubation (days)	40	100–150	100	20–60	270
Larval stages	4	15–17	17	8–29	4
Relative larval size	Large	Medium	Medium	Medium	Large
Larval life (days)	45	300–600	250–360	270–330	14–21
Intra and interspecific competition	Low	High	High	High	Medium
Source	(Jones 1988; Mikami & Greenwood 1997)	(Heydorn 1969; Cobb & Phillips 1980; Cobb et al. 1997; Phillips & Kittaka 2000; George 2005)	(Cobb et al. 1997; MacDiarmid & Kittaka 2000; George 2005)	(Cobb & Phillips 1980; Lyons et al. 1981; MacFarlane & Moore 1986; Briones-Fourzan & Conteras-Ortiz 1999; Phillips & Kittaka 2000; George 2005)	(Berry 1969; Farmer 1974; Farmer 1975; Cobb & Phillips 1980; Redant 1987; Relini et al. 1998)

in spring corresponds to a secondary pulse of juvenile recruitment. Two spawning peaks are apparent, the first in spring and a second in midsummer (Jones 1988).

Comparison of the reproductive seasonality of *Thenus* spp. with other lobsters (Newman & Pollock 1971; Silberbauer 1971; Aiken & Waddy 1980; Lyons et al. 1981) indicates that the annual cycle typical of tropical palinurid species is the most similar.

16.5 Molting and Growth

Growth studies of crustaceans are not common, largely because of the difficulties posed by discontinuous growth and the absence of a structural record of growth equivalent to that provided by the scales or otoliths of fish. A range of growth assessment techniques have been applied to various lobster species (Fogarty 1995; Stewart & Kennelly 2000), and specifically to *Thenus* spp. (Jones 1988; Courtney 2002), including size-frequency modal analysis (Pauly 1983), synthesis of intermolt/increment data (Mauchline 1976; Mauchline 1977), functional analysis (Fabens 1965), and progressive plotting (Jones 1988). A synthesis of these provides an accurate description of the growth rate of *Thenus* spp.

Physical and behavioral aspects of molting in *Thenus indicus* and *T. orientalis* have been examined from observations of captive lobsters (Jones 1988). Setagenesis has been described from observations of pleopods under magnification as per the methods of Lyle and MacDonald (1983).

16.5.1   Molt Staging

Molt-stage determination by setal development indicates great similarity in the molting process between *Thenus* spp. and other Palinura (Palinuridae and Scyllaridae). Molt stages and corresponding setagenic events are essentially the same as those described for *Panulirus marginatus* Quoy & Gaimard, 1825 (Lyle & MacDonald 1983). Stages recognized and their characteristics are listed below (see also Mikami & Kubala, Chapter 5).

- *Stage A, B, Postmolt*. The setal lumen is wide and filled with granular material. The epidermis is closely applied to the cuticle. The exoskeleton is pliable at first, hardening within 48 h.
- *Stage C, Intermolt*. The setal walls are now quite thick and the lumen narrow, with no granular filling. The epidermis is closely applied to the cuticle. The exoskeleton is rigid.
- *Stage D, Premolt*. Ecdysial sutures appear on branchiostegites within 72 h of ecdysis.
  - D_0: Epidermis begins to retract from cuticle around base of setae.
  - D_1': Separation between epidermis and cuticle widens, leaving a broad transparent zone.
  - D_1'': New setae appear in the transparent zone, although they are flaccid and irregular.
  - D_1''': The new setae have become rigid.
  - D_2: A translucent layer forms above the epidermis, surrounding the bases of the new setae.
  - $D_{3,4}$: New setae are well developed, their base articulations clearly apparent within the new cuticle.

16.5.2   Growth

Estimates of the age of *Thenus* spp. of different size have been generated from the known time of reproduction, estimates of larval duration, recruitment of juveniles to the fishery, and then modal analysis from size-frequency distributions. Age estimates are presented in Table 16.6.

Using a variety of methods as described by Jones (1988), growth curves were generated as per Figure 16.2. Based on the von Bertalanffy (1938) function, a singular "best fit" growth curve was generated for each species. The functions for these curves are:

- *T. indicus*: $L_t = 91 \times (1 - e^{-0.002 \text{ per day } x(t+79)})$
- *T. orientalis*: $L_t = 152 \times (1 - e^{-0.00075 \text{ per day } x(t+160)})$

where L_t = mm CL and t = number of days

Courtney's (2002) growth analysis of *Thenus* spp. was based on a much larger database and provides estimates of L_∞ that are considerably lower than those of Jones (1988). Although Jones (1988) provides anecdotal evidence of very large bay lobsters (exceeding 100 mm CL) that would support his larger L_∞ estimates, those of Courtney (2002) are more reliable on the basis of documented maximum size of both

TABLE 16.6

Mean CL (mm) and Estimated Age (days) for 0+ Year Class Individuals of *T. indicus* and *T. orientalis*. (From Jones 1988)

Month/Season	T. indicus		T. orientalis	
	Mean Carapace Length(mm)	Estimated Age (d)	Mean Carapace Length (mm)	Estimated Age (d)
December, summer	22.9	90	—	—
January, summer	30.0	120	29.4	120
February, summer	33.3	150	33.8	150
March, autumn	37.7	180	39.2	180
April, autumn	42.5	210	41.9	210

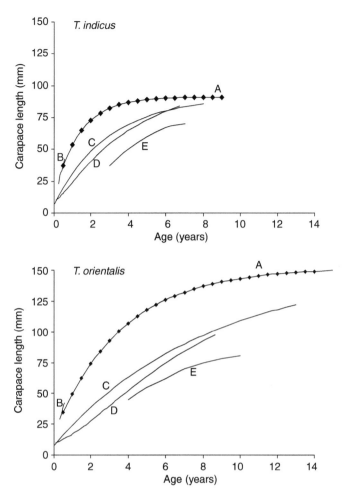

FIGURE 16.2 Definitive growth curves (A) for *T. indicus* and *T. orientalis* based on a synthesis of results from four growth-assessment techniques. For comparison, growth curves based on each assessment technique are included. Growth curves were generated from (B) modal progression analysis, (C) functional analysis, (D) intermolt/increment analysis, and (E) progressive plotting (From Jones, C.M. 1988. *The Biology and Behaviour of Bay Lobsters, Thenus* spp. (*Decapoda: Scyllaridae*) *in Northern Queensland*, Australia. Ph.D. dissertation: Brisbane, Australia: University of Queensland. 190 pp; used with permission.)

species. On this basis, the growth curves of Courtney (2002) (Figure 16.3) are likely to be the most accurate.

Theoretical growth of the two *Thenus* species is similar up to a size of approximately 70 mm CL and an age of about two years. Subsequently, however, the growth rate of *T. indicus* slows relatively quickly, while that of *T. orientalis* remains high for several more years up to a size of approximately 120 mm CL.

In comparison with growth functions documented for palinurid lobsters (Chittleborough 1976; Aiken 1980) and other scyllarids, especially *Scyllarides* spp. (see Bianchini & Ragonese, Chapter 9; Hearn et al., Chapter 14), the growth performance of *Thenus* spp. is very similar, albeit somewhat less than that of nephropids. The larger maximum size achieved by *T. orientalis* relative to *T. indicus*, which is comparable to that of the tropical palinurids (Cobb & Phillips 1980; Phillips et al. 1992) but smaller than that achieved by several *Scyllarides* species (Bianchini et al., Chapter 9; Hearn et al., Chapter 14), is likely to be attributable to the greater productivity of the preferred habitat of *T. orientalis*. These comparisons support the notion that *T. orientalis* represents an original descendent from a reef-dwelling palinurid stock, and that *T. indicus* is more recently evolved (see Webber & Booth, Chapter 2).

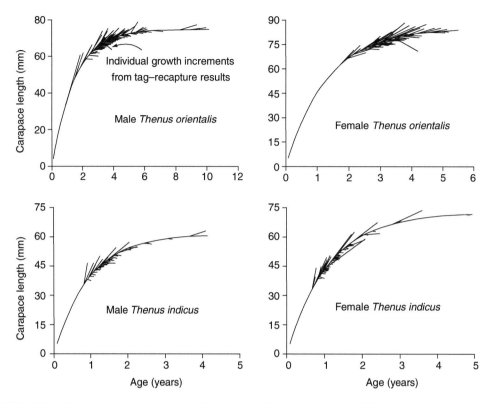

FIGURE 16.3   Growth curves for *T. indicus* and *T. orientalis*. (From Courtney, A.J. 2002). *The Status of Queensland's Moreton Bay Bug (Thenus spp.) and Balmain Bug (Ibacus spp.) stocks*. Brisbane, AU: Queensland Government, Department of Primary Industries; used with permission.)

16.6   Movement

The biology of *Thenus indicus* and *T. orientalis*, as discussed earlier, demonstrates a range of characteristics quite different from those of other achelate lobsters. In terms of locomotion, *Thenus* spp. is even more distinct, by virtue of its highly developed swimming ability (Jacklyn & Ritz 1986), which confers to these species great mobility. Such swimming ability appears to be further advanced than the typical tail-flick escape response documented for other lobsters, and is similar to that described for *Scyllarides latus* (Latreille, 1802) (Spanier et al. 1991; Spanier & Weihs 1992, 2004), *Ibacus peronii* Leach, 1815, and *I. alticrenatus* Bate, 1888 (Faulkes 2004).

The observed mobility of bay lobsters (Jones 1988) and the lack of complexity in their preferred habitat suggest that movement may be important to them. Homing activity in their relatively featureless environment is unlikely; however, nomadism and migration would confer considerable advantage to the species (Herrnkind 1980), allowing them to exploit the resources of a broad area.

Mobility of *Thenus* spp. has been examined by Jones (1988) and Courtney et al. (2001) through use of tag and recapture studies, monitoring of commercial catch levels and distribution, and observation of captive animals. These studies involved the tagging and release of more than 13,000 lobsters in several populations, with recapture rates of 7 to 10%. Accurate recapture information for many of these lobsters provided useful information for examination of movement. With an average time at liberty of 183 days, lobsters were recaptured an average of 24 km from the point of release. As movement would clearly not have been unidirectional, this statistic simply indicates that *Thenus* spp. can be very mobile, capable of moving large distances. The absence of any pattern to the direction of movement or season in which it occurred indicates that it is not likely to be migratory, as described for other species (Newman & Pollock 1971; Moore & MacFarlane 1984).

Observations of both *T. orientalis* and *T. indicus* held in large aquaria (Jones 1988) indicated that both species are characterized by substantial periods of activity involving walking and swimming (see Section 16.7 below). Homing movement, as defined by Herrnkind (1980), and territoriality were not evident. These observations suggest that movement direction and distance during periods of activity are random and that the translocation of tagged lobsters can be attributed to such random movement over the period at liberty.

Nomadic behavior has been described for several palinurid species, particularly in localities where cover and food supply are widely dispersed. Herrnkind (1980) has suggested that nomadism may be density dependent, only being expressed when biological resources of the localized environment become limited. The benthic environments of the shallow coastal waters of northern Australia are severely limited in their capacity to support populations of *Thenus* spp., particularly in view of the size of this lobster and its selective feeding preference. Low natural densities of this genus and other benthic species (Jones & Derbyshire 1988) supports this contention. It is not surprising then, that nomadism is well developed in *Thenus* spp. The adaptive values conferred by nomadism would include: optimal utilization of food and available habitat; ability to reutilize ephemeral habitats; and dispersal and mixing of genes. The latter value may be particularly important to *Thenus* spp. because of the organism's abbreviated larval life history. Although Herrnkind (1980) suggests that for palinurids nomadism may increase the probability of mortality by virtue of increased exposure to potential predators, it is unlikely to have such an effect on *Thenus* spp. for which predators are at a very low density.

It was evident from these studies that movement in *Thenus* spp. is of one type, defined by Herrnkind (1980) as nomadism and characterized by lack of direction, variability of distance, and temporal continuity. Both direct observations and indirect assessment techniques indicated that homing and migration were not displayed by either *Thenus* species. The simplicity of the bay lobster's physical habitat is sufficient to explain the absence of homing in this genus. There would be no advantage gained by moving from, and back to, a place within a featureless environment. Similarly, the advantages of migration, as outlined by Herrnkind (1980), have little applicability to *Thenus* spp. These advantages include: (1) the facilitation of larval release in areas where survival and dispersion are optimized; (2) physical protection during seasonal molting; (3) density reduction, after larval recruitment; and (4) optimal use of habitats subjected to seasonal oscillations. In populations of *Thenus* spp., nomadism would similarly achieve the first and third of these advantages, while the second and fourth would not be applicable.

The most significant aspect of movement in *Thenus* spp., in comparison with other lobsters, is that a large proportion of it is achieved by swimming. This one behavioral attribute, which sets this genus apart from all other lobsters, is responsible for its successful habitation of the coastal benthic environment.

16.7 Behavior

Within the typical habitat inhabited by *Thenus* spp., poor visibility, the low density of lobsters, and their nocturnal habit, make *in situ* observations impractical. Jones (1988) therefore employed naturalistic aquarium conditions to examine the behavior of *Thenus* spp., with particular attention paid to providing appropriate sediment characteristics, relative space per individual, water quality, photoperiod, food, and minimization of extraneous stimuli. Jones (1988) also assessed swimming activity by returning captive individuals to the sea near the site of their collection, and observing them while on self contained underwater breathing apparatus (SCUBA). Behavioral information collected included diel patterns of activity, food searching and feeding, concealment, sediment preference behaviors, locomotion, molting, and social activities.

16.7.1 Diel-Activity Pattern

Under simulated natural conditions using a 10:14 light:dark photoperiod, individual lobsters displayed a circadian activity rhythm consisting of four components, representing distinct time phases (Jones 1988). Each phase was characterized by the quantity and type of behavior displayed. Jones (1988) described these phases as: (1) day phase, from 0800 h through 1600 h; (2) evening phase, from 1600 h through

FIGURE 16.4   Photo of *Thenus orientalis* held in an aquarium. Lobster on left showing capacity to manipulate and open whole scallop (*Amusium* spp.). Lobster on right depicts typical buried stance, with body covered in sediment, eyes and antennules exposed.

FIGURE 16.5   Photo of *Thenus orientalis* held in an aquarium, showing typical resting stance, with body held off sediment, abdomen curled, and antennules elevated. This stance most often was displayed during night phase.

2000 h and including sunset; (3) night phase, from 2000 h until midnight; and (4) morning phase, from midnight to 0800 h and including dawn.

During the day phase, most lobsters remain buried in the sediment with only the eyes and antennules apparent (Figure 16.4). The mean activity level during this period is two to three min/h. Following the day phase, the evening phase is characterized by a significant increase in activity (mean 18 to 20 min/h), which consists mainly of searching/foraging activity to locate food. Its initiation is unrelated to changes in light intensity, as evidenced by the emergence of many individuals prior to sunset or sometime after. Ninety-five percent of lobsters emerged from the buried position and became active within 1 h of sunset. The method of searching/foraging results in the nomadic wandering of individuals and, as suggested previously, this type of movement would confer considerable advantages, the most immediate of which is the location of suitable food. Frequent bursts of swimming also occur during this phase, which do not appear to be initiated by any specific stimulus. They are generally of less than 10 sec duration, but can be much longer (see Section 16.7.5).

Despite the absence of any overt stimulus, activity levels drop significantly for a period of approximately four h during the night phase (mean activity level 11 to 13 min/h). During this phase lobsters are mostly still, remaining motionless but standing (Figure 16.5) or partially buried (Figure 16.6); however, they are not fully buried as they are during the day. Grooming activities are also often displayed during this night phase. The decrease in overt activity during the night phase may serve an energy-conserving function. Hindley (1975) suggested that *Penaeus merguiensis* conserved energy, in the absence of food, by alternately foraging and resting. A similar resting mechanism is evident in *Thenus* spp., but decreased

FIGURE 16.6 Photo of *Thenus orientalis* held in an aquarium in partially buried stance, often displayed during night phase.

activity is confined to a single period. If such a mechanism does operate, the activity pattern of preferred food species and predators is likely to be related.

The morning phase is marked by a second period of increased activity, consisting primarily of searching/foraging activity. As a result, the activity pattern for the hours of darkness is characterized by two well-defined peaks of searching/foraging activity, separated by a period of lesser activity. Despite the influence of food availability on activity, synchronization of this pattern cannot be attributed to any observed exogenous factor. Jones (1988) hypothesized that *Thenus* spp. activity is controlled by an endogenous clock, and this is further evidenced by the lack of association between cessation of activity and increase in light intensity at sunrise.

It has been suggested (Bregazzi & Naylor 1972; Rodriguez & Naylor 1972; Atkinson & Naylor 1973) that the degree of endogeneity in activity rhythms may be maximally developed in mobile but cryptic (e.g., burying) species. A selective advantage is conferred to such species by enabling them to anticipate environmental change and to return to their resting state prior to the onset of unfavorable conditions. Monitoring of the ambient environment is minimized, energy is conserved, and the timing of activity is optimized. This is contrary to the "exogenous control" school (Brown 1958; Naylor 2005) who suggest that external variables dictate the timing of activity. The adaptive significance of endogenous control to *Thenus* spp. would be considerable. Risk of predation by large, mobile, diurnal species would be minimized, while species of fish and invertebrates preferred as food would be most active and accessible.

Deviations from the typical periodicity and types of behavior observed can occur and Jones (1988) attributed them primarily to molting and to unfavorable conditions. A tail extension behavior, observed on occasions in premolt lobsters, was most likely employed to facilitate the separation of the old and new cuticles during ecdysis.

16.7.2 Food Searching and Feeding Behavior

According to Jones (1988), a large proportion (up to 70%) of the active behavior of *Thenus* spp. involves searching for food, which consists of slow forward motion (<1 min every 5 min) coupled with repeated changes in body orientation relative to the direction of movement. The sediment is investigated through a band of approximately three times the lobster's body width. As a lobster moves forward, the anterior two pairs of pereiopods are used to probe the sediment, while the antennules are continually raised and lowered as described by Lau (1987).

The dactyls of all walking legs are equipped with pores housing fine setae, which are likely to be chemosensory in function (Ache 1982; Derby et al. 2001; see Lavalli et al., Chapter 7 for a full description of currently known setal types on scyllarid pereiopods). The dactyls of the first two pairs of legs are significantly longer than those of the posterior pereiopods. Their epicuticular caps (the keratinized sheath covering the distal portion of the dactyl) are well developed and presumably serve a protective function in minimizing abrasion as probing of the sediment takes place. The posterior dactyls, which have relatively

TABLE 16.7

Relative Preference for Food Species Offered to Aquarium Captive *Thenus* spp. Lobsters

Most Preferred Food Species (100%)	Less Preferred Food Species (50%)	Least Preferred Food Species (<25%)	Food Species Avoided 0%
Scallops	Lizardfish	Flatfish	Urchin
Amusium balloti	*Synodus similis*	*Engyprosopon* spp.	*Maretia planulata*
Bernardi, 1861	McCulloch, 1921	*E. grandisquama*	Lamarck, 1816
A. pleuronectes	*Saurida* spp.(juv.)	Temminck &	
Linnaeus, 1758	*Trachinocephalus*	Schlegel, 1846	Cuttlefish
Chlamys leopardus	*myops* Gill, 1861	*Arnoglossus*	*Sepia* spp.
Reeve, 1853		*intermedius* Bleeker,	
		1865	
Goatfish			
Upeneus spp.			
U. sulphureus Cuvier,			
1829			
Shrimps			
Metapenaeopsis spp.			
Trachypenaeus spp.			

From Jones (1988). Percentages refer to frequency of response to food when offered; thus 100% indicates the food was chosen 100% of times it was offered.

Note: The relative abundance of these species in the bay lobster environment is documented in Jones & Derbyshire (1988).

small epicuticular caps, each possess two rows of coarse sensory hairs and were not observed by Jones (1988) to be used for sediment probing.

In his aquarium-based observations of *Thenus* spp., Jones (1988) noted that sediment probing is accompanied by frequent movement of the antennules in a vertical plane. While the antennules are moved in this fashion, their lateral filaments are repeatedly flicked, as described by Price and Ache (1977) and others (Daniel & Derby 1991; Berg et al. 1992; Mellon 1997). An increase in the frequency and magnitude of antennular flicking is always observed in response to the introduction of food (see Lavalli et al. Chapter 7 for further information on scyllarid antennular structure and function).

Jones (1988) observed that the introduction of fresh food (e.g., trawl by-catch fish and invertebrates) elicits a clear behavioral response in *Thenus* spp. including increased antennule flicking, reorientation of the body toward the food, and elevation of the first two pairs of walking legs. This initial response is typically followed by relatively rapid and directed movement toward the food. In some instances, a forward lunge, effected by the sudden extension of the abdomen, ensures that the food item is quickly and firmly grasped. Once contact is made with a food item, the walking legs are used to grasp and manipulate (see Johnston, Chapter 6 and Lavalli et al., Chapter 7 for more details on this phase of feeding). Large food items, equivalent in size to the lobster, are as vigorously pursued and grasped as smaller ones.

Although the eyes of *Thenus* spp. are relatively large, the importance of visual cues in food location may be less than that of chemical cues. Jones (1988) observed that when cleaned scallop shells (*Amusium* spp.) or deep-frozen whole scallops (i.e. with no aroma) are placed near a lobster, there is no response. However, the introduction of fresh or thawed scallops, presumably leaching a strong chemical signal, elicits an immediate response.

Observations of the feeding response to a variety of food species offered by Jones (1988) indicate a broad food preference hierarchy. Food items made available to lobsters included a variety of species from the most abundant macrofaunal groups commonly found associated with bay lobsters in shrimp trawl by-catch. The most common of these species are listed in Table 16.7. Preferred food species are ranked as: those species always chosen (100%) when offered, those species chosen in approximately 50% of offers, those chosen in <25% of offers, and those species apparently avoided. Clearly, plant and benthic infaunal elements, not provided in these observations, may also contribute to the diet of bay lobsters. A strong preference for scallops is supported by a well-developed ability to manipulate and open the shells. Johnson

(1980) pointed out the danger of inferences based on usage/availability data, and it should, therefore, be stressed that the food preference displayed by *Thenus* spp. indicates a ranking of selectivity of food species provided and not a measurement of food preference in nature. It is reasonable to assume, however, that the natural food preference of macrofaunal elements would approximate those described.

Observations of lobsters manipulating and successfully opening whole scallops and other bivalves in aquaria (Jones 1988) indicate a strongly developed behavioral capacity involving fine coordination of limbs (Figure 16.4). Similar complex coordination in opening bivalves was observed in several species of *Scyllarides* (Lau 1987; Spanier & Lavalli 1998; Malcom 2003; see Lavalli et al., Chapter 7 for a detailed description of limb coordination during feeding). Such behavior suggests either an innate or learned ability resulting from considerable exposure to bivalves as food in nature. Scallops are mobile creatures capable of short bursts of swimming. Although scallops provided as food in the laboratory were dead, the preference displayed for them suggests that they may also be preferred in nature. The observed ability to lunge forward, by a sudden extension of the tail, would be sufficient to attack a prey animal such as a live scallop, within 30 to 50 cm.

Given the low light intensity and soft sediment of the benthic environment inhabited by *Thenus* spp., it is not surprising that many of the cohabiting benthic species are reliant on camouflage for avoidance of predation. Among these are many of the preferred food species. In nature, it is likely that bay lobsters, using their well-developed chemosensory abilities, search out their preferred benthic food species, and having located them, lunge forward and grasp them with the pereiopods. This "search and attack" foraging mode is energy expensive and may occasionally be supported by less expensive foraging of infaunal elements. To optimize such a predatory food-seeking behavior in an environment characterized by low densities of preferred food species, a high degree of mobility would be advantageous, as discussed later.

16.7.3 Concealment

Concealment behavior to avoid predation is an adaptive response common to many crustaceans. Among the scyllarid lobsters, most species are inhabitants of hard substrates for which concealment might involve camouflage and shelter seeking within the physical complexity of the environment (Spanier & Lavalli 1998). In soft substrates, however, concealment can only be achieved by constructing burrows (as for nephropid lobsters, such as *Nephrops norvegicus* (Linnaeus, 1758) (Tuck et al. 1994) or *Homarus americanus* H. Milne-Edwards, 1837 (Lawton & Lavalli 1995), or immersion (burial) in the sediment (Bellwood 2002). This latter response is the concealment method applied by *Thenus* spp.

Burying is defined as the series of behaviors which enable lobsters to move into the substratum, with the sediment being in direct contact with and encasing the body (Bellwood 2002). In the case of *Thenus* spp., such behavior typically occurs at the cessation of a period of activity as described earlier. The direct observations of Jones (1988) of *Thenus* spp. in large aquaria furnished with a 10 cm depth of fine sand, showed that burying behavior is initiated with slow reverse movement coupled with slow and repeated extensions of the abdomen. The telson and uropods are fanned out and held against the sediment while the tail extensions take place so that a depression is created into which the tail and body are progressively pushed by the pereiopods. Between three and eight tail extensions are necessary to effectively cover the dorsal surface of the abdomen and the posterior third of the carapace. A sudden "shudder" effected by the pereiopods pushing backward and a series of rapid abdominal flexions complete the burying procedure, by covering the anterior exposed portion of the carapace and the antennae with a fine layer (<1 cm) of sediment. A fully buried lobster is entirely concealed except for the eyes and antennules (Figure 16.4). Burying is normally completed in less than two minutes. Contrary to some suggestions (George & Griffin 1972), the large platelike antennae are not involved in burying behavior. Despite the disparate sediment preferences displayed by the two species of *Thenus* observed (see Section 16.3.1), burying behavior was essentially the same for the two species, and similar to that described for brachyuran crabs by Bellwood (2002) and for *Ibacus* spp. by Faulkes (2006).

While buried, the antennules and eyes remain exposed and a relatively low frequency of antennular flicking is maintained. During hours of darkness (night phase) when lobsters are typically active, individuals are often observed burying themselves partially, for periods of relatively short duration. Partial burying is clearly distinguishable from normal burying and presumably provides suitable concealment during rest periods (Figure 16.6). Emergence from the buried position is facilitated by the extension of the pereiopods which causes the covering sediment to fall away to either side.

A respiratory (gill) current is maintained by the beating of the scaphognathites while in this resting state (Phillips et al. 1980; Atema 1985) — this current, while ventilating the gills, is also used to propel metabolites from the gills and urine and may be important in signaling as well as ventilation (Atema 1985). Jones (1988) confirmed this current in *Thenus* spp. by use of food dye — the operation of an afferent current beneath each eye and an efferent current flowing out at the base of the antennules. Water flow into the gills is over the entire ventral surface of the branchial chamber and its margin (the branchiostegite), because there is no sculpturing of the carapace to facilitate channeling. It is not clear whether there are any other specific morphological modifications that facilitate the respiratory current, as have been described for brachyuran crabs (Bellwood 2002). If a disturbance occurs the respiratory current immediately ceases, often for several seconds. Periodic reversal of the current occurs, presumably to flush detritus from the gill chamber.

16.7.4 Sediment Preference

The association between substrate characteristics and the distribution of commercially important benthic species is an important aspect of the fishing strategy for catching particular species (Williams 1958; Grady 1971; Rulifson 1981; Somers 1987; Snelgrove & Butman 1994; Perry 1999; Rangeley & Lawton 1999). Commercial trawl fishers in Australia recognize the two species of *Thenus* by their occurrence on different bottom types (Jones 1988). This suggests that a differential sediment preference operates for these species. Jones (1988) conducted sediment preference experiments to confirm this.

Results of sediment choice experiments indicate that *T. indicus* displays a significant preference for the finest sediment grade offered, which is dominated by particles of less than 63 μm in diameter (Jones 1988). Folk's (1965) nomenclature defines this sediment as mud/silt. *Thenus orientalis* displays a clear preference for sediments composed of moderate to coarse particle size, that is, medium to coarse sand.

Sediment/abundance correlations from wild-population studies (Jones 1993) indicated significant positive correlations for both species of *Thenus*. These correlations were consistent with maximum abundance of *T. indicus* on fine-grain sediments (mud) and maximum abundance of *T. orientalis* on coarse-grain sediments (sand). Thus, bay lobsters appear capable of distinguishing between sediments characterized by different particle size compositions, and they actively select habitats composed of preferred particle size ranges.

16.7.5 Locomotion

Locomotory ability within the family Scyllaridae includes walking, typical of most macrurous decapods; however, there is also evidence of highly specialized morphologies and behavioral adaptations which have led to an advanced swimming ability.

Jones (1988) observed that locomotion by walking in *Thenus* spp. is variable in direction, and in inter-leg coordination. Forward walking predominates, although sideways and backward movement also occur. The ability to walk in all directions is common among the macruran decapods (Evoy & Ayers 1982). For *Thenus* spp. the anterior two pairs of pereiopods are held above the substratum except when used for probing as described earlier (Section 16.7.2). The posterior three pairs of legs support the animal and are coordinated so that an alternating tripod gait is effected (Chapman 1969; Evoy & Ayers 1982). None of the pereiopods in either sex of *Thenus* spp. have fully formed and functional chelae and all are potentially available for locomotion. The utilization of legs for support varies from a maximum of ten to a minimum of three.

In addition to walking *Thenus* spp. is capable of swimming. Two types of swimming activity are recognized: "locomotion swimming" and "escape swimming." *Locomotion swimming* takes place at regular intervals throughout the active period and is characterized by its spontaneity and the apparent absence of an overt stimulus. Each period of swimming is of variable duration, usually less than one minute, although it can continue for up to 40 min without interruption (Jones 1988). Figure 16.8 shows *T. orientalis* in locomotion swimming mode in its natural habitat. From a normal standing position on the

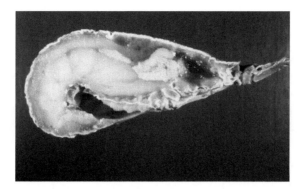

FIGURE 16.7   Photo of *Thenus orientalis* in sagittal section to illustrate the aerofoil shape of the body, which provides lift during swimming. The antennae (to right) are positioned at the trailing end of the aerofoil, and thus operate like the ailerons of an airplane wing.

FIGURE 16.8   Photo of *Thenus orientalis* in its natural habitat, depicting locomotory swimming. Note in particular, the lobster is swimming away from the camera to the left of field, the abdomen is in an extended position prior to flexion, the antennae are moving independently of each other to provide lateral balance, and the pereiopods are extended anteriorly for streamlining.

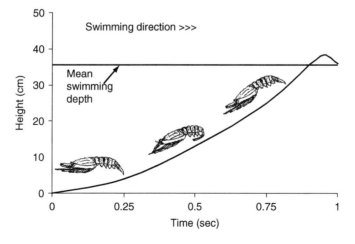

FIGURE 16.9   Diagrammatic representation of take-off into locomotion swimming mode of *Thenus* spp. (From Jones, C.M. 1988. *The Biology and Behaviour of Bay Lobsters, Thenus spp.* (*Decapoda: Scyllaridae*) *in Northern Queensland*, Australia. Ph.D. dissertation: Brisbane, Australia: University of Queensland. 190 pp. used with permission.)

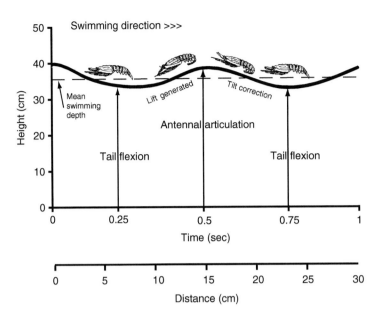

FIGURE 16.10   Diagrammatic representation of locomotion swimming of *Thenus* spp. (From Jones 1988; used with permission.)

sediment, take-off into the swimming mode is executed by sudden and repeated flexions of the abdomen, which propel the animal backward. Lift is generated by virtue of the longitudinal sectional shape of the lobster, which approximates an aerofoil (Figure 16.7) (Jacklyn & Ritz 1986; Spanier & Weihs 1992, 2004), and by the downward thrust of the tail (Spanier & Weihs 1992, 2004). As a lobster leaves the sediment surface, all pereiopods are extended anteriorly, improving streamlining (Figure 16.8 and Figure 16.9). The initial few flexions of the abdomen lift the lobster away from the sediment at an angle of between 30 and 45 degrees (Figure 16.9). Height above the sediment is controlled by the large flap-like antennae which are articulated in a dorsoventral plane, like the ailerons of an airplane wing. The height above the sediment maintained while swimming typically ranges from 12 to 54 cm, with a mean of 36 cm (Jones 1988) (Figure 16.9 and Figure 16.10). Because of the lift generated by the shape of the lobster, each flexion tilts the body so that the abdomen is higher than the carapace, and the lobster moves away from the sediment. To maintain a constant height, the antennae are depressed to correct this tilt (Spanier et al. 1991; Spanier & Weihs 2004). The resulting motion through the water is sinusoidal, the wavelength of which decreases as the frequency of the abdominal flexions—thus, the speed increases. This swimming motion is depicted in Figure 16.10 (see additional description in Lavalli et al., Chapter 7).

   Speed during locomotion swimming averages 29 cm/sec. Figure 16.11 shows the motion of a lobster swimming at twice the average speed. This motion is reminiscent of the hopping motion described by Jacklyn and Ritz (1986), although Jones (1988) only observed this when a lobster was perturbed.

   Due to the spatial constraints of the tank-based environment used, Jones (1988) found it difficult to gauge the directional stability of locomotion swimming in captive lobsters for distances greater than a few meters. Once the edge of the aquarium was encountered, observed lobsters usually bumped along the perimeter, unable to follow a desired course. However Jones (1988) also made observations of *Thenus* spp. in their natural environment by releasing previously captive lobsters. Such lobsters swim in a straight line at a constant depth for periods of several minutes, covering distances of between 20 and 80 m.

   The second type of swimming is *escape swimming*, characterized by forceful abdominal flexions and high-speed motion. Escape swimming is always in direct response to an overt stimulus. This stimulus usually involves direct physical interference, or the threat of such interference from another lobster or other external origin. For example, a shadow cast across a lobster can stimulate the escape swimming response. The force and frequency of abdominal flexions executed during an escape response are variable and appear to bear a relationship with the magnitude of the stimulus. Following the rapid abdominal

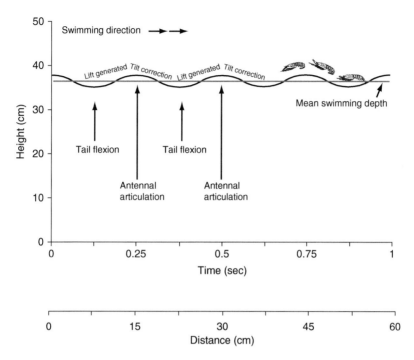

FIGURE 16.11 Diagrammatic representation of locomotion swimming of *Thenus* spp. at twice the mean swimming speed. (From Jones 1988; used with permission.)

flexions, a lobster will glide with the abdomen fully flexed and the pereiopods extended anteriorly to optimize streamlining. The motion described is an arc, the dimensions of which vary in accordance with the force of the abdominal flexion (Figure 16.12). Speed during escape swimming averaged approximately 1 m/sec. While this swimming mode is behaviorally different from the other swimming mode, it is not clear if these behavioral differences are due to distinct neural control systems, or if they are simply expressive extremes of one, flexible motor program (Faulkes 2004).

The morphological and behavioral adaptations which permit *Thenus* spp. to swim strongly and for sustained periods of time in the manner described are exceptional within the Crustacea. A similar swimming ability has been documented for the pelagic galatheid *Munida gregaria* (Fabricius, 1793) (Zeldis & Jillet 1982), and for another scyllarid lobster, *Scyllarides latus* (Spanier et al. 1991), but there appear to be few if any other macrurous decapods with the ability for sustained bursts of abdominal-flexion-mediated swimming.

Jacklyn & Ritz (1986) previously discussed the hydrodynamic advantages of scyllarid morphology in relation to observed swimming behaviors. The morphological characteristics of *Thenus* spp. are specialized adaptations for a mode of life in which swimming is of considerable significance. These characteristics include:

- Highly modified antennae which are broadly flattened, large in surface area relative to total body surface area, and smooth. The sculpturing and marginal spination are conducive to smooth flow of water from posterior to anterior. They are well innervated (Chacko 1967) and can be independently articulated to affect longitudinal inclination (pitch) and transverse inclination (roll) respectively, although control is clearly limited.
- A smooth, flat, and laterally expanded carapace which promotes stability and streamlining.
- A significant reduction in general body spination and sculpturing in comparison with many other scyllarid species, which also promotes streamlining.
- An aerofoil shape in sagittal section (Figure 16.7), which generates lift when exposed to a current flowing from posterior to anterior.

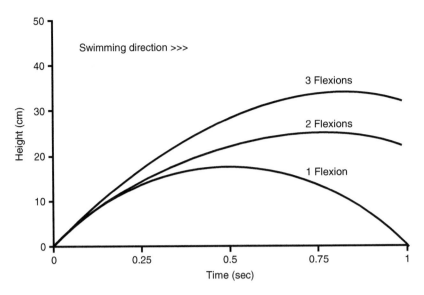

FIGURE 16.12 Diagrammatic representation of escape swimming motion of *Thenus* spp., after one, two or three tail flexions (From Jones, C.M. 1988. *The Biology and Behaviour of Bay Lobsters, Thenus spp. (Decapoda: Scyllaridae) in Northern Queensland*, Australia. Ph.D. dissertation: Brisbane, Australia: University of Queensland. 190 pp. used with permission.)

- Fine pereiopods which are minimally hirsute and can be extended anteriorly to promote streamlining.
- Eyes situated on the lateral extremes of the carapace on eyestalks which are relatively long. This enables the eyes to be extended clear of the deep orbits so that vision both anteriorly and posteriorly is possible. Posterior vision is advantageous to a backward swimming animal.
- Abdominal musculature that is particularly well developed in comparison to other lobster species. The abdomen accounts for over 40% of the total body weight, compared with less than 35% for most other lobsters (Ivanov & Krylov 1980; Jones 1988).

Where power for swimming is supplied by the abdominal muscles, directional control is provided by moving the antennae. Their muscle and nerve supply appears to be well developed (Chacko 1967), and movement in both horizontal and vertical planes is possible. From a dynamics standpoint, the control of body orientation with respect to direction (i.e. yaw) and roll requires very fine positional adjustment of these appendages. Jones (1988) observed the movements of the antennae, during episodes of swimming to be indelicate and quite sudden, indicating that their control was limited. Positional and directional stability are apparently maintained through use of the antennae, but maneuverability is not well developed. Jacklyn & Ritz (1986) suggested that the loops and rolls they observed in *T. indicus* were an indication of fine control and advanced maneuverability. The observations of Jones (1988), however, suggest the contrary, that looping and rolling are not deliberate movements, but an indication of poor control. The ability to execute such maneuvers may confer little advantage to an animal swimming in a largely featureless environment. Alternatively, they may be protean movements designed to confuse any potential predators as to the pathway traveled by the animal.

The natural habitat of *Thenus* spp. is typically flat and featureless, devoid of spatial complexity, and relatively poor in its carrying capacity for demersal macro-invertebrates and fish. Survival in such an environment for an animal with selective feeding preference necessitates mobility. Bay lobsters have achieved this by developing swimming ability and, in doing so, have filled a niche intermediate between the crawling benthos and the bottom-feeding, pelagic fish.

16.7.6 Molting

Although there is an abundance of literature regarding the physiological and morphological changes associated with lobster molting (Aiken 1980; Lipcius & Herrnkind 1982; Lipcius & Herrnkind 1987),

FIGURE 16.13   Photo of the exuvium of recently molted *Thenus orientalis* held in an aquarium. Note that the molted lobster extracts itself from the old shell through a split between the cephalothorax and the abdomen.

behavioral aspects of ecdysis have been less well documented (Tamm & Cobb 1978; Radhakrishnan & Vijayakumaran 1998). The behavioral mechanisms that enable a lobster to extricate itself from the old exoskeleton and to expand the new one prior to its hardening are important (see Mikami & Kuballa, Chapter 5 and Horne & Tarsitano, Chapter 8 for more details on the molt cycle), particularly in view of the vulnerability of the individual during this process. At no other time in the life history of the individual is the risk of mortality as high as during ecdysis.

Jones (1988) observed ecdysis of captive *Thenus* spp. on several occasions, and his findings are summarized here. Premolt lobsters are recognized by the appearance of ecdysial lines on the branchiostegites, as described for *Panulirus homarus* Linnaeus, 1758 (Radhakrishnan & Vijayakumaran 1998). Molting typically occurs within two days of detecting these lines and, in all instances, occurs during the active phase of the diel activity pattern, between 1600 and 0600 h. During the premolt phase (within two days of molting), activity is characterized by the absence of searching/foraging behavior and extended periods of inactivity while standing on the sediment surface. Within 4 h of ecdysis, "exercise" movements (Travis 1954) occur. The most conspicuous of these is the overextension of the abdomen so that the telson is held just above the anterior abdominal region. This pose is maintained for periods of up to 30 sec. Several less conspicuous movements, including slow abdominal flexions, and individual limb, eye, and antennal movements also take place. All presumably assist in the separation of the old and new cuticles.

Ecdysis in *Thenus* spp. occurs while lobsters stand upright on the sediment surface, there being no evidence of searching for a specific molting site, or behaving in a cryptic fashion (Aiken 1980; Lipcius & Herrnkind 1982; MacDiarmid 1989), or of lying on their side (Radhakrishnan & Vijayakumaran 1998). Following the premolt movements, *Thenus* spp. lower their abdomen to the sediment surface and shift their pereiopods into a bracing position where the anterior two pairs are held forward, the third and fourth pairs are extended laterally, and the posterior pair are held behind. General body motion ceases; however, contractions of the new, soft-shelled body beneath are evident. Within 30 sec, the ecdysial sutures along the branchiostegites and the dorsal membrane connecting the cephalothorax and the abdomen begin to tear (Figure 16.13). The head and thorax are then pulled upward and the legs extracted as the old carapace tilts forward along a hinge at the anterior end of the branchiostegites. At this time, several rapid flexions of the abdomen facilitate the final separation, and are maintained to propel the newly molted animal into a swimming mode. An arc of swimming is typically executed over a distance of one to two m before settling to the substratum. The time elapsed from assuming the bracing stance to this point ranges from 60 sec to approximately three min.

Frequent bursts of swimming for periods of up to 30 sec occur after ecdysis and are interspersed with longer periods of slow walking and quiescence. This activity pattern is maintained for a period of between 15 and 30 min. Newly molted lobsters partially bury themselves in the sediment (Figure 16.6). Considerable hardening of the shell is evident within three to four h after molting, although some compressibility is still noticeable up to several days afterwards. Normal behavior, including feeding, returns within 24 h of molting.

In the period prior to ecdysis, vulnerability to predation is likely to be normal, as typical mobility and escape responses operate. The period of maximum vulnerability, from assuming the bracing stance to emergence, is on the order of one to three min. Minimization of vulnerability is facilitated by the speed of ecdysis and the ability to recover mobility quickly. Despite the softness of the exoskeleton at the time of emergence, bay lobsters are capable of swimming and can effect a typical escape response.

16.7.7   Social Behavior

The often complex and ritualized nature of aggressive encounters between conspecific nephropid and palinurid lobsters (Atema & Cobb 1980; Thomas et al. 2003; Rutishauser et al. 2004) do not occur in bay lobsters (Jones 1988). Postural characteristics of approachment and avoidance involving chelae, antennae, and other appendages appear to have no counterpart in *Thenus* species and, unlike Barshaw and Spanier (1994) with *Scyllarides latus*, Jones (1988) was unable to recognize dominance or subordinance of individuals within small captive populations of up to 10 lobsters.

Jones (1988) observed that interactions that did occur between individual bay lobsters were characterized by indifference. The only instances in which aggression occurs involve possession of food items. Similar intraspecific interactions were reported by Barshaw and Spanier (1994) in *Scyllarides latus*. Such encounters are characterized by continuous maneuvering of the pereiopods to obtain the strongest grip on the food object, coupled with periodic and rapid tail flexions to effect an escape. These encounters are of short duration, usually less than 1 min. Probability of victory is unrelated to size, sex, or molt state, as it is for other lobsters (Jones 1988) and is more likely a function of mechanical advantage. The individual with the better grip, usually the first to have grasped a food object, is typically the victor.

During many hundreds of hours of observation of captive bay lobsters, neither courtship behavior nor mating were seen by Jones (1988). Mating of *T. orientalis* was observed on one occasion by Kneipp (1974) who described a very rapid behavioral sequence involving embracement of the ventral surfaces lasting only a few seconds. During this time, a spermatophoric mass was deposited on both sides of the female sternum posterior to the genital apertures. No courtship or precopulatory behavior was observed, although Kneipp (1974) suggested the possibility of pheromone involvement.

Production, presence, or operation of a sex pheromone for bay lobsters is not evident, although a sound-producing mechanism was suggested by Jones (1988). Sound vibrations were detected by a hand placed in the water within two m of a lobster, but the involvement of a stridulating organ like that of some palinurids (George & Grindley 1964; Mulligan & Fischer 1977) is not supported by morphological examination. The production of clicking sounds by grinding the mandibles together, as documented for *Jasus lalandii* (Atema & Cobb 1980), may be responsible. It is reasonable to assume that these sounds serve a communicatory function; however, their significance for sexual attraction between the sexes is uncertain.

Given the low natural densities of bay lobsters and the absence of communal behavior in aquarium studies, it is reasonable to assume that interactions between individual *Thenus* spp. lobsters in nature are rare. Bay lobsters are likely to be solitary animals and, therefore, it is not surprising that agonistic behavior is not well developed. Similarly, specific courtship and mating behaviors defined for several species of clawed and spiny lobsters appear to have no equivalent in either species of *Thenus*.

Given the solitary nature of bay lobsters, the functional significance of reproductive behavior is related more to the attraction of any mate, rather than selection of one among several potential mates. For this reason, the significance of sexual attraction mechanisms, pheromones, sound production, or otherwise may be greater than current evidence suggests.

16.8   The Australian *Thenus* Spp. Fishery

Several species of penaeid shrimp and scallop (*Amusium* spp.) constitute the target species of the trawl fishing industry of northern Australia. This industry is made up of several loosely defined, but distinguishable fisheries, each recognized by its target species, location, and season. Bay lobsters are caught as an incidental by-catch species in most of these fisheries (Jones 1984; Courtney et al. 2001; Courtney 2002).

The relative proportion of bay lobsters within the commercial catch varies, generally between 0 and 2.5% by weight (Jones 1988). Catches dominated by bay lobsters are infrequently reported. Although a diver-based fishery for *Thenus* spp. was reported from the Red Sea in the 1960s (Ben-Tuvia 1968), the exploitation of these lobsters in Australia and throughout most of southeast Asia is by way of trawlers (Holthuis 1991). The total production of bay lobsters in Queensland for 2001 was estimated at 386 metric tons, valued at AUD$4.6 million, but has been as high as 755 metric tons (Courtney 2002). More than 90% of the Australian catch is from Queensland waters.

Jones (1988) compared several population parameters measured for a *Thenus indicus* population in the southeast Gulf of Carpentaria between 1963, prior to the establishment of a shrimp trawl fishery, and 1983. The analysis indicates that significant changes occurred that are likely to be attributable to commercial fishing activities. The CPUE of lobsters in 1983 (1.8 lobsters ha^{-1}) was 44% less than that in 1963/64 (3.2 lobsters ha^{-1}). Of equal significance was a 15% reduction in the mean size of lobsters. The exploitable stock of *T. indicus* in the Gulf of Carpentaira is represented by a size range of approximately 25 to 70 mm CL and an age range of approximately two to five years. It is evident from examining catch-size composition that commercial exploitation is size selective toward the larger individuals. This selectivity has two sources: net efficiency and onboard selection. It is reasonable to assume that long-term commercial exploitation would cause a reduction in mean size of lobsters in the population, as has been documented for other lobster fisheries (Radhakrishnan et al. 2005).

The significance of both a reduction in CPUE and in size of individuals in the population is far greater than either alone. Based on the estimates given, the exploited standing stock of *T. indicus* in the Gulf of Carpentaria was reduced by 65% over a 20-year period.

The current fishery for bay lobsters in Queensland is managed and regulated. It is prohibited to take egg-bearing females at anytime, and a minimum legal size of 75 mm carapace width applies to both species. Courtney (2002) reported that CPUE has remained stable for the past two decades, although total catch has declined in later years due to decreased fishing effort. Unlike India where the fished populations appear to be significantly overexploited (Radhakrishnan et al. 2005), the *Thenus* fishery of northern Australia appears to be sustainable.

16.9 Summary

Thenus spp. represent a divergence from the familiar lobster stock, as typified by the palinurid genera *Panulirus, Palinurus*, and *Jasus* and the nephropid genera *Homarus* and *Nephrops*. It appears also to differ markedly from many of the other scyllarid genera, particularly in regard to morphology, distribution, and behavior.

The evolutionary sequence of scyllarid genera and species is comprehensively discussed by Webber and Booth (Chapter 2). They confirm the close affiliation of *Thenus* and the Scyllarinae, and the likelihood of a more recent specialization of *Thenus*. The most extraordinary of the various specializations include the capacities of swimming and of burying, both of which contribute to the successful habitation of the broad, sedimentary, inshore areas of the continental shelf in the Indo-West Pacific region.

Given the success and diversity of elasmobranch and teleost fish, echinoderms, mollusks, and brachyurans in the sedimentary benthic environment of the coastal shelf region, lobster species are conspicuously absent, particularly in subtropical and tropical latitudes. Consequently, lobsters are inhabitants of hard substrates associated with coral reefs, rocky shores, and boulder-strewn bottoms; thus, the presence of a lobster such as *Thenus* spp. on the depositional substrates of the continental shelf is exceptional in itself. Its specialized morphology and behavior make it all the more remarkable. Also significant is the recognition of two species of *Thenus* within Australian populations, and perhaps several more across the entire Indo-West Pacific distribution, all of which were previously considered a single species (Davie 2002). The similarity between the two Australian species in all biological respects suggests a very recent speciation event. Although at a broad level, the two species appear to be allopatric, there is some distributional overlap and consequently an ideal opportunity exists to examine species coexistence and the principles of competitive exclusion, behavioral isolation, and character displacement. Similarly, an opportunity exists for quantitative, *in situ* assessment of habitat/resource utilization in the context of optimal foraging theory.

Despite the morphological similarity of *T. indicus* and *T. orientalis*, examination of tissue chemistry (general proteins only) revealed clear heterogeneity. More intensive molecular assessment of the genus throughout its entire range is justified before confirmed descriptions of all *Thenus* species can be documented.

Despite having a good understanding of *Thenus* spp. biology, there are many aspects which require further clarification. Several aspects of reproduction remain enigmatic. Courtship and the nature of sexual attraction, the processes of copulation, sperm release and deposition, the timing of egg release, and the processes of fertilization, egg deposition and attachment, and larval development during incubation are all unclear at this time.

Documented information on the behavior of lobsters and its adaptive significance has primarily concerned lobsters of the families Nephropidae and Palinuridae. The strength of association between these groups and the Scyllaridae is borne out by the behavioral similarities observed in *Thenus* spp. At a broad level, several aspects of *Thenus* spp. behavior mirrored that described for species of the genera *Scyllarides*, *Panulirus*, *Jasus*, *Nephrops*, and *Homarus*. All genera are nocturnally active, spending much of their active time foraging for food. They are all quiescent and cryptic during daylight hours, although the method of concealment is dependent on the type of environment inhabited. Many aspects of reproductive and molting behavior are also shared between species of these genera. Similarities are not surprising; however, the differences are remarkable. Of these differences, the ability to sustain long bursts of swimming is singular for the Scyllaridae and most significant. Other aspects of *Thenus* spp. behavior are also notable, and more work is needed to determine if such behaviors are representative of or different from the other genera of the Scyllaridae.

The behavior of an animal is an expression of its adaptation to the environment inhabited. The environment of *Thenus* spp. is the shallow coastal benthos, characterized by soft, flat, unconsolidated sediments supporting a high diversity, but relatively low abundance of macro-invertebrate and demersal fish species (Jones & Derbyshire 1988).

According to the definitions of Hughes (1980) and Atema and Cobb (1980), bay lobsters can be considered searchers, which hunt and ambush their prey. Despite the relatively low abundance of food species in their environment, bay lobsters are selective feeders, which actively search out food. Their prey are not always small or easily captured. Swimming provides an efficient and economic means of locomotion that enables large areas of sea floor to be searched in the probability of locating preferred food items. By means of the advanced swimming ability, bay lobsters periodically relocate themselves throughout the nocturnal active period if searching/foraging activity meets with no success, that is, no food. The time spent searching and foraging between each relocation is variable and is probably related to food intake and rate of successful food capture.

Concealment behavior is common to most lobster genera (Atema & Cobb 1980). Its expression is a function of habitat type, but can be broadly divided into active burrowing (most nephropids), or passive sheltering (most palinurids and some scyllarids). Burying behavior displayed by *Thenus* spp. can be considered a modification of these typical concealment behaviors (Bellwood 2002). The mechanisms involved are similar to those described for *Ibacus* spp. (Faulkes 2006) and illustrate the remarkable adaptability of these genera to living in soft substrates.

Ecdysis for *Thenus* spp. involves a pattern of behavior similar to that described for *Panulirus* (Aiken 1980) and presumably, that of other lobsters (Dall 1977). However, the time taken to complete the process of molting is considerably shorter. Although palinurid species may not conceal themselves in dens while molting (Aiken 1980), the complexity of their habitat provides some physical protection. The vulnerability of a molting bay lobster with no habitat protection necessitates relatively rapid ecdysis.

Further, the nature of intraspecific behavior of *Thenus* spp., relative to other lobster species, may be attributable to environmental pressures. Their solitary existence is a product of the low carrying capacity of their environment. This has necessarily reduced the likelihood of intraspecific encounters and precluded the need of innate interactive behavior. The absence of ritualistic interactive behaviors in *Thenus* spp. may be an indication of specialization, as these behaviors would be redundant as bay lobsters adapted to their environment.

Acknowledgments

I thank Drs. Kari Lavalli and Ehud Spanier for their encouragement to contribute to this volume, and their careful revision of the manuscript. The reader will be aware that much of the information contained in the chapter was derived from my Ph.D. studies during the 1980s, and I am grateful for the opportunity to publish information on this fascinating genus which had not previously found its way into the scientific literature. From that perspective, I also acknowledge Drs. Jack Greenwood and Gerry Goeden who supervised the original study, and staff and colleagues from the Department of Primary Industries and Fisheries (Queensland), Northern Fisheries Centre, Cairns who contributed in various ways.

References

Ache, B.W. 1982. Chemoreception and thermoreception. In: Bliss, D.E. (ed.), *The Biology of Crustacea*, Vol. 3: pp. 369–398. New York: Academic Press.

Aiken, D.E. 1980. Molting and growth. In: Cobb, J.S. & Phillips, B.F. (eds.), *The Biology and Management of Lobsters, Vol. 1: Physiology and Behavior*: pp. 91–163. New York: Academic Press.

Aiken, D.E. & Waddy, S.L. 1980. Reproductive biology. In: Cobb, J.S. & Phillips, B.F. (eds.), *The Biology and Management of Lobsters, Vol. 1: Physiology and Behavior*: pp. 216–276. New York: Academic Press.

Atema, J. 1985. Chemoreception in the sea: adaptations of chemoreceptors and behavior to aquatic stimulus conditions. *Soc. Exp. Biol. Symp.* 39: 387–423.

Atema, J. & Cobb, J.S. 1980. Social behavior. In: Cobb, J.S. & Phillips, B.F. (eds.), *The Biology and Management of Lobsters, Vol. 1: Physiology and Behavior*: pp. 409–450. New York: Academic Press.

Atema, J. & Engstrom, D.G. 1971. Sex pheromone in the lobster, *Homarus americanus. Nature (London)* 232: 261–263.

Atkinson, R.J.A. & Naylor, E. 1973. Activity rhythms in some burrowing decapods. *Helgoländer Wissen Meersunters* 24: 192–201.

Barnard, K.H. 1950. Descriptive catalogue of South African Decapod Crustacea (crabs and shrimps). *Ann. South Afr. Mus.* 38: 1–837.

Barshaw, D.E. & Spanier, E. 1994. The undiscovered lobster: A first look at the social behaviour of the Mediterranean slipper lobster, *Scyllarides latus* (Decapoda, Scyllaridae). *Crustaceana* 67: 187–197.

Bellwood, O. 2002. The occurrence, mechanics and significance of burying behavior in crabs (Crustacea: Brachyura). *J. Nat. Hist.* 36: 1223–1238.

Ben-Tuvia, A. 1968. Report on the fisheries investigations of the Israel South Red Sea expedition, 1962. Rep. No. 33. *Bull. Sea Fish. Res. Stn. Isr.* 52: 21–55.

Berg, K., Voigt, R., & Atema, J. 1992. Flicking in the lobster *Homarus americanus*: Recordings from electrodes implanted in antennular segments. *Biol. Bull.* 183: 377–378.

Berry, P.F. 1969. The biology of *Nephrops andamanicus. Invest. Rep. Oceanogr. Res. Inst. Durban S. Afr.* 22: 1–55.

Boone, P.L. 1937. Scientific results of the world cruise of the yacht *Alav*, 1931. *Bull. Vanderbilt Mar. Mus.* 6: 54–71.

Branford, J.R. 1980. Notes on the scyllarid lobster *Thenus orientalis* (Lund, 1793) off the Tokar Delta (Red Sea). *Crustaceana* 38: 221–224.

Bregazzi, P.K. & Naylor, E. 1972. The locomotor activity rhythm of *Talitrus saltator* (Montagu) (Crustacea: Amphipoda). *J. Exp. Biol.* 57: 375–391.

Briones-Fourzan, P. & Conteras-Ortiz, G. 1999. Reproduction of the spiny lobster *Panulirus guttatus* (Decapoda: Palinuridae) on the Caribbean coast of Mexico. *Fish. Bull.* 19: 171–179.

Brown, F.A., Jr. 1958. Studies in the timing mechanism of daily, tidal and lunar periodicities in organisms. In: Buzzati-Traverso, A.A. (ed.), *Perspectives in Marine Biology*: pp. 269–282. Berkley: University of California Press.

Butler, M. 2003. Incorporating ecological process and environmental change into spiny lobster population models using a spatially-explicit, individual-based approach. *Fish. Res.* 65: 63–79.

Chacko, S. 1967. The central nervous system of *Thenus orientalis* (Leach). *Mar. Biol.* 1: 113–117.

Chapman, R.F. 1969. *The Insects: Structure and Function.* London, U.K.: The English Universities Press, Ltd.

Chilton, C. 1910. The Crustacea of the Kermadec Islands. *Trans. Proc. N. Z. Inst. Roy. Soc.* 43: 544–573.

Chittleborough, R.G. 1976. Growth of juvenile *Panulirus longipes cygnus* George on coastal reefs compared with those reared under optimal environmental conditions. *Aust. J. Mar. Freshw. Res.* 27: 279–296.

Cobb, J.S. & Phillips, B.F. 1980. *The Biology and Management of Lobsters.* New York: Academic Press.

Cobb, J.S., Clancy, M., & Booth, J.D. 1997. Recruitment strategies in lobsters and crabs: A comparison. *Mar. Freshw. Res.* 48: 797–806.

Cooper, R.A. & Uzmann, J.R. 1980. Ecology of juvenile and adult *Homarus.* In: Cobb, J.S. & Phillips, B.F. (eds.), *The Biology and Management of Lobsters, Vol. 2: Ecology and Management:* pp. 97–142. New York: Academic Press.

Courtney, A.J. 2002. *The Status of Queensland's Moreton Bay Bug (Thenus spp.) and Balmain Bug (Ibacus spp.) Stocks.* Brisbane, AU: Queensland Government, Department of Primary Industries.

Courtney, A.J., Cosgrove, M.G., & Die, D.J. 2001. Population dynamics of scyllarid lobsters of the genus *Thenus* spp. on the Queensland (Australia) east coast. I. Assessing the effects of tagging. *Fish. Res.* 53: 251–261.

Dall, W. 1977. Review of the physiology of growth and moulting in rock lobsters. *CSIRO Div. Fish. Oceanogr. Circ.* 7: 75–81.

Daniel, P.C. & Derby, C.D. 1991. Mixture suppression in behavior: The antennular flick response in the spiny lobster towards binary odorant mixtures. *Physiol. Behav.* 49: 591–601.

Davie, P.J.F. 2002. *Crustacea: Malacostraca, Ilocarida, Hoplocarida, Eucarida (Part 1).* Zoological Catalogue of Australia 19.3A., Canberra, AU: CSIRO Publishing / Australian Biological Resources Study (ABRS).

Derby, C.D., Steullet, P., Horner, A.J., & Cate, H.S. 2001. The sensory basis of feeding behavior in the Caribbean spiny lobster, *Panulirus argus. Mar. Freshw. Res.* 52: 1339–1350.

Elliot, J.M. 1977. *Some Methods for the Statistical Analysis of Samples of Benthic Invertebrates.* London, U.K.: Freshwater Biological Association.

Evoy, W.H. & Ayers, J. 1982. Locomotion and control of limb movements. In: Bliss, E. (ed.), *The Biology of Crustacea. Vol. 4:* pp. 62–105. New York: Academic Press.

Fabens, A.J. 1965. Properties and fitting of the von Bertalanffy growth curve. *Growth* 29: 265–289.

Factor, J. 1995. Introduction, anatomy, and life history. In Factor, J. (ed.), *The Biology of the Lobster Homarus americanus:* pp. 1–11. New York: Academic Press.

Farmer, A.S.D. 1972. A bilateral gynandromorph of *Nephrops norvegicus* (Decapoda: Nephropidae). *Mar. Biol.* 15: 344–349.

Farmer, A.S.D. 1974. Reproduction in *Nephrops norvegicus* (Decapoda: Nephropidae). *J. Zool.* 174: 161–183.

Farmer, A.S.D. 1975. Synopsis of biological data on the Norway lobster, *Nephrops norvegicus* (Linnaeus, 1758). *F.A.O. Fisheries Synopsis* 112: 1–97.

Faulkes, Z. 2004. Loss of escape responses and giant neurons in the tailflipping circuits of slipper lobsters, *Ibacus* spp. (Decapoda, Palinura, Scyllaridae). *Arthrop. Struct. Develop.* 33: 113–123.

Faulkes, Z. 2006. Digging mechanisms and substrate preferences of shovel nosed lobsters *Ibacus peronii* (Decapoda: Scyllaridae). *J. Crust. Biol.* 26: 69–72.

Fielder, D.R. 1964. The process of fertilization in the spiny lobster *Jasus lalandii* (H. Milne-Edwards). *Trans. R. Soc. South Aust.* 88: 161–166.

Fischer, W. & Bianchi, G. 1984. *FAO Species Identification Sheets for Fishery Purposes.* Western Indian Ocean (Fishing Area 51). Rome: FAO.

Fison, C.S. 1888. *Reports on the Oyster and Other Fisheries within the Ports of Moreton Bay and Maryborough.* Brisbane, AU: Qld Department of Ports and Harbours.

Fogarty, M.J. 1995. Populations, fisheries and management. In: Factor, J.R. (ed.). *The Biology of the Lobster Homarus americanus:* pp. 111–137. New York: Academic Press.

Folk, R.L. 1965. *Petrology of Sedimentary Rocks.* Hemphillis: University of Texas.

George, R.W. 2005. Evolution of life cycles, including migration, in spiny lobsters (Palinuridae). *N. Z. J. Mar. Freshw. Res.* 39: 503–514.

George, R.W. & Griffin, D.J.G. 1972. The shovel nosed lobsters of Australia. *Aust. Nat. Hist.* 40: 227–231.

George, R.W. & Grindley, J.R. 1964. *Projasus* — A new generic name for Parker's crayfish, *Jasus parkeri* Stebbing (Palinuridae: Silentes). *J. R. Soc. West. Aust.* 47: 87–90.

George, R.W. & Morgan, G.R. 1979. Linear growth stages in the rock lobster (*Panulirus versicolor*) as a method for determining size at first physical maturity. *Rap. Prog. Ver Reun., CIEM* 175: 182–185.

Goni, R., Renones, O., & Quetglas, A. 2001. Dynamics of a protected Western Mediterranean population of the European spiny lobster *Palinurus elephas* (Fabricius, 1787) assessed by trap surveys. *Mar. Freshw. Res.* 52: 1577–1587.

Grady, J.R. 1971. The distribution of sediment properties and shrimp catch on two shrimping grounds on the continental shelf of the Gulf of Mexico. *Proc. Gulf Carrib. Fish. Inst.* 23: 129–148.

Herrnkind, W.F. 1980. Spiny lobsters: Patterns of movement. In: Cobb, J.S. & Phillips, B.F. (eds.), *The Biology and Management of Lobsters, Vol. 1: Physiology and Behavior*: pp. 349–407. New York: Academic Press.

Herrnkind, W.F., Vanderwalker, J.A., & Barr, L. 1975. Population dynamics, ecology and behavior of spiny lobsters, *Panurilus argus*, of St. John, US Virgin Islands. (IV) Habitation, patterns of movement and general behavior. *Nat. Hist. Mus. Los Angeles County Sci. Bull.* 20: 31–46.

Heydorn, A.E.F. 1969. The rock lobster of the South African west coast, *Jasus lalandii* (H. Milne-Edwards). 2. Population studies, behaviour, reproduction, moulting, growth and migration. *S. Afr. Div. Sea Fish. Invest. Rep.* 71: 1–52.

Hindley, J.P.R. 1975. Effects of endogenous and some exogenous factors on the activity of the juvenile banana prawn, *Penaeus merguiensis. Mar. Biol.* 29: 1–8.

Holthuis, L.B. 1946. Biological results of the *Snellius* expedition. I. The Stenopodidae, Nephropsidae, Scyllaridae, Palinuridae. *Temminchia (Leiden)* 7: 1–178.

Holthuis, L.B. 1991. *Marine Lobsters of the World. An Annotated and Illustrated Catalogue of the Species of Interest to Fisheries Known to Date.* FAO Species Catalogue No. 125, Vol 13: pp. 1–292. Rome: Food and Agriculture Organization of the United Nations.

Hossain, M.A. 1974. On the squat lobster, *Thenus orientalis* (Lund) off Visakhapatnam (Bay of Bengal). *Curr. Sci.* 44: 161–162.

Hossain, M.A. 1978a. Appearance and development of sexual characters of sand lobster, *Thenus orientalis* (Lund) (Decapoda: Scyllaridae) from Bay of Bengal. *Bangladesh J. Zool.* 6: 31–42.

Hossain, M.A. 1978b. Telson setae and sexual dimorphism of the sand lobster *Thenus orientalis* (Lund). *Curr. Sci.* 47: 644–645.

Hossain, M.A. 1979. On the fecundity of the sand lobster, *Thenus orientalis* from Bay of Bengal. *Bangladesh J. Sci. Res.* 2: 25–32.

Hossain, M.A., Shyamasundari, K., & Hanumantha, R.K. 1975. On the landing of sand lobster, *Thenus orientalis* (Lund). *Seaf. Export J.* 7: 1–6.

Hughes, R.N. 1980. Optimal foraging theory in the marine context. *Oceanogr. Mar. Biol. Ann. Rev.* 18: 423–481.

Isarankura, A.P. 1971. *Assessment of Stocks of Demersal Fish off the West Coasts of Thailand and Malaysia.* Rome: FAO. 20 pp.

Ivanov, B.G. & Krylov, V.V. 1980. Length-weight relationships in some common prawns and lobsters (Macrura, Natantia and Reptantia) from the Western Indian Ocean. *Crustaceana* 38: 279–289.

Jacklyn, P.M. & Ritz, D.A. 1986. Hydrodynamics of swimming in scyllarid lobsters. *J. Exp. Mar. Biol. Ecol.* 101: 85–99.

Jefferies, W.B., Voris, H.K., & Man Yang, C. 1984. Diversity and distribution of the pedunculate barnacle *Octolasmis* Gray, 1825 epizoic on the scyllarid lobster, *Thenus orientalis* (Lund, 1793). *Crustaceana* 46: 300–308.

Johnson, D.H. 1980. The comparison of usage and availability measurements for evaluating resource preference. *Ecology* 61: 65–71.

Jones, C.M. 1984. Development of the bay lobster fishery in Queensland. *Aust. Fish.* 43: 19–21.

Jones, C.M. 1988. *The Biology and Behaviour of Bay Lobsters, Thenus spp. (Decapoda: Scyllaridae) in Northern Queensland, Australia.* Ph.D. dissertation: Brisbane, Australia: University of Queensland. 190 pp.

Jones, C. 1990a. Slipper lobsters in Australia. *Lobster Newslett.* 3(2): 3–4.

Jones, C.M. 1990b. Morphological characteristics of bay lobsters (Decapoda, Scyllaridae) (*Thenus* spp.) from Northern Australia. *Crustaceana* 59: 265–275.

Jones, C.M. 1993. Population structure of two species of *Thenus* (Decapoda: Scyllaridae), in northeastern Australia. *Mar. Ecol. Prog. Ser.* 97: 143–155.

Jones, C.M. & Derbyshire, K.D. 1988. Sampling the demersal fauna from a commercial penaeid prawn fishery off the central Queensland coast. *Mem. Queensland Mus.* 25: 403–415.

Kailola, P.J., Williams, M.J., Stewart, P.C., Reichelt, R.E., McNee, A., & Grieve, C. 1993. *Australian Fisheries Resource.* Canberra, AU: Bureau of Resource Sciences, Department of Primary Industries and Energy.

Kanciruk, P. 1980. Ecology of juvenile and adult palinuridae (spiny lobsters). In: Cobb, J.S. & Phillips, B.F. (eds.), *The Biology and Mangement of Lobsters, Vol. 2: Ecology and Management.* pp. 59–96. New York: Academic Press.

Keenan, C.P. & Shaklee, J.B. 1985. Electrophoretic identification of raw and cooked fish fillets and other marine products. *Food Technol. Aust.* 37: 117–124.

Kizhakudan, J.K., Thirumilu, P., Rajapackiam, S., & Manibal, C. 2004. Captive breeding and seed production of scyllarid lobsters — Opening new vistas in crustacean aquaculture. *Mar. Fish. Infor. Serv. T & E Ser.* 181: 1–4.

Kneipp, I.J. 1974. A Preliminary Study of Reproduction and Development in *Thenus orientalis* (Crustacea: Decapoda: Scyllaridae). BSc Honours thesis: Townsville: James Cook University.

Lara, G.V.R., Cervera, K.C., Mendez, J.C.E., Perez, M.P., Moguel, C.Z., & Ek, F.C. 1998. Spiny lobster (*Panulirus argus*) and queen conch (*Strombus gigas*) density estimation in the central area of Alacranes Reef, Yucatan, Mexico. *Proc. Gulf Carib. Fish. Inst.* 50: 104–127.

Lau, C.J. 1987. Feeding behavior of the Hawaiian slipper lobster, *Scyllarides squammosus*, with a review of decapod crustacean feeding tactics on molluscan prey. *Bull. Mar. Sci.* 41: 378–391.

Lawton, P. & Lavalli, K.L. 1995. Postlarval, juvenile, adolescent, and adult ecology. In: Factor, J.R. (ed.), *The Biology of the Lobster, Homarus americanus*: pp. 47–88. New York: Academic Press.

Lipcius, R.N. & Herrnkind, W.F. 1982. Molt cycle alterations in behavior, feeding and diel rhythms of a decapod crustacean, the spiny lobster, *Panulirus argus*. *Mar. Biol.* 68: 241–252.

Lipcius, R.N. & Herrnkind, W.F. 1987. Control and coordination of reproduction and molting in the spiny lobster *Panulirus argus*. *Mar. Biol.* 96 207–214.

Liu, H.C. 1976. The demersal fish stocks of the waters of north and north-west Australia. *Acta Oceanogr. Taiwan, Sci. Rep. Natl. Taiwan Univ.* 6: 128–134.

Lund, N.T. 1793. Slaegten *Scyllarus* iag hagelser til inskternes historie I. *Skr. Natur. Selsk. Kbh.* 2: 17–22.

Lyle, W.G. & MacDonald, C.D. 1983. Molt stage determination in the Hawaiian spiny lobster *Panulirus marginatus*. *J. Crust. Biol.* 3: 208–216.

Lyons, W.G. 1970. Scyllarid lobsters (Crustacea:Decapoda). *Mem. Hourglass Cruises* 1: 1–74.

Lyons, W.G., Barber, D.G., Foster, S.M., Kennedy, F.S., Jr., & Milano, G.R. 1981. The spiny lobster, *Panulirus argus*, in the middle and upper Florida Keys: Population structure, seasonal dynamics and reproduction. *FL Dep. Nat. Res. Mar. Res. Lab., Mar. Res. Rep.* 38: 38 pp.

MacDiarmid 1989. Moulting and reproduction of the spiny lobster *Jasus edwardsii* (Decapoda: Palinuridae) in northern New Zealand. *Mar. Biol.* 103: 303–310.

MacDiarmid, A.B. & Kittaka, J. 2000. Breeding. In: Phillips, B.F. & Kittaka, J. (eds.). *Spiny Lobster: Fisheries and Culture:* pp. 485–507. Oxford, U.K.: Blackwell Science Ltd.

MacFarlane, J.W. & Moore, R. 1986. Reproduction of the ornate rock lobster *Panulirus ornatus* in Papua New Guinea. *Aust. J. Mar. Freshw. Res.* 37: 55–65.

Malcom, C. 2003. Description of the Setae on the Pereiopods of the Mediterranean Slipper Lobster, *Scyllarides latus*, the Ridged Slipper Lobster, *S. nodifer*, and the Spanish Slipper Lobster, *S. aequinoctialis*. M.Sc. thesis: Texas State University at San Marcos, San Marcos, TX.

Matilda, C.E. & Hill, B.J. 1981. *Annotated Bibliography of the Portunid Crab, Scylla serrata (Forskal).* Brisbane, AU: Queensland Department of Primary Industries.

Matthews, D.C. 1954. A comparative study of the spermatophores of three scyllarid lobsters (*Parribacus antarcticus, Scyllarides squammosus* and *Scyllarus martensii*). *Q. J. Microsc. Sci.* 95: 205–215.

Mauchline, J. 1976. The Hiatt growth diagram for Crustacea. *Mar. Biol.* 35: 79–84.

Mauchline, J. 1977. Growth of shrimps, crabs and lobsters — An assessment. *J. Cons. Int. Expl. Mer* 37: 162–169.

Mellon, D.F. Jr. 1997. Physiological characterization of antennular flicking reflexes in the crayfish. *J. Comp. Physiol.* A180: 553–565.

Mikami, S. & Greenwood, J.G. 1997. Complete development and comparative morphology of larval *Thenus orientalis* and *Thenus* spp. (Decapoda: Scyllaridae) reared in the laboratory. *J. Crust. Biol.* 17: 289–308.

Moore, R. & MacFarlane, J.W. 1984. Migration of the ornate rock lobster *Panulirus ornatus* (Fabricus), in Papua New Guinea. *Aust. J. Mar. Freshw. Res.* 35: 197–212.

Morgan, G.R. 1980. Population dynamics of spiny lobsters. In: Cobb, J.S. & Phillips, B.F. (eds.), *The Biology and Management of Lobsters, Vol. 2: Ecology and Management.* pp. 189–217. New York: Academic Press.

Mulligan, B.E. & Fischer, R.B. 1977. Sounds and behavior of the spiny lobster *Panulirus argus* (Latreille, 1804) (Decapoda, Palinuridae). *Crustaceana* 32: 185–199.

Mutagyera, W.B. 1979. On *Thenus orientalis* and *Metanephrops andamanicus* (Macrura, Scyllaridae and Nephropidae) off Kenya coast. *East Afr. Agric. For. J.* 45: 142–145.

Naylor, E. 2005. Chronobiology: Implications for marine resource exploitation and management. *Sci. Marina* 69: 157–167.

Negrete-Soto, F., Lozano-Alvarez, E., & Briones-Fourzan, P. 2002. Population dynamics of the spiny lobster *Panulirus guttatus* (Latreille) in a coral reef on the Mexican Caribbean. *J. Shellfish Res.* 21: 279–288.

Newman, G.G. & Pollock, D.E. 1971. Biology and migration of rock lobster *Jasus lalandii* and their effect of availability at Elands Bay, South Africa. *Invest. Rep. Div. Sea Fish. S. Afr.* 94: 1–24.

Pauly, D. 1979. *Theory and Management of Tropical Multispecies Stocks. A Review, with Emphasis on the Southeast Asian Demersal Fisheries.* Manila: ICLARM.

Pauly, D. 1983. *Some Simple Methods for the Assessment of Tropical Fish Stocks.* Rome: FAO.

Peacock, N.A. 1974. A study of the spiny lobster fishery of Antigua and Barbuda. *Proc. Gulf Caribb. Fish. Inst.* 26: 117–130.

Perry, L. 1999. Sediment effects on the behavior and survival of young juvenile *Jasus edwardsii. Res. Crust. (Carcinol. Soc. Jap.).* 12: 4–5.

Phillips, B.F. & Kittaka, J. (eds.) 2000. *Spiny Lobsters: Fisheries and Culture.* Oxford, U.K.: Blackwell Science.

Phillips, B.F., Cobb, J.S., & George, R.W. 1980. General biology. In: Cobb, J.S. & Phillips, B.F. (eds.), *Biology and Management of Lobsters, Vol. 1: Physiology and Behavior.* pp. 1–82. New York: Academic Press.

Phillips, B.F., Palmer, M.J., Cruz, R., & Trendall, J.T. 1992. Estimating growth of the spiny lobsters *Panulirus cygnus, P. argus* and *P. ornatus. Aust. J. Mar. Freshw. Res.* 43: 1177–1188.

Prasad, R.R. & Tampi, P.R.S. 1968. On the distribution of palinurid and scyllarid lobsters in the Indian Ocean. *J. Mar. Biol. Assoc. India* 10: 78–87.

Price, R.B. & Ache, B.W. 1977. Peripheral modification of chemosensory information in the spiny lobster. *Comp. Biochem. Physiol.* 57A: 249–253.

Radhakrishnan, E.V. & Vijayakumaran, M. 1998. Observations on the moulting behaviour of the spiny lobster *Panulirus homarus* (Linnaeus). *Indian J. Fish.* 45: 331–338.

Radhakrishnan, E.V., Deshmukh, V.D., Manisseri, M.K., Rajamani, M., Kizhakudan, J.K., & Thangaraja, R. 2005. Status of the major lobster fisheries in India. *N. Z. J. Mar. Freshw. Res.* 39: 723–732.

Rangeley, R.W. & Lawton, P. 1999. Spatial scaling of habitat distributions in the American lobster. *J. Shellfish Res.* 18: 307 (abstr.).

Redant, F. 1987. *Reproduction and Seasonal Behavior of the Norway Lobster, Nephrops norvegicus, in the Central North Sea.* Copenhagen; Denmark: ICES.

Relini, L.O., Zamboni, A., Fiorentino, F., & Massi, D. 1998. Reproductive patterns in Norway lobster *Nephrops norvegicus* (L.), (Crustacea Decapoda Nephropidae) of different Mediterranean areas. *Sci. Mar. (Barcelona)* 62: 25–41.

Rodriguez, G. & Naylor, E. 1972. Behavioural rhythms in littoral prawns. *J. Mar. Biol. Assoc. U. K.* 52: 81–95.

Rulifson, R.A. 1981. Substrate preferences of juvenile penaeid shrimps in estuarine habitats. *Contrib. Mar. Sci.* 24: 35–52.

Rupert, E.E. & Barnes, R.D. 1994. *Invertebrate Zoology. 6th Edition.* Philadelphia, PA: Saunders College.

Rutishauser, R.L., Basu, A.C., Cromarty, S.I., & Kravitz, E.A. 2004. Long-term consequences of agonistic interactions between socially naive juvenile American lobsters (*Homarus americanus*). *Biol. Bull.* 207: 183–187.

Saeger, J., Martosubroto, P., & Pauly, D. 1976. *First Report of the Indonesian–German Demersal Fisheries Project.* (Results of a Trawl Survey in the Sunda Shelf Area. Jakarta: Indonesian – German Demersal Fisheries Project. 46 pp.

Seefried, M. 1983. *The Commercial Fishing Industry of Queensland.* Brisbane, AU: Queensland Department of Primary Industries.

Shindo, S. 1973. *General Review of the Trawl Fishery and the Demersal Fish Stocks of the South China Sea.* Rome: FAO.

Shirota, A. & Ratanachote, A. 1973. *Preliminary Observation on the Distribution and Catch of the Shovel-Nosed Lobster, Thenus orientalis Lund in the South China Sea.* Bagkok, Thailand: SEAFDEC. 8 pp.

Silberbauer, B.I. 1971. The biology of the South African rock lobster *Jasus lalandii* (H. Milne-Edwards). 2. The reproductive organs, mating and fertilization. *Invest. Rep. Div. Sea Fish. South Africa.* 93: 1–46.

Snelgrove, P.V.R. & Butman, C.A. 1994. Animal-sediment relationships revisited: Cause versus effect. *Oceanog. Mar. Biol.* 32: 111–179.

Somers, I.F. 1987. Sediment type as a factor in the distribution of commercial prawn species in the Western Gulf of Carpentaria, Australia. *Aust. J. Mar. Freshw. Res.* 38: 133–149.

Somerton, D.A. 1980. A computer technique for estimating the size of sexual maturity in crabs. *Can. J. Fish. Aqua. Sci.* 37: 1488–1494.

Spanier, E. & Almog-Shtayer, G. 1992. Shelter preferences in the Mediterranean slipper lobster: Effects of physical properties. *J. Exp. Mar. Biol. Ecol.* 164: 103–116.

Spanier, E. & Lavalli, K.L. 1998. Natural history of *Scyllarides latus* (Crustacea: Decapoda): A review of the contemporary biological knowledge of the Mediterranean slipper lobster. *J. Nat. Hist.* 32: 1769–1786.

Spanier, E. & Lavalli, K.L. 2006. *Scyllarides* spp. In: Phillips, B.F. (ed.), *Lobsters: Biology, Management, Aquaculture and Fisheries*: pp. 462–496. Oxford, U.K.: Blackwell Publishing.

Spanier, E. & Weihs, D. 1992. Why do shovel-nosed (slipper) lobsters have shovels? *Lobster Newslett.* 5(1): 8–9.

Spanier, E. & Weihs, D. 2004. Hydrodynamic aspects of locomotion in the Mediterranean slipper lobster, *Scyllarides latus.* In: *Proceedings of the 7th International Conference & Workshop on Lobster Biology and Management.* Tasmania, Australia, February 8–13, 2004, p. 61.

Spanier, E., Tom, M., Pisanty, S., & Almog, G. 1988. Seasonality and shelter selection by the slipper lobster *Scyllarides latus* in the southeastern Mediterranean. *Mar. Ecol. Prog. Ser.* 42: 247–255.

Spanier, E., Weihs, D., & Almog-Shtayer, G. 1991. Swimming of the Mediterranean slipper lobster. *J. Exp. Mar. Biol. Ecol.* 145: 15–31.

Stewart, J. & Kennelly, S.J. 2000. Growth of the scyllarid lobsters *Ibacus peronii* and *I. chacei. Mar. Biol.* 136: 921–930.

Tamm, G.R. & Cobb, J.S. 1978. Behavior and the crustacean molt cycle: Changes in aggression of *Homarus americanus. Science* 200: 79–81.

Thomas, C.W., Carter, C.G., & Crear, B.J. 2003. Feed availability and its relationship to survival, growth, dominance and the agonistic behaviour of the southern rock lobster, *Jasus edwardsii* in captivity. *Aquaculture* 215: 45–65.

Travis, D.F. 1954. The moulting cycle of the spiny lobster *Panulirus argus* (Latreille). I. Moulting and growth in laboratory maintained animals. *Biol. Bull.* 107: 433–450.

Tuck, I.D., Atkinson, R.J.A., & Chapman, C.J. 1994. The structure and seasonal variability in the spatial distribution of *Nephrops norvegicus* burrows. *Ophelia* 40: 13–25.

Tuck, I.D., Chapman, C.J., & Atkinson, R.J.A. 1997. Population biology of the Norway lobster, *Nephrops norvegicus* (L.) in the Firth of Clyde, Scotland — I: Growth and density. *ICES J. Mar. Sci.* 54: 125–135.

von Bertalanffy, L. 1938. A quantitative theory of organic growth. *Hum. Biol.* 10: 181–213.

Williams, A.B. 1958. Substrates as a factor in shrimp distribution. *Limnol. Oceanogr.* 3: 283–290.

Zeldis, J.R. & Jillet, J.B. 1982. Aggregations of pelagic *Munida gregaria* (Fabricius) (Decapoda, Anomura) by coastal fronts and internal waves. *J. Plankton Res.* 4: 839–858.

17

Fishery and Biology of Commercially Exploited Australian Fan Lobsters (Ibacus spp.)

James A. Haddy, John Stewart, and Ken J. Graham

CONTENTS

Abstract

This chapter summarizes the current knowledge of the fisheries and biology of four commercially exploited species of *Ibacus* (*I. alticrenatus* Bate, 1888, *I. brucei* Holthuis, 1977, *I. chacei* Brown & Holthuis, 1998, and *I. peronii* Leach, 1815) in Australia. These species are caught mostly off eastern Australia as by-product of demersal trawling, with annual landings (total for all species) of approximately 200 metric tons. Details are given for reproductive cycles, sizes at maturity, sex ratios, morphological data, mating, egg sizes, brood fecundities, egg-incubation periods, length-frequency distributions, molting, growth, movement, and life-history strategies. As *I. chacei* and *I. peronii* account for more than 95% of all *Ibacus* landed, the majority of current information is centered on these two species. Although there is considerable interspecies variation in relation to specific biological data, there are biological similarities among the four commercial species. Females grow to a larger size and reach sexual maturity at a larger size than males. Spawning occurs during the cooler months, with hatching occurring in spring/summer. Egg-incubation times range from two to four months, before hatching into larvae. Larvae develop through six to eight stages over two to four months before metamorphosing into a transparent nisto larva that does not feed. Nistos subsequently molt into juveniles. Growth rates do not vary until sexual maturity has been reached after four to six molts. The compilation of this chapter has highlighted that more research is required in order to sustainably manage the commercial exploitation of these species, and particular attention should focus on estimating mortality rates for use in egg- and yield-per-recruit modeling.

17.1 Introduction

There are eight recognized species of *Ibacus,* all confined to the Indo-West Pacific region (Holthuis 1991; Brown & Holthuis 1998). Four species (*Ibacus brevipes* Bate, 1888, *I. ciliatus* (Von Siebold, 1824), *I. novemdentatus* Gibbes, 1850, and *I. pubescens* Holthuis, 1960) have broadly tropical distributions between the South China Sea and northern Australia, with *I. novemdentatus* also occurring off east Africa. The remaining species (*I. alticrenatus* Bate, 1888, *I. brucei* Holthuis, 1977, *I. chacei* Brown & Holthuis, 1998, and *I. peronii* Leach, 1815) inhabit the more southern waters of Australia, with *I. alticrenatus* and *I. brucei* also inhabiting waters of New Zealand. A full account of the taxonomy, morphology, and distributions of the eight *Ibacus* species is contained in Brown & Holthuis (1998).

As with *Thenus,* the species of *Ibacus* inhabit relatively soft sandy or muddy substrates where they can be harvested by demersal trawlers, and all but the small deep water *I. brevipes* are commercially fished in at least some areas of their range. They are referred to as "fan lobsters" in the FAO species catalogue of the world's marine lobsters (Holthuis 1991), although in Australia they are generally called "bugs," a term also used locally for *Thenus.* The common name "Balmain bug" originally referred only to *I. peronii,* but has subsequently been adopted as the recommended marketing name for all species of *Ibacus* sold in Australia (Yearsley et al. 1999; Haddy et al. 2005). This account of the biology and fishery of Australian fan lobsters will focus on the four southern species that are commercially exploited off eastern and southeastern Australia, and which have been the subjects of a number of biological and fishery oriented studies (e.g., Stewart et al. 1997; Stewart & Kennelly 1997, 1998, 2000; Stewart 2003; Haddy et al. 2005). Additional abundance, size composition, and biological data were compiled from various exploratory and fishery surveys off Queensland, New South Wales (NSW), and Victoria.

17.1.1   *Ibacus alticrenatus* Bate, 1888

Common names include: deep water bug, white tailed bug, velvet bug, and prawn killer (New Zealand). The white tailed bugs is distributed from northeast Queensland (latitude 20°S), around the southern Australian coastline, to North West Cape (latitude 22°S) in Western Australia (Figure 17.1). It is also found around most of New Zealand, including the Chatham Islands. It has an overall depth range between 80 and 700 m, but is most commonly caught on the upper-continental slope between 200 and 400 m. Maximum size and weight is 65 mm CL (carapace length) and 140 g, respectively. Length–weight relationship (sexes combined): total weight $= 0.0007 \times CL^{2.9365}$ ($R^2 = 0.955; n = 513$).

17.1.2   *Ibacus brucei* Holthuis, 1977

Common names include: Bruce's bug, deepwater bug, and honey bug. Honey bugs are found along Australia's east coast, mainly between central Queensland (latitude 20°S) and central NSW (latitude 32°S) with a few isolated records as far south as latitude 35°S (Figure 17.1). The species has also been recorded from northern New Zealand waters off the Kermadec Islands and on the West Norfolk Ridge. Its overall depth range is between 80 and 560 m, but it is most abundant near the continental shelf break in depths between 150 and 250 m. Maximum size and weight is 74 mm CL and 190 g, respectively. Length–weight relationship (sexes combined): total weight $= 0.0003 \times CL^{3.1712}$($R^2 = 0.982; n = 1544$).

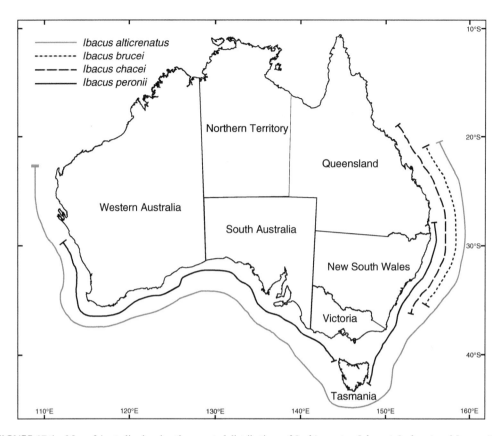

FIGURE 17.1 Map of Australia showing the reported distributions of *I. alticrenatus*, *I. brucei*, *I. chacei*, and *I. peronii*.

17.1.3 *Ibacus chacei* Brown & Holthuis, 1998

Common names include: smooth bug, Balmain bug, and garlic bug (Queensland). The name "garlic bug" alludes to the strong odor that emanates from mature males after capture; a similar odor is also associated with mature male *I. peronii*. This recently described species was previously confused with *I. peronii*, as the two species are similar in size and appearance and have overlapping geographic and depth distributions. However, *I. chacei* is confined to eastern Australia between northern Queensland (latitude 17°S) and southern NSW (latitude 36°S), although it is rarely caught south of Sydney (latitude 34°S) (Figure 17.1). Its overall depth range is between 20 and 330 m, but it is most abundant on the mid-continental shelf between 50 and 150 m. Maximum size and weight is 83 mm CL and 325 g, respectively. Length–weight relationship (sexes combined): total weight = $0.0004 \times \text{CL}^{3.0702} (R^2 = 0.992; n = 7597)$.

17.1.4 *Ibacus peronii* Leach, 1815

Common name: Balmain bug. This species is the best known of the Australian fan lobsters, and is widely distributed around the southern half of the continent from about the Queensland–NSW border (latitude 28°S) to central Western Australia (latitude 29°S), including the east coast of Tasmania and Bass Strait (Figure 17.1). The species is reported from southeast Queensland (Brown & Holthuis 1998), but none were found in trawl samples during recent studies (Courtney 2002). With an overall depth range of 4 to 288 m, *I. peronii* has the shallowest minimum depth reported for any *Ibacus* spp. and is mainly found close to shore in waters <80 m deep. Maximum size and weight is 89 mm CL and 375 g, respectively. Length–weight relationship (sexes combined): total weight = $0.0004 \times \text{CL}^{3.0886} (R^2 = 0.995; n = 425)$.

TABLE 17.1

Estimated Average Annual Landings (metric tons) of *Ibacus* spp. from Eastern and Southern Australia

	I. alticrenatus	*I. brucei*	*I. chacei*	*I. peronii*	**Total**
Queensland	0.5	1.5	110.0	0	112.0
NSW (north)	1.5	1.0	49.0	15.0	66.5
NSW (south)	1.0	0	0	3.5	4.5
Victoria	2.0	0	0	18.0	20.0
South Australia	0.5	0	0	2.5	3.0
Total	5.5	2.5	159.0	39.0	206.0

Derived from reported landings 2000 to 2004 from commercial catch databases of Queensland, New South Wales, Victoria, and South Australian fisheries agencies.

17.2 The Australian Fan Lobster Fishery

The annual total catch of *Ibacus* (all species) in Australia is approximately 200 metric tons and is valued in excess of AUD\$2 million at the first point of sale. Commercial catch statistics do not differentiate among the species, but broad categories are available for some regions. For example, catch data collected by the New South Wales Department of Primary Industries divides the reported catches into Balmain bugs (*I. chacei* and *I. peronii*) and the smaller "deepwater bugs" (*I. brucei* and *I. alticrenatus*). This information, together with research survey data and port and market observations, has been used to estimate the species composition of the landed catch (Table 17.1).

More than 85% of the Australian catch is harvested from southern Queensland and northern NSW waters (Table 17.1), and is taken almost totally as by-product in the eastern king prawn (*Penaeus plebejus* Hess, 1865) trawl fishery. This fishery encompasses continental shelf waters (20 to 200 m depth) from central Queensland to central NSW with more than 400 multirigged trawlers (12 to 20 m in length) targeting prawns at night (Kailola et al. 1993). Typical nightly catches for individual trawlers are between 50 and 100 kg of prawns, 5 to 10 kg of Balmain bugs, plus other assorted by-product. More than 90% of the bug catch in the king prawn fishery is *I. chacei*. Small, but regular, quantities of *I. peronii* are landed by prawn trawlers working inshore grounds off northern and central NSW, and from fish trawlers and seiners operating off southern NSW and Victoria. Again, daily catches are usually <10 kg. The relatively small total catch of *I. brucei* is caught in combination with king prawns, usually during winter months on outer-continental shelf grounds, while *I. alticrenatus* are taken on upper-continental slope grounds by prawn trawlers or fish trawlers targeting mixed catches.

Although *Ibacus* are mainly by-product species, increased consumer awareness, expanding markets and price increases have resulted in fishers specifically targeting these species (Haddy et al. 2005). Catch rates of the inshore species of *Ibacus* occasionally exceed 50 kg per night on grounds that have been unfished for some time, but fishers report that catches decline quickly to less than 10 kg per night. Fishers also report that catch rates of the two deep water species fluctuate widely and are largely unpredictable. Although catch rates as high as 50 kg/h (*I. brucei*) and 60 kg/h (*I. alticrenatus*) were achieved during exploratory trawling on unfished grounds off northern NSW (Gorman & Graham 1978), current commercial catch rates are normally around 5 to 10 kg/h. During research trawling on the southern NSW upper-continental slope in 1999 to 2001, catch rates of *I. alticrenatus* in 300 to 420 m ranged between zero and 50 kg/h (i.e. 0 to 1000 individuals per hour), but the average was less than 6 kg/h ($\bar{x} = 5.6 \pm 9.8; n = 44$) (Graham, unpublished data). Off western Victoria during the same period, *I. alticrenatus* was caught between 200 and 500 m with peak catch rates (12 to 36 kg/h) in 300 to 360 m ($\bar{x} = 23.2 \pm 7.6; n = 9$) (Graham, unpublished data).

As is the case with many shallow-water crustaceans, *I. chacei* and *I. peronii* appear to be largely inactive during daytime when they become less vulnerable to trawls, probably by burying in the seabed. During fishery surveys on northern NSW king prawn grounds, the mean catch number of *I. chacei* per 30 min tow was 18.9 ± 14.2 ($n = 44$) at night, but only 0.9 ± 1.0 for daytime tows ($n = 8$) (Graham et al. 1993a, 1993b). Survey catches of *I. peronii* on southern NSW grounds indicated similar behavior, with the average number per 60 min night-time tow (9.3 ± 7.8; $n = 96$) being almost double that from daytime tows (4.9 ± 4.4; $n = 96$) (Graham et al. 1995, 1996). In contrast, the deeper water *I. alticrenatus* and *I. brucei* do not exhibit such marked nocturnal behavior. The high catch rates for both species during exploratory trawling off northern NSW (60 and 50 kg/h; Gorman & Graham 1978) were recorded at night, whereas all catches of *I. alticrenatus* off southern NSW and western Victoria (up to 50 kg/h) were taken during daylight hours (Graham, unpublished data). In addition, daytime catch rates of *I. brucei* off northern NSW were as high as 15 kg/h (Gorman & Graham 1979).

The flattened morphology and burying behavior of *Ibacus* suggest that their vulnerability to trawl gear is probably low. However, in the king prawn fishery, most grounds are intensively trawled for much of each year and fishing pressure on *Ibacus* is continually high. The small-meshed (35 to 45 mm stretched mesh) prawn trawls employed across most of the fishery retain all size classes of bugs including, at times, large numbers of small juveniles (Graham et al. 1993a, 1993b; Graham & Wood 1997). In contrast, the proportion of juveniles to adults caught in larger-meshed fish trawls (minimum 90 mm stretched mesh) is relatively small (Graham et al. 1995, 1996) probably because the ground-ropes of the nets are not designed to fish as close to the seabed as prawn trawls and, even if caught, small bugs can readily fall through the large meshes in the front of fish trawls.

As *Ibacus* spp. are a by-product of demersal trawling, limiting fishing effort and changing mesh sizes are not appropriate management strategies. However, as discard mortality of trawl caught crustaceans is relatively low compared to fish (Hill & Wassenberg 1990), retention limits are appropriate management measures. Therefore to conserve stocks, fishery management regulations ban the retention of egg-bearing females of all species of *Ibacus*. Minimum legal sizes of 100 mm carapace width (CW, equivalent to approximately 50 to 55 mm CL) exist for *I. peronii* and *I. chacei* in NSW, but there are no minimum-size regulations for the other species in this state. A minimum legal size of 90 mm CW exists for *I. peronii* in Victoria, and a minimum CW of 100 mm CW exists for all species of *Ibacus* in Queensland. However, this "one-size limit fits all" approach in Queensland is inappropriate for the smaller species and is under review (Haddy et al. 2005). Although by-catch reduction devices in trawls are also mandatory in the NSW and Queensland prawn fisheries (Eayrs et al. 1997; Broadhurst 2000), these devices release few, if any, small bugs.

Although an increasing proportion of the catch from Queensland is exported, there is a strong market for bugs in Australia and the bulk of the catch is consumed domestically. The east coast prawn fleet typically lands its crustacean catch of prawns, bugs, and crabs already cooked (boiled), whereas the relatively small catches taken by southern fish trawlers are landed alive (green). However, wholesale market prices are more dependent on size rather than process. Sydney Fish Market auction sales data for 2004 and 2005 show that 77 metric tons of cooked *Balmain bugs* (*I. chacei* and *I. peronii*) averaged AUD\$12.18/kg, while 8 metric tons of green Balmain bugs sold for the almost identical price of AUD\$12.11/kg. During the same period, the market price for the smaller "deepsea bug" (*I. alticrenatus* and *I. brucei*) was substantially lower, averaging AUD\$6.87 per kg for 2.7 metric tons of product.

17.3 Reproductive Biology

17.3.1 General Description, Structure, and Development of Gonads

The internal and external reproductive morphologies of *Ibacus* spp. follow the basic descriptions for all decapod lobsters (Aiken & Waddy 1980) in that they are dioecious, possess the typical H-shaped gonads, and are easily sexed. Females possess paired genital openings on the coxae of the third pair of pereiopods, their fifth pereiopods are chelate, and their pleopods are enlarged with the endopod of the first pleopod having a large, broad, leaf-like appearance. The pleopods of mature females also bear long setae. In contrast, the pleopods of male *Ibacus* are relatively small, and the genital openings are located on the coxae of the fifth pair of pereiopods. The ovaries and testes are situated dorsal to the alimentary tract and

TABLE 17.2

Macroscopic and Histological Descriptions of Ovarian Development in *Ibacus* Spp.

Stage	Macroscopic Description	Histological Descriptions
1. Immature	Ovaries clear to white, small, straight and narrow. Individual oocytes not visible.	All oocytes in ovaries are previtellogenic and uniform in size. Maximum oocyte diameter is 0.2 mm.
2. Immature/regressed	Ovaries cream to yellow and small. Individual oocytes not visible.	Oocytes in ovaries display an increasing size gradient with the germinal strand and radiating mass of previtellogenic oocytes located in the center and larger vitellogenic oocytes surrounding the outer margin of the ovary. Approximately 30% of oocytes are vitellogenic with a maximum oocyte diameter of 0.44 mm.
3. Maturing	Ovaries yellow to orange enlarged throughout their length, but not convoluted. Individual oocytes are just visible through the ovary wall.	Previtellogenic oocytes restricted to the germinal strand. Approximately 80% of oocytes vitellogenic with a maximum oocyte diameter of 0.58 mm.
4. Mature	Ovaries bright orange, swollen, and convoluted filling all available space in the cephalothoracic region. Individual oocytes are clearly visible through the ovary wall.	All visible oocytes vitellogenic with a maximum oocyte diameter of 0.87 mm.
5. Spent	Ovaries cream to yellow/orange, large but not convoluted with flaccid and granular appearance. A few residual oocytes can sometimes be seen through the ovary wall.	Residual and atretic oocytes present, with a maximum oocyte diameter of 0.32 mm. Ovarian wall thick and contracted.

extend from the level of the eyes back to the abdomen. They consist of paired tubes joined by a transverse bridge located approximately one third of their length from the anterior end. The oviduct in females and the vas deferens in males pass from just posterior of the transverse bridge to the genital apertures. The proximal vas deferens is white, highly convoluted, and widens into a straight distal section that leads to the genital opening.

Female reproductive development has been classified into five ovarian stages that are easily assessed through the application of macroscopic descriptions (Table 17.2). Stewart et al. (1997) verified these macroscopic descriptions with histological examinations of *I. peronii* and *I. chacei* ovaries (Figure 17.2). Their results highlight that ovarian development is synchronous and mean oocyte size progressively increases with increasing maturity stage until batch spawning and ovipositioning occurs.

Limited work has been done on male reproduction; however, maturity can be classified into three categories based on the macroscopic appearance of the vas deferens: (1) immature: vas deferens thin, threadlike, and clear; (2) maturing: vas deferens slightly enlarged and clear to slightly opaque; and; (3) mature: vas deferens distended, swollen, and opaque white to yellowish. Stewart et al. (1997) assessed male *I. peronii* to be mature by the presence of spermatophores in histological sections of the vas deferens.

17.3.2 Temporal and Spatial Trends in Reproduction

The location and timing of spawning is best described using the proportion of ovigerous females in catches and/or seasonal variation in gonadosomatic indices (GSI). Table 17.3 details the seasonal occurrence and peak abundance of ovigerous *Ibacus* species inhabiting Australian waters. In the three most studied species, *I. brucei*, *I. chacei*, and *I. peronii*, all display repeatable cycles of reproduction. However, the location, onset, and duration of reproductive activity displays marked interspecific variability. For example, ovigerous *I. peronii* are present all year off the NSW coast, although ovigerous females are

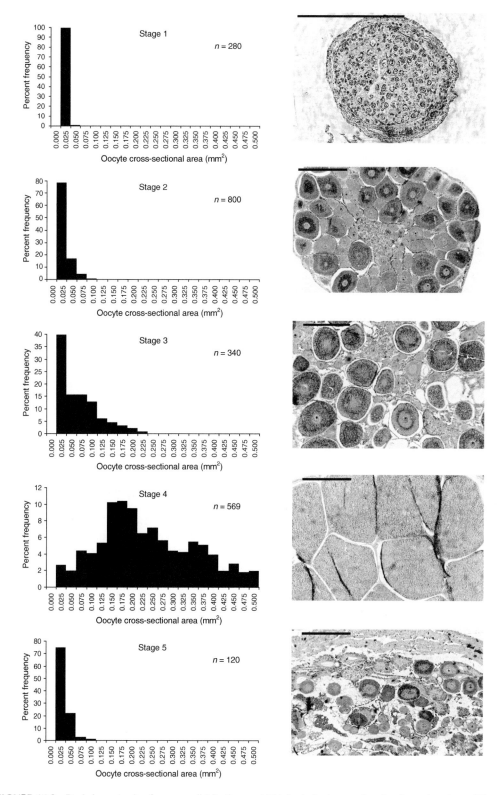

FIGURE 17.2 Pooled oocyte size frequency distributions and histological micrographs of each ovarian stage of *Ibacus peronii*. Scale bars = 0.5 mm. (Modified from Stewart, J., Kennelly, S.J., & Hoegh-Guldberg, O. 1997. *Crustaceana* 70: 344–367; used with permission of Brill Academic Publishers.)

TABLE 17.3

Summaries of Biological and Life-History Characteristics of Commercial Australian Fan Lobsters from the East Coast of Australia

Species	Ibacus alticrenatus	Ibacus brucei	Ibacus chacei	Ibacus peronii
Male maximum size (CL mm)	55	65	73	79
Female maximum size (CL mm)	65	74	83	89
CL–TW relationship	$TW=0.0007 \times CL^{2.937}$	$TW=0.0003 \times CL^{3.171}$	$TW=0.0004 \times CL^{3.070}$	$TW=0.0004 \times CL^{3.089}$
Female PM L_{50}	39	47	55	50
Female PM (% CL_{max})	60.0	63.5	66.3	56.2
Egg size (mm diameter)	0.94–1.29	0.73–1.01	1.02–1.37	0.97–1.45
Fecundity range ($\times 10^3$)	1.7–14.8	2.0–61.3	2.1-28.8	5.5–36.7
CL–BF relationship ($\times 10^3$)	$BF=0.00005 \times CL^{3.1315}$	$BF=1.3354 \times CL-50.882$	$BF=0.8849 \times CL-43.405$	$BF=0.943 \times CL-45.296$
Ovigerous season (peak)	Apr–Oct (Jul)	Apr–Aug (May)	Jun–Jan (Sep)	All year (Jul)
Egg-incubation period (months)	3–4	2–3	3–4	3–4
Larval duration (months)	4–6	—	—	3
Number of phyllosomal stages	7	—	—	6

CL = carapace length (mm); TW = total weight (g); PM = physiological maturity (CL mm); BF = brood fecundity. Data obtained from Haddy et al. 2005, Stewart et al. 1997, and Stewart & Kennelly 1998.

most abundant in July to September (winter/early spring), with a secondary peak in abundance during December. Similarly, the presence of ovigerous *I. chacei* off the Queensland coast is also prolonged (June to January), with a peak abundance period between July and September and a minor increase in December. In contrast, catches of ovigerous *I. brucei* off Queensland are restricted to the months of April to August, with peak abundance in May. Despite these differences, the general trend is for ovipositioning to occur during the colder months (late autumn to early spring; May to September) with hatching occurring during the spring/summer period.

Seasonal changes in GSI values and proportions of gonad stages of female *I. chacei* indicate that ovarian recrudescence is relatively rapid, with GSI values and the proportion of females possessing mature ovaries rapidly increasing within a one to two month period between February and April (Figure 17.3). The coordination of vitellogenic activity with environmental variables is poorly understood in *Ibacus*. However, it is possible that a decreasing photoperiod (i.e., after the summer solstice) may initiate this activity when water temperatures are still relatively high. The onset of vitellogenic activity in *I. chacei* is further defined geographically, with the southern boundary of vitellogenesis occurring between 26°S and 27°S (Haddy et al. 2005). The mechanisms controlling this southern limit for breeding are unknown, as mature-sized females commonly occur as far south as Sydney (34°S). In contrast, vitellogenesis does not appear to be geographically restricted for the other three species, with mature and ovigerous females occurring throughout their distributions.

The presence of mature ovaries in ovigerous females and a secondary ovigerous abundance peak late in the season in *I. chacei* and *I. peronii* suggest that some individuals spawn more than once a year. Haddy et al. (2005) noted that the majority of *I. chacei* undergoing successive ovarian maturation was caught early in the spawning season, and that this occurred across a wide range of sizes (55 to 72 mm CL). In

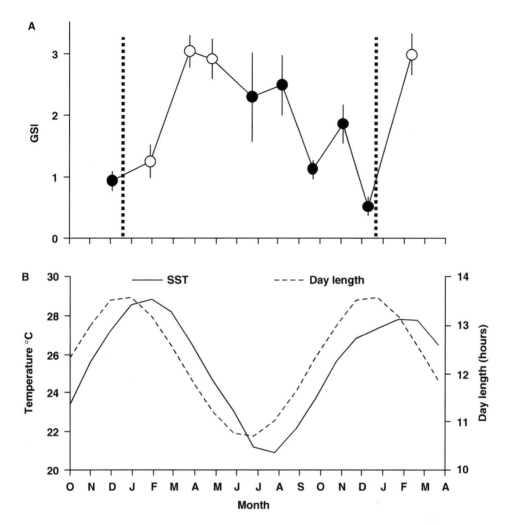

FIGURE 17.3 Seasonal changes in (A) mean gonadosomatic index (GSI ±S.E.) from adult female *Ibacus chacei* (>54 mm CL), and; (B) sea surface temperature (SST) and daylength, off Central–Southern Queensland from October 2001 to March 2003. ● indicates samples where ovigerous females were present. Dotted lines indicate the timing of the summer solstice. (From Haddy, J.A., Courtney, A.J., & Roy, D.P. 2005. *J. Crust. Biol.* 25: 263–273; used with permission.)

contrast, there is currently no evidence to suggest that *I. alticrenatus* or *I. brucei* produce multiple broods, as all ovigerous females examined have possessed reproductively inactive ovaries (Stages 2 and 5).

Few studies have examined the seasonal reproductive biology of male *Ibacus*. However, Stewart et al. (1997) found spermatophores in the vas deferens of *I. peronii* from NSW and Victoria throughout the year. They concluded from this that males are reproductively active throughout the year with spermatogenesis occurring continuously after maturation.

17.3.3 Sex Ratio, Size at Maturity, Mating, Egg Size, Fecundity, and Incubation Periods

The sex ratios of the four commercially caught species of *Ibacus* off eastern Australia are approximately 1:1. Sex ratios (female to male) have been reported as being 1:1.28 for *I. alticrenatus*, 1:1.05 for *I. brucei*, 1:1.07 for *I. chacei*, and 1:1.07 for *I. peronii*. (Stewart 1999; Haddy et al. 2005).

The size at sexual maturity of *Ibacus* can be assessed at two levels: (1) physiological maturity — through assessments of gonadal development; and (2) functional maturity — through observations of the sizes of ovigerous females and the presence of well-developed setae (≥10 mm long, see Stewart et al.,

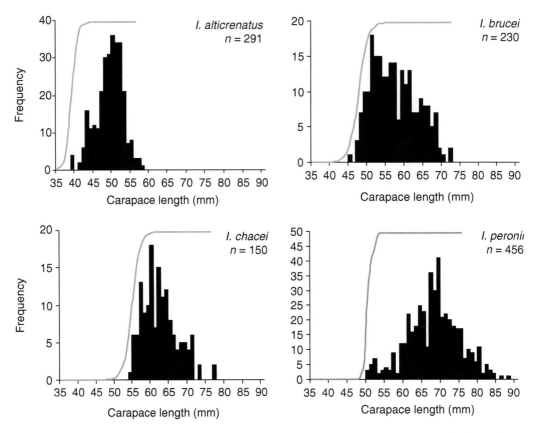

FIGURE 17.4 Length-frequency distributions of ovigerous females and logistic size at maturity curves for *Ibacus alticrenatus*, *I. brucei*, *I. chacei*, and *I. peronii*. Data pooled from Haddy, J.A., Courtney, A.J., & Roy, D.P. 2005. *J. Crust. Biol.* 25: 263–273, Graham (unpublished data), and Stewart (unpublished data); used with permission.

1997) on pleopods (Figure 17.4). Females initiate ovarian maturation at carapace lengths of 38 mm for *I. alticrenatus*, 44 mm for *I. brucei*, 50 mm for *I. chacei*, and 48 mm for *I. peronii*, and attain 50% physiological maturity at carapace lengths of 39, 47, 55, and 50 mm, respectively. However, there is currently some concern that functional maturity may not be attained until the molt that occurs after reaching physiological maturity. The estimated sizes at physiological maturity range between 56.2 and 66.3% of the maximum reported sizes (Table 17.3). Stewart et al. (1997) investigated the potential for geographical variation in the size at maturity of *I. peronii* from Victoria and northern NSW, but found no difference in the size at which ovarian maturity occurred at these locations. Although very limited data is available on males, they typically mature at smaller sizes than females, with male maturity in *I. alticrenatus*, *I. brucei*, *I. chacei*, and *I. peronii* occurring as small as 37, 43, 47, and 38 mm CL, respectively (Haddy & Graham, unpublished data).

Mating has not been observed in *Ibacus*, but observations of *I. chacei* and *I. peronii* in the field and in aquaria allow some insight into mating mechanics. Molting is not a prerequisite; thus, mating occurs when the female is hard-shelled (Stewart et al. 1997; Haddy et al. 2005). Spermatophoric masses are persistent, gelatinous, an opaque white color, and are deposited as two elongated strips, approximately 20 to 30 mm long, close to the genital openings (Stewart et al. 1997; Haddy et al. 2005). It is thought that fertilization occurs externally and relatively soon after mating while the eggs are transported from the genital apertures to the setae.

As for many decapod crustaceans, the eggs on ovigerous females can be classified into three main categories based on color and the stage of development of the larvae inside the egg. Newly deposited (early stage) eggs are spherical and bright orange. During development, the eggs progressively increase in size and develop two black eyespots (mid-stage). As hatching approaches, the eggs turn a clear-brown color and the two eyespots enlarge (late stage). During this development, the size of the egg progressively increases (Figure 17.5). Brood fecundity is highly variable both within and among the four species, ranging between ~ 1700 to 14,800 eggs for *I. alticrenatus*, 2000 to 61,300 eggs for *I. brucei*, 2100 to 28,800 eggs for *I. chacei*, and 5500 to 36,700 eggs for *I. peronii* (Stewart & Kennelly 1997; Haddy et al. 2005). The brood fecundity–carapace length (BF–CL) relationships for *I. brucei*, *I. chacei*, and *I. peronii* are positive and linear, while in *I. alticrenatus*, this relationship is slightly curvilinear (Figure 17.6; Table 17.3). Although the egg-carrying capacity of *Ibacus* is size dependent, considerable interspecific variation exists. The main reasons for these differences relate to life history strategies and the sizes of eggs produced (Stewart & Kennelly 1997; Haddy et al. 2005). For example *I. brucei* is a medium-sized bug that produces a small egg (0.73 to 1.01 mm) and, as a result, the brood fecundity of *I. brucei* is relatively high. In

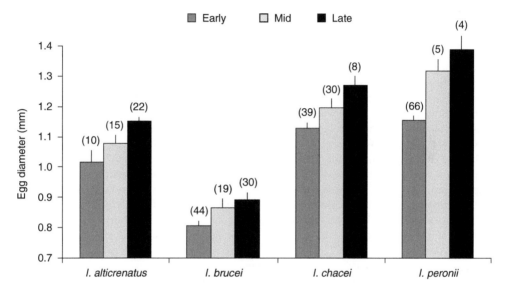

FIGURE 17.5 Mean egg size (±95% confidence intervals) for each egg-development stage for *Ibacus alticrenatus*, *I. brucei*, *I. chacei*, and *I. peronii*. (*n* values are given in parenthesis) (From Haddy, J.A., Courtney, A.J., & Roy, D.P. 2005. *J. Crust. Biol.* 25: 263–273, and Stewart (unpublished data); used with permission.)

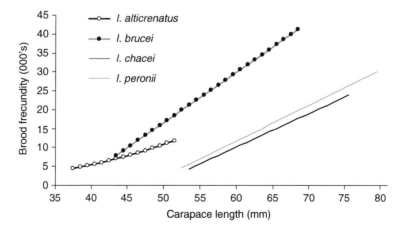

FIGURE 17.6 Brood fecundity–carapace length regression relationships for *Ibacus alticrenatus*, *I. brucei*, *I. chacei*, and *I. peronii*. (Data compiled from Haddy, J.A., Courtney, A.J., & Roy, D.P. 2005. *J. Crust. Biol.* 25: 263–273, and Stewart, J. & Kennelly, S.J. 1997. *Crustaceana* 70: 191–197; used with permission of Brill Academic Publishers.)

contrast, *I. alticrenatus, I. chacei,* and *I. peronii* produce relatively large eggs (1.0 to 1.4 mm in diameter) and possess a relatively lower fecundity than *I. brucei.*

Observations of brood fecundity in relation to egg development show that individuals with early-stage eggs generally possess higher brood fecundities than similarly sized individuals with mid-or late-stage eggs, indicating that egg losses occur during incubation (Haddy et al. 2005). However, the degree of egg loss during the egg-incubation period is currently unknown. Egg incubation times have been estimated to vary between approximately two to three months for *I. brucei* and three to four months for *I. alticrenatus, I. chacei,* and *I. peronii.* Incubation periods within species are likely to be temperature dependent, and there is some evidence that *I. peronii* from cooler Victorian waters have longer incubation periods than those in NSW (Stewart 1999).

17.4   Molting and Growth

Molting and growth of *Ibacus* has been studied using tag-recapture data for *I. peronii* and *I. chacei* (Stewart & Kennelly 2000), modal progression analysis for *I. chacei* (Haddy et al. 2005), and larval and juvenile rearing for *I. peronii* (Marinovic et al. 1994). Studies on the larval development of *Ibacus* spp. indicate that they undergo six to eight phyllosoma stages during two to four months before metamorphosing into a postlarval stage called a nisto (Ritz & Thomas 1973; Takahashi & Saisho 1978; Phillips et al. 1981; Atkinson & Boustead 1982; Marinovic et al. 1994). Observations of *I. peronii* showed that larvae hatch as first-stage phyllosomas that unfold their swimming legs after 20 min to become actively swimming phyllosomas that are strongly attracted to light (Stewart et al. 1997). Larval *I. peronii* have been reared through six phyllosomal instars during 71 to 97 days (mean = 79 days), before molting into a nonfeeding, transparent nisto (Marinovic et al. 1994). During this stage they become increasingly pigmented before molting into juveniles after 17 to 24 days (mean = 19 days), at an average size of 12.2 mm CL (Marinovic et al. 1994). Similarly, Atkinson & Boustead (1982) reported that late-stage *I. alticrenatus* phyllosomas (molt 7) metamorphosed into the nisto stage before molting into juveniles (14.7 mm CL) after 25 to 29 days. Marinovic et al. (1994) also reported that juvenile *I. peronii* in captivity molted after 40 days and increased in size to approximately 17 mm CL, a size close to the smallest mode apparent in length-frequency distributions of *I. peronii* retained in trawl nets (Figure 17.7).

Growth in *Ibacus* is dependent on molt frequency and molt increment. Strong modes in the length-frequency distributions of *Ibacus* (Figure 17.7) suggest that after their size at first capture by trawl gear (recruiting to the fishery), *I. peronii* and *I. chacei* each molt five more times before attaining physiological maturity (50 to 55 mm CL). These five molts occur during a period of approximately two years. In contrast, *I. alticrenatus* appears to molt three times and *I. brucei* four times after being recruited to the fishery before attaining their sizes at sexual maturity (39 and 47 mm CL, respectively). Molt frequencies of captive *I. peronii* in NSW suggest that after maturation, molting is a seasonal event (October to January) with males molting slightly earlier than females. This is also supported by tag-recapture data in which 97% of recaptured animals that were at liberty between October and January in any year had molted (Stewart & Kennelly 2000). Molting, however, may not necessarily occur annually, as five recaptured tagged males were at liberty for more than two years and one was at liberty for more than three years without having increased in size. Thirteen recaptured females were at liberty for more than one year without increasing in size, but none for more than two years. It is assumed that animals that did not increase in size had not molted, as none had indentations in their carapace margins where the tag penetrated. Such indentations were commonly observed in animals that had molted.

Molt increments for *I. brucei, I. chacei,* and *I. peronii* have been determined from modal peaks in pooled length-frequency distributions (Haddy et al. 2005) and tag-recapture data (Stewart & Kennelly 2000). Results indicate that early juvenile growth in *Ibacus* Spp. is characterized by relatively large molt increments (20 to 35% of their premolt size) and frequent molting. However in larger, mature individuals, both the relative size of the molt increments (11 to 15%) and the molt frequency is significantly lower. The mirroring of modal size classes in pooled length frequency distributions between sexes (Figure 17.7) suggests that there are no differences in molt increments between sexes at immature sizes in all four commercial species. Stewart & Kennelly (2000), using tag-recapture data, also found no differences in

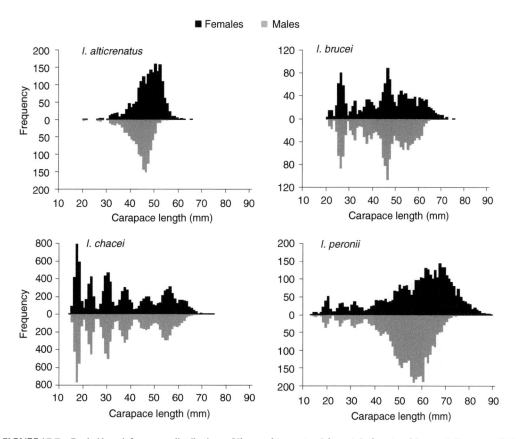

FIGURE 17.7 Pooled length frequency distributions of *Ibacus alticrenatus*, *I. brucei*, *I. chacei*, and *I. peronii*. Data compiled from numerous length frequency databases from the New South Wales Department of Primary Industries and the Queensland Department of Primary Industries. Note variable Y-axis ranges.

molt increments between sexes at all sizes of *I. peronii*, but females attained larger sizes than males because they molted more frequently. These findings suggest that for the four species of *Ibacus*, early growth follows a similar pattern, with few differences in growth between sexes, and each having equal molt increments and molt intervals. It seems likely that after males attain sexual maturity, their molting becomes less frequent than females, resulting in females attaining larger sizes.

The growth models for *I. peronii*, determined by Stewart & Kennelly (2000), suggest the potential for this species to live for more than 15 years in NSW waters. This was recently confirmed by several long-term recaptures of tagged *I. peronii* (up to 3417 days at liberty) that have supported the predicted growth model (Stewart 2003). There is also some evidence that *I. peronii* from Victoria do not live as long as those in NSW, with the longest time at liberty of tagged animals being only 846 days and no long-term recaptures reported. The growth model also suggested that the maximum size of *I. peronii* in Victoria was attained after eight years. Similarly *I. chacei* reaches its recorded maximum size in five to seven years (Stewart & Kennelly 2000; Haddy et al. 2005).

17.5 Movements

Contrasting patterns of movement have been described for *Ibacus peronii* and *I. chacei* off the east coast of Australia. Stewart & Kennelly (1998) tagged 3892 *I. peronii* and 716 *I. chacei* using standard plastic T-bar tags (Hallprint Pty Ltd). A total of 557 *I. peronii* (14.3%) were recaptured and demonstrated a nomadic pattern of movement, with individuals being recaptured close to their place of release (mean of 0.35 km)

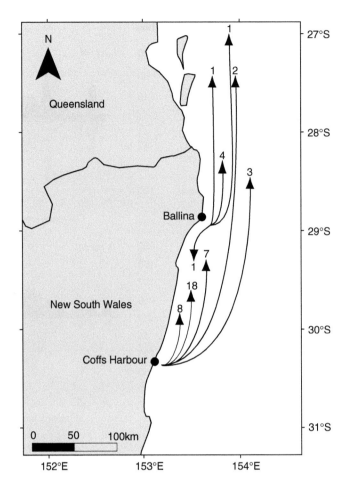

FIGURE 17.8 Summary of tagged *Ibacus chacei* that were recaptured more than 10 km from their release locations. Numbers are the number of animals represented by each arrow. (Modified and reprinted from Stewart, J. & Kennelly, S.J. 1998. *Fish. Res.* 36: 127–132, with permission from Elsevier.)

and with no specific directional movement. In contrast, recaptured *I. chacei* (94 animals or 13.1% of those tagged), demonstrated a clear northward migration from their release locations, traveling up to 310 km in 655 days (Figure 17.8). This northward movement was observed across a wide size range (30 to 70 mm CL) and was consistent throughout the year from all release locations. The average minimum rate of movement (calculated as the shortest distance between release and recapture location divided by time at liberty) was 0.15 km/day, but ranged up to 0.71 km/day. To date *I. chacei* is the only member of the Scyllaridae known to undertake such extensive migrations.

17.6   Comparisons of Life-History Strategies and Recruitment

The different patterns of movement exhibited by *Ibacus peronii* and *I. chacei* provide some insight into their life-history strategies. These two species overlap in their distributions, but are generally separated by depth, with *I. peronii* mainly inhabiting waters up to 80 m deep and *I. chacei* mainly inhabiting waters from 50 to 150 m. The deeper waters inhabited by *I. chacei* along the center of the continental shelf are strongly influenced by the south flowing, East Australian Current (EAC). This current is at its strongest in summer, often exceeding three knots across the shelf and five knots further offshore. Currents close inshore are generally much weaker (<1 knot) and more variable in direction, frequently flowing northward during winter (Cresswell 1974, 1984). The small localized movement pattern of *I. peronii* in these inshore waters could be expected to facilitate localized larval retention and settlement. Catches of small (<20 mm CL) juvenile *I. peronii* off NSW were taken in waters 10 to 50 m deep (November to December) (Graham & Wood 1997). In contrast, because *I. chacei* inhabits depths strongly influenced by the EAC, it is hypothesized that this species has evolved an "upstream" migratory behavior almost identical to the eastern king prawn (Montgomery 1990). *Ibacus chacei* migrate north of latitude 27°S to reach locations where vitellogenesis and spawning occurs, facilitating the dispersal of larvae southward via the EAC. Research surveys found the main concentrations of juvenile *I. chacei* (<20 mm CL) off NSW were in depths of 40 to 85 m (November to December) (Graham et al. 1993a, 1993b; Graham & Wood 1997) and off Queensland in depths of 80 to 100 m (January to February) (Haddy et al. 2005).

Settlement patterns are poorly understood in the deeper water *I. alticrenatus* and *I. brucei*, and these species may have evolved different life-history strategies. *Ibacus brucei* inhabits the outer-continental shelf where the EAC flows at its strongest and produces oceanic eddies as broad as 200 nautical miles in diameter. In such environments, larval mortality is believed to be high due to low larval retention (DeMartini & Williams 2001), and it is likely that *I. brucei* has evolved a higher fecundity to compensate for these potentially high larval losses. Juvenile *I. brucei* (<23 mm CL) have been recorded in depths between 138 and 210 m (November to February) (Brown & Hothuis 1998; Haddy et al. 2005). In contrast, *I. alticrenatus* mainly inhabits the continental slope between 200 to 400 m and possesses a relatively low fecundity. Therefore, it appears that this species has developed an alternative strategy to account for the potentially high larval losses in oceanic environments. Juvenile *I. alticrenatus* (<24 mm CL) are rarely caught and have only been reported from deep water (400 to 686 m; January). This suggests that this species possesses a deep water settling behavior, as is the case in many deep water spiny lobsters (Kittaka et al. 1997; Haddy et al. 2003). This behavior would be advantageous for larvae dispersed into oceanic environments as it would assist in reducing mortality levels.

17.7 Conclusions

Biological knowledge of the genus *Ibacus* is at a relatively low level. Recent studies have provided some understanding of basic biological parameters such as growth rates, reproduction, and life histories for the commercially caught *I. alticrenatus*, *I. brucei*, *I. chacei*, and *I. peronii*. Unfortunately, most of what we know of the Ibacinae has been drawn from *I. chacei* and *I. peronii* and there is a need for further research on the lesser and noncommercial species in order to better understand this genus. *Ibacus chacei* and *I. peronii* are heavily fished on east coast trawl grounds and there is a need for further assessments to be done in order to sustainably manage their exploitation. The stock structure of *Ibacus* species needs to be assessed, in particular for *I. peronii*, as there is a potential for genetically distinct populations to occur across Australia. Estimates of natural and fishing mortality rates, the development of basic egg- and yield-per-recruit models, and a better understanding of discard mortality rates would assist in assessment of management measures.

References

Aiken, D.E. & Waddy, S.L. 1980. Reproductive biology. In: Cobb, J.S. & Phillips, B.F. (eds.), *The Biology and Management of Lobsters,* Vol. 1: pp. 215–276. New York: Academic Press.

Atkinson, J.M. & Boustead, N.C. 1982. The complete larval development of the scyllarid lobster *Ibacus alticrenatus* Bate, 1888 in New Zealand waters. *Crustaceana* 42: 275–287.

Broadhurst, M.K. 2000. Modifications to reduce bycatch in prawn trawls: A review and framework for development. *Rev. Fish Bio. Fish.* 10: 27–60.

Brown, D.E. & Holthuis, L.B. 1998. The Australian species of the genus *Ibacus* (Crustacea: Decapoda: Scyllaridae), with the description of a new species and addition of new records. *Zool. Med. Leiden* 72: 113–141.

Courtney, A.J. 2002. The status of Queensland's Moreton Bay bug (*Thenus* spp.) and Balmain bug (*Ibacus* spp.) stocks. *Queensland Department of Primary Industries, Information Series*, QI02100. p. 18.

Cresswell, G.R. 1974. Ocean currents measured concurrently on and off the Sydney area continental shelf. *Aust. J. Mar. Freshw. Res.* 25: 427–438.

Cresswell, G.R. 1984. *The East Australian Current on the northern NSW continental shelf.* CSIRO Marine Laboratory Research Report, 1981–84: 153–154.

DeMartini, E.E. & Williams, H.A. 2001. Fecundity and egg size of *Scyllarides squammosus* (Decapoda: Scyllaridae) at Maro Reef, northwestern Hawaiian Islands. *J. Crust. Biol.* 21: 891–896.

Eayrs, S., Buxton, C., & McDonald, B. 1997. *A Guide to Bycatch Reduction in Australian Prawn Trawl Fisheries*: pp. 54. Launceston, Australia: Australian Maritime College.

Gorman, T.B. & Graham, K.J. 1978. *FRV Kapala Cruise Report No. 47* for cruises 78–02, 03, 05 and 06. New South Wales State Fisheries. pp. 8.

Gorman, T.B. & Graham, K.J. 1979. *FRV Kapala Cruise Report Nos. 52 & 53* for cruises 78–16 to 78–22. New South Wales State Fisheries. pp. 20.

Graham, K.J. & Wood, B.R. 1997. *Kapala Cruise Report No. 116.* Sydney: NSW Fisheries. pp. 86.

Graham, K.J., Liggins, G.W., Wildforster, J., & Kennelly, S.J. 1993a. *Kapala Cruise Report No. 110.* Sydney: NSW Fisheries. pp. 69.

Graham, K.J., Liggins, G.W., Wildforster, J., & Kennelly, S.J. 1993b. *Kapala Cruise Report No. 112.* Sydney: NSW Fisheries. pp. 74.

Graham, K.J., Liggins, G.W., & Wildforster, J., & Wood, B. 1995. *Kapala Cruise Report No. 114.* Sydney: NSW Fisheries. pp. 63.

Graham, K.J., Liggins, G.W., & Wildforster, J. 1996. *Kapala Cruise Report No. 115.* Sydney: NSW Fisheries. pp. 63.

Haddy, J.A., Roy, D.P., & Courtney, A.J. 2003. The fishery and reproductive biology of barking crayfish, *Linuparus trigonus* (Von Siebold, 1824) along Queenslands East Coast. *Crustaceana* 76: 1189–1200.

Haddy, J.A., Courtney, A.J., & Roy, D.P. 2005. Aspects of the reproductive biology and growth of Balmain bugs (*Ibacus* spp.) (Scyllaridae). *J. Crust. Biol.* 25: 263–273.

Hill, B.J. & Wassenberg, T.J. 1990. Fate of discards from prawn trawlers in Torres Strait. *Aust. J. Mar. Freshw. Res.* 41: 53–64.

Holthuis, L.B. 1991. *Marine Lobsters of the World. An Annotated and Illustrated Catalogue of the Species of Interest to Fisheries Known to Date.* FAO Species Catalogue No. 125, Vol. 13: pp. 197–206. Rome: Food and Agriculture Organization of the United Nations.

Kailola, P.J., Williams, M.J., Stewart, P.C., Reichelt, R.E., McNee, A., & Grieve, C. 1993. *Australian Fisheries Resources.* Canberra, Australia: Bureau of Resource Sciences and Fisheries Research and Development Corporation.

Kittaka, J.A., Fernando, A., & Prescott, J.H. 1997. Characteristics of palinurids (Decapoda; Crustacea) in larval culture. *Hydrobiologia* 358: 305–311.

Marinovic, B., Lemmens, J.W.T.J., & Knott, B. 1994. Larval development of *Ibacus peronii* Leach (Decapoda: Scyllaridae) under laboratory conditions. *J. Crust. Biol.* 14: 80–96.

Montgomery, S.S. 1990. Movements of juvenile eastern king prawns, *Penaeus plebjeus*, and identification of stock along the east coast of Australia. *Fish. Res.* 9: 189–208.

Phillips, B.F., Brown, P.A., Rimmer, D.W., & Braine, S.J. 1981. Description, distribution and abundance of late larval stages of the Scyllaridae (slipper lobsters) in the South-Eastern Indian Ocean. *Aust. J. Mar. Freshw. Res.* 32: 417–437.

Ritz, D.A. & Thomas, L.R. 1973. The larval and post larval stages of *Ibacus peronii* Leach (Decapoda, Reptantia, Scyllaridae). *Crustaceana* 24: 5–16.

Stewart, J. 1999. *Aspects of the biology of balmain and smooth bugs, Ibacus spp. (Decapoda: Scyllaridae) off Eastern Australia.* Ph.D. thesis: University of Sydney, Australia.

Stewart, J. 2003. Long-term recaptures of tagged scyllarid lobsters (*Ibacus peronii*) from the east coast of Australia. *Fish. Res.* 63: 261–264.

Stewart, J. & Kennelly, S.J. 1997. Fecundity and egg-size of the Balmain bug *Ibacus peronii* (Leach, 1815) (Decapoda, Scyllaridae) off the east coast of Australia. *Crustaceana* 70: 191–197.

Stewart, J. & Kennelly, S.J. 1998. Contrasting movements of two exploited scyllarid lobsters of the genus *Ibacus* off the east coast of Australia. *Fish. Res.* 36: 127–132.

Stewart, J. & Kennelly, S.J. 2000. Growth of the scyllarid lobsters *Ibacus peronii* and *I. chacei. Mar. Biol.* 136: 921–930.

Stewart, J., Kennelly, S.J., & Hoegh-Guldberg, O. 1997. Size at sexual maturity and the reproductive biology of two species of scyllarid lobster from New South Wales and Victoria, Australia. *Crustaceana* 70: 344–367.

Takahashi, M. & Saisho, T. 1978. The complete larval development of the scyllarid lobster, *Ibacus ciliatus* (Von Siebold) and *Ibacus novemdentatus* Gibbes in the laboratory. *Mem. Fac. Fish. Kagoshima Univ.* 27: 305–353.

Yearsley, G.K., Last, P.R., & Ward, R.D. 1999. *Australian Seafood Handbook*: p. 469. Australia: CSIRO Publishing.

18

Slipper Lobster Fisheries — Present Status and Future Perspectives

Ehud Spanier and Kari L. Lavalli

CONTENTS

Abstract

Several species of slipper lobster, especially of the genera *Scyllarides*, *Thenus*, and *Ibacus*, are of commercial importance as food, although prices of slipper lobsters are usually lower than those of local spiny lobsters. Others may serve as baits for additional fisheries or are of interest to the aquarium trade. Many scyllarid lobsters are caught as a by-catch of other fisheries, including that of spiny lobsters. Trawl nets are generally used for fishing on soft bottoms, while traps and trammel nets are used for hard bottoms. Reliable fisheries statistics and stock assessments are not available for most scyllarid species. The majority of slipper lobster fisheries are characterized by the absence of or insufficient regulations and negligible or inadequate reinforcement of these regulations. Several populations of slipper lobsters have, therefore, been depleted and some even became rare in part of their distribution range due to overexploitation. In response to overfishing of spiny lobsters, fishermen in various locations tend to quickly shift their efforts to slipper lobsters, which further endanger these species. Future perspectives focus on extending biological knowledge on major commercial scyllarid species, identifying stocks, and estimating stock size as well as spawning stock. Fishing gear should be improved to reduce mortality and catch of sublegal and berried females. There is also a need for studies to increase health and safety, maximize economic efficiency, and improve postharvest practices for slipper lobsters. Proper management actions are needed to regulate recreational, artisanal, and commercial fisheries, to establish sustainable fisheries, and to preserve present populations of slipper lobster species and their habitats. Urgent measures should be taken in cases of overexploited and declining populations in an attempt to recover the population or at least stop its decline. This should include appropriate legislation, control and reinforcement, establishment of marine protected areas, and deployment of artificial habitats. Enhancement by laboratory-reared lobsters should be carefully considered where aquaculture of the species is available. Future research also should consider possible shifts in marine ecosystems in response to recent global and man-made changes and their effects on slipper lobsters populations.

18.1   Introduction

Although traditionally slipper lobsters were considered secondary to clawed and spiny lobsters as targets for fisheries, many species had commercial value. Holthuis (1991) indicated that out of 71 species of Scyllaridae recognized by that time, 30 species had some interest to fisheries even though the economic value was frequently minor. Yet, except for most representatives of the genus *Scyllarus*, which are too small to have a considerable economic value as food for humans, all the rest of the species of the family Scyllaridae were treated by Holthuis (1991) as "species of interest to fisheries." Species of considerable economic importance mentioned by Holthuis (1991) were *Scyllarides brasiliensis* Rathbun, 1906, *S. latus* (Latreille, 1802), *S. squamosus* (H. Milne Edwards, 1837), *Ibacus alticrenatus* Bate, 1888, *I. ciliatus* (Von Siebold, 1824), *I. novemdentatus* Gibbes, 1850, *I. peronii* Leach, 1815, *Parribacus antarcticus* (Lund, 1793), and *Thenus orientalis* (Lund, 1793). However in the 15 years since Holthuis' important monograph was published, we are witnessing several trends in fisheries of populations of scyllarids of commercial value. Some of these developments were indicated in certain scyllarid species more than a decade ago and include overfishing, lack of fisheries management and regulation, deficient information on yield and stock assessment, and a shift in fisheries to scyllarid species due to overexploitation of other commercial species in the area (e.g., slipper lobster fisheries replacing spiny lobster fisheries in Hawaii; see DeMartini & Williams 2001; Polovina 2005; DiNardo & Moffitt, Chapter 12; in the Galápagos, Hearn et al., Chapter 14; and in India, Radhakrishnan et al., Chapter 15).

The prices of slipper lobsters are usually lower than those of local spiny lobsters; this is the case in the Philippines, where the price of *Thenus orientalis* is lower than that of spiny lobsters (Motoh & Kuronuma 1980) and in the Galápagos where *Scyllarides astori* is also priced lower (Hearn et al., Chapter 14). Yet in Israel, the price of the local *S. latus* used to be higher than that of imported American clawed lobster, *Homarus americanus* H. Milne Edwards, 1837 (Spanier & Lavalli 1998). However, recent increases in the costs of importation have made the imported clawed lobster more expensive (USD\$26 per lobster) compared to the local slipper lobsters (USD\$22 per lobster) (Spanier, personal observations).

Although most species of slipper lobsters are used for food for humans, some are used as bait for other species (see Section 18.3). In addition, it is interesting to note that some species of slipper lobsters that are of little importance as food are of commercial interest in the aquarium trade because of their unusual form and bright colors; these include, for example, *Arctides regalis* Holthuis, 1963 (Holthuis 1991) and *Scyllarus americanus* (S. I. Smith, 1869) (Sharp et al., Chapter 11).

In his invited review to the 7th International Conference and Workshop on Lobster Biology and Management held in Hobart, Tasmania, Bruce Phillips (2005) outlined eight topics associated directly with and important for lobster fisheries. These include: stocks, fishing gear and effect of fishing, recreational fishing, post-harvest practice, economics, aquaculture and enhancement, ecosystem management, and marine protected areas. Obviously, other areas of knowledge, such as larval and juvenile distribution and ecology, behavior, genetics, and other biological facets may also be associated with fisheries, albeit indirectly. While his review refers mainly to clawed and spiny lobsters, these topics may be useful to consider when dealing with slipper lobsters.

18.2   Fisheries Statistics and Stock Assessment

Due to the nature of slipper lobster fisheries, where most are a by-product of fisheries of other target species and others are the result of small and artisanal fishermen groups or SCUBA divers and recreational fishermen, reliable information about landings, size, and sex of the fished lobsters is limited or completely unavailable in most species. Holmyard (2005), reviewing the world landing of all species of lobsters in an FAO publication, states that of the world's annual production of lobsters, the American lobster, *Homarus americanus*, accounts for ~48.5% of the landings, the European lobster, *Homarus gammarus* (Linnaeus, 1758), accounts for 3%, the spiny lobsters (*Panulirus* spp.) account for 42.4%, and rock lobsters (*Jasus* spp.) account for ~6.1%. Obviously there is no reference to slipper lobster landings. Diaby (2004), reviewing trends in the world lobster production stated that it increased steadily from 157,000 metric tons in 1980 to more than 230,000 metric tons in 1997, before stabilizing at 227,000 metric tons in 2001. He added to the above four categories of lobsters a fifth one: "others," but there is no indication of the proportion of slipper lobsters in this last category. Slipper lobster landings appear only in a few of the 24 major fishing areas of the world, as reported in the FAO Fisheries Database and Statistics (see the plot of regional fishery characteristics, capture production by species items at http://www.fao.org/fi/statist/statist.asp) (data available in Table 18.1).

In the proceedings of the Sixth International Conference and Workshop on Lobster Biology and Management, held in Key West, Florida, USA in 2000, published in *Marine and Freshwater Research*, Vol. 52(8) in 2001 (pp. 1033–1675), there was no reference to stock assessment, management, or marine protected areas for slipper lobsters, and out of a total of 68 printed papers only one referred to slipper lobsters. A slight increase in interest in slipper lobsters was apparent in the proceedings of the recent 7th International Conference and Workshop on Lobster Biology and Management, held in Hobart, Tasmania, Australia in 2004, published in the *New Zealand Journal of Marine and Freshwater Research*, Vol. 39 (2 + 3) in 2005 (pp. 227–783). Out of a total of 50 printed papers, three referred to slipper lobsters.

Despite the lack of global quantitative data on slipper lobsters fisheries, this volume supplies several examples of regional data and trends, quantitative in nature, for particular species of slipper lobster. This partial information enables us to postulate some interim conclusions and outline future recommendations that may be applicable to areas where slipper lobsters are/were fished, but where no quantitative data exists on stock size, or commercial and recreational fishing pressure. Chapters 11, 12, 14, 15, 16, and 17 supply valuable information on fisheries in the Florida Keys and Gulf of Mexico, the northwestern Hawaiian Islands, the Galápagos Islands, waters off India, and waters around Australia. Yet, the knowledge gained and conclusions derived from the study of certain species should be treated carefully when applied to other, even closely related, species due to differences in habitats, life histories, behaviors, and the effects of the fisheries themselves.

Sharp et al. (Chapter 11), reviewing the Florida slipper lobster fishery, stated that scyllarid lobsters are of limited economic importance compared to Caribbean spiny lobster, *Panulirus argus* (Latreille, 1804), a species that supports one of the state's most valuable fisheries (average annual landings of 17 tons of slipper lobster with a value of USD$82,000 vs. 2722 tons of spiny lobster with estimated value of USD$21 million over the same time period). These numbers reemphasize the point raised in the previous section that slipper lobsters yield lower market prices than spiny lobsters (in Florida ~USD$4.8/kg vs. ~USD$7.7/kg for spiny lobsters). Florida scyllarids are primarily landed as by-catch from other fisheries — in trawls as by-catch of the shrimp fishery (mainly *Scyllarides nodifer* (Stimpson, 1866)), or as by-catch in spiny lobster traps (*S. nodifer*, *S. aequinoctialis* (Lund, 1793) and *Parribacus antarcticus*). Sharp et al. further point out that despite the landing information indicating that scyllarids and especially *S. nodifer* continue to support a limited by-catch fishery, the status of these stocks is completely unknown. Slipper lobsters are caught in Florida also by divers and in the recreational fisheries. As mentioned, small live-market fisheries (few hundred lobsters annually for slipper lobsters) for the marine aquarium trade exist in Florida. Divers land *S. nodifer*, *S. aequinoctialis*, *P. antarcticus*, and *Scyllarus americanus* for this purpose.

DiNardo & Moffitt (Chapter 12) describe the interrelationship between the fisheries of the green spiny lobster, *Panulirus penicillatus* (Olivier, 1791), the endemic Hawaiian spiny lobster *Panulirus marginatus* (Quoy & Gaimard, 1825), the Hawaiian scaly slipper lobster, *Scyllarides squammosus*

TABLE 18.1

Capture Production (Metric Tons) of Slipper Lobsters (Bold) During 1997 to 2003 in the Major FAO Fisheries Area Presented Alongside Production Data of Other Lobster Groups (Species Group 43)

Fishing Area	English Name	Scientific Name	1997	1998	1999	2000	2001	2002	2003
Atlantic Southeast (C-47)	Cape Rock lobster	*Jasus lalandii* (H. Milne Edwards, 1837)	1879	2076	2097	2058	1974	3029	2844
	Tristan da Cunha rock lobster	*Jasus tristani* Holthuis, 1963	321	376	336	316	425	301	534
	Southern spiny lobster	*Palinurus gilchristi* Stebbing, 1900	892	864	429	305	1053	651	89
	Slipper lobsters nei	***Scyllaridae***	**0**	**1**	**1**	**1**	**0**	—	—
	Lobsters nei	*Reptantia*	25	31	68	10	1062	1008	
Indian Ocean, Western (C-51)	Tropical spiny lobsters nei	*Panulirus* spp.	2547	2822	2459	2239	2243	2370	2429
	St.Paul rock lobster	*Jasus paulensis* (Heller, 1862)	295	308	345	192	183	334	383
	Natal spiny lobster	*Palinurus delagoae* Barnard, 1926	10	6	7	8	10	7	90
	Spiny lobsters nei	*Palinuridae*	233	239	204	228	213	270	2
	Slipper lobsters nei	***Scyllaridae***	**74**	**67**	**20**	**9**	**26**	**20**	**14**
	Mozambique lobster	*Metanephrops mozambiqus* Macpherson, 1990	156	192	152	180	141	130	145
	Lobsters nei	*Reptantia*	—	—	—	62	2	—	1
Indian Ocean, Eastern	Australian spiny lobster	*Panulirus cygnus* George, 1962	9896	10400	13065	14605	11353	9050	11477
	Tropical spiny lobster nei	*Panulirus spp.*	1321	1051	1361	1432	1800	1911	2315
	Southern rock lobster	*Jasus novaehollandiae* (Holthuis, 1963)	4888	4615	4655	4756	4677	4403	4255
	Flathead lobster	***Thenus orientalis***	**177**	**533**	**37**	**23**	**58**	**282**	**181**
	Slipper lobster nei	***Scyllaridae***	**0**	**0**	**0**	**0**	—	—	—
	Metanephrops nei	*Metanephrops spp.*	—	—	—	39	105	88	—
Pacific, Western Central	Tropical spiny lobster nei	*Panulirus* spp.	4564	3627	4281	4323	5200	5801	6562
	Flathead lobster	***Thenus orientalis***	**2785**	**3005**	**1700**	**2289**	**1797**	**2005**	**2222**
	Slipper lobsters nei	***Scyllaridae***	**736**	**866**	**98**	**96**	**128**	**127**	**135**
	Metanephrops nei	*Metanephrops spp.*	—	—	—	—	—	9	—
	Lobsters nei	*Reptantia*	200	100	200	500	700	1127	1000
Pacific, Southwest	Green rock lobster	*Jasus verreauxi* (H. Milne Edwards, 1851)	158	124	123	152	113	111	138
	Red rock lobster	*Jasus edwardsii* (Hutton, 1875)	5000	2707	2818	2788	2551	2495	2553
	Slipper lobsters nei	***Scyllaridae***	**158**	**160**	**7**	**1**	**4**	**2**	**0**
	New Zealand lobster	*Metanephrops challengeri* (Balss, 1914)	1093	909	925	1004	1033	1025	826
World production	All species of lobsters		233, 381	217, 126	228, 602	227, 596	221, 749	224, 883	224, 079

Only areas where slipper lobsters appeared were included in this table (adapted from: FAO Fisheries Database and Statistics internet page (http://www.fao.org/fi/statist/statist.asp, October 2005) — plot of regional fishery characteristics, capture production by species items). nei = not explicitly identified.

(H. Milne Edwards, 1837), and that of the Hawaiian ridgeback slipper lobster, *S. haanii* (De Haan, 1841) in the uninhabited and remote Northwestern Hawaiian Islands (NWHI). The fisheries statistics supplied for 1985 to 1999 refer just to "spiny lobster" and "slipper lobsters" in general. It should be mentioned that in addition to the above two dominant Hawaiian scyllarids, other species of scyllarids are found in this region (see DiNardo & Moffitt, Chapter 12, Table 12.1). Gear modifications and improved markets in Hawaii in 1985 to 1986 led to an increase in scaly slipper lobster landings in a fishery that initially targeted Hawaiian spiny lobsters. Catches of scaly slipper lobster were characterized by fluctuations. They remained high from 1985 to 1987 with peak of ~45% (1,245,000) of the total reported catch of lobsters, followed by a general decline from 1988 to 1996 (only ~11% of the total reported catch in 1995), and then increased significantly from 1997 to 1999 (~63% of the total reported catch of lobsters in 1999). Several approaches used to model NWHI lobster populations and establish levels of sustainable yield are described in Chapter 12. The fishery was closed in 2000 as a precautionary measure because of a lack of confidence in the models used to assess the stocks. Shortly thereafter, the Northwestern Hawaiian Islands Coral Reef Ecosystem Reserve was established and commercial fishing was prohibited, perhaps indefinitely.

Hearn et al. (Chapter 14) state that in 2002, the total catch of the Galápagos slipper lobster, *Scyllarides astori* Holthuis, 1960 was estimated to be 13.8 metric tons (according to Murillo et al. 2003), whereas for 2003, that value decreased slightly to 12.8 metric tons (according to Molina et al. 2004) live weight. They added that since 2002, approximately three metric tons of tails (corresponding to 8.4 metric tons live weight) of whole *S. astori* are exported annually to continental Ecuador.

Radhakrishnan et al. (Chapter 15) indicate that according to Kagwade et al. (1991), in Veraval, India, the average annual catch of *T. orientalis* was 148.3 metric tons during 1980 to 1985, with a wide fluctuation thereafter, decreasing from 126 metric tons in 1987 to 85 metric tons in 1988, followed by a peak of 215 metric tons in 1990. However, the mean annual landings during 1991 to 2000 was only 97.7 metric tons (Radhakrishnan et al. 2005), and there has been a steady decline ever since with catches at 22, 18, 7, and 6 metric tons during the years 2001 to 2004. A similar trend was indicated by Radhakrishnan et al. (Chapter 15) in the fisheries of this species in the waters off Mumbai, India. While Kagwade et al. (1991) reported a catch of 184.9 metric tons in the late 1970s, the fishery declined dramatically to 50 metric tons in 1989, 2 metric tons in 1994 and finally collapsed in 1996. Analysis of length composition of male and female *T. orientalis* from Mumbai waters for 1980 to 1985 clearly indicated that the species was overexploited.

Jones (Chapter 16) reports that the exploitation of *Thenus* spp. in Australia and throughout most of Southeast Asia is as by-catch by trawlers and that catches dominated by these lobsters are infrequently reported. He adds, however, that more than 90% of the Australian catch is from Queensland waters and that the total production of these lobsters in Queensland for 2001 was estimated at 386 metric tons and valued at AUD$4.6 million (Courtney 2002), but has been as high as 800 metric tons. The exploitable stock of *T. indicus* Leach, 1816 is represented by a size range of approximately 25 to 70 mm carapace length (CL) and an age range of approximately two to five years.

Haddy et al. (Chapter 17) estimates that approximately 200 metric tons of commercially exploited *Ibacus* spp. are landed in Australia per year, with the majority being caught off eastern Australia. *Ibacus chacei* (Brown & Holthuis, 1998) and *I. peronii* Leach, 1815 account for more than 95% of all *Ibacus* landed.

There is limited or no information on other slipper lobster fisheries. For example, Holthuis (1991) reported that two to three metric tons of *Scyllarides latus* are taken annually in Israel. Yet attempts for more recent landing estimations, which is not uncommon in the markets and in the seafood restaurants in season in Israel, failed because most fishing is done by numerous SCUBA divers who do not report to the fisheries department (Spanier personal observations). There is presently no assessment of stocks for this species in areas where commercial fishing exists (e.g., Israel, Turkey, Algeria). But it is obvious that stocks were almost depleted in other parts of the historic distribution range of this species, such as the Mediterranean coasts of Europe (Spanier 1991; Spanier & Lavalli 1998).

18.3   Fishing Gear

The fishing gear used for slipper lobsters largely depends on the habitats of the given species and their associated substrates, although many scyllarid lobsters are caught as by-catch of fisheries of other species including that of spiny lobsters, shrimp, and even finfish fisheries. In general, trawl nets are used for fishing of soft-bottom species (e.g., *Thenus* spp., see Chapters 15 and 16 and *Ibacus* spp., see Chapter 17) and traps and trammel nets are used for hard-bottom species (e.g., *Scyllarides* spp.). SCUBA is used by many recreational fishermen, but also is popular, along with hooka diving, in the professional fisheries for scyllarids. Divers take them by hand (usually with or without gloves) some time during the day using torches to locate them, or they spear them, or capture them with dip nets.

A review of scyllarid species that are listed in Holthuis (1991) of having at least some commercial value reveals that *Arctides antipodarum* Holthuis, 1960 (rocky bottoms) is sometimes caught in lobster traps set for species of *Jasus* in Australia and northern New Zealand, but it is also taken by divers. *Scyllarides aequinoctialis*, which typically dwells on sand and rock, is taken in traps set for other lobsters (see Chapter 11), but also is captured with fixed gill nets and seines. Its meat also serves as bait in spiny lobster, *Palinurus argus* (Latreille, 1804) pots (Holthuis 1991). *Scyllarides astori* (rocky substrates) are caught in traps and by trammel nets for spiny lobsters and other species, but mainly are captured by divers (for details see Chapter 14). *Scyllarides deceptor* Holthuis, 1963 (sandy bottoms) is occasionally taken by trawl. *Scyllarides delfosi* Holthuis, 1960 (muddy bottoms) is sometime taken by trawls, dredges, or traps. *Scyllarides elisabethae* (Ortmann, 1894) (mud and fine sand) is occasionally taken by trawlers. *Scyllarides haanii* (rocky bottoms) is usually taken by lobster pots (see also Chapter 12). *Scyllarides herklotsii* (Herklots, 1851) (sand, rock, and mud substrates) is usually taken in vertical nets or sometime in trawls. *Scyllarides latus* (rocky habitats, but also soft bottoms during migration; see Spanier et al. 1988; Spanier & Lavalli 1998) is taken mainly by hand (SCUBA), trammel nets, trawls, and lobster pots. The typical trap for *S. latus* was developed in Israel following the model of the Azores Islands (Figure 18.1). This trapezoidal-shaped wooden trap is equipped with a galvanized tin funnel opening on top. It is effective when soaked in seawater for several days before use and baited with live bivalves, catching up to 17 lobsters at a given trap haul. Apparently there is a subsequent attractant with lobsters already in the traps being a magnet for additional lobsters; this may be due to a "guide effect" similar to that reported in juvenile spiny lobsters (e.g., Zimmer-Faust & Spanier 1987; Spanier & Zimmer-Faust 1988; see Lavalli et al., Chapter 7 for more information). *Scyllarides nodifer* (sandy bottom sometime mixed with mud) is mostly obtained in traps for other species (Chapter 11). *Scyllarides squammosus* (hard substrates) is mostly taken by hand at night, but also by plastic traps, particularly the Fathoms Plus® trap (see Chapter 12 for illustrations of this trap). *Evibacus princeps* S. I. Smith, 1869 (mainly soft bottoms) is taken in fairly large quantities by trawling in the Gulf of Panama. *Ibacus alticrenatus*,

FIGURE 18.1   Wooden trap with galvanized tin upper opening used in Israel to catch *Scyllarides latus* (Photo by S. Breitstein; used with permission).

I. ciliatus, and *I. peronii* (soft bottom) are taken by trawlers (see Chapter 17), while *I. novemdentatus* (soft substrates) is taken by trawlers in Japan and Taiwan and also by long-line fishing in Japan (Holthuis 1991). *Parribacus antarcticus* (reef dweller) is taken by nets set for spiny lobsters (see Chapters 11 and 12), while *P. caledonicus* Holthuis, 1960 (reefs) is caught by divers (George 1971). *Parribacus japonicus* Holthuis, 1960 (shore-reef dweller) is caught by gill net and *P. perlatus* Holthuis, 1967 (shallow water rocks) is taken by hand. *Scyllarus arctus* (Linnaeus, 1758) (rock or muddy substrates) is taken as by-catch of other fisheries by gill nets, trawls, dredges, traps, seines, and SCUBA. *Scammarctus* (formerly *Scyllarus*) *baeti* (Holthuis, 1946), *Remiarctus* (formerly *Scyllarus*) *bertholdii* (Paulson, 1875), *Petrarctus* (formerly *Scyllarus*) *brevicornis* (Holthuis, 1946), *Eduarctus* (formerly *Scyllarus*) *martensii* (Pfeffer 1881), and *Petrarctus* (formerly *Scyllarus*) *rugosus* (H. Milne Edwards, 1837) (all soft bottoms) are caught in Taiwan by "baby shrimp trawls" as by-catch (Chan & Yu 1986). *Thenus orientalis* (sand and mud) is taken as by-catch in the net of trawlers. In Queensland, Australia the shrimp fisheries lands *Thenus* as a by-catch but ranks *Ibacus* as a targeted food item (Grant 1978; see also Chapter 16). *Thenus indicus* is also taken by trawlers (Chapters 16), although there is at least one report of a diver-based fishery for *Thenus* in the Red Sea (Ben-Tuvia 1968).

18.4 Catch Per Unit Effort

A few reports exist on catch per unit effort (CPUE) of commercially important scyllarids. There are considerable differences depending on the species, the fishing gear, and previous fishing efforts.

Ivanov & Krylov (1980) recorded catches of 19.2 and 22.6 kg/h of *Ibacus novemdentatus* in the western Indian Ocean using trawl nets. Haddy et al. (Chapter 17) report by-catches of *Ibacus* spp. associated with the prawn fishery as usually < 10 kg/day. However, catch rates as high as 50 kg/h for *I. brucei* (Holthuis 1977) and 60 kg/h for *I. alticrenatus* were achieved during exploratory trawling on unfished grounds off northern New South Wales in Australia (Gorman & Graham 1978).

Studying spatial distributions of *T. indicus* and *T. orientalis*, Jones (1993) reports that both species of lobsters are highly aggregated. The mean catch per hour and mean catch per hectare reported by him, using standard trawl fishery gear, was 16.8 and 1.83, respectively for *T. indicus*. For *T. orientalis* mean catch per hectare was 2.01. Kagwade et al. (1991) reported a catch rate of 1.7 kg per unit effort (trawler) of *T. orientalis* in certain areas of India.

Even less information exists on CPUE in *Scyllarides* species. Groeneveld et al. (1995) conducted a trap-fishing survey off the east coast of South Africa. An average abundance index of 446.8 g per trap was estimated for both *Scyllarides elisabethae* and *Palinurus delagoae* Barnard, 1926 combined. The spiny lobsters contributed 301.9 g/trap (67.6%) and *S. elisabethae* 144.9 g/trap (32.4%). Hearn et al. (Chapter 14) report CPUEs ranging from 7.4 to 38.8 kg live weight per diver day of *Scyllarides astori* in some of the Galápagos Islands in 2002 and 2003 (for details see Table 14.2 in Chapter 14).

DiNardo & Moffitt (Chapter 12) present annual metrics of spiny and slipper lobster CPUEs from research surveys in two Islands in NWHI.

18.5 Management and Regulations

Most slipper lobster fisheries are characterized by the absence of or insufficient regulations and absent or inadequate reinforcement of whatever regulations exist. However, there are a few places where proper regulations have been established and where such regulations are effectively reinforced. Some of the regulations were adopted in response to population changes and mortality associated with fisheries activities. Sometimes the establishment of these regulations is too little, too late for the dwindling populations to sufficiently rebound, as witnessed for *Scyllarides latus* in the Azores (Martins, personal communications) and Italy (Bianchini et al. 2001; Bianchini & Ragonese, Chapter 9), and *Thenus orientalis* in India (Radhakrishnan et al., Chapter 15).

Sustainable yields for Hawaiian slipper lobsters in the NWHI were estimated using surplus production, and catch quotas were adopted as a management tool in 1992 (for details, DiNardo & Moffitt, Chapter 12).

In the Galápagos Marine Reserve there is a preliminary (and not yet enforced) zonation scheme for *Scyllarides astori* that prohibits extractive activities in 18% of the coastal waters. Additionally, the Galápagos Five-Year Fishing Calendar identifies the need to incorporate specific regulations for 2004, including closed seasons and a minimum landing size (for details, see Hearn et al., Chapter 14). Sadly, these regulations have not been established yet. In Australia, fishery-management regulations ban the retention of egg-bearing females of all species of *Ibacus* and minimum legal size carapace widths were established for the respective species in various regions (for details, see Haddy et al., Chapter 17). In several countries, there are different regulations for different states/provinces in the same country, and regulations often are not strictly enforced (e.g., fishing of *T. orientalis* in India; see Radhakrishnan et al., Chapter 15). The total closure of an area to fishing of slipper and other species of lobsters is usually a last-ditch management tool to prevent the complete extermination of a population. It can also be done as a preliminary measure when it is clear that populations models are not tracking stock status well, most likely due to incorrect biological parameters, as in the case of NWHI where the fishery was closed in 2000 (for details, see DiNardo & Moffitt, Chapter 12), or as a response to a dramatic collapse of the yield due to overfishing (e.g., *S. latus* in the Azores and along many of the Mediterranean coasts of Europe, and *T. orientalis* in several maritime states in India). Since the slipper lobster fishery is frequently associated with fisheries targeting other species, such as the case of the combined trap fishery for *Scyllarides elisabethae* and *Palinurus delagoae* off the east coast of South Africa (Groeneveld et al. 1995), there is a need to take both lobster species into consideration in any management strategy.

18.6   Recreational Fishing

Due to increases in the standard of living and easy access to equipment (boats, traps, diving gear), recreational fishing has become, in recent years, a major threat to many slipper lobster populations. Not all countries allow fishing of lobsters for recreational purpose, but it is permitted in Australia, New Zealand, South Africa, the United States (Phillips 2005; Sharp et al., Chapter 11; DiNardo & Moffitt, Chapter 12), as well as many others; few countries reinforce regulations on recreational fishermen or even require reporting of catches. There is usually a built-in conflict between commercial and recreational fishermen. The first group wants to limit the activities of the second, while the second group wants to increase their share in the catch. It is easier to control some relatively small number of commercial fishermen (who are generally more organized, sometimes even into collectives, and who operate from permanent bases) than the many recreational fishermen who are spread over large geographical ranges, are usually not organized, and who may include tourists, foreigners, or otherwise. For these reasons, it is also very difficult to obtain constant and reliable biological and fishery information from recreational fishermen (although there are exceptions), and this lack of information interferes with our ability to estimate stocks of slipper lobsters. Yet the political power of recreational fishers may be considerable as Phillips (2005) states, "The politicians have perhaps noted that there are more votes to be gained from large numbers of recreational fishers than the smaller number of commercial fishers!" The situation can be politically complex in countries where there are also groups of artisanal fishermen, in addition to the two sectors mentioned above. Clearly new ideas are needed in this arena.

18.7   Postharvest Practice and Economics

Slipper lobsters, as with spiny lobsters, are marketed live or frozen as whole lobsters or tails. Although legs are not used as food, as in some other species of lobsters, fishermen attempt to supply the slipper lobsters (especially to restaurants and hotels) in an esthetic form with all their appendages intact, since those intact lobsters usually gain a better market price. Because many of the scyllarid lobsters are found in subtropical or tropical countries with elevated air temperatures, there is a danger of spoiling the lobster meat on the way from the fishing grounds to the consumer, if fishermen lack adequate refrigeration facilities on board or other means of holding the lobsters live. *Scyllarides latus* are usually not kept alive in seawater on board because they tend to harm each other with the sharp points of their pereiopod dactyls in such crowded conditions. In the Mediterranean, fishermen usually keep them in wet sacks until they reach the port. Along

FIGURE 18.2 Live *Scyllarides latus* kept tied to a line in the water until marketing in southwest Turkey (Photo by O. Ben-Eliyahu; used with permission).

the Aegean coast of Turkey, *S. latus* is kept in a small group tied to a line (in a similar manner that lobsters were tethered in the study of Lavalli & Spanier 2001), with a space between each lobster, for many hours in seawater (usually over the dock) until marketing (Spanier, personal observations; see Figure 18.2). For economic reasons, fishermen prefer larger specimens and occasionally release undersized and berried lobsters. However, the survival probability of these released individuals is unclear. There is clearly a need for studies on discard mortality to increase the health and safety of such discards (or to eliminate trapping undersized or ovigerous females altogether), to maximize economic efficiency, and to improve postharvest practices of slipper lobster fisheries.

18.8 Effects of Fisheries, Aquaculture, and Enhancement

There are several examples of the effect of extensive fisheries on stocks of slipper lobsters in this volume (e.g., Chapters 11, 12, 14, 15, and 17). When these fisheries are not regulated or management tools are not applied properly, the results may be devastating for slipper lobster populations. An extreme situation is described in India by Radhakrishnan et al. (Chapter 15), where no viable stock of *Thenus orientalis* remains in the sea off Mumbai. The authors suggest two possible solutions: (1) total conservation of the remaining residual population by banning any landing of the species, and (2) the return of lobsters captured by the fisheries to the sea until the stock is revived. Since success has been achieved in aquaculture of this species, as well as other species of scyllarids (e.g., Kizhakudan et al. 2004; see also Mikami & Kubala, Chapter 5

FIGURE 18.3   (See color insert following page 242) *Scyllarides latus* in an artificial reef at 20-m depth off Haifa, Israel, 2005. (Photo by S. Breitstein; used with permission.)

and Pessani & Mura, Chapter 13), captive breeding and hatchery production, followed by sea ranching of hatchery-produced juveniles may also be a viable way in which to recover severely depleted stocks. The behavior of hatchery-reared lobsters may be different from that of wild lobsters, and this can strongly affect their survival in the sea. Laboratory-reared, postlarval clawed lobsters, *Homarus americanus*, do not seek shelter immediately when released to cobble habitats, which increases predation upon them (Castro et al. 2001); however, older juveniles do successfully inhabit shelters and grow to recruit into the fishery. Grow-out of older juveniles, combined with sensitizing of the animals to predator odors, has resulted in a successful restocking effort in Norway with *Homarus gammarus* (see Agnalt et al. 2004 for a review of this enhancement effort). Maladaptive hatchery behavior has been linked to high-predation rates in other species (e.g., Olla et al. 1998). Another problem pointed out in enhancement programs for spiny lobsters (e.g., Oliver et al. 2005) is the timing and location of the release of the hatchery-reared juveniles to the field. Absence of sufficient biological knowledge may lead to the release of the juveniles in the wrong location and time of the day or year, resulting in unfavorable current regimes, insufficient food or shelter, and exposure to high predation. In other species of commercial slipper lobsters where no success has been achieved in aquaculture, such as in the genus *Scyllarides* (where there is no information about nistos or juveniles), enhancement can be done by using adult specimens. A small-scale stock enhancement by restocking of adult *S. latus* was done in Sicily (e.g., Bianchini et al. 1997, 1998; Bianchini & Ragonese, Chapter 9). However to assess the success of such enhancement programs on reviving stock, it should be done with a considerable number of lobsters and stocked individuals should be tracked for much longer time periods in a given region in order to gain significant information concerning survival rates from these efforts. Even if such restocking programs are effective in reviving local populations, one should be careful in implementing such an approach. Restocking of a depleted region with considerable numbers of lobsters from another region (or country) can cause a dramatic decline of the lobster populations in the source area.

18.9   Ecosystem Management and Marine Protected Areas

As Spanier et al. (1988, 1990) have pointed out, rocky outcrops that are the preferable habitat for *Scyllarides latus*, are limited in the southeastern Mediterranean. Thus, one should carefully manage these ecosystems, which also serve as preferred grounds for other commercial species, such as groupers, sea breams, and octopuses. The fisheries in these areas should be carefully regulated and destructive fishing methods, such as rough-bottom trawls and the use of explosives should be completely banned. This is true also for other essential habitats of hard-bottom scyllarids, such as coral reefs. When the amount of such preferred habitats is limited, Spanier (1991) suggests constructing artificial reefs that will supply the needed habitats (see Figure 18.3 for an example of such artificial reefs). These man-made structures should be constructed according to the behavioral–ecological adaptations of the given species and the phases in the life cycles most likely to use them (Spanier 1994; for details of scyllarid preferences, see Lavalli et al., Chapter 7). Different artificial structures can be designed and deployed for the settling nistos ("collectors") as has been done with pueruli of spiny lobsters (Phillips & Booth 1994; Quinn & Kojis 1995), for early benthic juvenile stages, and for adults (e.g., "casitas"). Artificial habitats to enhance juvenile and adult lobster stock has been utilized to enhance damaged stocks of the American clawed lobsters *H. americanus* (e.g., Castro et al. 2001) and to enhance natural populations when shelter-providing habitat is scarce in a variety of spiny lobsters species (e.g., Briones-Fourzán et al. 1994; Butler & Herrnkind 1997; Losada-Tosteson & Posada 2001).

In addition, special ecosystem management is needed when considering the reproductive population. This may include a complete prohibition of fishing the slipper lobster in the reproductive season, a ban on fishing ovigerous females, and limitations on fishing in selected artificial and natural sites that may serve as sanctuaries or refugia for the reproductive population. Although there were and are attempts to create fishery exclusion zones or Marine Protected Areas (MPAs) for slipper lobsters in Hawaii (see DiNardo & Moffitt, Chapter 12), the Galápagos (see Hearn et al., Chapter 14), the Azores Islands, and the Mediterranean (Diaz et al. 2005), the long-term effects of such MPAs is not clear. It is possible, depending on the local political situation, to declare a large area as a MPA regarding certain species of slipper (and other) lobsters. However, it is often not practical to actually control such a vast area against illegal fishers. Thus, MPAs should be carefully selected with regards to their location, suitable size for the purpose at hand, and methods need to be developed to monitor the effectiveness of such measures. Many questions remain to be addressed in the establishment of these MPAs. For example, how does the closure of an area to slipper lobster fishing (commercial?, recreational? or both?) affect lobster populations (in terms of lobster sizes, numbers, densities, sex ratios, size/age of sexual maturity, and fecundity) in the protected areas compared to similar unprotected areas in the vicinity and at a distance from the MPA? How does the closure of an area to lobster fisheries affect the lobster fisheries itself (lobster yield, sexual maturity, and fecundity) in areas adjacent to the protected area, as well as areas that are further away from it? To what extent do adult lobsters from the protected areas move to unprotected areas and vice versa? These questions may be more complex since more than one species of lobster may be included in the no-take zone (e.g., in Hawaii and the Galápagos, the protected areas include species of slipper and spiny lobsters).

The selection of a MPA is crucial for its effectiveness. Lipcius et al. (2001) showed that if one wished to increase recruitment of *Palinurus argus* via establishment of a MPA, that MPA had to be one that incorporated transport process of larvae. Unfortunately these processes are poorly known for most species of scyllarid lobsters. Another consideration is that in a MPA, the fisheries for natural predators of lobsters, such as fish, will be limited or nonexistent. This situation may result in a higher predation rate in the MPA, especially of sensitive phases such as recently settled juvenile lobsters, relative to nonprotected areas. The availability of suitable shelters or burial substrate for such stages should also be considered; as such shelters will help to reduce predation (Diaz et al. 2005). The "trade-off" between "human predation" (fishing) and natural predation of lobsters should be taken into account when considering the establishment of such protected areas.

18.10   Future Perspectives

Phillips (2005) stated that "Good management of lobster species comments on the depth or lack thereof of the biological information available on the species comprising the fishery. The greater the depth of the knowledge, the better the management system that can be devised and the more successfully it can be operated." Yet researchers of slipper lobsters are often faced with a frustrating situation — they have to make urgent management decisions regarding species where not enough biological information is available, and they cannot postpone management activities to future times when more knowledge will have accumulated. These situations have already been faced since quite a few populations of slipper lobster species have been depleted, some even to the point of the lobsters becoming rare in part of their distribution range (e.g., *S. latus* was declared marine species in need for protection in the western Mediterranean — see Spanier 1991). Although considerable knowledge on the biology, ecology, behavior, and fisheries of several species of slipper lobsters has been gained in the recent decade and is reflected in this volume, there are still many gaps of basic information for most species and, especially, the commercial species of the genera *Scyllarides*, *Thenus*, and *Ibacus*. In the absence of sufficient biological and fisheries information on slipper lobsters, one may try to draw conclusions from other families of lobsters (especially spiny lobsters) where knowledge is based on a more considerable body of research. But this information cannot substitute the direct knowledge for a given scyllarid species. Moreover, one has to be careful in projecting biological knowledge between species that inhabit different habitats, such as soft (e.g., *Thenus* spp.) and hard (e.g., *Scyllarides* spp) substrates, and may differ in their life cycles (naupliosomas, larvae, postlarvae, juveniles, and adults).

Future biological studies should focus on the following subjects in the major commercial scyllarid species — larval ecology and dispersion, potential bottlenecks affecting nisto recruitment to the benthos, juvenile biology including factors that increase or decrease recruitment into the adult population, reproductive biology, behavior, behavioral ecology of the various phases, physiology, genetics, and diseases. In the field of fisheries and management, future efforts should be directed at identifying stocks, estimating stock size, and determining spawning stock. Fishing gear should be improved to reduce discard mortality and catch of sublegal, and berried females, and management procedures should be enacted to regulate both recreational and commercial fisheries. Proper management actions are needed to establish sustainable fisheries (see Polovina 2005) and preserve present populations of slipper lobster species with considerable stocks. Urgent measures should be taken in cases of overexploited and declining populations of scyllarid species in an attempt to recover the population or at least stop its decline. This should include appropriate legislation, control and reinforcement, establishment of MPAs, and deployment of artificial habitats for different phases where such habitats are limited, damaged, or severely degraded by human activities. Also enhancement by hatchery-reared lobsters should be carefully considered where aquaculture of the species is available.

Due to overfishing of spiny lobsters and other species, fishermen in various locations tend to quickly shift their efforts to slipper lobsters (e.g., DeMartini & Williams 2001; Polovina 2005; DiNardo & Moffitt, Chapter 12; Hearn et al. Chapter 14), which are also becoming more popular in various world markets. This is an alarming trend that if not properly and immediately managed will result in a devastating and irreversible outcome to many slipper lobster populations. This trend with lobster species is similar to what has happened with fish species, where it seems that we are "fishing down marine food webs" (Pauly et al. 1998). If one compares global fish-catch data against food-web models since the 1950s, it appears that global fish harvests have shifted down the food web, away from preferred, higher trophic level (predator) species to species of lower trophic levels (plankton eaters). This shift has occurred as populations of predator fish have been decimated by overfishing, forcing fishers to harvest less preferred species at a lower trophic level. Likewise, we are seeing the shift from the preferred spiny lobster species to the less preferred slipper lobsters, as spiny lobsters become less plentiful. When dealing with shifting of the human fisheries from the depleting stocks of spiny lobster to still-existing populations of slipper lobsters, one should wonder if natural predators have not adopted the same strategy, thereby further increasing the pressure on slipper lobsters populations.

The present oceans are undergoing man-made and global changes. How do these changes affect slipper lobster populations? Phillips (2005) points to the increasing number of construction projects and operation of underwater sea pipelines (for gas, oil) that may act as a physical barrier to lobster movement and migration and, in turn, may affect lobster survival, reproduction, or catchability. Other marine man-made activities can change bottom topography, water chemistry, temperature, and currents, and can produce noise and vibrations that may affect lobster survival, physiology, and behavior. Transport of lobsters (or organisms associated with them, such as competitors, predators, and parasites) by ballast water of ships or man-made canals to biogeographic regions where they were absent in the past (e.g., Galil et al. 1989) can also affect fisheries. Shifts in marine ecosystems in response to changes in the ocean and atmospheric climate (e.g., Polovina 2005) can affect lobster populations via changes in behavioral, physiological, and distributional patterns and may impact fisheries.

The constant accumulation of abandoned and lost fishing gear such as "phantom nets" or "ghosts pots" (e.g., Parrish & Kazama 1992) that keep on catching slipper lobsters without benefiting the fisheries presents another challenge for study.

Last, but not least, intensive international cooperation is needed regarding exchange of knowledge and information, as well as joint biological and fisheries associated research and management. The need for international cooperation is required for two reasons: (1) the extensive geographical distribution of certain species (e.g., *T. orientalis*: Iran, see Fatemi 2001; India, see Chapter 15; Australia, see Chapter 16) that spans over the territorial waters of many countries with overlapping populations or fragments of metapopulations; and (2) the existence of species that are similar within one genus in different countries (e.g., *Scyllarides* spp.) where partial biological and fisheries knowledge has been accumulated on each. The pooling of this fractioned scientific information can enable the better understanding of the genus and development of improved management tools.

References

Agnalt, A.L., Jørstad, K.E., Kristiansen, T., Nøstvold, E., Farestvelt, E., Næss, H, Paulsen, O.I., & Svåsand, T. 2004. Enhancing the European lobster (*Homarus gammarus*) stock at Kvitsøy Islands: Perspectives of rebuilding Norwegian stocks. In: Leber, K.M., Kitada, S., Blankenship, H.L., & Svasand, T. (eds.), *Stock Enhancement and Sea Ranching: Developments, Pitfalls and Opportunities,* 2nd Edition: pp. 415–426. Oxford, U.K.: Blackwell Publishing Ltd.

Ben-Tuvia, A. 1968. Report on the fisheries investigations of the Israel South Red Sea expedition, 1962. Rep. No.33. *Bull. Sea Fish. Res. Sta. Haifa* 52: 21–55.

Bianchini, M.L., Ragonese, S., Greco, S., Chessa, L., & Biagi, F. 1997. Valutazione della fattibilità e potenzialità del ripopolamento attivo per la magnosa, *Scyllarides latus* (Crostacei Decapodi). *Final Rep. MiRAAF (Pesca Marittima)*: 1–145.

Bianchini, M.L., Greco, S., & Ragonese, S. 1998. Il progetto "Valutazione della fattibilità e potenzialità del ripopolamento attivo per la magnosa, *Scyllarides latus* (Crostacei Decapodi)": Sintesi e risultati. *Biol. Mar. Medit.* 5: 1277–1283.

Bianchini, M.L., Bono, G., & Ragonese, S. 2001. Long-term recaptures and growth of slipper lobsters, *Scyllarides latus*, in the Strait of Sicily (Mediterranean Sea). *Crustaceana* 74: 673–680.

Briones-Fourzán, P., Lozano-Álvarez, E., & Eggleston, D.B. 1994. The use of artificial shelters (casitas) in research and harvesting of Caribbean spiny lobster in Mexico. In: Phillips, B.F., Cobb, J.S., & Kittaka, J. (eds.), *Spiny Lobster Management*: pp. 340–361. Oxford: Fishing News Books.

Butler, M.J. & Herrnkind, W.F. 1997. A test of recruitment limitation and the potential for artificial enhancement of spiny lobster (*Panulirus argus*) populations in Florida. *Can. J. Fish. Aquat. Sci.* 54: 452–463.

Castro, K.M., Cobb, J.S., Wahle, R.A., & Catena, J. 2001. Habitat addition and stock enhancement for American lobsters *Homarus americanus. Mar. Freshw. Res.* 52: 1253–1261.

Chan T.W. & Yu, H.P. 1986. A report on the *Scyllarus* lobsters (Crustacea: Decapoda: Scyllaridae) from Taiwan. *J. Taiwan Mus.* 39: 147–174.

Courtney, A.J. 2002. *The Status of Queensland's Moreton Bay Bug (Thenus spp.) and Balmain Bug (Ibacus spp.) Stocks.* Brisbane, AU: Queensland Government: Department of Primary Industries.

DeMartini, E.E. & Williams, H.A. (2001). Fecundity and egg size of *Scyllarides squammosus* (Decapoda: Scyllaridae) at Maro Reef, Northwestern Hawaiian Islands. *J. Crust. Biol.* 21: 891–896.

Diaby, S. 2004. Trends in the US and world lobster production, import and export. *US Forest & Fishery Product Division Report.* 15 pp.

Diaz, D., Zabala, M., Linares, C., Hreu, B., & Abelló, P. 2005. Increased predation of juvenile European spiny lobsters (*Palinurus elephas*) in a marine protected area. *N. Z. J. Mar. Freshw. Res.* 39: 447–453.

Fatemi, S.M.R. 2001. Lobster fisheries in the Sea of Oman. In: Goddard, S., Al-Oufi, H., McIlwain, J., & Claereboudt, M. (eds.), *Proceedings of the 1st International Conference on Fisheries, Aquaculture and Environment in the NW Indian Ocean*: pp. 8–14. Muscat, Sultanate of Oman: Sultan Qaboos University.

Galil, B., Pisanty, S., Spanier E., & Tom, M. 1989. The Indo-Pacific lobster *Panulirus ornatus* (Fabricius 1798) (Crustacea: Decapoda): A new Lessepsian migrant to the Eastern Mediterranean. *Israel J. Zool.* 35: 241–243.

George, R.W. 1971. Report of a first tour of a spiny lobster survey in Fiji, Western Samoa, American Samoa and Tonga, 20 October–23 December 1970. South Pacific Islands Fisheries Development Agency. *FAO FI:SF/SOP/Reg.* 102/4:1–9.

Gorman, T.B. & Graham, K.J. 1978. *FRV Kapala Cruise Report No. 47* for cruises 78–02, 03, 05 and 06. New South Wales State Fisheries. pp. 8.

Grant, E.M. 1978. *Guide to Fishes,* 4th Edition. Brisbane: Department of Harbours and Marine. 768 pp.

Groeneveld, J.C., Cockcroft, A.C., & Cruywagen, G.C. 1995. Relative abundances of spiny lobster *Palinurus delagoae* and slipper lobster *Scyllarides elisabethae* off the east coast of South Africa. *S. Afr. J. Mar. Sci.* 16: 19–24.

Holmyard, N. 2005. *Lobster Supply Stable but Prices on the Increase.* FAO GLOBEFISH. September, 1p.

Holthuis, L.B. 1991. *Marine Lobsters of the World. An Annotated and Illustrated Catalogue of the Species of Interest to Fisheries Known to Date.* FAO Species Catalogue No. 125, Vol. 13: pp. 1–292. Rome: Food and Agriculture Organization of the United Nations.

Ivanov, B.G. & Krylov, V.V. 1980. Length-weight relationship in some common prawns and lobsters (Macrura, Natantia and Reptantia) from the western Indian Ocean. *Crustaceana* 38: 279–289.

Jones, C.M. 1993. Population structure of two species of *Thenus* (Decapoda: Scyllaridae), in northeastern Australia. *Mar. Ecol. Prog. Ser.* 97: 143–155.

Kagwade, P.V. Manickaraja, M., Deshmukh, V.D., Rajamani, M., Radhakrishnan, E.V., Suresh, V., Kathirvel, M., & Rao, G.S. 1991. Magnitude of lobster resources of India. *J. Mar. Biol. Assoc. India* 33: 150–158.

Kizhakudan, J., Thirumilu, K., Rajapackiam, P., & Manibal, C. 2004. Captive breeding and seed production of scyllarid lobsters — opening new vistas in crustacean aquaculture. *Mar. Fish. Infor. Serv. T & E Ser.* 181: 1.

Lavalli, K.L. & Spanier, E. 2001. Does gregariousness function as an antipredator mechanism in the Mediterranean slipper lobster, *Scyllarides latus*? *Mar. Freshw. Res.* 52: 1133–1143.

Lipcius, R.N., Stockhousen, W.T., & Eggleston, D.B. 2001. Marine reserves for Caribbean spiny lobster: Empirical evaluation and theoretical metapopulation recruitment dynamics. *Mar. Freshw. Res.* 52: 1589–1598.

Losada-Tosteson, P. & Posada, J.M. 2001. Using tires as shelters for the protection of juvenile spiny lobsters, *Panulirus argus*, as a fishing gear for adults. *Mar. Freshw. Res.* 52: 1445–1450.

Molina, L., Chasiluisa, C., Murillo, J.C., Moreno, J., Nicolaides, F., Barreno, J.C., Vera, M., & Bautil, B. 2004. Pesca Blanca y pesquerías que duran todo el año, 2003. In: Murillo, J.C. (ed.), *Evaluación de las Pesquerías en la Reserva Marina de Galápagos. Informe Compendio 2003*: pp. 103–139. Santa Cruz, Galápagos: Fundación Charles Darwin y Parque Nacional Galápagos.

Motoh, H. & Kuronuma, K. 1980. *Field Guide for the Edible Crustacea of the Phillipines*: i, ii: pp. 1–96. Tigbauan, Ilioilo, Phillippines: SEAFDEC.

Murillo, J.C., Chasiluisa, C., Molina, L., Moreno, J., Andrade, R., Bautil, B., Nicolaides, F., Espinoza, E., Chalén, L., & Barreno, J.C. 2003. Pesca blanca y pesquerías que duran todo el año en Galápagos, 2002. In: Murillo, J.C. (ed.), *Evaluación de las pesquerías en la Reserva Marina de Galápagos. Informe Compendio 2002*: pp. 97–124. Santa Cruz, Galápagos, Ecuador: Fundación Charles Darwin y Servicio Parque Nacional Galápagos.

Oliver, M.D., Stewart, R., Mills, D., MacDiarmid, A., & Gardner, C. 2005. Stock enhancement of rock lobsters (*Jasus edwardsii*): Timing of predation on naive juvenile lobsters immediately after release. *N. Z. J. Mar. Freshw. Res.* 39: 391–397.

Olla, B., Davis, M.W., & Ryer, C.H. 1998. Understanding how the hatchery environment represses or promotes the development of behavioral survival skill. *Bull. Mar. Sci.* 62: 531–550.

Parrish, F.A. & Kazama, T.K. 1992. Evaluation of ghost fishing in Hawaiian lobster fishing. *Fish. Biol.* 90: 720–725.

Pauly, D., Christensen, V., Dalsgaard, J., Foroese, R., & Torres, F. 1998. Fishing down marine food webs. *Science* 279: 860–863.

Phillip, B. 2005. Lobsters: The search for knowledge continues (and why we need to know!). *N. Z. J. Mar. Freshw. Res.* 39: 231–241.

Phillips, B.F. & Booth, J.D. 1994. Design, use, and effectiveness of collectors for catching the puerulus stage of spiny lobsters. *Rev. Fish. Sci.* 2: 255–289.

Polovina, J.J. 2005. Climate variation, regime shift, and implication for sustainable fisheries. *Bull. Mar. Sci.* 76: 233–244.

Quinn, N.J. & Kojis, B.L. 1995. The use of Witham collectors to increase production in lobster (*Panulirus argus*) mariculture. *Proceedings of the 30th Caribbean Food Crops Society of Annual Mtg.* 109–122.

Radhakrishnan, E.V., Deshmukh, V.D., Manisseri, M.K., Rajamani, M., Kizhakudan, J.K., & Thangaraja, R. 2005. Status of the major lobster fisheries in India. *N. Z. J. Mar. Freshw. Res.* 39: 723–732.

Spanier, E. 1991. Artificial reefs to insure protection of the adult Mediterranean slipper lobster, *Scyllarides latus* (Latreille, 1803). In: Boudouresque, C.F., Avon, M., & Gravez, V. (eds.), *Les Espéces Marines á Protéger en Méditerranée,* GIS 1991: pp. 179–185. France: Posidonia Publication.

Spanier, E. 1994. What are the characteristics of a good artificial reef for lobsters? *Crustaceana* 67:173–186.

Spanier, E. & Lavalli, K.L. 1998. Natural history of *Scyllarides latus* (Crustacea Decapoda): A review of the contemporary biological knowledge of the Mediterranean slipper lobster. *J. Nat. Hist.* 32: 1769–1786.

Spanier, E. & Zimmer-Faust, R.K. 1988. Some physical properties of shelter that influence den preference in spiny lobsters. *J. Exp. Mar. Biol. Ecol.* 122: 137–149.

Spanier, E., Tom, M., Pisanty, S., & Almog, G. 1988. Seasonality and shelter selection by the slipper lobster *Scyllarides latus* in the southeastern Mediterranean. *Mar. Ecol. Prog. Ser.* 42: 247–255.

Spanier, E., Tom, M., Pisanty, S., & Almog-Shtayer, G. 1990. Artificial reefs in the low productive marine environment of the Southeastern Mediterranean. *P. S. Z. N. I. Mar. Ecol.* 11: 61–75.

Zimmer-Faust, R.K. & Spanier, E. 1987. Gregariousness and sociality in spiny lobsters: Implications for den habitations. *J. Exp. Mar. Biol. Ecol.* 105: 57–71.

Index